History of Mathematics Education

Springer's History of Mathematics Education Series aims to make available to scholars and interested persons throughout the world the fruits of outstanding research into the history of mathematics education, provide historical syntheses of comparative research on important themes in mathematics education; and establish greater interest in the history of mathematics education.

More information about this series at http://link.springer.com/series/13545

Nerida F. Ellerton • M. A. (Ken) Clements

Toward Mathematics for All

Reinterpreting History of Mathematics in North America 1607–1865

Springer

Nerida F. Ellerton
Department of Mathematics
Illinois State University
Bloomington, IL, USA

M. A. (Ken) Clements
Department of Mathematics
Illinois State University
Normal, IL, USA

ISSN 2509-9736 ISSN 2509-9744 (electronic)
History of Mathematics Education
ISBN 978-3-030-85723-3 ISBN 978-3-030-85724-0 (eBook)
https://doi.org/10.1007/978-3-030-85724-0

This Springer imprint is published by the registered company Springer Nature Switzerland AG
The registered company address is: Gewerbestrasse 11, 6330 Cham, Switzerland

Dedicated to the memories of our parents, Lydia Ruth Gersch (Meyer), Paul Johannis Gersch, Brenda Margaret Clements (Fleming), and John Albert Clements who gave us opportunities that they themselves never had.

Preface to this Book

In November 2019, when we had almost finished our first draft of this book, we learned that another book on the history of mathematics in North America, authored by David E. Zitarelli, had just been released. That book was published by the MAA Press and was endorsed as an imprint of the American Mathematical Society. Its title was *A History of Mathematics in the United States and Canada (Volume 1: 1492–1900)*. Not much later we found that David Roberts' (2019), *Republic of Numbers* had appeared. We had not been aware that these books were being prepared.

This was the second time in recent years that this had happened to us. The first time occurred in 2016 when, just as we were completing the first draft of our book *Samuel Pepys, Isaac Newton, James Hodgson and the Beginnings of Secondary School Mathematics: A History of the Royal Mathematical School at Christ's Hospital 1673–1868* (which would be published by Springer in 2017), we learned that at the end of 2015 Clifford Jones had had his book *The Sea and the Sky: The History of the Royal Mathematical School of Christ's Hospital* published.

On that first occasion we took the opportunity to read Clifford's book carefully, and that caused us to change parts of what we had written. A similar thing happened this time, although we were saddened to hear of David Zitarelli's passing late in 2018. We would have enjoyed discussing with him how he framed his history. We were particularly impressed with his well-documented account of developments relating to mathematics at Harvard College in the seventeenth century. One thing which quickly became clear to us was that he had prepared his book from a mathematician's perspective, and we had prepared ours from a different vantage point. We were more interested than he in accounting for the history of mathematics in the 13 colonies and in the United States (to 1865) as that mathematics was experienced by people across the full spectrum of ages and places—by young children (aged less than 10), by children aged between 10 and 15, by pre-college students aged between about 15 and 18, by college students, and by others, including adults outside of formal education institutions. And, of course, like both David Zitarelli and David Roberts, we were also interested in the perspectives of "mathematicians"—scholars who worked in the field and had an enduring interest in any of the branches of mathematics.

When people look at the same set of events from different vantage points they see mostly similarities, but also differences. As we were writing this book we were struck by the fierce inequalities of opportunity which developed with respect to the possibilities of learning Western Mathematics. On a related matter, we were also interested to note that at the end of the seventeenth century only the most elementary aspects of the Hindu-Arabic numeration system, which we regard as the most transformative theoretical position in the history of mathematics, was not known to the majority of inhabitants in North America. As you read that sentence did you think—"That can't be true"? Well, it *is* true, and that fact provides an interesting starting point for the history to be presented in this book.

In 1992 the second author of this book (Ken Clements) authored a book titled *Mathematics for the Minority: Some Historical Perspectives on School Mathematics in Victoria*. In that book he argued that the history of school mathematics in the state of Victoria (Australia) had been controlled by politicians and mathematicians—initially British politicians and British mathematicians. After the Federation of Australian states was achieved and the nation of Australia was born in 1901 academics at the University of Melbourne, and education administrators employed by the Victorian government, believed that the main task of school mathematics was to prepare students for scientific, technical and mathematical studies at the University, at the Working Man's College, and at the Ballarat School of Mines—which at that time were the State's only higher education institutions. In other words, school mathematics was aimed at the needs and aspirations of a minority. In this present book we present a similar line of argument. This book tells the story of how European-background forms of mathematics were translated from "home" to the education institutions being established in the North American New World. By and large, the main idea was to make the mathematics studied in the New World identical with the mathematics taught in respectable "home" education institutions. This continued to be the case throughout the period 1607–1865—from the beginnings of the first permanent European-background settlement at Jamestown, Virginia, through to the end of the Civil War.

The world of publishing has changed considerably over the past few decades, and that has had an impact on how we have asked authors to prepare chapters for books in Springer's History of Mathematics Education series. In the past, authors and series editors could assume that a whole book, or at least quite a few chapters in it, would be read by interested persons. But now, e-books and individual chapters of a book in digital form are readily available, and that has affected how we have written individual chapters of this book. Thus, for example, a careful reader of this book might notice that, occasionally, there is repetition of points made in earlier chapters. Obviously, because some readers will have access to just one of the chapters in the book, it made sense for us to repeat material covered in earlier chapters. We have attempted to limit such repetition to cases where what is being repeated represents essential knowledge if the present chapter is to be understood as a stand-alone document. Another sign of the times is that there is a reference list at the end of *each* chapter, *and* a *composite* reference list after all nine chapters have been presented. The reason for that is simple: readers who have access to just one chapter are likely to want to have access to a fully documented statement setting out the works to which reference is made in the chapter.

We wish to thank librarians, archivists and the staff at the Phillips Library at the Peabody Essex Museum, Salem, Massachusetts, the Butler Library at Columbia University, New York, the Clements Library at the University of Michigan, the Houghton Library at Harvard University, the Library of Congress (in Washington DC), the Wilson Library at the University of North Carolina at Chapel Hill, the Beinecke Library at Yale University, the Winterthur Museum in Delaware, the Lilly Library at the University of Indiana, the Special Collections Research Center in the Swem Library at the College of William and Mary and at the Rockefeller Library (both in Williamsburg, Virginia), the New York Public Library, the British Library (London), Guildhall Library, London Metropolitan Archives, the Royal Observatory and the National Maritime Museum at Greenwich, the Bodleian Libraries at the University of Oxford, the Cambridge University Library, the Pepys Library at Magdalene College within the University of Cambridge, the State Library of Victoria (Australia), and the Milner Library at Illinois State University, for locating relevant manuscripts, artifacts, and books for us.

We could not have been more pleased with the cooperation we received by our publisher, Springer Nature, especially Melissa James and Nick Melchior. We always felt that we were being supported in the best possible ways. We would also like to thank Dr George Seelinger, the Head of the Mathematics Department at Illinois State University (in which we both worked until our recent retirements) for encouraging us in our research endeavors.

References

Clements, M. A. (1992). *Mathematics for the minority: Some historical perspectives on school mathematics in Victoria*. Geelong, Australia: Deakin University.

Jones, C. (2015). *The sea and the sky: The history of the Royal Mathematical School of Christ's Hospital*. Horsham, England: Author.

Roberts, D. L. (2019). *Republic of numbers: Unexpected stories of mathematical Americans through history*. Baltimore, MD: Johns Hopkins University Press.

Zitarelli, D. A. (2019). *A history of mathematics in the United States and Canada*. Volume 1: 1492–1900. Washington, DC: American Mathematical Society.

Preface to the Series

Books in Springer Nature's series on the history of mathematics education comprise scholarly works on a wide variety of themes, prepared by authors from around the world. An important aim of the series is to develop and report syntheses of historical research which have already been carried out in different parts of the world with respect to important themes in mathematics education—like, for example, "Historical Perspectives on how Language Factors Influence Mathematics Teaching and Learning," and "Historically Important Theories Which Have Influenced the Learning and Teaching of Mathematics."

The mission for the series can be summarized as:

- To make available to scholars and interested persons around the world the fruits of outstanding research into the history of mathematics education;
- To provide historical syntheses of comparative research on important themes in mathematics education; and
- To establish greater interest in the history of mathematics education.

In this present book we offer a history of mathematics in North America between 1607 and 1865, as told from a mathematics-for-all vantage point. As far as we know, no other writers have addressed that theme. As the text proceeds, readers are invited to think about how mathematics in North America (excluding Canada and Alaska) emerged during a period when curricula of education institutions were controlled by what we have called the "classics stranglehold." Of special interest are the profound effects this background had on fundamental questions like: "What should be the intended mathematics curricula in schools?" "Should the intended curricula be the same for all learners?" And "Who should be responsible for bringing about changes to implemented mathematics curricula in schools and colleges?"

We hope that the series will continue to provide a multi-layered canvas portraying rich details of mathematics education from the past, while at the same time presenting historical insights that can support the future. This is a canvas which can never be complete, for today's mathematics education becomes history for tomorrow. A single snapshot of mathematics education today is, by contrast with this canvas, flat and unidimensional—a mere pixel in a detailed image. We encourage readers both to explore and to contribute to the detailed image which is beginning to take shape on the canvas for this series.

Any scholar contemplating the preparation of a book for the series is invited to contact Nerida Ellerton (ellerton@ilstu.edu) in the Department of Mathematics at Illinois State University or Melissa James, at the Springer Nature New York office.

List of Publications in the Springer Nature History of Mathematics Education Series

Barbin, É., Guichard, J-P., Moyon, M., Guyot, P., & Morice-Singh, C. (2017). *Let history into the mathematics classroom*

De Bock, D. (Ed.), (2022). *Modern mathematics–An international movement?*

De Bock, D., & Vanpaemel, G. (2019). *Rods, sets and arrows: The rise and fall of modern mathematics in Belgium*

Ellerton, N. F., & Clements, M. A. (2017). *Samuel Pepys, Isaac Newton, James Hodgson and the beginnings of secondary school mathematics: A history of the Royal Mathematical School within Christ's Hospital 1673–1868*

Garnica, A. V. M. (2019). (Ed.), *Oral history and mathematics education*

Kanbir, S., Clements, M. A., & N. F. Ellerton (2018). *Using design research and history to tackle a fundamental problem with school algebra*

Owens, K. D., Lean, G. A. Paraide, P., & Muke, C. (2017). *History of number: Evidence from Papua New Guinea and Oceania*

Ravn, O., & Skovsmose, O. (2018). *Connecting humans to equations: A reinterpretation of the philosophy of mathematics*

Ellerton, N. F., & Clements, M. A. (2022). *Toward mathematics for all: Reinterpreting history of mathematics in North America 1607–1865*

Paraide, P., Owens, K. D., Clarkson, P. C., Owens, C., & Muke, C. (2022). *Mathematics education in Papua New Guinea: A case study of colonial and postcolonial influences on mathematics education*. New York, NY: Springer.

Recent Springer Books on the History of Mathematics Education by Nerida F. Ellerton and M. A. (Ken) Clements

(2012). *Rewriting the history of mathematics education in North America 1607–1861*

(2014). *Abraham Lincoln's cyphering book and ten other extraordinary cyphering books*

(2015). *Thomas Jefferson and his decimals 1775–1810: Neglected years in the history of U.S. school mathematics*

(2017). *Samuel Pepys, Isaac Newton, James Hodgson and the beginnings of secondary school mathematics: A history of the Royal Mathematical School within Christ's Hospital 1673–1868*

(2022). *Toward mathematics for all: Reinterpreting the history of mathematics in North America 1607–1865*

Also Note:

Clements, M. A., Keitel, C. Bishop, A. J., Kilpatrick, J., & Leung, F. (2013). From the few to the many: Historical perspectives on who should learn mathematics. In M. A. Clements, A. J. Bishop, C. Keitel, J. Kilpatrick, & F. Leung (Eds.), *Third international handbook of mathematics education* (pp. 7–40). New York, NY: Springer.

Ellerton, N. F., & Clements, M. A. (2022). Australian school mathematics and "colonial echo" influences, 1901–1975. In D. De Bock (Ed.), *Modern mathematics–An international movement?* New York, NY: Springer.

Singh, P., & Ellerton, N. F. (2013). International collaborative studies in mathematics education. In M. A. Clements, A. J. Bishop, C. Keitel, J. Kilpatrick, & F. Leung (Eds.), *Third international handbook of mathematics education* (pp. 827–860). New York, NY: Springer.

Mathematics in the eighteenth century . . . did not originate generally in the schools of this country . . . If we except such mechanical features as the elementary operations of arithmetic and algebra and consider the progress of real mathematics, neither the elementary schools of this country nor the colleges were much concerned in that period with the subject.

(Smith & Ginsburg, 1934, p. 16)

I should rejoice to see . . . Euclid honourably shelved or buried "deeper than did ever plummet sound" out of the schoolboys' reach.

(Statement by J. J. Sylvester, 1870, p. 261).

Entrance requirements for 1786 at Columbia College (New York) were specified as follows:

"No candidate shall be admitted into the College . . . unless he shall be able to render into English Caesar's Commentaries of the Gallic War; the four orations of Cicero against Catiline; the four first books of Virgil's *Aeneid*, and the gospels from the Greek; and to explain the government and connection of the words, and to turn English into grammatical Latin, and shall understand the four first rules of arithmetic, with the rule of three."

(Quoted in Broome, 1903, p. 34)

[When Benjamin Silliman (aged 13) entered Yale College (in 1792)] "the entrance requirements might also have been appropriate for a 'school of Plato.' Candidates for admission to the Freshman Class were examined in Cicero's Select Orations, Virgil, Sallust, the Greek Testament, Dalzel's Collectanea Græca Minora, Adam's Latin Grammar, Goodrich's Greek Grammar, Latin Prosody, Writing Latin, Barnard's or Adams' Arithmetic, Murray's English Grammar, and Morse's Worcester's or Woodridge's Geography. Jacob's Greek Reader and the four Gospels were admitted as a substitute for *Græca Minora* and the Greek Testament."

(Fulton & Thomson, 1947, p. 9)

Contents

List of Figures

List of Tables

Abstracts

Abstract for the Book

The 104 men and boys who arrived in Jamestown, Virginia, in 1607 heralded the first permanent settlement by European-background persons on territory now part of the United States of America. This book provides a history of mathematics in North America (excluding Canada, Alaska, and Mexico) between 1607 and the end of the Civil War, in 1865. The position taken is that all people engaged actively, in some way, with mathematics, and therefore the history of mathematics should tell the story of how mathematics emerged, for all, in the New World.

In this book, we take a mathematics-for-all perspective by considering the history of early-childhood mathematics, elementary-school and secondary-school mathematics, college mathematics, mathematics employed outside of formal education institutions, and attempts to create "new" forms of mathematics, mainly by "mathematicians." Furthermore, the mathematics developed by and applied within different social groups—like, for example, females, Native Americans, African-American slaves, rural families, and persons engaged in specific employment areas (such as business, teaching, navigation, surveying, building construction, astronomy, and local and other forms of government administration)—should also be part of the story.

Today, in the 2020s, it is assumed that everybody should be offered the opportunity to learn mathematics. However, it was not until well into the twentieth century that "mathematics for all" became a recognizable and achievable goal in much of North America. Before then, the geographical location of schools in relation to children's homes, the availability (or non-availability) of plantation workers and of teachers capable of teaching mathematics, the attitudes within families—especially parental attitudes—to schooling, economic circumstances of families, and social and psychological presuppositions and prejudices about mathematical ability or giftedness all influenced greatly the amount and type of mathematics a person would have the opportunity to learn. Moreover, in many societal subcultures, the perceived difference between two social functions of mathematics—its utilitarian, modeling function and its capability to sharpen the mind and induce logical thinking—generated mathematics curricula and forms of teaching in local schools which met the needs of some learners more than others.

This book identifies a historical progression towards the achievement of mathematics for all: from schooling for all to quantitative literacy for all, to basic mathematics for all, to secondary mathematics for all, to college mathematics, to mathematics research, and to mathematical modelling in order to solve real-life problems. As much as has been possible, arguments have been based on data available in primary sources, and on interpretative analyses of those data.

Abstracts for the Nine Individual Chapters

Abstract for Chapter 1: "The Scope of the Book"

This book presents a history of mathematics between 1607 and 1865 in that part of North America which is the present United States of America (excluding Alaska), and this first chapter begins with some discussion of the meanings which could be given to the title of the book. During most of the seventeenth century the number of European-background settlers was always small in comparison with the number of Native American peoples, and the struggles by the settlers to survive meant that any desire to study higher forms of mathematics, or to conduct research in mathematics, was virtually non-existent. It was difficult for them even to provide ways and means by which young children could learn the Hindu-Arabic methods of counting or calculating. Products of technology like paper, slate, and ink were not readily available, and very few mathematics-knowledgeable teachers were available. The situation improved during the period 1700–1865, but even during the first half of the nineteenth century most young children did not have ready access to mathematics textbooks. In this introductory chapter, issues associated with the education of Native American children, and of children of indentured European-background workers and African American slaves are also considered. Toward the end of the chapter, six research questions are stated, and summaries of what will be investigated in the remaining eight chapters of the book are given.

Abstract for Chapter 2: "Young Children's Introduction to Mathematics in North America Between 1607 and 1865"

In this chapter we consider the mathematics studied by young children—not yet 10 years of age—the eastern colonies during the seventeenth and eighteenth centuries. We draw special attention to the hornbook—the artifact which most influenced intended, implemented and attained curricula for young European-background children during the period 1607–1799—and provide details on what is possibly the earliest extant hornbook constructed and used in North America during that period. During the seventeenth century there was little opportunity for most young children to advance their understandings of mathematics. Evidence is put forward that as late as the beginning of the nineteenth century most children aged less than 10 years were not given any opportunity to study any form of mathematics beyond counting verbally and learning to read and write the Hindu-Arabic numerals. It was not until the early 1820s that the idea began to be accepted by some North American educators. that all young children from about the age of 6 should learn to read and write Hindu-Arabic numerals, and to develop other elementary arithmetical concepts and skills. Once that idea was put forward, initially by Warren Colburn, it steadily gathered momentum among scholars, educators, and the society at large.

Abstract for Chapter 3: "The Influence of the Cyphering Tradition on North American Elementary- and Middle-School Mathematics Between 1607 and 1865"

Commercially-published textbooks do not offer the most important data for those interested in the histories of mathematics and mathematics education in North America during the period 1607–1865. In fact, until well into the nineteenth century most North American schoolchildren who were learning mathematics did not own a mathematics textbook, and many teachers of mathematics did not own one either. By contrast, almost all students aged from 10 to 16 years who studied any branch of mathematics prepared handwritten cyphering books, and often their teachers made available to them the cyphering books that they had prepared in their own school days. In this chapter we summarize our previous work on the cyphering tradition, drawing attention to theoretical bases, and also to the way the tradition controlled both the implemented and the attained mathematics curricula in grammar schools and in other pre-college institutions. Summaries of curriculum content and of the teaching and learning patterns which were an inherent part of the cyphering tradition are given. The discussion is based on our analyses of about 1500 extant North American cyphering books from the period.

Abstract for Chapter 4: "Mathematics Textbooks and the Gradual Decline in the Use of Middle- to Advanced-Level *Abbaco* Arithmetic 1607–1865"

This chapter focuses on the influence of textbooks and textbook authors on the teaching and learning of middle- to more advanced-level abbaco arithmetic in North America during three sub-periods—from 1607 to 1776, from 1776 to 1825, and from 1825 to 1865. During the first sub-period, from 1607 to 1776, there were relatively few students who concentrated on learning any form of mathematics beyond low-level abbaco arithmetic. Those who prepared cyphering books copied statements of rules, cases and model examples from "parent" cyphering books or directly from textbooks. During the second sub-period, from 1776 to 1825, textbooks by North American authors were increasingly used to assist students preparing cyphering books, the most popular authors being Thomas Dilworth, Nicolas Pike, Nathan Daboll, Daniel Adams, Michael Walsh, Stephen Pike, and Warren Colburn. Although algebra and geometry were more studied than in the previous sub-period, any movement away from traditional *abbaco* arithmetic to other forms of mathematics tended to be resisted in the schools. The third sub-period, 1825–1865 witnessed a struggle between those who wanted to revolutionize and expand the teaching and learning of mathematics in the United States of America and those who clung to the content and pedagogical approaches associated with traditional *abbaco* arithmetic intended curricula. In this chapter we concentrate on showing that although initially in school mathematics textbooks were used to complement cyphering, ultimately they came to play a more decisive role.

Abstract for Chapter 5: "The Struggle for Algebra"

This chapter focuses on the emergence of algebra in the intended and implemented curricula of U.S. schools between 1607 and 1865. Before 1776 only a tiny proportion of school-age children, in what is now the mainland part of the United States, studied algebra. The chapter begins by providing evidence that until about 1820 the study of mathematics other than abbaco arithmetic was not something seriously engaged in by most young people in North America. Very few textbooks on any branch of mathematics other than arithmetic were suitable for school children, and relatively few cyphering books which focused on mathematics other than arithmetic were prepared. That changed in the early 1820s, after the first public high schools were opened, and after colleges began to require prospective students to demonstrate a knowledge of algebra. Nevertheless, even in the 1850s less than 10% of school-age North American children studied any of algebra, geometry, trigonometry, surveying, navigation, or calculus. The cyphering tradition was strongly linked to both abbaco arithmetic and algebra, but algebra was much less studied. In 1730 a Dutch-language textbook, by Pieter Venema, on arithmetic and algebra, was published in New York, but at that time there was little demand for it and a second edition never appeared. Documentary evidence—never before available to historians—from a "precursor" document prepared by Venema in New York in 1725, is discussed and analyzed. In that document Venema demonstrated how algebra could be used to prove and to generalize. Venema was ahead of his time and offered North American mathematics education an opportunity which it failed to grasp. Venema was ahead of his time and offered North American mathematics education an opportunity which it failed to grasp.

Abstract for Chapter 6: "Pre-College Geometry, Mensuration, Trigonometry, Surveying, and Navigation 1607–1865"

This chapter analyzes pre-college education developments in geometry, mensuration, trigonometry, surveying, and navigation between 1607 and 1865, in the 13 colonies and then in the United States of America. Although throughout that period relatively few students prepared cyphering books which focused on anything other than *abbaco* arithmetic. Some school students did study one or more of algebra, geometry, trigonometry, astronomy, navigation, and surveying, but most of those who did had not previously studied topics like angles, decimals, fractions, logarithms, or elementary mechanics, and therefore it was extremely difficult for them to make good progress. Evidence will be presented showing that some students nevertheless managed to succeed. In particular, data from a cyphering book prepared by Thomas Willson in Pennsylvania in 1789 will be examined in detail, and the analysis will suggest what implemented curricula in post-*abbaco* forms of mathematics were like at that time. It has often been argued that so far as mathematics education was concerned much was achieved in the schools of that time, because there was an over-emphasis on mere memorization. In this chapter it is argued, however, that that contention rests on the untested assertion that students who prepared cyphering books did not understand and could not apply what they entered in their cyphering books. An important aim for the cyphering tradition was that students who prepared manuscripts would consult them if and when they felt the need to do so later in their lives.

Abstract for Chapter 7: "College Mathematics, 1607–1865"

Throughout the period 1607–1865 most families had very few books other than a Bible in their homes, and most people did not know much mathematics beyond reading, writing, and counting with Hindu-Arabic numerals. Between 1636 and 1865 only a tiny proportion of the population of that part of North America which is now mainland United States of America attended college and, of those who did, most had not previously studied mathematics beyond low-level *abbaco* arithmetic, elementary algebra, and the first few books of Euclid's Elements. It is not surprising, therefore, that the period did not produce more than three or four scholars who, by European standards, might be considered to have been "outstanding" mathematicians. The U. S. college curriculum had its origin in the classical curriculum traditions of the medieval universities of Europe and especially of Cambridge and Oxford Universities. However, many of those who attended North American colleges did study what we have called "applied mathematics"—embracing fields like astronomy, surveying, mensuration and navigation— while they were at college, and we argue that this aspect of the implemented curriculum had been successfully translated mainly from Great Britain.

Abstract for Chapter 8: "Different Perspectives on Mathematics in North America1607–1865"

It would be unreasonable to expect the inhabitants of North America to have produced great works of mathematics—judging by European standards—during the period 1607–1865. At that time a New World began to be constructed in North America by the European "invaders"—houses, schools, and towns were built, administrative structures were created, and lands were cleared for farming. But very few books other than bibles and, perhaps, almanacs were to be found in homes or schools, and most of the relatively few settlers who knew enough mathematics to teach it had other things to do. It is not surprising, therefore, that the 258-year period did not produce more than three or four mathematicians who, by the European standards of the time, might be regarded as "outstanding." Between 1775 and 1820 U.S. college curricula drew their inspiration from the classical curricular traditions of the medieval universities of Europe and especially of Cambridge and Oxford Universities. However, many students who attended the North American colleges did enroll in "applied mathematics" subjects—embracing fields like astronomy, surveying, mensuration, and navigation. Interest in those forms of mathematics had been successfully translated mainly from Great Britain.

Abstract for Chapter 9: "Toward Mathematics for All: Answers to Research Questions, Limitations, and Possibilities for Further Research"

This final chapter begins by answering the six research questions which were posed towards the end of the first chapter. Those questions were:

1. What were the intended, implemented and attained mathematics curricula for young children aged less than 10 years (in North America) (a) during the seventeenth century? And (b) during the period 1700–1865? And, to what extent did the answers to those questions vary across North America, and in different groups of children (e.g., boys versus girls, European-background children versus Native American children, and European-background children versus African-American children)?
2. What were the intended, implemented and attained mathematics curricula for North American children aged between 10 and 15 years during (a) the seventeenth century, and (b) the period 1700–1865? And,

to what extent did the answers to those questions vary across different parts of North America, and across different groups?

3. What were the intended, implemented and attained mathematics curricula for North American pre-college students aged between about 15 and 18 years during (a) the seventeenth century, and (b) the period 1700–1865? And, to what extent did the answers to those questions vary across different parts of North America, and across different groups?
4. What were the intended, implemented and attained mathematics curricula for North American college students during (a) the seventeenth century, and (b) the period 1700–1865? And, to what extent did the answers to those questions vary across different parts of North America, and across different groups?
5. What perspectives on the status of mathematics in college curricula were held in the North American colonies during the period 1607–1865?
6. What are the implications of the answers to the first five questions (above) for those investigating the history of mathematics in North America? What future research is needed, and to what extent will it be feasible to conduct that research?

While carrying out the research for this book we came to recognize that authors of general histories of mathematics have tended to view the history of mathematics in terms of whether an event or person(s) associated with an event contributed to a "weight-bearing link" (Parshall, Historical contours of the American mathematical research community. In: Stanic GMA, Kilpatrick J (eds) A history of school mathematics, vol 1. National Council of Teachers of Mathematics, Reston, VA, 2003, p. 114) with the present state of knowledge for key areas of mathematics. In other words, they have looked back from the present situation with respect to high-level mathematics in an attempt to identify persons who, and events which, progressed mathematics toward what is now regarded as important. In this book, however, we have considered the history of mathematics in North America (excluding Alaska and Canada) between 1607 and 1865 from a more inclusive, bottom-up, mathematics-for-all perspective (Clements et al., From the few to the many: Historical perspectives on who should learn mathematics. In: Clements MA, Bishop AJ, Keitel C, Kilpatrick J, Leung F (eds) Third international handbook of mathematics education. Springer, New York, NY. https://doi.org/10.1007/978-1-4614-4684-2_1, 2013). In this final chapter the above questions are answered from analyses provided in the preceding eight chapters. The chapter closes with a discussion of limitations of the research, and how a consideration of those limitations draws attention to various questions which need to be the subject of further research.

About the Authors

Nerida F. Ellerton was professor within the Mathematics Department at Illinois State University between 2002 and 2018. She holds two doctoral degrees—one in Physical Chemistry and the other in Mathematics Education. Between 1997 and 2002 she was Dean of Education at the University of Southern Queensland, Australia. She has taught in schools and at four universities, and has also served as consultant in numerous countries, including Australia, Bangladesh, Brunei Darussalam, China, Malaysia, the Philippines, Thailand, the United States of America, and Vietnam. She has written or edited 17 books and has had more than 150 articles published in refereed journals or edited collections. Between 1993 and 1997 she was editor of the *Mathematics Education Research Journal*, and between 2011 and 2016 she was Associate Educator for the *Journal for Research in Mathematics Education*. Since 2012 Springer has published six books which she co-authored with Ken Clements. She and Ken are joint editors of Springer's *History of Mathematics Education* Series.

M. A. (Ken) Clements's masters and doctoral degrees were from the University of Melbourne, and at various times in his career he has taught, full-time, in primary and secondary schools, for a total of 15 years. He has taught in six universities, located in three nations, and in 2019 he retired after being professor within the Mathematics Department at Illinois State University for 15 years. He has served as a consultant and as a researcher in Australia, Brunei Darussalam, India, Malaysia, Papua New Guinea, South Africa, Thailand, the United Kingdom, the United States of America, and Vietnam. He served as co-editor of the three *International Handbooks of Mathematics Education*—published by Springer in 1996, 2003 and 2013—and with Nerida Ellerton, co-authored a UNESCO book on mathematics education research. He has authored or edited 36 books and more than 200 refereed articles on mathematics education and is honorary life member of both the Mathematical Association of Victoria and the Mathematics Education Research Group of Australasia. He married Nerida Ellerton in 2005, and between them they have 7 children, 19 grandchildren, and 3 great-grandchildren.

Chapter 1
The Scope of This Book

Abstract This book presents a history of mathematics between 1607 and 1865 in that part of North America which is the present United States of America (excluding Alaska), and this first chapter begins with some discussion of the meanings which could be given to the title of the book. During much of the seventeenth century the number of European-background settlers was always small in comparison with the number of Native American peoples, and the struggles by the settlers and the Indigenous inhabitants to survive meant that any desire to study higher forms of mathematics, or to conduct research in mathematics, was virtually non-existent. It was difficult for them even to provide ways and means by which young children could learn the Hindu-Arabic methods of counting or calculating. Products of technology like paper, slate, and ink were not readily available, and very few mathematics-knowledgeable teachers were available. The situation improved during the period 1700–1865, but even during the first half of the nineteenth century most young children did not have ready access to mathematics textbooks. In this introductory chapter, issues associated with the education of Native American children, and of children of indentured European-background workers and African-American slaves are also considered. Toward the end of the chapter, six research questions are stated, and summaries of what will be investigated in the remaining eight chapters of the book are given.

Keywords *Abbaco* sequence for arithmetic • *Abbaco* tradition • Counting systems • Decimal systems of counting • Definitions of mathematics • Glendon Lean • Hindu-Arabic numeration system • History of mathematics • History of mathematics education • Indigenous counting systems • Jamestown

> History, as nearly no one seems to know, is not merely something to be read. And it does not refer merely, or even principally, to the past. On the contrary, the great force of history comes from the fact that we carry it within us, are unconsciously controlled by it in many ways, and history is literally present in all that we do.
>
> (Baldwin, 1998, pp. 722–723)

What is Mathematics?

The title of this book, "*Toward Mathematics for All: Reinterpreting History of Mathematics in North America 1607–1865,*" demands comment. First, this will be *a* reinterpretation of history rather than "*the* history." We hold that there is no such thing as a unique history of any discipline for any period of time. Whoever writes a history writes it from a particular vantage point. Historians "see" different things

N. F. Ellerton, M. A. (Ken) Clements, *Toward Mathematics for All*, History of Mathematics Education,
https://doi.org/10.1007/978-3-030-85724-0_1

when they write about the same events and eras from different vantage points. That draws attention to the idea that a *history* is not a unique set of events occurring during a particular period of time, but rather an account of a set of events and relationships between events as seen, constructed, and interpreted by the person, or persons, writing the history. In that sense, all historical accounts must be subjective.

What is true of history is also true of mathematics (Stedall, 2012). Richard Courant's and Herbert Robbins's (1941) book, *What is Mathematics?* has been the subject not only of much praise by scholars but also of muted criticism (see, for example, Blank, 2001). It seems to us that some authors have colonized the meanings of "mathematics" and "mathematicians" whenever they have discussed histories of mathematics. For them, the word "mathematician" should be applied only to academics teaching "high-level mathematics" in colleges or to persons conducting high-level research which involves mathematical analysis in research institutions (Stedall, 2012). Correspondingly, some would prefer to reserve the word "mathematics" for high-level studies and research carried out by "mathematicians."

David Zitarelli (2019), in his recent book *A History of Mathematics in the United States and Canada*, addressed the meaning of "mathematician" directly when he wrote:

> What is a mathematician? A modern mathematician, after all, would chafe at the notion of someone who did not produce one mathematical work labeled in such a way. ... Up to the time of the watershed year 1876, a mathematician in America was someone sufficiently steeped in the subject to be able to teach advanced parts of the subject and, moreover, to apply these topics to related fields. ... However, I claim that David Rittenhouse is a mathematician by today's standards, because he published papers on mainstream mathematics that were entirely new to him. Other figures defy this easy distinction, such as Isaac Greenwood; even though he presented his own approach to topics novel to American students at the time, they were not original, and so I label him a mathematical practitioner. Generally, the only four individuals (up to 1876) I call mathematicians are Rittenhouse, Nathaniel Bowditch, Robert Adrain, and Benjamin Peirce. All others were mathematical practitioners. (p. 55)

Later in his book, Zitarelli (2019) distinguished between mathematical "enthusiasts," mathematical "practitioners," and "mathematicians" (p. 118). For example, he described Benjamin Franklin as a "mathematical enthusiast of the first rank" (p. 78) and commented that although "Franklin and Jefferson may have contributed little directly to mathematics, they certainly appreciated the subject's importance and took pride in their ability to apply it" (p. 79). This book will attempt to present a history of mathematics in North America (excluding Canada) for the 258-year period 1607–1865, and not a "history of mathematicians." That latter task has been well tackled by others (see, e.g., Bell, 1945; Roberts, 2019).

Unlike some commentators (e.g., Kline, 1972; Parshall, 2003; Smith & Ginsburg, 1934), we will not restrict the meaning of the word "mathematics" to the findings of "research" carried out by "mathematicians." From our perspective, 2- to 3-year-old children learning to quantify a collection of objects are engaged in a form of important mathematics; so too are 11-year-old children learning to measure angles with a protractor; so are 16-year-olds as they reflect on what it means to prove when they first meet the traditional *reductio ad absurdum* proof that there is no rational number which, when squared, equals 2; so too are 18-year-olds struggling to cope with the intricacies of elementary differential calculus; so too are adults who have left school but are attempting to work out the implications for their family finances of a mortgage reduction from 4 percent to 3.5 percent. And, of course, so too was Andrew Wiles as he attempted, ultimately successfully, to prove Fermat's Last Theorem (Stedall, 2012). This book is concerned with the history of mathematics in North America, as seen from the democratized perspective just outlined.

Although we can agree with David Zitarelli's (2019) definition of a mathematician as "someone who contributed an original piece of mathematics" (p. 118), we wonder what the word "original" means in that context. We do *not* agree with those who would think that school students studying "mathematics," or subjects like "arithmetic," "algebra," "trigonometry," "geometry," or "calculus," are not engaged in mathematics. A person playing a piano may not be a musician, but that person is engaged in making music. A person studying history may not be a historian but is nevertheless engaged with history. A middle-school school student coming to recognize the truth of the associative property for the multiplication of rational numbers would not normally be regarded as a mathematician but *is* engaged with mathematics.

In this book a *mathematical* task will be regarded as one which requires the use of calculations, or algebra (including functions, graphs), or formal logical reasoning, or geometry, or trigonometry, or limits, or calculus, or anything else commonly recognized as being "mathematical." Furthermore, mathematics can be either "pure" or "applied." Applied mathematics is to be associated with tasks which are concerned with developing and using mathematics to pose, model, and solve, and also to extend and generalize real-world-related problems—like, for example, in business, or surveying, or navigation, or astronomy (including space exploration), or, at the present time, with information technology.

This book offers a history of mathematics from a vantage point which includes mathematics formally investigated by research mathematicians, by "applied mathematicians," and by persons in families, in schools, in colleges, and in society in general who are attempting to "mathematize" problems that they want to solve. Although we have enjoyed, and profited from, reading David Zitarelli's (2019) *A History of Mathematics in the United States and Canada*, we recognize that David's concept of mathematics is very different from ours.

Karen Hunger Parshall (2003) had this to say about the "historiographical" point of view on the history of mathematics that Morris Kline embraced

> From the historiographical point of view that Kline adopted in his study, mathematical results merited inclusion in the historical narrative provided they formed a weight-bearing link in that great chain of mathematical ideas that stretches across time from the present to the past. (Notice here the direction of time's arrow!) For Kline, the history of mathematics is the story of how *contemporary* mathematical theories evolved; it is a technically oriented, intellectual history of ideas. This sort of historiographical framework suggests historical questions such as "How did X use Y's mathematical work to advance theory Z?" and "How did A do B without knowing C?" Answers to these and other questions provide important insights into the development of mathematical theory; or, to put it another way, the historiographical perspective that generates these kinds of questions illuminates important aspects of the history of mathematics. But do other crucial facets of that history remain obscure from the viewpoint?
>
> (Parshall, 2003, pp. 114–115)

We plead guilty to narrowing the meaning of the symbols "in North America 1607–1865" in our title so that the words have a different meaning from what they usually have. Of course, Canada is part of North America, but in this book we conveniently confine "North America" to all parts of the present mainland United States of America (except Alaska) and recognize that the extent of the territory described varied during the period 1607–1865. It will never refer to any parts of what are now called Canada, Alaska, or Mexico. It will often refer to the colonial settlements largely on the eastern coast of North America which were outside of Canada (with Florida being included after 1822), and to the present mainland states of the United States of America.

The date 1607 has been chosen because it denotes the year when the first permanent European settlement in "North America" began. The early settlers had left behind the houses, castles, churches, schools, universities, systems of administration, and other cultural artifacts of their homelands to take on the challenges they found in Jamestown (Ames, 1957). The year 1865 was a less obvious choice as an upper bound. For us, it represents a time when a new meaning was being given to the word "mathematics" in the United States of America. More on that will be discussed in Chapters 4 through 6 of this book. Here it suffices to notice that 1607–1865 is a 258-year span of time that has as its upper bound a year that marks the end of the Civil War. Even in 1865 only a small proportion of children in the United States of America were given the opportunity to study formal mathematics beyond counting and the four operations on Hindu-Arabic numerals. In one sense, "mathematics for all" was a long way from being achieved—but, in another sense, a pathway toward it was being established, and the methods being used to create it, and the identities of those who would create it, would indelibly affect not only direction but also the terrain over which that pathway would go.

From a historiographical perspective, the most important difference in the history which will be presented in this book from other histories of mathematics is that it is intended to throw light on the discontinuities and challenges faced by *all* who have walked, or are now walking, or who would soon begin to walk, on the "mathematics-for-all" pathway. From that perspective, this history is written from an *education* vantage point. We recognize, though, that the perspective on history that we offer is a *beginning*—much more will need to be done.

Mathematics Studied in North America in the Seventeenth Century

In May 1607, 104 English males (mostly men, but a few youths) arrived in North America to start a settlement. They decided to establish several forts, which they called "Jamestown," in what is now the State of Virginia. Jamestown was named after King James I of England, and "Virginia" after the company which financed the venture (Egloff & Woodward, 1992; Wecter, 1937). There had been numerous earlier, failed, attempts by Europeans to establish footholds in this New World—for example, at St Augustine in today's Florida in 1565, and the Roanoke Colony in today's North Carolina in 1585—but the 1607 event would result in the first *permanent* British settlement being established in North America (Morison, 1971; Price, 2003). During the seventeenth century not only did the Jamestown settlement survive, but other "colonies" were established along the east coast (Ames, 1957), for example—in New Hampshire, Massachusetts Bay, Connecticut, Rhode Island, New York, New Jersey, Delaware, Pennsylvania, Maryland, Virginia, North Carolina, South Carolina, and Georgia.

The total number of European-background people—including indentured servants—living in the colonies grew to about 250,000 by the beginning of the eighteenth century with "the women and children comprising at least two-thirds of the population" (*Ames, 1957, p. 6). During the seventeenth century the number of Native Americans fell but the number of black slaves brought from Africa steadily increased (Berlin, 1998; Blackburn, 1997; Dexter, 1887; Guasco, 2014; Wareing, 1985; Wells, 1975). Most European-background families were engaged in a struggle to survive (Ames, 1957; University of Michigan, 1967)—coping with the heavy demands of clearing the land, building, planting, harvesting, trading, performing household chores, defending territory and buildings, and establishing churches, businesses and legal and administrative structures (Eggleston, 1888). Locally-appointed councils created and interpreted the rules by which different communities operated. The Church was important in all of the colonies, and participation in its establishment and forms of worship was an important societal expectation. Although schools were established, and often supported by locally-arranged mandatory taxation, attendance at these schools was irregular because the labor of all but the youngest of the children was needed to assist in the struggle to survive. It became common for boys to go to school in winter, but not at other times. Nevertheless, it was true that some of the European-background settlers had attended high-class educational institutions in their homelands, before moving to North America, and they wanted their children to receive a higher education—and that explains why several "Latin" grammar schools, and higher-education colleges were established (Andrews, 1912; Cremin, 1970; Cubberley, 1920; Dauben & Parshall, 2014).

In 1642, the Massachusetts Bay Colony passed the first law in the New World requiring children to be taught to read and write. In 1647, Massachusetts passed another law requiring all towns of 50 families to have an elementary school and every town of 100 families to have a "Latin" school (Cremin, 1970; Cubberley, 1920). But passing laws to make attendance at school compulsory for children in a certain age-group, and making those laws effective were two different things, and it was many years before schools were attended regularly by all children in European-background families. In almost all cases, Native American children, children of indentured servants, and children of African American slaves were not welcomed in the "public" schools. The politics associated with the decisions which created these situations has been treated extensively elsewhere (see, e.g., Cremin, 1970; Smith, 1947), and is not a subject of attention in this book.

The summary in the above paragraphs suggests why most of the early settlers did not regard the formal study of mathematics as a sensible thing for themselves or for their children. Certainly, some families wanted their children to be well educated, and that motivated the establishment of higher-level education institutions. But these were more the result of settlers wanting to ensure that there was a reasonable local supply of medical doctors, lawyers and, especially, clergymen than of any serious appreciation of the value of higher education. As in Europe, the thinking was that any decent institution of higher learning should focus on the classics—definitely Latin, also some Greek, and perhaps a little Hebrew, should be part of the intended curriculum. Also, school learning was to complement the family and church so far as religious teaching was concerned. In all European-background communities, learning to read the Bible was regarded as extremely important. By contrast, mathematics beyond, perhaps, knowing how to count and measure in local situations was seen, by most, as largely irrelevant. Any idea of offering courses involving high-level mathematics, or conducting and reporting mathematics research, was rarely contemplated.

Eggleston (1888) summarized the position of education in the British colonies around 1700 in the following way:

> The schools were few and generally poor. Boys, when taught at all, learned to read, write and "cast accounts." Girls were taught even less. Many of the children born when the colonies were new grew up unable to write their names. There were few books at first, and no newspapers until after 1700. There was little to occupy the mind except the Sunday sermon. (p. 95)

For most European-background settlers there was neither time nor opportunity to pursue formal studies of any of algebra, geometry, or applied subjects like surveying, or navigation.

Estimates of the number of Native Americans already living, in 1607, in those parts which would become known as the "British colonies" have varied greatly—from 1 to 5 million. Whatever the number was, it fell as the seventeenth century progressed as a result of the introduction of devastating European diseases and race wars. The number of European-background persons grew from 104 at Jamestown in 1607 to about 250 thousand in the colonies in 1700 (Marshall, 2001; United States Census Bureau, 2004). For much of the seventeenth century, if not all of it, the number of Native Americans exceeded the number of European-background persons.

Terminology

We are not concerned, specifically, to provide extensive details in relation to the settlement of Jamestown in 1607. Rather "Jamestown" and "1607" will be used symbolically, denoting, respectively, that part of North America which today is part of the mainland of United States (not including Canada, or Alaska, or Hawaii, etc.), and the time when permanent settlement of Europeans in the New World (of "North America") first occurred. In this chapter we will be especially interested in the "mathematics" in this New World—not only the mathematics brought to the New World by the settlers, but also the forms of mathematics known and used by Native Americans at that time.

Our definition of the term "mathematics" for this chapter is inclusive in the sense that we are giving equal weight to mathematics and mathematics education. By the term "mathematics" we will include all aspects related to quantification, or counting, of discrete sets of objects, and ways of facilitating such quantification. It will also include methods of locating objects, and reasoning in space, and all aspects related to measurement of quantities, as well as to words and methods by which related concepts are defined and related, and the reasoning which permits theorems to be provided and proved.

We defined mathematics in this inclusive way in an attempt to make clear what we are investigating in this chapter. In the first half of the seventeenth century European educational institutions were still coming to grips with groundbreaking new mathematical ideas being put forward by mathematicians like the Frenchmen François Viète (1540–1603) and René Descartes (1596–1650), the Dutchman, Simon Stevin (1554–1620), the Scot, John Napier (1550–1617), and the Englishman, James Harriot (Struik, 2012). But such developments were a long way from the minds of most of the settlers in Jamestown or of other European-background settlers in what would become the British colonies. What mattered most for them was getting enough food and clothes in order to survive with dignity, and to establish peaceful relationships with local Native Americans.

In this book some attention will be given to the "spatial," "time," and "measurement" aspects of mathematics—the history of the development of these concepts, and how there are important cultural differences, has provided an ongoing agenda for researchers (see, e.g., Harris, 1981, 1991; Núñez & Cooperrider, 2013). Paul Libois, the radical Belgian mathematician and mathematics educator, referred to different kinds of geometrical spaces—a Euclidean space (x, y, z), a Galilean space (x, y, z, t) and other spaces like (x, y, z, t, p, T), where t denotes time, p pressure and T temperature. According to Libois (1951). the space of Euclid "was obtained through abstraction starting from (essentially) the consideration of solid bodies, imagined independently from time, and fixed with respect to an immovable body (the Earth)," but the other spaces were obtained from abstraction derived from real-life "optical, electrical and magnetic phenomena" (quoted from De Bock and Vanpaemel's (2019) translation, p. 15), and for Libois this suggested an educational approach for mathematics starting with naïve observations of "real" physical objects and proceeding via paths which involved increasing levels of abstraction. In other words, mathematics was not only what was arrived at through abstraction but included the path toward abstraction. The distinction is important in the history of mathematics in North America between 1607 and 1865, as deep thinkers like Benjamin Franklin, Thomas Jefferson, and Abraham Lincoln—persons not always regarded as mathematicians—consciously attempted to create abstract systems from realities, and then to apply those abstract systems to solve problems which confronted them. That will be discussed further in Chapter 8 of this book.

Indigenous Counting Systems and the Coming of the Hindu-Arabic Numeration System

Despite Tobias Dantzig's (1930) assertion to the contrary, there is considerable evidence that all well-formed groups of people have developed ways of counting (Bishop, 1988; Owens, Lean, Paraide & Muke, 2018; Silverman, 2006). That was obviously true in North America in the seventeenth century. More than a century ago, W. C. Eels (1913) reported that his research had revealed 306 different number systems employed by North American Indians and, of those, 146 were essentially decimal (i.e., base 10), 106 were essentially quinary or quinary decimal (i.e., base 5), 35 were vigesimal (i.e., base 20) or quinary-vigesimal, 15 were quaternary (i.e., essentially base 4), 3 were ternary (base 3), and 1 was octonary (base 8). Eels admitted that some of his classifications could have been wrong "due to inadequate data" (p. 293*n*).

Glendon Angove Lean's (1992) research, carried out between 1970 and 1990 in Papua New Guinea and Oceania, uncovered over 800 different languages and over 800 different counting systems (Owens et al., 2018)—many of which were still being used in villages in 1990. Although many of the counting systems documented by Lean (1992) were of the decimal variety, those decimal systems—originating in most cases from the number of fingers and thumbs on two hands—had subtle differences. Both Eels and Lean recognized that different base 10 structures existed—for example, in one structure "16" might be thought of, and expressed as, $10 + 5 + 1$, and in another as $10 + 2 + 2 + 2$, etc.

Eels (1913) and Lean (1992) found that there were many systems which employed bases other than 10, and there were also some "body-count" systems (with no "base"). Often systems had bases related to fingers and toes. For example, counting fingers and toes probably gave rise to vigesimal systems, and were often found—although not always—among groups which did not normally wear moccasins or other forms of "shoe" which covered feet. An interesting case came from the now-extinct Yuki language in California, which had an octal system because the speakers counted using the spaces between their fingers rather than the fingers themselves (Ascher, 1992). In 1752 a former William and Mary College mathematics professor, the Reverend Hugh Jones, argued that a base 8 number system was superior to decimal systems for arithmetical computations. His 47-page manuscript on that theme, *The Reasons and Rules and Uses of Octave Computation or Natural Arithmetic,* is now held in the British Museum.

Lean (1992) found that none of the indigenous counting systems that he identified had a name for "zero." Specific numerical and linguistic treatments of fractions were not found either (although indigenous languages always included expressions for sharing, or splitting, etc., which thereby enabled what might be regarded as fraction concepts to be identified and discussed).

During the period 1607–1865 there were large groups of Native Americans to be found in many regions within North America—the Iroquois (including the Mohawks, Senacas, Oneida, Onondago, Cayoga), the Navajo, the Apache, the Cheyenne, the Sioux (including the Lakota, Dakota, and Nakota), the Hopi, the Seminoles, and the Commanches, were just a few of these groups. Each had its own language and its own counting system. The groups' counting systems helped them to keep track of what they owned and what they measured, and to provide answers to practical issues arising from how they lived. Most did not know, or care about, the counting systems of others (Eels, 1913). Worksheets colorfully summarizing the number systems of 68 different indigenous groups in North America can be found at http://www.native-languages.org/numbers.htm.

It would be unhelpful to provide further details, here, for indigenous counting systems—that is not the main theme we are addressing. Rather, it is important to note that when the European "settlers" arrived in Jamestown in 1607 they brought with them another counting system—one which had not been used by any of the Native American societies up to that time. That system was the Hindu-Arabic numeration system, with its numerals 0, 1, 2, 3, 4, 5, 6, 7, 8 and 9, and its ingenious place-value system for representing numbers greater than 9. It also had well-developed ways of predicting how many objects were in sets of objects by applying standard algorithms for "addition, subtraction, multiplication and division." This system had initially been developed in India during the first seven centuries of the Common Era (CE) and had then been adopted and utilized across Arab nations before it found its way into European nations during the period 900–1200 CE (Danna, 2019; Ifrah, 2000; Menninger, 1969; Smith & Karpinski, 1911; Wardley & White, 2003; Høyrup,

2014). Its power had been displayed as it transformed local, national, and international commerce.

Historical perspective suggests that the Hindu-Arabic system of numeration was the most transformative mathematical development of all time, and the rapidity of its spread across India, then across Arab nations, and then across European nations testified to the recognition, by merchants in many parts of the "Old World," that it was a key to wealth and success (Danna, 2019; Høyrup, 2014; Ifrah, 2000). But before 1607 it was unknown to the Native American peoples. The invaders spoke various strange languages, but the leaders of the various Native American communities had no reason to suspect that over the next several hundred years there would be as much pressure, and sometimes more pressure, placed on them to change from their traditional counting systems to this "new" Hindu-Arabic numeration system as there would be for them to change from the languages that they used in everyday conversations.

Who Used the Hindu-Arabic Numeration System in North America, 1607–1699?

During the period 1607–1620 the only persons in North America to use the Hindu-Arabic numeration system would have been the settlers at Jamestown. Although we do not know how many of the original settlers were able to use the system freely, we do know that as the seventeenth century progressed more and more Europeans who knew how to use the system crossed the Atlantic and settled at various points along the East coast of North America. We also know that in 1635 the Boston Latin School was established in New England, and New College—which would become Harvard College—was established at nearby Newtown(e) (now Cambridge) in 1636. Although the early European education institutions did not give special attention to arithmetic—their focus was on community living and, for older children, on Latin—they provided basic education in religion and reading complemented by writing and a small amount of arithmetic. The forms of education which were implemented differed markedly from those in today's schools because there was very little paper or ink available, there were rarely any textbooks, there were no written examinations, and most teachers lacked sound understandings of what they were expected to teach. At the former New College, which became Harvard College in 1639, students from well-to-do families prepared to become lawyers, medicos and, especially, clergymen, and throughout much of the day students were expected to converse in Latin (Morison, 1935)—although English tended to be used for instruction in mathematics (Zitarelli. 2019).

During the seventeenth century most of the children of European-background free settlers would have been expected to learn to read, write, and say numerals expressed as combinations of some of the Hindu-Arabic numerals 1, 2, 3, 4, 5, 6, 7, 8, 9 and 0 (Finegan, 1917; Kilpatrick, 1912). Of one thing we can be certain: any forms of mathematics studied in the early "schools" would not have been known by more than a handful of indigenous persons living in Native American communities.

The *Abbaco* Sequence for Arithmetic

The "intended curriculum" for most mathematical programs in North American schools during the period from 1607 through 1865 derived from what has been called the *abbaco* sequence (Ellerton & Clements, 2012). That sequence was a well-ordered set of topics associated with courses in business arithmetic which had been standardized in European reckoning schools. For many years it was accepted by scholars that the *abbaco* sequence was initially developed in India, then further developed in Arabic nations, and finally translated into European city states—largely through Leonardo of Pisa's (Fibonacci's) *Liber Abbaci*, which was written around 1200 CE (see, e.g., Smith & Karpinski, 1911; Yeldham, 1926, 1936). In recent years however, Jens Høyrup (2005) has shown that between 900 and 1200 CE there were features of the *abbaco* tradition, and also aspects of algebra, already to be found in parts of Western Europe, and especially in Spain.

The *abbaco* sequence began with "numeration tables" which provided summaries of the Hindu-Arabic numeration base 10, place-value system. It then moved on to algorithms for the four operations (addition, subtraction, multiplication, and division) on whole numbers (Yeldham, 1936). Then came elementary measurement (including units) in which the Hindu-Arabic numerals were used to indicate measurements of amounts of quantities. Part of this measurement section was concerned with a topic known as "reduction." Then followed loss and gain, ratio and proportion (called the "rules of three"), currency exchange, equation of payments, barter, interest (simple and compound), tare and tret(t), discount, and brokerage. At the most advanced level would come topics like vulgar (i.e., "common") fractions, commission, alligation (i.e., the arithmetic of mixing quantities), fellowship (i.e., the arithmetic of partnerships), false position, progressions, involution and evolution, permutations and combinations, and mensuration.

The *abbaco* sequence usually did not include formal study of any of algebra, Euclidean geometry, or trigonometry, and only a small proportion of students, almost all of them from well-to-do families, ever got to study those branches of mathematics—usually in "high-class" grammar schools and colleges. For most students, however, the emphasis was on learning rules and cases in the *abbaco* sequence and on applying those to problems which might arise in business contexts (Ellerton & Clements, 2012, 2014).

In the seventeenth and eighteenth centuries, pre-college students rarely owned a textbook, and only a small proportion of them proceeded to the more advanced *abbaco* topics. In fact, only a few of the teachers had ever studied the more advanced topics themselves. The method of instruction was almost always consistent with what has been called the "cyphering tradition," by which most male students aged

10 years or more, and about 20 percent of female students in that age bracket, prepared handwritten "cyphering books" (see Chapter 3 of this book).

Differences in the Opportunity to Learn *Abbaco* Arithmetic in North America, 1607–1865

Mathematicians have always been interested in identifying and documenting the careers of females who were exceptionally gifted in mathematics—probably because many members of society have long questioned the idea, sometimes put forward, that mathematics should be regarded as a quintessential male subject. There has been much written about the contention that females, considered as a group of people, are not as talented as males in mathematics and the physical sciences, but are more talented than males in language studies, needlework and sewing (see e.g., Cohen, 1993; Harris, 1997; Patterson, 2012). But, of course, there are categories of people other than those distinguished by gender which, historically, have been associated with lower levels of participation, or lack of participation, in higher mathematics. One can think of class (working class versus upper class, etc.), region (rural versus urban), race differences, and so on.

This book is concerned with the history of mathematics in North America between 1607 and 1865. It will be assumed that the word "mathematics" embraces all aspects related to measurement of quantities, as well as spoken words and written symbols and other aspects of the language by which physical objects and concepts were defined, quantified, related, and communicated during the period under consideration. We also include "higher mathematics"—the kinds of mathematics studied in the upper echelons of departments of mathematics in colleges and universities, and also the kinds of mathematics that researchers investigate—within the ambit of our discussion. We will argue that historians investigating the development of mathematics in North America between 1607 and 1865 need to recognize that many groups of people within North America have always been, and continue to be, severely disadvantaged with respect to the opportunities that they have been given to study mathematics, and especially higher forms of mathematics. By contrast, certain other groups have been advantaged.

We begin by creating 16 subdivisions based on a subdivision in time (two periods, one between 1607 and 1699, and the other between 1700 and 1865); four subdivisions based on race and servitude (European-background persons who were not indentured servants, European-background persons who were indentured servants, Native Americans, and African-American persons); and two subdivisions based on gender (male or female).

In summary, the framework draws attention to:

- *Two subdivisions based on time:* we distinguish between the amount of participation in mathematics in North America (a) between 1607 and 1699, and (b) between 1700 and 1865.

- *Four subdivisions based on different groups studying, or teaching, or researching mathematics in the following categories*: (a) European-background persons who were not indentured servants; (b) European-background persons who were indentured servants; (c) Native Americans; and (d) African-Americans.
- *Two subdivisions based on gender*: (a) male persons and (b) female persons.

These subdivisions can give rise to $2 \times 4 \times 2 = 16$ distinguishable groups—for example, one might consider "the 1607–1699 group of Native American males," or "the 1700–1865 group of African-American females."

The reader might wonder whether the number of European-background whites who were indentured servants was sufficiently large to warrant their being separated into a unique category. The answer is definitely "Yes" (Chessman, 1965). Economic historians and economists have reported data indicating that the number of indentured servants increased in all 13 colonies in the seventeenth century (Galenson, 1984). There are data indicating that between the years 1630 and 1776, one-half to two-thirds of Caucasian immigrants to the 13 colonies came as indentured servants (Ames, 1957; Smith, 1947; Whaples, 1995).

Of the 16 groups, only 6 had significant percentages of persons—i.e., significantly more than 0%—who received a formal education that took account of more than a very elementary level in the *abbaco* sequence. Our *estimated* percentages of students in the 6 groups who received such an education are shown between parentheses at the end of each line in the following list:

1. 1607–1699 European-background males who were not indentured servants (40%);
2. 1607–1699 European-background males who were indentured servants (10%);
3. 1700–1865 European-background males who were not indentured servants (70%);
4. 1700–1865 European-background males who were indentured servants (30%);
5. 1700–1865 European-background females who were not indentured servants. (20%);
6. 1700–1865 European-background females who were indentured servants. (10%)

We emphasize that the percentages shown merely represent our estimates—research has not been done which would reveal the actual percentages. It was only in rare circumstances that a Native American or an African American person had the opportunity to study *abbaco*-type arithmetic in common schools at any time between 1607 and 1865. That fact needs to be recognized in any evaluation of Kamens and Benavot's (1991) claim, a claim repeated by Jeremy Kilpatrick (2014), that in

U.S. common schools, arithmetic was made a compulsory school subject by 1790. The meaning of "compulsory" in that assertion is problematic. In fact, although Kamens and Benavot acknowledged that there was no U.S. national curriculum for common schools in 1790, their analysis assumed that this was "not a serious drawback" (p. 171). We disagree. For example, they do not take account of the fact that throughout the whole of the period 1607–1865 most boys who attended schools in rural districts, did so in winter months only; furthermore, attendance rates and intended curricula differed from state to state. We find Kamens and Benavot's (1991) analysis of curricula in U.S. common schools of the seventeenth and eighteenth centuries seriously lacking in specific detail and their main conclusions highly questionable.

We estimate that only about 20 percent of all white European-background males living in North America during the seventeenth century had ever studied, or would study, *abbaco* arithmetic beyond the most elementary level, and that during the eighteenth century the corresponding percentage was about 35. During the seventeenth century much less than 10 percent of European-background females living in North America would have studied *abbaco*-arithmetic beyond the most elementary level, and during the eighteenth century the percentage was never likely to have risen to above 20 (Ellerton & Clements, 2012, 2014).

The remarkable thing is that even in the 1790s certainly less than 10 percent and probably well less than 5 percent of those belonging to all the other 10 categories, had ever studied arithmetic beyond the most elementary *abbaco* level. The fierce inequalities of educational opportunity which might be associated with that statement have never been adequately addressed by researchers in education, history, or mathematics.

The 16-subgroup structure for analysis outlined in the above paragraphs offers a basis for a research agenda so far as the history of mathematics and mathematics education in North America is concerned. Consider, for a moment, what other categories might be added (e.g., rural versus urban, North versus South, English-speaking versus non-English-speaking, students doing apprenticeships versus students still at day-school, students living in big cities versus those living in remote frontier regions). One might reflect, too, on the extent to which the situations would differ if we were wishing to provide a basis for comparing the histories of mathematics and mathematics education in Great Britain, or France, or Spain, or Germany, or The Netherlands, or, more generally, in Western Europe.

An examination of the conjectures we have just made should make it clear that we contend that during the seventeenth century relatively few people living in the 13 colonies studied mathematics beyond *abbaco* arithmetic or other elementary forms of "Western" mathematics. That was largely because most had neither the opportunity nor the desire to do so. We do not know how many would have liked to study *abbaco*-type mathematics in the various groups but were not given the opportunity to do so, but it is likely that that number would have been small. What is interesting is our conjecture that after 1607 the situation improved—if that is the right word—only slightly over the next 200 years. That conjecture is consistent with the summary presented by David Eugene Smith and Jekuthiel Ginsburg (1934). And, incidentally, a similar situation prevailed in Great Britain with respect to the mathematics education of the young—Howson and Rogers (2014) have reported that in 1824 less than 50 percent of those attending British schools were taught arithmetic.

It is not surprising, then, that by 1865, in North America, there was a massive problem facing anyone who did not have a European background and wanted to study mathematics, at any level (Drake, 1963). In the state of Virginia, for instance, there were more African-American slaves and their children than there were European-background persons who were not indentured servants, and hardly any of the African Americans had attended school (Drake, 1963; Wareing, 1985). Research is needed which establishes benchmarks and progressions in learning so far as participation in mathematics of different racial groups in North America is concerned. Compared with what prevailed in France and Germany, for example, and contrary to a claim made by Kamens and Benavot (1991) and accepted by Kilpatrick (2014), we believe that in 1865 the United States, as a nation, had a lot of "catching up" to do, at all levels of mathematics education (Kline, 1972; Parshall, 2003; Smith & Ginsburg, 1934). If our conjectures are reasonably accurate then there is no way we would expect that by 1900 more than a tiny proportion of North American mathematicians children" would reach the same level of research quality in mathematics as that reached by European mathematicians. In 1865, and even in 1900, most girls, Native Americans, and African-Americans (and working-class children, etc.) had much less opportunity than "corresponding children of the same age in some Western European nations to advance in any mathematical studies (Vickers, 2008).

The Main Aims for This Book

This book offers an overview of a history of mathematics in the 13 colonies during the colonial period and in the United States of America during the period 1776–1865. Throughout the book the word "mathematics" will be taken to mean mainly the Hindu-Arabic *abbaco* sequence for arithmetic if we are referring to children (up to the age of 15 years). For students, between 15 and 18 years, it will refer to more advanced topics in the Hindu-Arabic *abbaco* sequence, to measurement, and sometimes to algebra, trigonometry, geometry, and sometimes (though rarely) to calculus. At the college and research levels, we will be referring to the mathematics studied or taught or researched by students and teachers.

In Chapters 5 through 8 we will argue, like Parshall (2003) and Smith and Ginsburg (1934), that internationally-recognized research in mathematics by North American scholars did not appear until the early 1800s, and that there was not a great deal of this before 1865. One of the issues considered in this book is why it took so long for an internationally-recognized mathematics research sub-culture to appear in North America.

We shall assume that the terms *intended* mathematics curriculum, *implemented* mathematics curriculum, and *attained* mathematics, as introduced by Ian Westbury (1980), are well defined. The "intended curriculum" corresponds to the sequence of mathematical topics, and approaches, which schools, textbook authors, local education authorities, and teachers expect students to learn for a well-defined period (like, for example, over a period of one year, or over a period of, say, four years). It also includes the idea of preferred teaching methods of the schools, textbook authors, and teachers for delivering the intended content. By contrast, the "implemented curriculum" will refer to the content actually taught, and to the ways it was taught. The "attained curriculum" will refer to what the students learned and retained about the content of the implemented curriculum.

The Six Research Questions

We now state the following six main questions which will be investigated in this book:

1. What were the intended, implemented and attained mathematics curricula for young children (aged less than 10 years) in North America (a) during the period 1607–1820? and (b) the period 1820–1865?, and to what extent do the answers to those questions vary across North America, and in different groups of children (e.g., boys versus girls, European-background children versus Native American children, and European-background children versus African-American children)?
2. What were the intended, implemented and attained mathematics curricula for North American children aged between 10 and 15 years during (a) the seventeenth century, and (b) the period 1700–1865, and to what extent do the answers to those questions vary across North America, and across different groups?
3. What were the intended, implemented and attained mathematics curricula for North American pre-college children aged between about 15 and 18 years during (a) the seventeenth century, and (b) the period 1700–1865, and to what extent do the answers to those questions vary across North America, and across different groups?
4. What were the intended, implemented and attained mathematics curricula for North American college students during (a) the period 1607–1776? and (b) the period 1776–1865, and to what extent do the answers to those questions vary across North American colleges, and across different groups?
5. What perspectives on the purposes and status of mathematics in college curricula were held in the North American colonies during the period 1607–1865?

6. What are the implications of the answers to the first five questions (above) for those investigating the history of "higher" mathematics in North America? What future research is needed, and to what extent will it be feasible to conduct that research?

Research mathematicians reading this book might be disappointed with those six questions because only one of them—the fifth—refers, albeit indirectly—to the history of mathematics research in North America. We have worked from the perspective that "mathematics" encompasses more much than merely research in mathematics or the teaching of higher-level mathematics in advanced colleges. We believe that for the period between 1607 and 1865 the history of mathematics in North America should be as much concerned with the history of the development of structures by which people of all ages were enabled to learn mathematics—that is to say, with the history of mathematics education—as with changes in the mathematics which was studied or researched in higher-education institutions. That is not to say that serious research in mathematics did not take place in North America during the period 1607–1865. Identifying that research is regarded as something within the scope of this book.

One might ask why anybody should write a book on the history of mathematics in North America between 1607 and 1865? What use could such a history possibly be for today's readers? Is this book nothing more than an academic exercise? Well, no, we hope that this book will be important for those who want to gain an insight into why mathematics came to be identified, by so many, with white, male privilege. The quotation from James Baldwin (1998)—after the abstract and keywords at the start of this chapter—is relevant to what we are trying to say through the pages of this book. Please read Baldwin's statement again, now, and also, read it once more after you've finished reading Chapter 9, the last chapter of this book.

The Concept of "School" in this Book

Before moving on it will be useful to define the concept of "school" as it was used in North America during the seventeenth and eighteenth centuries. The word "school" will be taken to include "academies," "apprenticeship schools," "common schools," "dame schools," "evening schools," "grammar schools," "local schools," "private schools," "public schools," "subscription schools," and "writing schools" (Clements & Ellerton, 2015; Cremin, 1970, 1977), as well as more specialized establishments like "dance schools," "elocution schools," and "navigation schools' and "French ladies' colleges." A narrower interpretation of the word "school" than what is implied by that collection of terms is also relevant—so that any formal education environment in which at least one "teacher" regularly met with at least one "student," at an agreed place, for the purpose of helping the student(s) to learn facts, concepts, and skills, from at least one of reading, writing, or arithmetic, will be regarded as having been a school (Ellerton & Clements, 2012). This definition implies that for the purposes of this book a school did not need to offer formal tuition in any form of mathematics.

Higher-level colleges—such as King's College (now called Columbia University), Harvard, William and Mary, and Yale—will *not* be regarded as "schools." During the seventeenth and eighteenth centuries, and also during the early nineteenth century, such higher-level institutions were usually called "colleges" and were sharply distinguished from "schools."

Outline of Chapters in this Book

There are nine chapters. In this first, introductory, chapter we have provided necessary definitions, and offered conjectures which were intended to define a research agenda for scholars already investigating, or intending to investigate, the history of mathematics in North America. We also put forward six research questions which will be addressed and answered.

Chapter 2 will offer a summary of the mathematics studied by young children (aged less than 10 years) in North America during the period 1607–1865, and Chapter 3 will offer a summary of intended, implemented and attained mathematics curricula in North America during the same period for children aged between 10 and 16 years. Chapter 4 will do likewise, only with respect to those who proceeded as far as more advanced *abbaco* arithmetic topics, or for those who studied elementary forms of algebra, trigonometry, geometry (and perhaps applied topics like navigation and surveying) at the pre-college level. With each of Chapters 2, 3, and 4, findings will be linked to the conjectures we made after we defined 16 categories of people earlier in this chapter. Chapter 5 will be concerned with the introduction and development of algebra in curricula after 1607, and Chapter 6 will focus on creative *applied* mathematics-related developments which occurred and were reported by education establishments. Chapter 7 will address issues associated with college mathematics during the period 1607–1865, and Chapter 8 will identify persons who developed distinctive ways of looking at, and using, mathematics during the same period.

In the final chapter (Chapter 9), tentative answers will be given to each of the six research questions. The statements of these tentative answers will lead directly to a consideration of questions which might fruitfully be addressed by future researchers, and of difficulties that those carrying out such future research might be expected to experience.

We think of this book as representing our final words to those who will carry out needed research in the future. We hope the book will be rich in the sense that it will pass on to readers what we have learned over the past 15 to 20 years as we have researched the history of North American mathematics and mathematics education. At times it has been an exhilarating experience for us, chasing rare references, artifacts, and documents, reflecting on what others have written, and reporting the conclusions that we have reached. Some might think it is unfortunate that so few scholars have contributed to the enterprise, but a more positive view is that the field "is ripe unto harvest."

References

Andrews, C. M. (1912). *The colonial period.* New York, NY: Henry Holt and Company.

Ascher, M. (1992). Ethnomathematics: A multicultural view of mathematical ideas. *The College Mathematics Journal, 23*(4), 353–355. https://doi.org/10.2307/2686959

Baldwin, J. (1998). *Collected essays.* New York, NY: Library of America.

Bell, E. T. (1945). *The development of mathematics* (2nd ed.). New York, NY: McGraw-Hill Book Company.

Berlin, I. (1998). *Many thousands gone: The first two centuries of slavery in North America.* Cambridge, MA: Harvard University Press.

Bishop, A. J. (1988). *Mathematical enculturation.* Dordrecht, The Netherlands: Reidel. https://doi.org/10.1007/978-94-009-2657-8

Blackburn, R. (1997). *The making of new world slavery: From the Baroque to the modern 1492–1800.* London, England: Verso.

Blank, B. E. (2001). *What is mathematics: An elementary approach to ideas and methods. Book review. Notices of the AMS, 48*(11), 1325–1329.

Clements, M. A., & Ellerton, N. F. (2015). *Thomas Jefferson and his decimals; neglected years in the history of U.S, school mathematics.* New York, NY: Springer.

Chessman, R. (1965). *Bound for freedom.* New York, NY: Abelard-Schuman.

Cohen, P. C. (1993). Reckoning with commerce. Numeracy in 18th-century America. In J. Brewer, & R. Porter (Eds,). *Consumption and the world of goods* (pp. 320–334). New York, NY: Routledge.

Courant, R., & Robbins H. (1941). *What is mathematics? An elementary approach to ideas and methods.* London, England: Oxford University Press.

Cremin, L. A. (1970). *American education: The colonial experience 1607–1783.* New York, NY: Harper & Row.

Cremin, L. A. (1977). *Traditions of American education.* New York, NY: Basic Books.

Cubberley, E. P. (1920). *The history of education.* Boston, MA: Houghton Mifflin Company.

Danna, R. (2019). *The spread of Hindu-Arabic numerals in the European tradition of practical mathematics (13th–16th centuries).* Cambridge, England: University of Cambridge Department of History.

Dantzig, T. (1930). *Number: The language of science.* London, England: The Macmillan Company. https://doi.org/10.2307/2224269

Dauben, J. W., & K. H. Parshall (2014). Mathematics education in North America to 1800. In A. Karp & G. Schubring (Eds.), *Handbook on the history of mathematics education* (pp. 175–185). New York, NY: Springer.

Dexter, F. B. (1887). Estimates of population in American colonies. *Proceedings of the American Antiquarian Society* (pp. 22–50). Boston, MA. American Antiquarian Society.

Drake, S. (1963). *The American dream and the Negro: 100 years of freedom*. Chicago, IL: Roosevelt University.

De Bock, D., & Vanpaemel, G. (2019). *Rods, sets and arrows: The rise and fall of modern mathematics in Belgium*. Cham, Switzerland: Springers. https://doi.org/10.1007/978-3-030-20599-7

Eels, W. C. (1913). Number systems of the North American Indians. *The American Mathematical Monthly*, *20*(10), 293–299. https://doi.org/10.1080/00029890.1913.11997985

Eggleston, A. (1888). *A history of the United States and its people*. New York, NY: American Book Company.

Egloff, K., & Woodward, D. (1992). *First people: The early Indians of Virginia*. Charlottesville, VA: The University Press of Virginia.

Ellerton, N. F., & Clements, M. A. (2012). *Rewriting the history of mathematics education in North America, 1607–1861*. New York, NY: Springer. https://doi.org/10.1007/978-94-007-2639-0

Ellerton, N. F., & Clements, M. A. (2014). *Abraham Lincoln's cyphering book and ten other extraordinary cyphering books*. New York, NY: Springer. https://doi.org/10.1007/978-3-319-02502-5

Finegan, T. E. (1917). Colonial schools and colleges in New York. *Proceedings of the New York State Historical Association, 16,* 165–182.

Galenson, D. W. (1984). The rise and fall of indentured servitude in the Americas: An economic analysis. *The Journal of Economic History, 44*(1), 1–26. https://doi.org/10.1017/S002205070003134X

Greenwood, I. (1729). *Arithmetick, vulgar and decimal, with the application thereof to a variety of cases in trade and commerce*. Boston, MA: Kneeland & Green.

Guasco, M. (2014). *Slaves and Englishmen: Human bondage in the early modern Atlantic world*. Philadelphia, PA: University of Pennsylvania Press. https://doi.org/10.1017/S0310582200011482

Harris, M. (1997). *Common threads: Women, mathematics and work*. Stoke-on-Trent, England: Trentham Books.

Harris, P. (1981). Measurement in tribal Aboriginal communities. *The Australian Journal of Indigenous Education, 9*(2), 53–61. https://doi.org/10.1017/S0310582200011482

Harris, P. (1991). *Mathematics in a cultural context; Aboriginal perspectives on space, time and money*. Geelong, Australia: Deakin University.

Howson, A. G., & Rogers, L. (2014). Mathematics education in the United Kingdom. In A. Karp & G. Schubring (Eds.), *Handbook on the history of mathematics education* (pp. 257–282). New York, NY: Springer.

Høyrup, J. (2005). Leonardo Fibonacci and *abbaco* culture: A proposal to invert the roles. *Revue d'Histoire des Mathématiques, 11*, 23–56. https://doi.org/10.1007/978-1-4614-9155-2_13

Høyrup, J. (2014). Mathematics education in the European Middle Ages. In A. Karp & G. Schubring (Eds.), *Handbook on the history of mathematics education* (pp. 109–124). New York, NY: Springer.

Ifrah, G. (2000). *The universal history of numbers from prehistory to the invention of the computer*. New York, NY: John Wiley & Sons. Inc.

Jones, H. (c. 1752). *The reasons and rules and uses of octave computation or natural arithmetic*. London, England: Author (held in the British Museum).

Kamens, D. H., & Benavot, A. (1991). Elite knowledge for the masses. The origins and spread of mathematics and science education in national curricula. *American Journal of Education, 99,* 137–180.

Kilpatrick, J. (2014). Mathematics education in the United States and Canada. In A. Karp & G. Schubring (Eds.), *Handbook on the history of mathematics education* (pp. 323–334). New York, NY: Springer. https://doi.org/10.1007/978-1-4614-9155-2_16

Kilpatrick, W. H. (1912). *The Dutch schools of New Netherland and colonial New York.* Washington, DC: United States Bureau of Education.

Kline, M. (1972). *Mathematical thought from ancient to modern times.* New York, NY: Oxford University Press.

Lean G. A. (1992). *Counting systems of Papua New Guinea and Oceania.* PhD dissertation, Papua New Guinea University of Technology (Lae, Papua New Guinea).

Libois, P. (1951). *Les espaces*. Liège, Belgium: Thone.

Marshall, P. J. (Ed.). (2001). *Oxford history of the eighteenth century*. Oxford, England: Oxford University Press.

Menninger, K. W. (1969). *Number words and number symbols: A cultural history of numbers*. Boston, MA: MIT Press. https://doi.org/10.2307/2799719

Morison, S. E. (1935). *The founding of Harvard College*. Cambridge, MA: Harvard University Press.

Morison, S. E. (1971). *The European discovery of America*. New York, NY: Oxford University Press.

Núñez, R., & Cooperrider, K. (2013). The tangle of space and time in human cognition. *Trends in Cognitive Sciences, 17*(5), 220–229. https://doi.org/10.1016/j.tics.2013.03.008

Owens, K., Lean, G., Paraide, P. & Muke, C. (2018). *History of number: Evidence from Papua New Guinea and Oceania*. New York, NY: Springer. https://doi.org/10.1007/978-3-319-45483-2

Parshall, K. H. (2003). Historical contours of the American mathematical research community. In G. M. A. Stanic & J. Kilpatrick (Eds.), *A history of school mathematics* (Vol. 1, pp. 113–158). Reston, VA: National Council of Teachers of Mathematics.

Patterson, E. C. (2012). *Mary Somerville and the cultivation of science, 1815–1840*. Boston, MA: Kluwer.

Price, D. A. (2003). *Love and hate in Jamestown: John Smith, Pocahontas, and the start of a new nation*. New York, NY: Alfred A. Knoff.

Roberts, D. L. (2019). *Republic of numbers: Unexpected stories of mathematical Americans through history*. Baltimore, MD: Johns Hopkins University Press.

Silverman, D. J. (2006). Faith and boundaries: Colonists, Christianity and community among the Wampanoag Indians of Martha's Vineyard, 1600–1871. *History, 91*(304), 583–584. https://doi.org/10.1111/j.1468-229X.2006.379_7.x

Smith, A. E. (1947). *Colonies in bondage: White servitude and convict labor in America 1607–1776*. Chapel Hill, NC: University of North Carolina.

Smith, D. E., & Ginsburg, J. (1934). *A history of mathematics in America before 1900*. Chicago, IL: The Mathematical Association of America. https://doi.org/10.1090/car/005 https://doi.org/10.1090/car/005

Smith, D. E., & Karpinski, L. C. (1911). *The Hindu-Arabic numerals*. Boston, MA: Ginn & Co.

Stedall, J. (2012). *The history of mathematics: A very short introduction*. New York, NY: Oxford University Press.

Struik, D. J. (2012). *A concise history of mathematics* (4th ed.). New York, NY: Dover Publications.

United States Census Bureau. (2004). *Colonial and pre-Federal statistics*. Suitland, MD. Author.

University of Michigan. (1967). *Education in early American*. Ann Arbor, MI: The University of Michigan Library.

Vickers, D. (2008). *A companion to colonial America*. New York, NY: Wiley-Blackwell.

Wardley, P., & White, P. (2003). The Arithmeticke Project: A collaborative research study of the diffusion of Hindu-Arabic numerals. *Family and Community History, 6*(1), 5–17. https://doi.org/10.1179/fch.2003.6.1.002

Wareing, J. (1985). *Emigrants to America: Indentured servants recruited in London 1718–1733*. Baltimore, MD: Genealogical Publishing Co.

Wecter, D. (1937). *The saga of American society: A record of social aspiration 1697–1937*. New York, NY: Charles Scribner's Sons.

Wells, R. V. (1975). *Population of the British colonies in America before 1776: A survey of census data*. Princeton, NJ: Princeton University Press.

Westbury, I. (1980). Change and stability in the curriculum: An overview of the questions. In H. G. Steiner (Ed.), *Comparative studies of mathematics curricula: Change and stability 1960 –1980* (pp. 12–36). Bielefeld, Germany: Institut für Didaktik der Mathematik-Universität Bielefeld.

Whaples, R. (1995). Where is there consensus among American economic historians? *The Journal of Economic History, 55*(1), 139–154. https://doi.org/10.1017/S0022050700040602

Yeldham, F. A. (1926). *The story of reckoning in the Middle Ages*. London, England: George A. Harrap.

Yeldham, F. A. (1936). *The teaching of arithmetic through four hundred years (1535–1935)*. London, England: George A. Harrap.

Zitarelli, D. A. (2019). *A history of mathematics in the United States and Canada. Volume 1: 1492–1900*. Washington, DC: American Mathematical Society. https://doi.org/10.1090/spec/094

Chapter 2
Young Children's Introduction to Mathematics in North America Between 1607 and 1865

Abstract In this chapter we consider the mathematics studied by young children—not yet 10 years of age—the eastern colonies during the seventeenth and eighteenth centuries. We draw special attention to the hornbook—the artifact which most influenced intended, implemented and attained curricula for young European-background children during the period 1607–1799—and provide details on what is possibly the earliest extant hornbook constructed and used in North America during that period. During the seventeenth century there was little opportunity for most young children to advance their understandings of mathematics. Evidence is put forward that as late as the beginning of the nineteenth century most children aged less than 10 years were not given any opportunity to study any form of mathematics beyond counting verbally and learning to read and write the Hindu-Arabic numerals. It was not until the early 1820s that the idea began to be accepted by some North American educators. that all young children from about the age of 6 should learn to read and write Hindu-Arabic numerals, and to develop other elementary arithmetical concepts and skills. Once that idea was put forward, initially by Warren Colburn, it steadily gathered momentum among scholars, educators, and the society at large.

Keywords *Abbaco* system for arithmetic • Andrew White Tuer • Antiquarian Society of America • Battledore • Dame schools • George A. Plimpton • Girls and mathematics • Harvard • Hornbook • Implemented curriculum • Indentured servants • Intended curriculum • Native American children • New Amsterdam • Pestalozzi • Slaves • Warren Colburn

Educating Young Children in North America, 1607–1799

Most treatises on the history of North American education for the period 1607–1799 have had little to say about the mathematics formally studied by young children less than 10 years of age (see, e.g., Ames, 1957; Eggleston, 1888; Littlefield, 1904; Monaghan, 2007). Indeed, for that same period no well-researched history of the teaching and learning of mathematics with respect to young children in North America has ever been published. In their chapter on "From Discovery to an Awakened Concern for Pedagogy: 1492–1821," in the National Council of Teachers of Mathematics' (1970) *Thirty-Second Handbook,* Phillip Jones and Arthur Coxford (1970) quoted, with approval, the claim that even around 1800 the intended curriculum for day schools comprised "spelling, reading, and writing" (see p. 13). Other than making that statement, Jones and Coxford were silent on the subject.

N. F. Ellerton, M. A. (Ken) Clements, *Toward Mathematics for All*, History of Mathematics Education,
https://doi.org/10.1007/978-3-030-85724-0_2

In Alexander Karp's and Gert Schubring's (2014) 634-page edited collection, *Handbook on the History of Mathematics Education*, there is no specific chapter on, or systematic discussion of, the history of mathematics education for children less than 10 years of age. Nor was there any focused discussion on that aspect of mathematics education in the *International Journal for the History of Mathematics Education,* which was published between 2006 and 2015.

William Heard Kilpatrick (1912), in his detailed study of the Dutch schools of New Amsterdam and colonial New York, drew attention to the difference between intended, implemented and attained curricula as far as mathematics for young children was concerned (Westbury, 1980). After pointing out that although in 1636 the official Dutch program for its colonies required teachers to instruct "the youth in reading, writing, cyphering, and arithmetic" (p. 220), Kilpatrick (1912) provided evidence that in many schools in New Amsterdam (later to be renamed New York), arithmetic was not taught because it was not sufficiently well known by the teachers. He concluded that very little arithmetic was found "in the schools of Holland America" (p. 221). However, he qualified that statement by adding that arithmetic was much more likely to appear in central New Amsterdam schools than in "the outlying Dutch villages" (p. 221). According to Kilpatrick, arithmetic was "a commercial subject, and formed a part of the curriculum only where the demands of a trade made it desirable" (p. 221). His analysis made it clear that girls were rarely if ever taught arithmetic. He concluded: "The Dutch of America followed early seventeenth-century traditions of the fatherland: reading and writing for both girls and boys, with but little arithmetic save in the more commercial atmosphere of the capital and at Albany" (pp. 221–222). However, "the religious part of the program was much stressed" (p. 222).

Further evidence for the lack of attention given to arithmetic, especially so far as young children was concerned, in schools in colonial North America can be found by considering the use of hornbooks in North America. That is a subject to which we now turn.

Hornbook Education

During the first three decades of the twentieth century the well-known philanthropist George Arthur Plimpton (1855–1936), who was head of Ginn & Co., the New York-based educational publishing house, worked with his friend, David Eugene Smith (1860–1944), a noted mathematician and mathematics educator at Columbia University, New York, to build two of the world's finest collections of books, manuscripts and artifacts related to the histories of mathematics and mathematics education. Plimpton and Smith subsequently donated their collections to the Rare Book and Manuscript Library at Columbia University and those collections remain in that Library today. The Plimpton collection comprises 317 medieval and Renaissance manuscripts or artifacts, including Plimpton 322, a famous clay tablet from around 1800 BCE (Britton, Proust, & Schnider, 2011). Plimpton 322 consists of a table of triples of numbers, which today's mathematicians recognize as "Pythagorean triples."

Despite his success in building a fabulous collection of education manuscripts and artifacts, there was one artifact that George Plimpton desired to have but was unable to purchase for his collection. He set himself the task of locating and purchasing a hornbook which had been constructed in colonial North America during the seventeenth or eighteenth century. He did not succeed in finding one (Plimpton, 1912, 1928). In October 1916 he gave an address to the Antiquarian Society on "The Hornbook and its Use in America." In that address, the text of which is readily available today (see Plimpton, 1916), he described a hornbook in the following way:

> The hornbook, in point of fact, is not a book at all. Originally it was a piece of board with a handle ... On the face of the hornbook was a piece of vellum or paper upon which the lesson was inscribed. This was protected by a sheet of translucent horn. This protection was of course necessary to keep the lesson from the possible stain of a pair of dirty little hands ... (p. 3)

Plimpton (1916) showed actual hornbooks, from his collection, to those present at his talk.

At the time of his address, all of the hornbooks in Plimpton's collection had originated in Europe, but he made it clear that he still wanted to find a hornbook which had been constructed in North America. He stated:

> We are interested especially in the hornbook in America, and to what extent do we find the hornbooks were used in this country. There is ample evidence that the hornbook in the colonies, as in the old country, was the favorite device for starting children upon the ladder of learning. The Pilgrim Fathers, of course, came from Holland, and at that time the hornbook was the prevailing method of teaching the children there. The Dutch were such clever handicraftsmen that we find many of the early hornbooks were actually made in Holland. (p. 7)

Plimpton added:

> Most of the hornbooks that were used in the early days in this country were undoubtedly imported, but I am able to show what is very likely the oldest hornbook made on this continent. It is a Mexican hornbook, probably of the early seventeenth century. (p. 7)

Plimpton stated that he had been able to find only one hornbook, and that "it was doubtful whether any had ever been made in this country" (p. 36). Today, the Plimpton collection in the Rare Book and Manuscript Library includes an artifact from about the 1820s which is noted as possibly a hornbook which *might* have been constructed in North America. It is in poor condition, having lost its horn, and there are doubts about its origins.

Florian Cajori's (1890) Misleading Definition of the Hornbook

Florian Cajori (1859–1930), one of the most respected North American writers on histories of mathematics and mathematics education, described a hornbook in the following way:

> It consisted of one sheet of paper about the size of an ordinary primer, containing a cross (called "criss-cross"), the alphabet in large and small letters, followed by a small regiment of monosyllables, then came a form of exorcism and the Lord's Prayer, and, finally, the *Roman numerals* (original emphasis). (Cajori, 1890, p. 11)

Immediately after that passage Cajori asserted that it was on the strength of the Roman numerals that he ventured "to propose the hornbook for the honor of being the first mathematical primer used in this country" and that "hornbooks were quite common in England and in the English colonies in America down to the time of George II," but "they disappeared entirely in this country before the Revolution" (p. 11).

The hornbook shown in Figures 2.1a and 2.1b illustrates some of the elements of Cajori's description. However, we disagree with Cajori on two points:

1. Our intuition was that most hornbooks did *not* show Roman numerals. We believed that the fact that the hornbook shown in Figures 2.1a and 2.1b did show Roman numerals made that hornbook different—or, most likely—older than most other surviving hornbooks. We examined, page by page in Andrew White Tuer's (1896) two-volume masterpiece on the hornbook and found that altogether there were 128 drawings or photographs showing the faces of different hornbooks, and none of those 128 faces showed Roman numerals (although quite a few showed Hindu-Arabic numerals).
2. The hornbook shown in Figures 2.1a and 2.1b did not have a cross at the start of the first row.

Use of the Hornbook in Colonial North America

The hornbook shown in Figures 2.1a and 2.1b was purchased via an online auction together with a written communication from the last owner stating that it had previously been owned by his aunt who, after working for many years as a teacher to Native Americans in and around Michigan, had been given the hornbook as a parting gift.

In order to establish that this hornbook had been constructed in North America, small samples of wood from the back of the hornbook (see Figure 2.1b) were submitted, on two separate occasions, to the Center for Wood Anatomy Research

Figure 2.1(a) The upper face of a hornbook (c. 1700) held in the Ellerton-Clements collection of hornbooks. The outer dimensions of the rectangular faces are about 7" by 5". The "lesson" is covered by a translucent horn and is kept in place by metal strips which are tacked to the wood. The wood has been verified as "very old" American white pine.

Figure 2.1(b) The "back" of the hornbook shown in Figure 2.1a.

at the Forest Products Laboratory, Wisconsin (which is a Division of the United States Department of Agriculture). On both occasions, the official analyses reported that the wood in the hornbook was "very old" American "white pine." Since white pine is native to eastern North America, it is almost certain that the hornbook was handmade in North America. It was likely constructed in the 17th century, and its text includes the Roman numerals as well as the Hindu-Arabic numerals (Clements & Ellerton, 2015; Miter, 1896; Tuer, 1896). Other hornbooks that we have examined (in the Rare Book and Manuscript Library at Columbia University, and in Indiana University's Lilly Library) do not include Roman numerals. The hornbook shown in Figures 2.1a and 2.1b might be the only extant hornbook constructed and used in North America during the colonial period—although it is possible that hornbooks in the Smithsonian and in the Library of Congress mentioned later in this chapter were constructed in North America. The hornbook shown in Figures 2.1a and 2.1b is probably the *oldest* extant hornbook constructed in North America.

It should be obvious that the "lesson" printed on the paper beneath the horn of a hornbook could not be extensive—after all, it was directed at young children, and there was very little space (the rectangular faces ranged in size from about 4" by 3" to 7" by 5") for anything other than "basics." Essentially, a hornbook was an artifact designed to enable young learners (up to the age of 10) to see how the printed alphabet looked. With the hornbook shown in Figures 2.1a and 2.1b, the letters of the alphabet were presented in both upper- and lower-case. The vowels were also displayed separately. Numerals were shown in both Hindu-Arabic and Roman forms.

Many hornbooks did not show numerals at all (Bailey, 2013; Folmsbee, 1942; Tuer, 1896), and the extent to which hornbooks were used to help young North American children learn the alphabet and to write and read numerals has not been carefully studied by anyone other than Tuer (1896), whose research made clear that parents who wanted their young children to advance to higher levels of education realized that their children needed to be given much practice in reciting the alphabet and in counting to 10 (Bailey, 2013). They also wanted them to learn to read and write simple text which included the Hindu-Arabic numerals (Clements & Ellerton, 2015).

With most hornbooks the first row of the lesson began with a symbol of a cross and, as Cajori (1890) pointed out, for that reason first rows were called "criss-cross" rows. A pupil was expected to "cross" himself or herself before beginning a lesson (Plimpton, 1916). When we first received the hornbook shown in Figures 2.1a and 2.1b we looked for a form of a cross at the start of the first row but were unable to see one. That puzzled us until we read George Littlefield's (1904) comment that often very early Puritan missionaries in North America, like John Eliot (see Cogley, 1999), deliberately omitted the cross when constructing hornbooks, for they did not want to expose Native American children to "forms of idolatry." By the early 1700s most very young children in North America whose parents or grandparents were of European background, but were not indentured servants, attended schools of one kind or another, at least for a part of the year. Most boys in rural areas attended school during winter months only because, from a young age, they were required to work on the farm during the other months.

Parents wanting their children to take advantage of the available forms of education expected them to memorize all the information in the lesson on a hornbook. They also wanted their children to learn to read the Bible and a primer, and to be introduced to the psalter and catechism. Often, girls were also taught to sew, knit, weave, and embroider (Bailey, 2013, Cremin, 1970; Earle, 1899; Edmonds, 1991; Eggleston, 1888; Monaghan, 2007; Ring, 1993; Swan, 1977; Woloch, 1992). Most young European-background children, other than the children of indentured servants, attended so-called "dame schools" or schools supported by local taxation. In the eighteenth century some children attended schools assisted by the Society for the Propagation of the Gospel in Foreign Parts—an English missionary agency, formed by the Church of England in 1701—which helped provide teachers with paper, and funds. During the seventeenth and eighteenth centuries, wealthy families, especially those in the South, tended to employ governesses or tutors for their children, and some sent their children to European boarding schools—although the potential dangers associated with Atlantic Ocean crossings limited the number of parents prepared to do that (Cremin, 1970; Kraushaar, 1976). The seventeenth and eighteenth centuries witnessed the emergence of expensive "preparatory schools," especially in New England, New York, Pennsylvania, and Virginia (Cremin, 1970).

A Research Imperative

This chapter comes with an invitation to interested persons to attempt to answer the question: "In 17th and 18th-century colonial North America to what extent were hornbooks the main learning aid for introducing very young learners to numerical notations, and to number sense?" We think it is important that the extent to which hornbooks were made available to, and used by, young Native-American children, by African-American children, and by the children of European-background indentured servants, be researched. We say that because although much has been written about the unwillingness of many European-background settlers to allow children of slaves and servants to attend schools (Goodell, 1853), there were settlers who did want all children to be properly educated and we suspect that allowing children to have a hornbook would have been seen as representing a good start. We believe that the absence of the symbol of a cross at the beginning of the criss-cross row in the hornbook shown in Figures 2.1a and 2.1b points to the likelihood that that hornbook was used to educate some Native American people (Littlefield, 1904) and, as already stated, we believe that the hornbook shown in Figures 2.1a and 2.1b dates to around 1700 (or before that).

Curiously, none of the persons who have written about the history of mathematics, or education, in North America have paid any attention to such a fundamentally important "bottom-up history" question. Smith and Ginsburg (1934) stated that an "occasional hornbook with the numerals" was made available to pupils (p. 8), but with most groups (e.g., with groups of Native-Amerian children taught by missionaries) data relating to the actual situation are probably not available, and therefore it will be difficult to research the question well.

Andrew White Tuer's Pioneering Work on the History of Hornbooks

In 1896 Andrew White Tuer, a British scholar, published his magnificent *History of the Horn-Book.* After describing the long history of the hornbook as a key learning aid for young European children, Tuer proceeded to lament, specifically, the dearth of extant hornbooks, not only in Great Britain but also in North America. He stated that he had not seen, or heard of, the existence of any hornbook which had been constructed in North America. Ironically, most of the relatively few hornbooks in the United States of America today originated in Europe and were purchased in London when hornbooks in Tuer's collection were auctioned. We (Ellerton and Clements) own four genuinely-old hornbooks, three of which were purchased from England. The fourth—which is our treasure—is the one shown in Figures 2.1a and 2.1b.

Hornbooks Currently Held in the United States of America

During the 17th and 18th centuries the first introduction to mathematics for some young school children in Europe and North America was via hornbooks—with the children being asked to learn to read and write the alphabet and the Hindu-Arabic numerals. Note, however, our use of the expression "some young school children," for unlike the hornbooks shown in Figures 2.2, 2.3, and 2.4, only about half of the 50 hornbooks that we have physically handled have shown the Hindu-Arabic numerals, and a much smaller proportion of them had attached abacuses. It seems that, primarily, hornbooks were regarded as an aid to reading and recognizing the letters of the alphabet—with some makers adding the Hindu-Arabic numerals for those who might be interested.

Except in wealthy families in which private governesses or tutors were employed, or in families in which children were sent to preparatory schools for the well-to-do, more often than not those guiding young children's learning in colonial North America did not have a strong grasp of numerical concepts. Quite simply, most of the "dames" who conducted "dame schools" or the often poorly-educated persons who taught in the local schools supported by taxation, had not proceeded far with arithmetic, and often they did not know anything about it beyond being able to count and read to twelve or so Hindu-Arabic numerals (Cremin, 1970).

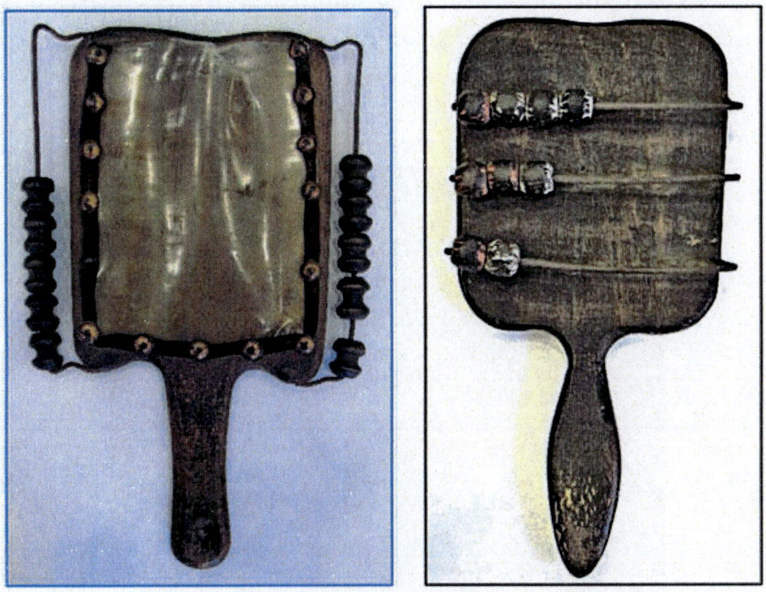

Figure 2.2 These hornbooks originated from Europe (and were in the Tuer collection). (Courtesy Lilly Library, Indiana University, Bloomington, Indiana).

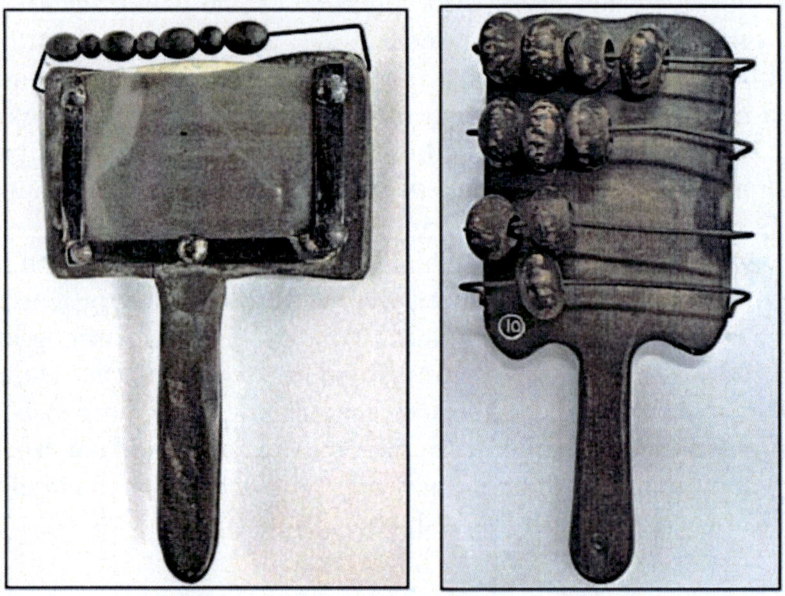

Figure 2.3 These two hornbooks were originally from Great Britain (in the Tuer collection). (George A. Plimpton collection, Rare Book and Manuscript Library, Columbia University).

Figure 2.4 This hornbook, held by the Library of Congress in Washington DC, also has an attached abacus.

Hornbooks, and the Mathematics Education of Young Children in North America in the Seventeenth and Eighteenth Centuries

Many of those who administrated local governments in the British colonies believed that it was unwise to give the poor a formal education because, ultimately, that might endanger governing authorities. As A Member of the Royal Institution (1812), a british author, wrote:

> It is not proposed that the children of the poor be educated in an expensive manner or even taught to read and cypher. Utopian schemes for the universal diffusion of general knowledge will soon ... confound that distinction of ranks and classes of society on which the general welfare hinges and the happiness of the lower orders, no less than that of the higher, depends. ... There is a risk of elevating, by an indiscriminate education, the minds of those doomed to the drudgery of daily labour above their condition, and thereby rendering them discontented and unhappy in their lot. (p. 46)

The modern reader surely cringes when reading such a statement, but it represents not only the feelings of many of Europeans of the time, but also of many of the leaders in colonial North America (Howson, 1982, 2010; Wecter, 1937).

Dame Schools

Primarily, dame schools served a year-round child-minding function, but children who attended them were often taught to read, write, and count. Often, Native-American or African-American children, and children of European-background indentured servants, were not permitted to attend dame schools or schools supported by local taxation (Chessman, 1965; Monaghan, 2007). De Bellaigue's (2007) analysis of middle-class private schools in England from 1800 to 1867 suggests that some British dame schools provided non-trivial academic preparations but, according to Geoffrey Howson and Leo Rogers (2014), the dame schools in Great Britain "had low educational aims, were run by ill-educated persons, and were attended mainly by children of the working class" (p. 259).

Parents would "drop off" their children at the home of the "dame" early in the morning, and the children would stay all day. Fees were low and the children were required to help the dame with chores. The dame would spend much time sewing, weaving, cleaning, cooking, and telling stories, especially Bible stories, to the children. Some dames also attempted to help their charges learn to recite the alphabet, to count, and to say, and remember, the Lord's Prayer and other religious statements (Harper, 2010). In most cases the dame herself knew very little arithmetic beyond being able to count and to read and write the Hindu-Arabic numerals. Although that represented a significant amount of arithmetical knowledge for a young child, there are no strong data on how well, and how many, dames were able to pass on a minimal level of arithmetical knowledge to the children for whom they had responsibility (Monaghan, 2007). Paper was scarce and expensive, and writing materials were not usually available—but many children brought hornbooks to school (Clements & Ellerton, 2015).

With the young children in well-to-do families, a wider range of learning aids was sometimes available. It has been argued that dame schools provided the glue which held together early North American European-background communities. They generated respect for others and enhanced the cultural awareness of those who attended (Earle, 1899).

Figure 2.5 shows Andrew White Tuer 's (1896) depiction of a dame-school in a well-to-do part of London. The quality of the children's shoes, their dress, and the elaborate furnishings, suggest that the children were from families which were not poor, and that the same was true of the dame herself. The presence of a naughty boy, with a dunce's hat, serving time standing on a chair, suggests that it was not uncommon for discipline to be enforced. In a less affluent neighborhood, in Great Britain or in North America, the children may not have had shoes, even in winter months. Can you see the six hornbooks depicted in Figure 2.5?

Arithmetic in dame schools rarely went beyond familiarizing children with the Hindu-Arabic numerals. Children were expected to learn to count, to recognize and sometimes to write numerals. Although mental addition and mental subtraction of small counting numbers might have been part of the intended curriculum in some dame schools, and in most local taxation-supported schools there can be no guarantee that anything beyond counting and reading numerals would have been part of the implemented curricula. Teachers in taxation-supported schools usually did not know much arithmetic, and there were no easy ways to ensure that what they taught their students was accurate. However, some direction arose from the fact that parents who wanted their children subsequently to attend higher-level middle-schools or grammar schools knew that, often such schools would not permit children to begin cyphering, or even to be admitted, unless they could read and write simple texts, count, and perform simple calculations mentally (Ellerton & Clements, 2012).

Most hornbooks used in North America between 1607 and 1799 were imported from Europe—especially from Great Britain, Holland, and the Germanic states. Only a few were made in North America. In the 17th century, John Eliot used locally constructed hornbooks in his work with Native Americans (Anderson, 1962; Littlefield, 1904). As stated earlier, between 1910 and 1916, George Plimpton made a huge—but ultimately unsuccessful—effort to find a hornbook which had been constructed in North America before 1800 (Plimpton, 1916). He stated that he had found it very difficult to find *any* surviving hornbooks—even hornbooks originally imported from Europe.

In his analysis of the history of the hornbook in Europe and North America, Plimpton concluded that although he knew of no extant hornbooks which had been constructed before the 16th century, there was evidence that, in fact, the hornbook was used well before then. He stated that in "an arithmetical manuscript of

Figure 2.5 Andrew White Tuer's (1896, p. 115) depiction of a dame school in England.

Sacrobosco, dated about 1400, we come unexpectedly upon the picture of a hornbook" (p. 2). Plimpton (1916) also pointed out that William Shakespeare had declared of a teacher (in *Love's Labour Lost*, which is thought to have been written in the 1590s): "Yes, yes, he teaches boys the hornebook" (p. 5).

Tuer (1896) lamented the fact that very few hornbooks had survived anywhere. He stated that the British Museum had only three, and the Bodleian Library at Oxford, one. There were several single specimens of hornbooks in private libraries in America, but these were all "English" hornbooks. Tuer stated that he had been "so fortunate as to pick up from time to time 24 specimens of the hornbook," and that "the best examples have been found in England" (p. 5).

Tuer not only referred to the dearth of hornbooks in North America but added that he had not seen, or heard of, the existence of any hornbook which had been constructed in North America. Ironically, most of the hornbooks presently in the United States of America were purchased in London from Tuer's estate (Ward, 2007). The Library of Congress—which has three hornbooks, two in ivory and one in wood—claims that it is possible that its wooden hornbook was constructed in North America.

In Figures 2.1a through 2.4 we have shown some of the hornbooks currently held in the United States of America. Each of the hornbooks in Figures 2.2 through 2.4 incorporates an abacus. In Figure 2.2, the reverse faces of two elaborate hornbooks from the impressive collection in the Lilly Library at Indiana University are shown. The hornbooks shown in Figure 2.3—which are held in the Plimpton collection in the Rare Book and Manuscript Library at Columbia University—also show reverse faces. All four of the hornbooks shown in Figures 2.2 and 2.3 are of European origin.

Figure 2.4 shows a hornbook held in the Library of Congress in Washington, DC. The Library's Internet description for that hornbook states: "Wood hornbook, eighteenth century, possibly American." Then comes:

> With alphabet in lower and uppercases, followed by vowels, ligatures, and the Lord's prayer. Paper text covered with translucent horn tacked to the face. Two-line abacus with 12 beads in cutout at top. Carved indentions on verso.

Despite the Library's claim that the hornbook is "possibly American," no evidence has been provided that the wood of the hornbook was from a tree native to America and, without that, one must conclude that it is more likely than not that it is of European origin. This comment would also apply to small hornbooks held in the Smithsonian's "Richard Lodish American School Collection" in the National Museum of American History.

We have observed that the five hornbooks shown in Figures 2.2, 2.3 and 2.4 have attached abacuses, and noted that most hornbooks did not have attached abacuses. Indeed, only about half of all hornbooks showed the Hindu-Arabic numerals. The hornbook pictured in Figures 2.1a and 2.1b did not have an attached abacus, but it did have Roman numerals as well as the Hindu-Arabic numerals. We have not seen another hornbook with Roman numerals.

Battledores

As we have said, after about 1800 the use of hornbooks as an aid to educating young children seems to have disappeared, both in Europe and in North America (Tuer, 1896). They were often replaced by what became known as "battledores," which were simple rectangular cards with the alphabet, vowels, and religious messages printed on both sides. Sometimes, Hindu-Arabic numerals were also shown. There was no wood attachment or handle and, typically, a battledore was folded into three sections (Bailey, 2013; Welsh, 1902).

Battledores were easier, and cheaper, than hornbooks to make, but were more likely to be lost or damaged. With the advent, during the period 1800–1840, of cheaper textbooks for helping young children to read, spell, and get correct answers for mental arithmetic tasks, they steadily became less used. The era of the battledore did not last nearly as that for the hornbook—it finished around 1850.

Summary: The Influence of the Hornbook

The above summary has been primarily concerned with how the hornbook might have helped young children less than 10 years of age to be daily made aware of what the Hindu-Arabic numerals looked like. Some teachers had a hornbook which included the numerals, and a few even had hornbooks with attached abacuses, aimed at helping children to count and to read numerals. Nevertheless, instruction based on a hornbook offered a very restricted view of what arithmetical learning was all about. Even if a hornbook did show Hindu-Arabic numerals and also had an attached abacus—which was unlikely—that of itself would not have inspired many youngsters to believe that arithmetic was anything other than something which would help them to remember things about numbers. But many hornbooks did not even show the Hindu-Arabic numerals.

It is difficult to imagine that young children aged between about 4 and 9 years would have been well prepared for any future mathematical studies merely by becoming acquainted with symbols which, somehow, represented numerals for counting. From a mathematics education perspective, instruction based on a hornbook would rarely have given learners a desire to want to go further in mathematics. In any case, most dames and teachers in small taxation-supported schools, did not know enough mathematics themselves to be in a position to assist young children to be confident when dealing with number-related contexts and to enjoy working with numbers.

Warren Colburn's Challenge to Those Who Defined Intended and Implemented Mathematics Curricula for Young Children

Modern thinking suggests that it was only a matter of time before someone would propose that *all* children aged less than 10 years should be expected to learn more mathematics than merely learning to count and to read and write numerals according to the Hindu-Arabic system. It has become received tradition that in North America the change came about in the early 1820s as a result of the enormous influence of Warren Colburn (1793–1833), a Harvard graduate who, in 1821, published a little book aimed at assisting teachers to help young children (from around six years of age) to learn the first principles of numeration, counting, and the four operations. The title of Colburn's (1821) book, *An Arithmetic on the Plan of Pestalozzi, with some Improvements*, acknowledged the influence of Johann Heinrich Pestalozzi, a Swiss/German educator (Biber, 1831).

Colburn's work on school arithmetic can be closely associated with his time as a student at Harvard College (between 1816 and 1820). In 1821 John Farrar, the Hollis Professor of Mathematics and Natural Philosophy at Harvard, wrote:

> Having been made acquainted with Mr. Colburn's treatise on Arithmetic and having attended an examination of his scholars who had been taught according to his system, I am well satisfied that it is the most easy, simple, and natural way of introducing young persons to the first principles in this science of numbers. The method here proposed is the fruit of much study and reflection. The author has had considerable experience as a teacher, added to a strong interest in the subject, and a thorough knowledge not only of this but of many of the higher branches of mathematics. This little work is therefore seriously recommended to the notice of those who are employed in this branch of early instruction, with the belief that it only requires a fair trial in order to be fully approved and adopted. (Recommendation by John Farrar, in Colburn, 1821, p. ii)

The strength of Farrar's recommendation is enhanced, perhaps, by the fact that Colburn had been Farrar's student at Harvard.

Between 1820 and 1823 Colburn taught at a private school for girls in Boston. His *First Lessons in Arithmetic,* published in 1821, was widely used for more than half a century (Karpinski, 1980). Colburn published a *Sequel to the First Lessons* in 1822 and subsequently a series of school readers and a textbook on algebra (see Chapter 5). He was one of the founders of the American Institute of Instruction, and at the first national conference of that body, held in 1830, he read his now-famous pedagogical treatise on the teaching and learning of arithmetic (Monroe, 1911).

Writing 75 years after the publication of Colburn's first textbook, George H. Martin (1897) stated that that book had been "an efficient force in raising the standard of instruction" (p. 140). Martin, maintained that previous to Colburn's pioneering *Arithmetic* . . .

> all arithmetic work had been unintelligent ciphering. This book came into the schools as refreshing as a northwest wind, and as stimulating. It was eagerly seized upon by the more intelligent teachers. Its use was a mark of an intelligent teacher, a sign of life from the dead. Embodying the principles of the new education, it wrought a revolution in the teaching of arithmetic, and it determined the character of all subsequent text-books. (p. 40)

It was while he was a student at Harvard that Colburn's interest in mathematics education first piqued. Colburn and his friend, and fellow student, James Carter, went for daily walks and on those walks they began to explore how the philosophies of Pestalozzi could be applied to the education, especially the mathematics education, of young children (Edson, 1856). Colburn and Carter articulated the need for boys and girls, aged from about 6 years, to be given a systematic course in arithmetic. They emphasized that learning mathematics should be an active process, not something requiring mere memorization. The teacher was continually to ask leading questions which directed the students' minds toward generalizations of numerical properties. Although all of that might not sound very radical today, around 1820 it called for a revolution in thinking about the purpose, and methods, of educating the young.

In 1820, George B. Emerson, the proprietor of a ladies' college in Boston, employed Colburn to teach arithmetic to young girls at his school (Barnard, 1851). Colburn remained at the school for three years and during that time the young girls in his classes became a laboratory for testing the ideas that he and Carter had developed.

The first page of Colburn's Arithmetic began with the following questions which would become famous in the history of school mathematics in the United States of America:

1. How many thumbs have you on your right hand? How many on your left? How many on both together?
2. How many hands have you?
3. If you have two nuts in one hand and one in the other, how many have you in both?
4. How many fingers have you on one hand?
5. If you count the thumb with the fingers, how many will it make?
6. If you shut your thumb and one finger and leave the rest open, how many will be open?
7. If you have two cents in one hand and two in the other, how many have you in both?
8. James has two apples, and William has three; if James gives his apples to William, how many will William have?

(Colburn, 1821, p. 2)

Colburn's publishers provided plates showing pictures to illustrate his questions, and teachers were encouraged to use them, and to teach groups of students in a manner which required individuals to answer questions immediately after they were asked.

Pestalozzi's (and Colburn's) move toward group teaching, based on leading questions and illustrations (or, better still, real objects), called for revolutionary change in thinking about the education of young children. Unlike the assumption which had prevailed for centuries, by virtue of the *abbaco* tradition, young learners were *not* to be introduced to large numbers (thousands, millions, even billions) at the very beginning of their mathematical studies. Rather, they were to be encouraged to reflect on general ideas based on patterns that they noticed when they were answering questions.

It could be argued that Colburn's interpretation of Pestalozzi's education theory was restrictive in the sense that it called for teachers to make use of a particular set of questions. in a particular order. There was the built-in, but largely untested, assumption that if the teacher led the pupils through those questions then the pupils would learn associated principles "through induction." Thus, for example, a teacher might ask her pupils for the values of 8 and 6, 18 and 6, 28 and 6, 38 and 6, etc., and then ask them for the values of 8 and 7, 18 and 7, 28 and 7, etc. With respect to the question, "What cost three yards of tape, at two cents a yard?" Colburn advised teachers to get pupils "to observe that three yards will cost three times as much as one yard; and say, 'if one yard cost two cents, three yards will cost three times two cents,'" etc. Colburn created a fixed sequence of questions which were supposed to suit most students.

Colburn (1835) actually wrote that "no man ever actually learned mathematics in any other method than by analytic induction; that by learning the principles by the examples he performs; and not by learning principles first, and then discovering by them how the examples are to be performed" (p. ix)

In the 1820s Colburn only taught in schools for a few years—after 1824 he became the owner of a factory, and he stopped teaching in schools. He authored a sequel (see Chapter 4 of this book) to his first arithmetic and also an algebra (see Chapter 5) and language textbooks. However, it was his first book—*An Arithmetic on the Plan of Pestalozzi*—which was the most successful. It was so successful, in fact, that it became one of the most influential texts in the history of education in the United States of America (Cajori, 1890, Doar, 2006). One hundred and seventeen years after it first appeared, Education professors Clifford Breed, Frederick Overman and James Woody (1938) claimed that American teachers, inspired by Colburn's interpretations of what Pestalozzi's ideas meant for the mathematics education of young children in the United States of America, began to use inductive, mental, methods which invited students to construct their own mathematical meanings (see, also, Monroe, 1912). This not only resulted, so the tradition goes, in students beginning to learn arithmetic at about 6 years of age (rather than at 10 or 11), but also in more students understanding more arithmetic than ever before. Furthermore, females began to study arithmetic in much greater numbers than at any previous time.

According to the standard interpretation, a new breed of Colburn-inspired teachers quickly energized the teaching and learning of arithmetic. Among other things, these teachers adopted more inductive approaches so that, right across the young nation, teachers began to ask young learners to figure out answers mentally to carefully sequenced sets of questions, to generalize, and then to articulate what they had learned in well-formed sentences. In this way, the teacher played a far more important role in facilitating arithmetic learning than ever before.

We believe that available data do *not* support this standard interpretation, but it is not our intention here to question the importance of Colburn's (1821, 1822) texts or of other arithmetic texts written by Colburn-inspired authors in the 1820s and 1830s (see, e.g., Goodrich, 1833; Ray, 1834; Smith, 1827). Certainly, we have reported data which indicate that most U.S. teachers who taught arithmetic to students aged 10 years or more continued to use the cyphering approach throughout the period 1820–1840 (Ellerton & Clements, 2012). Although Colburn may have helped change the thinking of generations of teachers who taught young children, there is little evidence that he had a strong influence on teachers of older students. He did write an arithmetic textbook, and an algebra textbook, aimed at older students (Karpinski, 1980), but these later books were not as influential as the arithmetics he directed at young children and their teachers.

There are first-hand testimonies to the distinctiveness of Colburn's approach to arithmetic education, but often those making such testimonies point out that Colburn's approach was opposed by many teachers. For example, James Freeman Clarke, who used one of Colburn's arithmetics at the Boston Latin School in the 1820s, commented, many years later, that "this admirable book was soon banished from the schools by the pedants, who thought that whatever was interesting must be bad" (quoted in Clarke & Hale, 1892, p. 278).

Before 1820, the standard type of school was of the one-room variety in which children aged from about 4 to about 16 (or 17 or 18), would be found. The children tended to work by themselves on material set by teachers during times when the teacher talked with individual students during "recitations." One-room schools continued to be commonplace across the United States during the period 1820–1865, but the most common form of teaching changed in a noticeable way. Increasingly, and especially in urban schools, children of roughly the same age would be asked to work together in small groups (Monaghan 2007). Recitations, and student learning became less individualized and between 1820 and 1865 the so-called "blab approach" (Braden, 1983; Ellerton & Clements, 2014), and other approaches to teaching arithmetic by which the teacher rarely explicitly taught more than one or two children at the same time, were gradually replaced by methodologies which had children working together on common themes and taking recitations in groups (Downs, 1978). In the large towns, students began to be placed in age-related grades in which whole-class teaching became the norm—but that was not feasible in most one-room rural schools.

Changes in the Modes of Implementing Mathematics Curricula in U.S. Schools

The relatively few modern writers who have written about the history of mathematics education in the United States have all noticed the strong influence of Warren Colburn (see, e.g. Kilpatrick, 2013). Although we acknowledge Colburn's influence on school mathematics, we believe that there were more general causal influences at work, and that these were particularly important from 1840 onward.

We have identified the following four factors—which we believe were not entirely independent of each other—which, after about 1840, precipitated changes to methods and expectations for one-room school education (Ellerton & Clements, 2012):

1. The introduction, from the mid-1840s onwards, of written examinations which were set externally by local education officials, meant that teachers increasingly felt a need to "teach to the test." Data from examinations were used, by local school authorities and by state officials, to check the quality of teaching and learning in elementary schools, both in the towns and in one-room rural schools (Caldwell & Courtis, 1925).
2. From the 1820s onwards, the steady growth of the U.S. population generated a corresponding growth in the number of public higher-level schools across the nation. That resulted in an expectation that students in elementary schools, including those in one-room rural schools, be prepared for the written entrance examinations for higher-level schools in reading, writing, spelling and arithmetic. Once again, teachers increasingly felt pressured to "teach to the test." The one-room school teachers realized that their future employment as teachers depended on how well their students did on the tests, and this led to a narrowing of the curriculum. Somewhat ironically, however, relatively few of the students who attended rural one-room schools would proceed to the high schools or private academies. Myrna Grove (2000) estimated that as late as 1900, "only five percent of one-room school graduates proceeded to urban high schools" (p. 75).
3. The establishment of state normal schools—or, in modern parlance, "teachers colleges"—across the nation from about 1840 (Harper 1935, 1939) affected profoundly the type of education offered in the one-room schools. Horace Mann—Secretary of the Massachusetts Board of Education—was largely responsible for establishing a climate of opinion favoring the establishment of the early normal schools. In the teacher-education courses that these normal schools offered young, prospective teachers, and in the summer programs that they offered practicing teachers, there can be no doubt that instructors in these normal schools used theories based on the writings of Johann Heinrich Pestalozzi to discredit individualized approaches to school education which had been the cornerstone of education practice in early North American one-room schools. The normal schools pushed for an emerging group-recitation, blackboard approach (Barnard 1851; Ellerton & Clements, 2012).
4. The increasing availability of textbooks, in reading, spelling, and arithmetic, many of which were especially written for use in U.S. public schools (Ellerton & Clements, 2012), also affected how teachers taught and how students learned. In the normal schools, prospective and

practicing teachers were trained to use these textbooks and, in time, texts written with school children in mind—such as those by William Holmes McGuffey (for reading), Noah Webster (for spelling) and Joseph Ray (for arithmetic)—came to be used in virtually every school, including one-room schools, across the nation (Ellerton & Clements, 2012; Hildreth 1936). This supported the use of group recitation methods and allowed for an easy switch from a fully individualized approach—the kind which Abraham Lincoln experienced (Ellerton & Clements, 2014)—to one in which students worked together to solve problems from grade-appropriate textbooks which the students themselves owned.

These new influences took effect from the early 1820s. Warren Colburn's books and those of other authors (e.g., Ray, 1834; Smith, 1827) who, almost certainly, wrote arithmetic textbooks modeled on Colburn's, prepared the way for profound changes which occurred after 1840. Those changes will be more fully outlined in the next two chapters.

References

A Member of the Royal Institution. (1812). *A vindication of Mr. Lancaster's system of education*. London, England: Longman & Co.

Anderson, C. (1962). *Technology in American education: 1650–1900* (Report No. OE-34018). Washington, DC: Office of Education, U.S. Department of Health, Education, and Welfare.

Bailey, M. L. (2013). Hornbooks. *Journal of the History of Childhood and Youth, 6*(1), 3–14.

Barnard, H. (1851) *Normal schools, and other institutions, agencies and means designed for the professional education of teachers*. Hartford, CT: Case, Tiffany & Co.

Biber, E. (1831). *Henry Pestalozzi, and his plan of education*. London, England: John Souter School Library.

Braden, W. W. (1983). *Theo tradition in the South*. Baton Rouge, LA: Louisiana State University Press

Breed, F. S., Overman, J. R., & Woody, C. (1938). *Child-life arithmetics. Grade Five*. Chicago, IL: Lyons & Carnahan.

Britton, J. P., Proust, C, & Shnider, S. (2011), Plimpton 322: A review and a different perspective. *Archive for History of Exact Sciences*, 65(5), 519–566. https://doi.org/10.1007/s00407-011-0083-4

Cajori, F. (1890). *The teaching and history of mathematics in the United States* (Circular of Information No. 3, 1890). Washington, DC: Bureau of Education.

Caldwell, O. W., & Courtis, S. A. (1925). *Then and now in education, 1845–1923*. Yonkers-on-Hudson, New York, NY: World Book Company.

Chessman, R. (1965). *Bound for freedom*. New York, NY: Abelard-Schuman.

Clarke, J. F., & Hale, E. E. (1892). School days in New England. In K. Munroe & M. H. Catherwood (Eds.), *School and college days* (Vol. VII, pp. 265–280). Boston, MA: Hall and Locke Company.

Clements, M. A., & Ellerton, N. F. (2015). *Thomas Jefferson and his decimals 1775–1810: Neglected years in the history of U.S. school mathematics*. New York, NY: Springer. https://doi.org/10.1007/978-3-319-02505-6

Cogley, R. W. (1999). *John Eliot's mission to the Indians before King Philip's War*. Cambridge, MA: Harvard University Press.

Colburn, W (1821). *An arithmetic on the plan of Pestalozzi, with some improvements*. Boston, MA: Cummings and Hilliard.

Colburn, W. (1822). *Arithmetic; being a sequel to first lessons in arithmetic*. Boston, MA: Cummings, Hilliard, & Co

Colburn, W. (1835). *Colburn's first lessons, intellectual arithmetic upon the inductive method of instruction*. Hallowell, ME: Glazier, Masters and Smith.

Cremin, L. A. (1970). *American education: The colonial experience 1607–1783*. New York, NY: Harper & Row.

De Bellaigue, C. (2007). *Educating women: Schooling and identity in England and France 1800–1867*. Oxford, England: Oxford University Press. https://doi.org/10.1093/acprof:oso/9780199289981.001.0001

Doar, A. K. (2006). *Cipher books in the Southern Historical Collection*. Master of Science thesis, Wilson Library, University of North Carolina at Chapel Hill.

Downs, R. B. (1978). *Friedrich Froebel*. Boston, MA: Twayne.

Earle, A. M. (1899). *Child-life in colonial days*. New York, NY: Macmillan.

Edmonds, M. J. (1991). *Samplers and samplermakers: An American schoolgirl art, 1700–1850*. New York, NY: Rizzoh.

Edson, T. (1856). *Memoir of Warren Colburn, written for the American Journal of Education*. Boston, MA: Brown, Taggart & Chase.

Eggleston, A. (1888). *A history of the United States and its people*. New York, NY: American Book Company.

Ellerton, N. F., & Clements, M. A. (2012). *Rewriting the history of school mathematics in North America 1607–1861*. New York, NY: Springer. doi:https://doi.org/10.1007/978-94-007-2639-0

Ellerton, N. F., & Clements, M. A. (2014). *Abraham Lincoln's cyphering book and ten other extraordinary cyphering books*. New York, NY: Springer. https://doi.org/10.1007/978-3-319-02502-5

Folmsbee, B. (1942). *A little history of the hornbook*. Boston, MA: The Horn Book Inc.

Goodell, W. (1853). *The American slave code in theory and practice*. New York, NY: American & Foreign Anti-Slavery Society.

Goodrich, S. G. (1833). *Peter Parley's method of teaching arithmetic to children: With numerous engravings*. Boston, MA: Carter, Hendee, & Co.

Grove, M. (2000). *Legacy of one-room schools*. Morgantown, PA: Masthof Press.

Hildreth, G. (1936). *Learning the three R's: A modern interpretation*. Minneapolis, MN: Educational Publishers Inc.

Jones, P. S., & Coxford, A. F. (1970). From discovery to an awakened concern for pedagogy: 1492–1821. In National Council of Teachers (Ed.), *Thirty-second handbook* (pp. 11–23). Washington, DC: National Council of Teachers of Mathematics.

Harper, C. (1935). *Development of the teachers college in the United States with special reference to Illinois State University*. Bloomington, IL: McKnight & McKnight.

Harper, C. (1939). *A century of public teacher education: The story of the state teachers colleges as they evolved from the normal schools*. Washington, DC: American Association of Teachers Colleges.

Harper, E. P. (2010). Dame schools. In T. Hunt, T. Lasley, & C. D. Raisch (Eds.), *Encyclopedia of educational reform and dissent* (pp. 259–260). Thousand Oaks, CA: SAGE Publications.

Howson, G. (1982). *A history of mathematics education in England*. Cambridge, England: Cambridge University Press. doi:https://doi.org/10.1017/CBO9780511897481

Howson, G. (2010). Mathematics, society, and curricula in nineteenth-century England. *International Journal for the History of Mathematics Education, 5*(1), 21–51.

Howson, G., & Rogers, L. (2014). Mathematics education in the United Kingdom. In A. Karp & G. Schubring (Eds.), *Handbook on the history of mathematics education* (pp. 257–282). Dordrecht, The Netherlands: Springer. doi:https://doi.org/10.1007/978-1-4614-9155-2_13

Karp, A., & Schubring, G. (Eds.). (2014). *Handbook on the history of mathematics education*. New York, NY: Springer. doi:https://doi.org/10.1007/978-1-4614-9155-2

Karpinski, L. C. (1980). *Bibliography of mathematical works printed in America through 1850*. New York, NY: Arno Press.

Kilpatrick, J. (2013). Warren Colburn and the inductions of reason. In K. Bjarnadóttir, F. Furinghetti, J. Prytz, & G. Schubring (Eds.), *"Dig where you stand" 3. Proceedings of the Third International Conference on the History of Mathematics Education* (pp. 219–232). Uppsala, Sweden: Uppsala University.

Kilpatrick, W. H. (1912). *The Dutch schools of New Netherland and colonial New York*. Washington, DC: United States Bureau of Education.

Kraushaar, O. F. (1976). *Private schools from the Puritans to the present*. Bloomington, IA: Phi Delta Kappa Educational Foundation.

Littlefield, G. E. (1904). *Early schools and school-books of New England*. Boston, MA: The Club of Odd Volumes.

Martin, G. G. (1897, *The evolution of the Massachusetts public school system: A historical sketch*. New York, NY: D. Appleton & Co.

Miter (1896, September 8). Our London letter. *The American Stationer, 40*(10), 367–368 and 379. https://doi.org/10.15281/jplantres1887.10.117_367

Monaghan, E. J. (2007). *Learning to read and write in colonial America*. Amherst, MA: University of Massachusetts Press.

Monroe, W. S. (1911). Warren Colburn. In P. Monroe (Ed.), *A cyclopedia of education* (Vol. 2, p. 48). New York, NY: Macmillan.

Monroe, W. Walter S. (1912). A chapter in the development of arithmetic teaching in the United States. *The Elementary School Teacher, 13*(1), 17–24.

Plimpton, G. A. (1912, August 1). Hornbooks. *The Independent, 72*(3322), 264–268.

Plimpton, G. A. (1916). *The hornbook and its use in America*. Worcester, MA: American Antiquarian Society. https://doi.org/10.2307/3474145

Plimpton, G. A. (1928). The history of elementary mathematics in the Plimpton Library. Science, 68(1765), 390–395. https://doi.org/10.1126/science.68.1765.390

Ray, J. (1834). *Introduction to Ray's eclectic arithmetic. The little arithmetic, elementary lessons in intellectual arithmetic on the analytic and inductive methods of instruction, being an introduction to the author's Eclectic Arithmetic*. Cincinnati, OH: Truman, Smith and Co.

Ring, B. (1993). *Girlhood embroidery: American samplers and pictorial needlework, 1650–1850*. New York, NY: Knopf Publishers.

Smith, D. E., & Ginsburg, J. (1934). *A history of mathematics in America before 1900*. Chicago, IL: The Mathematical Association of America. https://doi.org/10.1090/car/005

Smith, R. C. (1827). *Practical and mental arithmetic on a new plan, in which mental arithmetic is combined with the use of the slate*. Boston, MA: Richardson, Lord & Goodrich.

Swan, S. B. (1977). *American women and their needlework 1700–1850*. New York, NY: Holt, Rinehart and Winston.

Tuer, A. M. (1896). *History of the horn-book*. London, England: Leadenhall Press.

Ward, H. E. (2007). To all who know their ABCs, greeting: A history of the ABCs, Lilly Library, Indiana University. *Indiana Libraries, 26*(3), 58–61.

Wecter, D. (1937). The saga of American society: A record of social aspiration 1697–1937. New York, NY: Charles Scribner's Sons.

Welsh, C. (1902). Hornbooks and battledores *The Literacy Collector, 3*(2), 33–37. https://doi.org/10.1038/scientificamerican02011902-37ebuild

Westbury, I. (1980). Change and stability in the curriculum: An overview of the questions. In H. G. Steiner (Ed.), *Comparative studies of mathematics curricula: Change and stability 1960–1980* (pp. 12–36). Bielefeld, Germany: Institut für Didaktik der Mathematik-Universität Bielefeld.

Woloch, N. (1992). *Early American women*. Belmont, CA: Wadsworth Publishing Company.

Chapter 3
The Influence of the Cyphering Tradition on North American Elementary- and Middle-School Mathematics Between 1607 and 1865

Abstract Commercially-published textbooks do not offer the most important data for those interested in the histories of mathematics and mathematics education in North America during the period 1607–1865. In fact, until well into the nineteenth century most North American schoolchildren who were learning mathematics did not own a mathematics textbook, and many teachers of mathematics did not own one either. By contrast, almost all students aged from 10 to 16 years who studied any branch of mathematics prepared handwritten cyphering books, and often their teachers made available to them the cyphering books that they had prepared in their own school days. In this chapter we summarize our previous work on the cyphering tradition, drawing attention to theoretical bases, and also to how the tradition controlled both the implemented and the attained mathematics curricula in grammar schools and in other pre-college institutions. Summaries of curriculum content and of the teaching and learning patterns which were an inherent part of the cyphering tradition are given. The discussion is based on our analyses of about 1500 extant North American cyphering books from the period.

Keywords *Abbaco* sequence • Abraham Lincoln's cyphering book • Algebra education • Ciphering book • Cyphering book • Cyphering tradition • Fibonacci • Harvard College • Phillips Library in Salem, MA • Rockefeller

Mathematics Studied by Children Aged Between 10 and 16

One of the questions raised, but *not* fully addressed, in the first two chapters was: "Why did it take so long for a world-class mathematician to appear in North America?" We have argued, elsewhere (Ellerton & Clements, 2011a), that the first North American scholar to earn genuine respect among European mathematicians was Nathaniel Bowditch, whose translation of, and commentary on, Pierre-Simon Laplace's extraordinarily complex *Mécanique Céleste* between 1815 and 1830 prompted Sylvestre François Lacroix to proclaim, in 1832, that he was astonished at Bowditch's achievement; Adrien-Marie Legendre, also in 1832, stated that Bowditch's translation and commentary should be regarded as "a new edition"; and Charles Babbage (again in 1832) maintained that Bowditch's work was "a proud circumstance for America" (see Cajori, 1890, p. 105). From our perspective (Ellerton & Clements, 2011a), the historical lineage of the U.S. mathematics research community began with the remarkable Nathaniel Bowditch—see

N. F. Ellerton, M. A. (Ken) Clements, *Toward Mathematics for All*, History of Mathematics Education,
https://doi.org/10.1007/978-3-030-85724-0_3

Chapter 8 of this book for an elaboration of Bowditch's career and summary of his achievements—and was continued through the efforts of Bowditch's distinguished protégé, Benjamin Peirce.

Why, then, did it take so long for an internationally recognized mathematician to emerge in North America? We believe that the history of mathematics research in North America was built on an ethnomathematical foundation which had the "cyphering tradition" as its cornerstone. This tradition was built around an inherited *abbaco* curriculum which proved to be ideal for some colonial contexts but was not conducive to the development of highly creative mathematicians. We do not deny that in exclusive colleges like Harvard, Yale, William and Mary, King's (which would be renamed Columbia), New Jersey (which would be renamed Princeton) and, in the nineteenth century, in the U.S. Military Academy at West Point, some students were able to study mathematics well beyond the highest level in the *abbaco* sequence. At each of those colleges some students were required to study higher levels of geometry, algebra, trigonometry, and calculus, and they also studied sophisticated forms of applied mathematics associated with navigation, astronomy, and surveying. But, almost all of the students who went on to tackle those areas of mathematics had received their elementary grounding in the subject via the cyphering tradition—even though in most cases the mathematics that they studied and recorded in their cyphering books had not been primarily aimed at preparing them for higher studies in pure or applied mathematics (Burton, 1833).

During the period 1607–1865 many persons who would become famous Americans prepared their own cyphering books—including George Washington (Crackel, Rickey & Silverberg, 2017), Abraham Lincoln (Ellerton, Aguirre-Holguin & Clements, 2014), and Nathaniel Bowditch (Bowditch, 1840). Although a youthful Abraham Lincoln did not study mathematics beyond middle-level *abbaco* arithmetic in the one-room schools in and around Pigeon Creek, Indiana, that he attended between 1819 and 1826, analyses of the mathematics he entered into his cyphering book over that period suggest that he could have gone on to study higher mathematics if he had wanted, and had had the opportunity, to do so (Dunlap, 1959; Ellerton et al., 2014; Wickersham, 1886). But the possibility of studying higher mathematics would not have occurred to young Abraham—even though, later in his life he would, of his own volition, saturate himself with the logic of Euclid, and would consciously apply that logic when preparing famous speeches (like those he made when debating Stephen Douglas, or when preparing his Gettysburg address—see Hirsch & Van Haften, 2010, 2016).

Of fundamental importance to understanding the roots of twenty-first century North American mathematics is the history of the cyphering tradition in North America. The cyphering tradition defined the mathematics which was taught and learned in North-American education institutions from the early days of European settlement to its gradual demise from the 1840s. Our major data source has been the Ellerton-Clements cyphering book collection which is the largest collection of North American cyphering books and is now held by the Library of Congress. We also examined manuscripts in the collection held in the Phillips Library of the Peabody Exeter Museum (Salem, MA). That collection includes a particularly impressive body of *eighteenth-century* cyphering books.

Numerous other mathematicians and mathematics educators have contributed to the literature on the history of U.S. mathematics—Amy Ackerberg-Hastings, Eric Temple Bell, Carl Boyer, Florian Cajori, Patricia Cline Cohen, Joseph Dauben, Louis Karpinski, Victor Katz, Peggy Kidwell, Jeremy Kilpatrick, David Klein, Karen Hunger Parshall, Fred Rickey, David Roberts, Lao Genevra Simons, Nathalie Sinclair, David Eugene Smith, George Stanic, John Stillwell, Dirk Struik, George Stanic and Frank Swetz are names which immediately spring to mind—but none of their publications has focused on the cyphering traditiion (see, for example, Parshall, 2003). The theme being addressed in this chapter does not have a well-established scholarly literature.

In our book, *Rewriting the History of School Mathematics in North America, 1607–1861: The Central Role of Cyphering Books* (Ellerton & Clements, 2012) we addressed, among other things, the following four questions:

1. What was a cyphering book?
2. Which North American students prepared cyphering books?
3. What were the key components of the "cyphering tradition"?
4. What evidence do we have to support our contention that the cyphering tradition framed implemented mathematics curricula in North American pre-college education institutions during the period 1607–1861?

In this chapter, we will not only identify what we believe to be the most important manuscripts in the Ellerton-Clements cyphering book collection, but also comment on why we think they are important. We will pay special attention to matters related to the above four questions.

In Chapter 2 our focus was on the lack of opportunity during the period under consideration for most North American children aged less than 10 years to learn to read, write and use the Hindu-Arabic numerals. Here we focus on the forms of mathematics which children in European-background families whose parents were not indentured servants were able to study once they had reached the age of 10 years. For such children, the opportunity to learn was largely dependent on whether they had access to teachers who themselves had gained the relevant mathematical knowledge. The intended, implemented, and attained mathematics curricula of those who were given the opportunity to learn were framed by the "cyphering tradition."

The Cyphering Tradition

Definition of a Cyphering Book

In this chapter we will often refer to "cyphering books." Before the Declaration of Independence, the word used was "cyphering," with a "*y*," but after 1776 the influence of Noah Webster, and others, resulted in the gradual adoption of a different spelling—so that "cyphering" often became "ciphering," with the *y* being replaced by an *i*. We have defined a *cyphering book* as a *handwritten* manuscript with the following four properties:

1. Either the contents were written by a student who, through the act of preparing it, was expected to learn and be able to apply whatever content was under consideration; or, the book was prepared by a teacher who wished to use it as a model which could be followed by students preparing their own cyphering books.
2. All entries in the book appeared in ink—either as handwriting or as illustrations. Headings and sub-headings were presented in a decorative, calligraphic style.
3. The book was dedicated to setting out rules and cases associated with a sequence of mathematical topics, including notes for each topic and relevant word problems. The problems were in arithmetic, especially business arithmetic, or in algebra, or geometry, or trigonometry, or were applications of mathematics in the fields of gauging, navigation, surveying, military strategy, etc.
4. The topics covered were sequenced so that they became progressively more difficult. The content also reflected the expectation that, normally, no child less than 10 years of age would be assigned the task of preparing a cyphering book.

(Ellerton & Clements, 2012, pp. 3–4)

Other terms occasionally used include "copybook," "sum book" and, especially in recent times, "cipher book." The contemporary term, however, was "cyphering book" (or "ciphering book").

Most North American cyphering books dealt with just one branch of mathematics—usually arithmetic—but sometimes it was algebra or geometry, or surveying, or navigation, etc. Occasionally, a cyphering book had entries from several areas of mathematics. Cyphering books were usually made up of unlined, rectangular folio-sized sheets of paper, which formed "quires" (sometimes called "signatures"). In the 18th and 19th centuries, "rag" paper was often used, and in North America pages usually had dimensions about 12.5" by 8" (i.e., approximately 32 cm by 20 cm). The quires were routinely sewn together to form manuscripts. In most cases, protective covers were added. Typically, the first page of a manuscript was beautifully decorated, and the name of the owner as well as the year and location in which the manuscript began to be prepared, were indicated (Burton, 1833; Cohen, 2003; Ellerton & Clements, 2011b, 2014).

Most cyphering books dealt with several arithmetical topics. Although, usually, the treatment of the same topic progressed on successive pages, occasionally topics were revisited, and extended, later in the same cyphering book. Cyphering books were intended to serve as books for future reference by those who prepared them. It was part of the cyphering tradition that entries in a book should be well written and entirely accurate. Before being entered into a cyphering book each entry should have been approved by a teacher, or private tutor, or by some other authority.

Mathematics Content in the Cyphering Books—The *Abbaco* Sequence

During the period 1200–1850 most Western European and North American boys aged 10 years, or more, who learned mathematics at school or with a private tutor were expected to prepare handwritten cyphering books. The "implemented curriculum" was well represented by what was written in the cyphering books. By contrast, textbooks might be thought of as providing an "author-intended curriculum." Usually, an implemented curriculum would have included less mathematics than a textbook-defined intended curriculum. With the advent, in the fifteenth century of the Common Era (CE), of commercially-printed mathematics textbooks, entries in cyphering books were increasingly copied from textbooks.

A "cyphering tradition" has been associated with the so-called *abbaco* sequence of mathematics topics (Ellerton & Clements, 2012; Van Egmond, 1976, 1980). From about the thirteenth century onwards, sharp rises in the numbers of international trading and banking companies in Western European city-states prompted the formation of vernacular schools across Europe in which the commercially-oriented *abbaco* sequence was a major part of the intended curriculum. Students were expected to learn to apply *abbaco* topics to business and accounting problems. Beautiful calligraphic writing was required of students and calculational accuracy was of paramount importance. All of this was supervised by "reckoning masters" (Heal, 1931; Karpinski, 1925) who were employed to educate the sons of the merchant class. In the "reckoning schools" students prepared handwritten "cyphering books" and the quality and appearance of these books came to be recognized not only as important indicators of a student's potential for gainful employment or future study but also of the teaching power of the associated reckoning master.

The "intended curriculum" for most mathematical programs in North American schools and colleges during the period from 1607 through 1865 derived from the *abbaco* sequence. That sequence was a well-ordered set of topics associated with business arithmetic in Arabic and Western European nations (Ellerton & Clements, 2012). For many years it was accepted by scholars that the tradition was initially developed in India, then further developed in Arabic nations, and finally in European city states after it had been introduced into Europe by Arab immigrants and through Leonardo of Pisa's (Fibonacci's) *Liber Abbaci*—which was written around 1200 CE (see, e.g., Gies & Gies, 1969; Jackson, 1906; Long, McGee & Stahl, 2009; Wardhaugh, 2012; Yeldham, 1936). In recent years however, Jens Høyrup (2005, 2008) has shown that even before 1200 CE features of the tradition, including some algebra, had already appeared in parts of Western Europe, and especially in Spain.

The *abbaco* sequence began with Hindu-Arabic base 10 terminology and its associated place-value system, and then moved to algorithms for the four operations on whole numbers, "compound operations" (i.e., elementary measurement, including units), reduction, practice, and the rules of three. Then followed loss and gain, equation of payments, barter, interest (simple and compound), tare and tret, discount, and brokerage. At a more advanced level would come topics like decimals, vulgar fractions, commission, alligation (the arithmetic of mixing quantities), fellowship (the arithmetic of partnerships), false position, arithmetical and geometrical progressions, involution and evolution, permutations and combinations, and mensuration. The *abbaco* sequence usually did not include formal study of algebra or Euclidean geometry, or trigonometry, but about 10 percent of the cyphering books in the E-C collection include entries relating to one or more of those branches of mathematics (see Table 5.1 in Chapter 5). Typically, emphasis was on learning *abbaco* rules and cases and applying those to problems which might arise in business contexts (Ellerton & Clements, 2014).

Pedagogy, Cyphering, Recitation, and the Cyphering Tradition

As we saw in Chapter 2, before the first half of the nineteenth century most North American school-age children did not attend school throughout the year. Some European-background children, usually males, but rarely children of indentured servants, studied arithmetic (Chessman, 1965). But, less than half of the children attending school between 1607 and 1865 actually prepared cyphering books. Most European-background girls who attended did not prepare cyphering books.

In the New England colonies or states, most of the boys in European-background families but whose parents were not indentured servants who attended school did so during winter months only, and for them the implemented mathematics curricula tended to be confined to *elementary abbaco* arithmetic. Teachers did not stand at the front of the room and teach, and most pupils, even those studying mathematics, did not own a mathematics textbook (Ellerton & Clements, 2014). Before the 1840s, written examinations of any kind were not used. Most teachers had never studied any mathematics beyond *abbaco* arithmetic (Cajori, 1890; Ellerton & Clements, 2012).

In schools, the cyphering tradition came to be associated with a form of pedagogy which fostered individual, yet supervised, learning. This included the following components:

1. The teacher prescribed during a one-on-one recitation what an individual student would do. Sometimes the teacher would even write headings and problems to be solved into a student's cyphering book (Dickens, 1850; Walkingame, 1785). Each student was expected to prepare for the next recitation session by memorizing rules he or she had been asked to learn.
2. The student would work individually on finding solutions to set exercises, often on a slate or on scraps of paper. These would be shown to the teacher during another recitation session.
3. If the tentative solutions were approved by the teacher, the student would then be told to complete his or her cyphering book entries for

that topic. That would require the student to write introductory statements for the given topic, to state rules and cases, to copy model examples, and to provide solutions to exercises. Sometimes the teacher lent a student an older cyphering book, or a textbook, to show what was required. Entries would be made with a quill and home-made ink (Ellerton & Clements, 2012; Meriwether, 1907).

4. Finally, the student would show the teacher his or her completed cyphering-book entries for the topic under consideration. It was expected that headings would be in fine calligraphy, and penmanship would be of a good quality. A cyphering book was to be a guidebook for life—a text which could be consulted if and when the need arose—and it was important that all entries were correct.

In each of the recitation components, the teacher would talk to individual students on a one-to-one basis. During the recitations, some teachers merely listened to students as they verbalized rules that they had memorized. Often, teachers asked probing questions designed to test and extend their students' understandings (Adams, 1848; Babcock, 1829; Cobb, 1835; Colburn, 1821; Emerson, 1835; Sterry & Sterry, 1795). According to Lyman Cobb (1835), a diligent teacher would meet and talk with each learner at least once every day, probing the student's understanding of the principles upon which the rules were founded. Cobb (1835) advised:

> The teacher should not permit him [the learner] to commence a new sum, or engage in a new rule, until he is fully and thoroughly acquainted with the principles of the rule in which he has been working. Young scholars are generally anxious to make rapid progress in passing through the Arithmetick. This propensity, however laudable, should not be indulged at the expense of a partial knowledge of the subject. The teacher should endeavour to convince the scholar that, in order to make his progress advantageous, he must perfectly understand each rule and its principles. (p. 2)

Thus, the cyphering tradition incorporated a form of pedagogy which combined individual yet supervised learning, with follow-up evaluative sessions, usually through one-on-one recitations.

Figure 3.1 shows a reproduction of an etching by Abraham Bosse, a French-Huguenot artist who, in the 1630s, depicted the kind of schoolrooms in middle-class Western European neighborhoods at that time. The master was shown speaking to an individual boy. He might have been checking what the boy had written in his cyphering book or, perhaps, asking questions, or setting new work for the student to do. Another student was waiting to see the master, and others were working, individually, preparing their cyphering books or other workbooks for when it would be their turn for recitation. In the North American colonies, almost all school rooms were less elaborate than the one shown in Figure 3.1, and often teachers were not qualified to teach arithmetic beyond numeration and the four operations with whole numbers (Ellerton & Clements, 2012). But this same individualistic approach to pedagogy and learning prevailed.

Figure 3.1. Le Maître d'Ecole (c. 1635), by Abraham Bosse, (1611–1678). This etching is held by the Metropolitan Museum of Art, New York (public domain image).

Early in the seventeenth century, the Huguenots, the Dutch, and some other well-defined groups within Western Europe (e.g., in Italy, and in England), valued arithmetic highly and regarded it as a vital component of the education of all children. It was only natural that those who emigrated to North America would want their children to learn the same arithmetical content, and to learn it in the same ways that they had learned it in the Old World.

Educational Rationale for the Cyphering Tradition

The entries in most cyphering books featured two genres. The first is what we have called the **IRCEE** ("**I**ntroduction, **R**ules, **C**ases, **E**xamples, **E**xercises") genre, by which, for any particular topic, there was an **I**ntroduction—usually from two to four lines introducing the topic about to be covered—then would come a statement of **R**ules (given in sentence form) which would apply to problems for the topic); and, then for each rule, **C**ases would be stated (also given in sentence form) for which the rules would apply; then would follow one or more worked **E**xamples; and finally several related **E**xercises would be set for the student to do. The **PCA** ("**P**roblem. **C**alculation, **A**nswer") genre applied to expectations for the setting out of solutions to exercises: the **P**roblem would be stated, in writing, then would follow **C**alculations (usually without any explanations for any of the steps involved in the calculations), and finally the **A**nswer to the problem would be written and underscored (or "Ans." or "Answer" would be written). The **PCA** genre was evident in almost all cyphering books (Cohen, 2003; Ellerton & Clements, 2009).

In Figure 3.2, which shows a page from a 1789–1790 cyphering book prepared in Philadelphia by Thomas Willson, a rule associated with the volume of a "cylindroid" is shown. Note the calligraphic headings, the introduction (in which a cylindroid is defined), the rule, and the model example. The **PCA** genre is well illustrated with the statement of the problem, the unexplained calculations, and the statement of the answer at the bottom of the page. In Figure 3.2, ink from the reverse side of the page has "seeped through"—that was a common phenomenon.

Although this theoretical rationale was implicitly known and understood, it was never explicitly described. It was taken for granted and was assumed to be an integral part of the implemented curriculum by all involved—especially by teachers and students.

Figure 3.3 shows a model developed by the present writers which depicts the cyphering tradition's societally-oriented, structure-based, and problem-based form of school mathematics:

- It was "societally-oriented" because the "model" problems were deliberately chosen by the instructors so that they would likely be relevant to the present and future social situations of the students.
- It was "structure-based" since each problem was chosen because it offered the opportunity to help the student to recognize that a particular problem required mathematics of a particular kind, with a special structure, if it were to be readily solved.
- It was "problem-based" because each problem was chosen because it offered the opportunity to help the student to recognize that in order to solve a problem it was necessary to identify the kind of mathematics which would be required, and then to apply that mathematics to gain a solution. The students were expected to learn how to solve problems independently.

The following educational rationale (taken from Ellerton and Clements, 2009) summarizes our historical and theoretical base for our research into the cyphering tradition:

1. The fundamental aim of teachers of arithmetic, algebra, geometry, trigonometry, navigation, surveying, etc., was to help their students become independent problem solvers in their chosen vocations, even if this required individualized teaching/learning processes for different students.
2. Students were to be invited to recognize important structural similarities in carefully chosen problem situations. The aim was for them to learn how rules, and associated algorithms, could be applied to arrive at solutions to the problems.

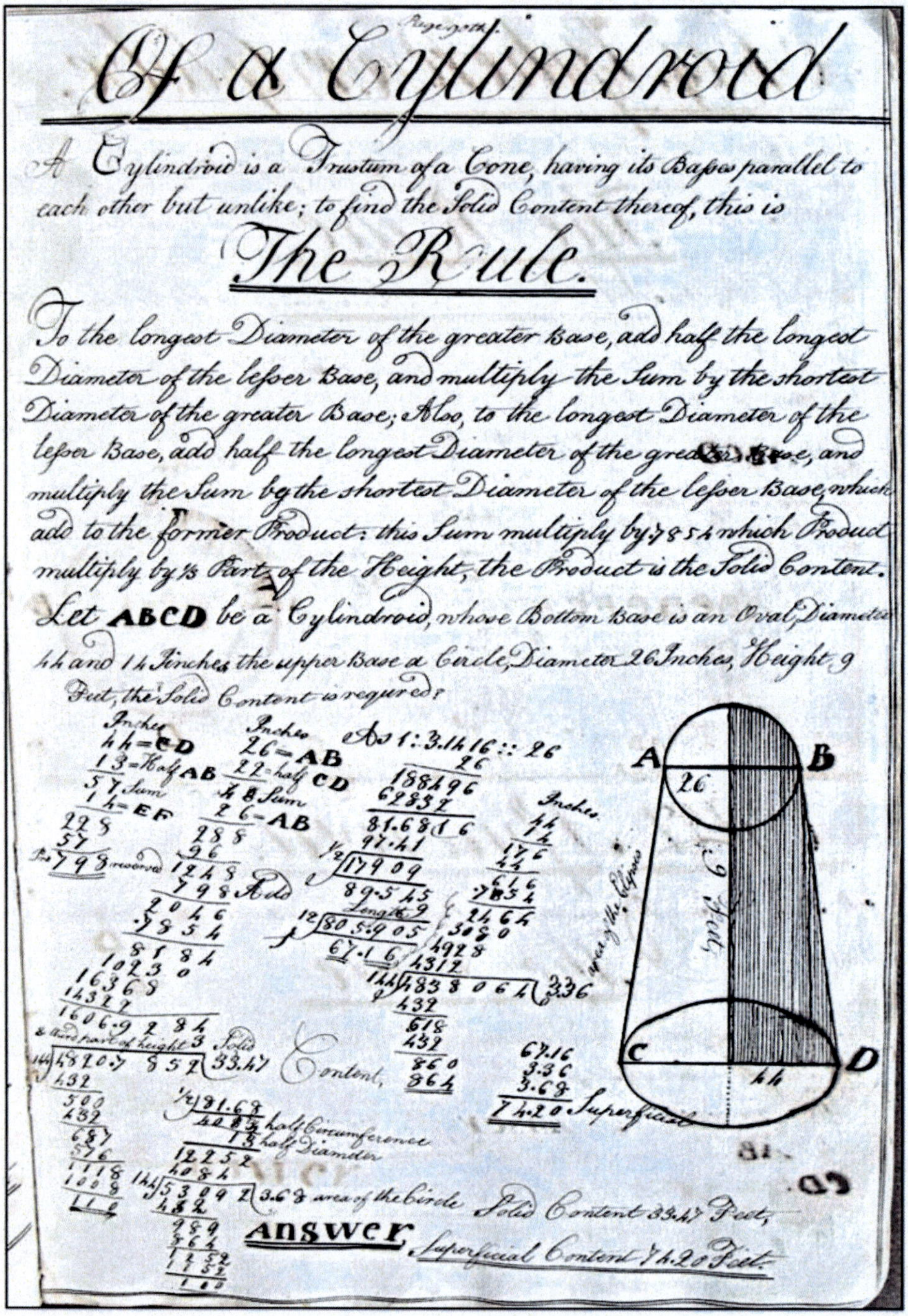
Of a Cylindroid

A Cylindroid is a Frustum of a Cone, having its Bases parallel to each other but unlike; to find the Solid Content thereof, this is

The Rule.

To the longest Diameter of the greater Base, add half the longest Diameter of the lesser Base, and multiply the Sum by the shortest Diameter of the greater Base; Also, to the longest Diameter of the lesser Base, add half the longest Diameter of the greater Base, and multiply the Sum by the shortest Diameter of the lesser Base, which add to the former Product: this Sum multiply by .7854 which Product multiply by 1/3 Part of the Height, the Product is the Solid Content.

Let ABCD be a Cylindroid, whose Bottom Base is an Oval, Diameter 44 and 14 Inches the upper Base a Circle, Diameter 26 Inches, Height 9 Feet, the Solid Content is required?

Answer. Solid Content 33.47 Feet, Superficial Content 7.420 Feet.

Figure 3.2. **IRCEE** and **PCA** genres in Thomas Willson's (1789–1790) cyphering book.

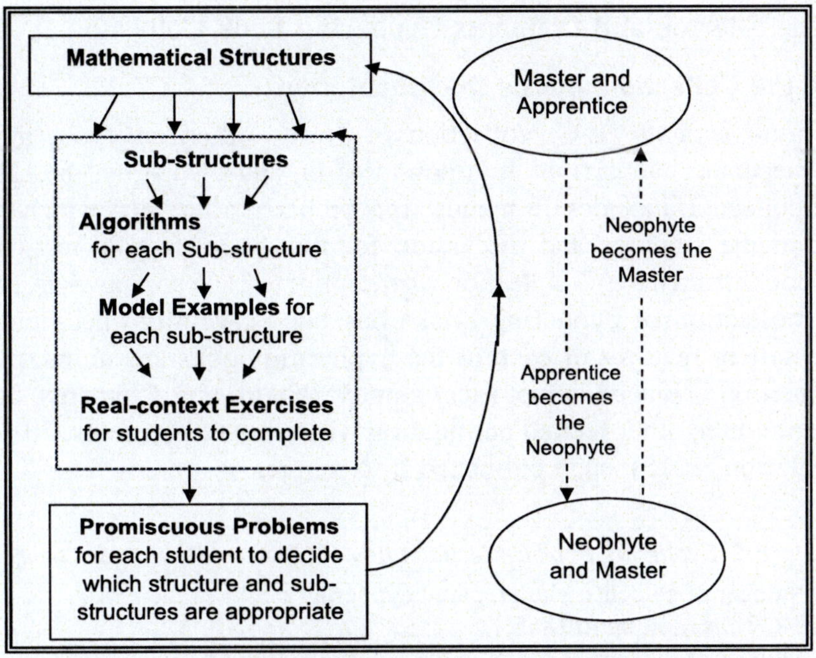

Figure 3.3. A "societally-oriented, structure-based, problem-based" theoretical base (from Ellerton & Clements, 2009).

3. Problems were expected to illustrate well-defined problem categories and sub-categories, and for each sub-category it was expected that at least one "model" problem would be chosen and solved for the students.
4. Then students should be asked individually to solve problems ("exercises") which were structurally identical to the model problems that they had already seen.
5. Before students would be permitted to enter solutions to the exercises in their cyphering books their solutions should be checked by their masters for accuracy.
6. Finally, solutions should be handwritten into the cyphering books, with teachers requiring students to display their best possible penmanship.
7. After repeating this sequence with all the rules and cases associated with several main categories of problems, students should be set miscellaneous problems (often called "promiscuous questions"), with each problem embodying one of the structures that they had just been studying. The problems should relate to real-life situations which were either meaningful to students at that time or might be expected to be meaningful to them in the future.

The Ellerton and Clements Cyphering Book Collection

The Extent of the Collection and its Documentation

The Ellerton-Clements (E-C) collection of North American cyphering books comprises 549 separate manuscripts. It is now held in, and owned by, the Library of Congress. The collection includes 13 manuscripts prepared after 1865 which have not been included in the analyses and discussion for this book. Details related to the 536 cyphering books from the E-C collection written prior to 1865 are shown in Table 3.1.

The E-C collection of cyphering books has been carefully documented in a book in which salient features of each of the cyphering books are summarized and the overall collection is placed in historical context (Ellerton & Clements, 2021). In addition, there are three other related publications (Ellerton & Clements, 2012, 2014; Clements & Ellerton, 2015), each published by Springer.

Table 3.1
Data Related to the E-C Cyphering Book Collection up to 1865 (from Ellerton & Clements, 2021)

1.	Number of students who prepared cyphering books in the E-C collection up to 1865:	507
2.	Number of cyphering books in the E-C collection up to 1865:	536
3.	Number of male students who prepared cyphering books up to 1865 (excluding students whose names were not given):	369
4.	Number of female students who prepared cyphering books up to 1865 (excluding students whose names were not given):	84
5.	Number of students who prepared cyphering books up to 1865 but whose names were not given (or whose gender could not be inferred from the name given):	53

Numbers of Students Who Prepared Cyphering Books in the E-C Collection prior to 1865 and Who Indicated they were from Massachusetts, New York, Pennsylvania, Virginia, or Maryland

Massachusetts:	76 Students,	58 Males,	13 Females,	5 Unknowns	(93 Manuscripts)
New York:	72 Students,	56 Males,	10 Females,	6 Unknowns	(65 Manuscripts)
Pennsylvania:	104 Students,	79 Males,	15 Females,	10 Unknowns	(114 Manuscripts)
Virginia:	18 Students	14 Males,	4 Females,	0 Unknowns	(20 Manuscripts)
Maryland:	15 students,	12 Males,	3 Females,	0 Unknowns	(16 Manuscripts)

The E-C cyphering book collection is much larger and more representative of U.S. cyphering books than any other collection. It includes cyphering books from many colonies (or states) from 1667 on, whereas other collections tend to have cyphering books which were prepared in a limited number of neighboring geographical regions.

Entries in the bottom part of Table 3.1 suggest that slightly more than one-half of the cyphering books in the E-C collection were prepared in New England, or New York, or Pennsylvania (Ellerton & Clements, 2021). From an analysis of the cyphering books in the E-C cyphering book collection we have concluded that between 1667 and 1865 there were as many cyphering books prepared in what is now the state of Ohio as there were in all of Alabama, Georgia, Louisiana, Mississippi, North Carolina, Kentucky, and Tennessee combined (Ellerton & Clements, 2021).

In particular, it seems to be almost certain that only a small proportion of children living in Southern colonies or states prepared cyphering books—we say that despite the existence of an impressive collection of "Southern" cyphering books now held in the Wilson Library at the University of North Carolina (for details, see Doar, 2006). In a Federal government report issued toward the end of the nineteenth century, A. D. Mayo (1898) blamed weaknesses in Southern education on the tendency of wealthy landowners from Virginia and the Carolinas to look after their own children educationally, but to offer only "meagre provision for the mass of white people" (p. 715). According to Mayo, this resulted in "widespread illiteracy" (p. 715). Data from our research would suggest that in the South, innumeracy became a problem because relatively few Southern children prepared cyphering books (Ellerton & Clements, 2012). And, of course, poverty and lack of opportunity associated with race would also have been a major factor (Brewer & Porter, 1994). There is no strong evidence that any of the 507 students who prepared the cyphering books in the E-C collection were of African-American heritage—although a certain Daniel F. Masten, who prepared his cyphering book in New York between 1832 and 1843, may have been an exception to that statement (Ellerton & Clements, 2021).

So far as gender differences in participation in school mathematics were concerned, our analysis indicates that slightly less than 20 percent of the students who prepared cyphering books in the E-C collection were female. Most of the manuscripts prepared by females were from the period 1800–1865 (Ellerton & Clements, 2021). Thus, although most teachers who taught mathematics to learners aged 10 years or more were male, a small proportion of girls did cypher. That said, there can be no doubt that a far greater proportion of boys than of girls studied any form of school mathematics beyond the most elementary level.

Figure 3.4 shows a page from a manuscript prepared, in the late 1760s, by Sally Halsey of New Jersey. Sally's manuscript is part of the E-C cyphering book collection. Sally was certainly not the first North American female to produce a cyphering book, for in a footnote on page 105 of Robert Middlekauff's (1963), *Ancients and Axioms: Secondary Education in Eighteenth-Century New England,* reference was made to Alice Chase's (1755) cyphering book, which was held by the Rhode Island Historical Society. Middlekauff claimed that for arithmetic, "girls memorized the same rules that boys did, solved the same problems, and kept copybooks with rules, problems, and computations arranged under the same rubrics" (p. 105). According to Middlekauff, "algebra, geometry, and other branches of mathematics were not for them" (p. 105). But, few girls, even among those girls who went to school, studied as much arithmetic as Sally Halsey or Alice Chase. Patricia Cline Cohen (1982), in her *A Calculating People: The Spread of Numeracy in Early America*, claimed that before the 1820s "females rarely progressed beyond a few years of schooling" (p. 129).

The Wilson Library at the University of North Carolina holds a magnificent composite cyphering book which was prepared between 1776 and 1782 by sisters Martha and Elisabeth Ryan of Bertie County, North Carolina (for details, see Chapter 4 of Ellerton and Clements, 2014).

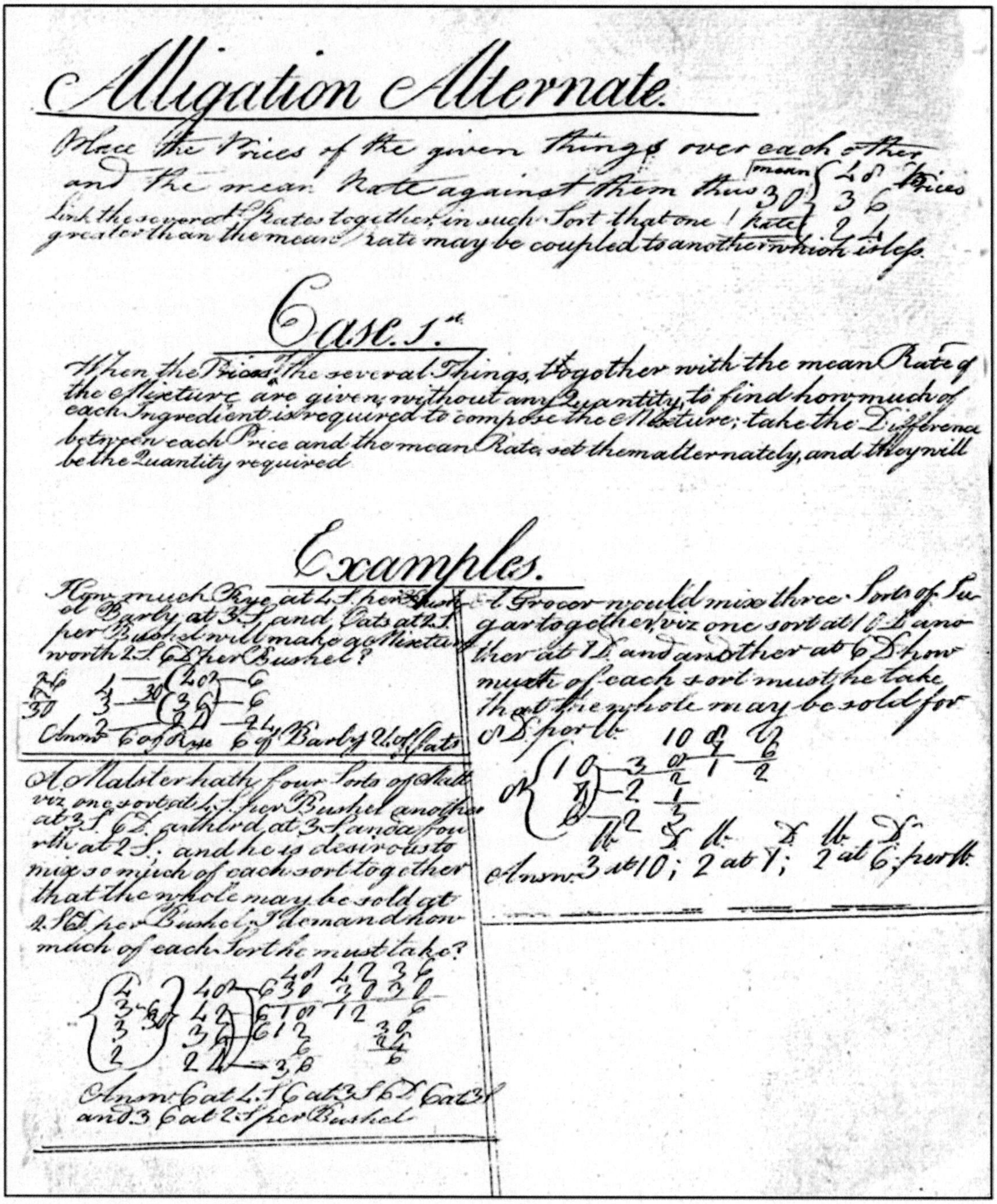
Alligation Alternate.

Place the Prices of the given things over each other and the mean Rate against them thus mean 30 { 40 36 24 Prices

Link the several Rates together in such Sort that one Rate greater than the mean rate may be coupled to another which is less.

Case 1st.

When the Prices of the several Things, together with the mean Rate of the Mixture are given, without any Quantity, to find how much of each Ingredient is required to compose the Mixture; take the Difference between each Price and the mean Rate, set them alternately, and they will be the Quantity required.

Examples.

How much Rye at 4 S. per Bushel Barly at 3 S. and Oats at 2 S. per Bushel will make a Mixture worth 2 S. 6 D. per Bushel?

Answ. 6 of Rye 6 of Barly 24 of Oats

A Malster hath four Sorts of Malt viz one sort at 4 S. per Bushel another at 3 S. 6 D. a third at 3 S. and a fourth at 2 S. and he is desirous to mix so much of each sort together that the whole may be sold at 2 S. 6 D. per Bushel. I demand how much of each Sort he must take?

Answ. 6 at 4 S. 6 at 3 S. 6 D. 6 at 3 S. and 36 at 2 S. per Bushel

A Grocer would mix three Sorts of Sugar together viz one sort at 10 D. another at 7 D. and another at 6 D. how much of each sort must he take that the whole may be sold for 8 D. per lb.

Answ. 3 at 10; 2 at 7; 2 at 6 per lb.

Figure 3.4. A page prepared by a female (Sally Halsey, of New Jersey) in the late 1760s.

Although Cyphering Books Were Special and Private Creations, They Sometimes Included Important Historical Documents

Well into the nineteenth century there was a shortage of paper in North America, especially outside of large Eastern coastal cities like New York, Boston, and Philadelphia. Ink was not easy to get—and as a result, the ink used for cyphering was usually of a home-made variety. This resulted in cyphering books becoming much-valued documents, and often students who prepared them would refer to them, in writing, as "My book." They might also include handwritten ditties (like "Abraham Lincoln, his hand and pen; He will be good, but God knows when"—see Ellerton et al., 2014, p. 139), and often family trees appeared on the last few pages.

Future employers, and educational institutions were likely to want to see the cyphering books of prospective employees or students, so those who prepared cyphering books often took great pains to make their cyphering books as attractive as possible. Calligraphic headings and handwriting tended to be elaborate and beautiful. The decorative tendency was also partly a response by students to pressure from their teachers. A teacher's re-appointment could depend on the perceived quality of cyphering books, for they would be made available to parents to inspect at special school functions held toward the ends of semesters. It has been claimed that next to the Bible and an almanac, a cyphering book was often regarded as the most precious document in a family home (Ellerton & Clements, 2012). Perhaps that is why so many cyphering books survive today.

Because paper was expensive and often hard to get, cyphering books were sometimes used, by families and communities, to record important information which one would not normally expect to find buried in a child's "school" book which dealt with arithmetic. In fact, quite a few of the cyphering books in the E-C collection contain historical documents relating to key events in the history of the United States of America. We now have a brief look at two such cases.

A 1774 protest letter. An important historical document was buried in a cyphering book prepared by Amos Lockwood at Newport, Rhode Island, in 1774. This was a copy of a protest letter (against representatives of the British Government) signed by William Ellery (who would later sign the Declaration of Independence), Joseph Wanton, Henry Ward, John Collins, and John Mawdsley—all of whom were leading citizens in Newport. This protest letter, which was handwritten in brown ink, was on the last page of Amos's cyphering book (see Figure 3.5). It was dated "January 15th, 1774," and it was expected that it would be read to those who attended a local protest meeting.

The "protest letter" was hidden within an impressive 14" by 9" cyphering book which dealt with mathematical tables, computations, accounting exercises, practice letters, etc. Some, but not all, of the pages in the manuscript featured attractive penmanship.

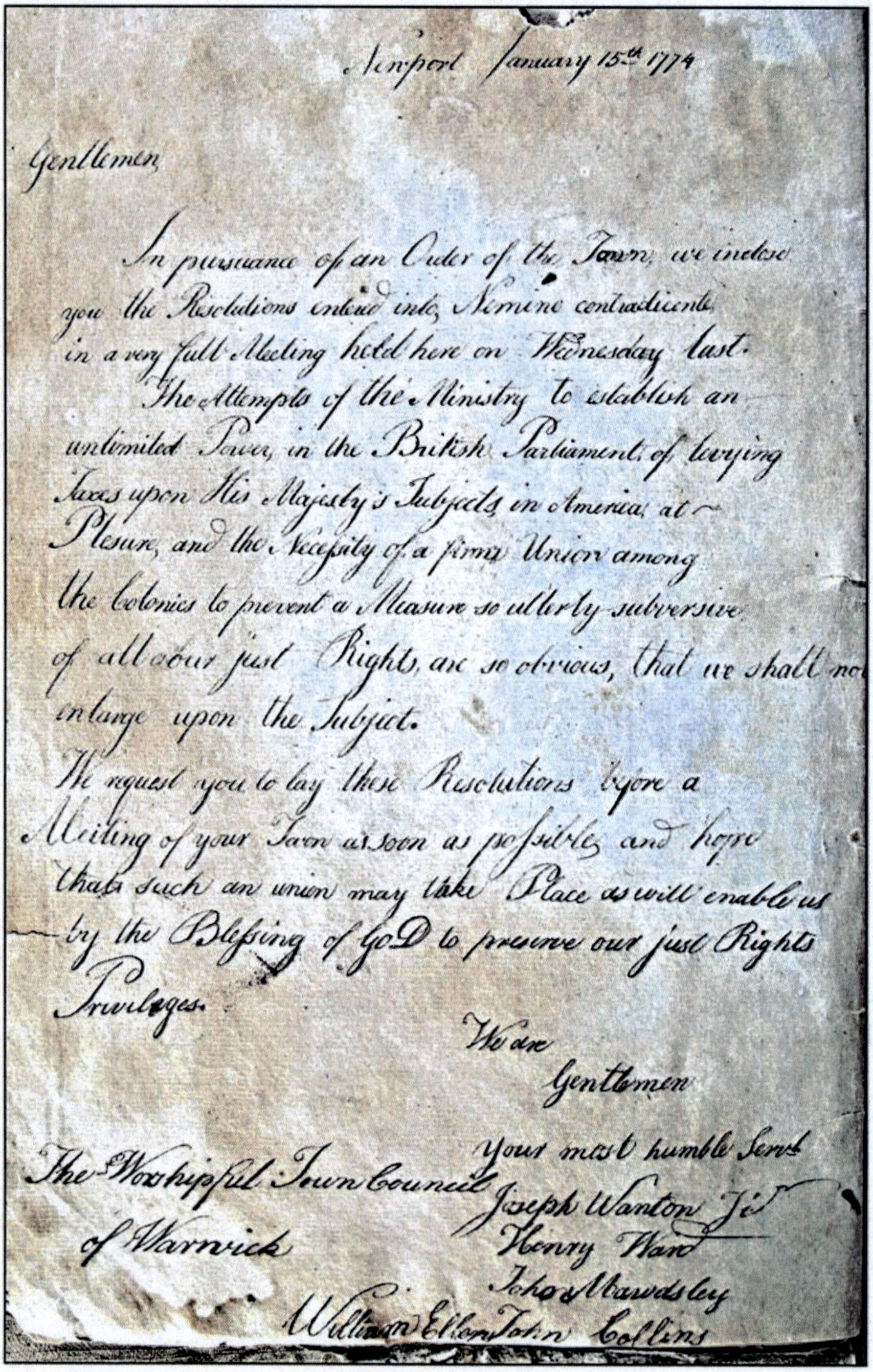

Newport January 15th 1774

Gentlemen

In pursuance of an Order of the Town, we inclose you the Resolutions entered into, Nemine contradicente, in a very full Meeting held here on Wednesday last.

The Attempts of the Ministry to establish an unlimited Power, in the British Parliament of levying Taxes upon His Majesty's Subjects in America, at Plesure, and the Necessity of a firm Union among the Colonies to prevent a Measure so utterly subversive of all our just Rights, are so obvious, that we shall not enlarge upon the Subject.

We request you to lay these Resolutions before a Meeting of your Town as soon as possible, and hope that such an union may take Place as will enable us by the Blessing of GOD to preserve our just Rights Privileges.

We are
Gentlemen
Your most humble Servts

The Worshipful Town Council of Warwick

Joseph Wanton Jr
Henry Ward
John Mawdsley
William Ellery John Collins

Figure 3.5. A protest letter hidden in a 1774 cyphering book.

A Revolutionary War Oath of Allegiance. A Pine/Chichester cyphering book includes a page on which there is a signed Revolutionary War oath of allegiance to King George III. The document was probably prepared in 1778. The oath was copied by James Pine Chichester, a teenager who lived in Huntington, New York. James was the son of Eliphalet and Mary Pine Chichester. Eliphalet fought against the British during the Revolutionary War.

An introductory remark reads, in part: "The oath that the inhabitense (sic.) was (sic.) obliged to take while prisoners under the British government during the Revulution (sic.) War for our Independence. I do certify that Pine Chichester, aged 19, of Huntington township has voluntarily swore (sic.) before me to bear faith and true allegiance to his Majesty King Georg[e] the Third ..." The page is signed, "William Tryon, Governor." Tryon served as New York's colonial governor from 1771 through 1780. Huntington, which was on Long Island and was not far away from New York City, became the headquarters of the North American British Army for some years after 1776. Figure 3.6 shows the relevant part of the manuscript.

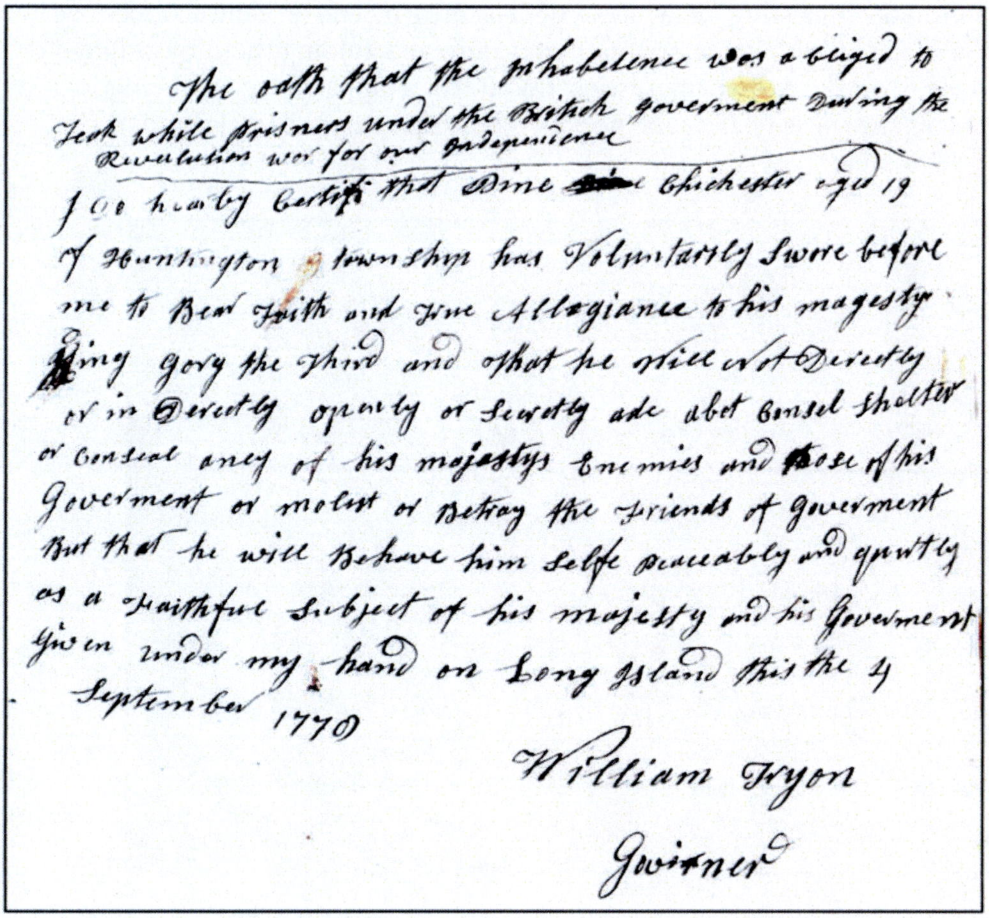
The oath that the inhabetence was obliged to
Teak while prisners under the British goverment During the
Revolution war for our Independence
I do hearby Certifi that Pine Chichester aged 19
of Huntington township has Voluntarily Swore before
me to Bear faith and True Allegiance to his majesty
King gorg the Third and that he will not Derectly
or in Derectly openly or Secretly ade abet Consel shelter
or Conseal aney of his majestys Enemies and those of his
Goverment or molest or Betray the friends of goverment
But that he will Behave him selfe peaceably and quietly
as a faithful Subject of his majesty and his Goverment
Given under my hand on Long Island this the 4
September 1770

William Tryon
Governer

Figure 3.6. A 1778 oath of allegiance, hidden in a cyphering book.

Some Special Cyphering Books

The Oldest Extant North American Cyphering Book

A seven-page manuscript, which may have been prepared in Maine in the late 1660s, is probably the nation's oldest extant cyphering book. The document became part of the E-C cyphering book collection in June 2013 when we purchased it through an online auction. Figure 3.7. shows a page from this manuscript.

The manuscript had been advertised in the following way (original wording has been retained):

> This is a curious and quite possibly ultra-rare early (c. 17th century) American mathematics lesson book, handwritten on watermarked, hand-laid paper! Measuring 8" by 6," and spanning 16 pages, 7 of which contain manuscript handwriting. Language is Olde English, with handwriting characteristics undoubtedly mid 1600s–early 1700s. Recovered from a Camden, Maine, auction; assumed American through association, as other documents were all American in the sale. No further identifying characteristics other than watermark consisting of a crest with a crown, and a central cross, as shown. Paper with some staining, creasing, and folding, held by original strings loosely. A few interesting doodles accompany the mathematical rules, as shown as well. Quite possibly the earliest cypher style manuscript we have seen.

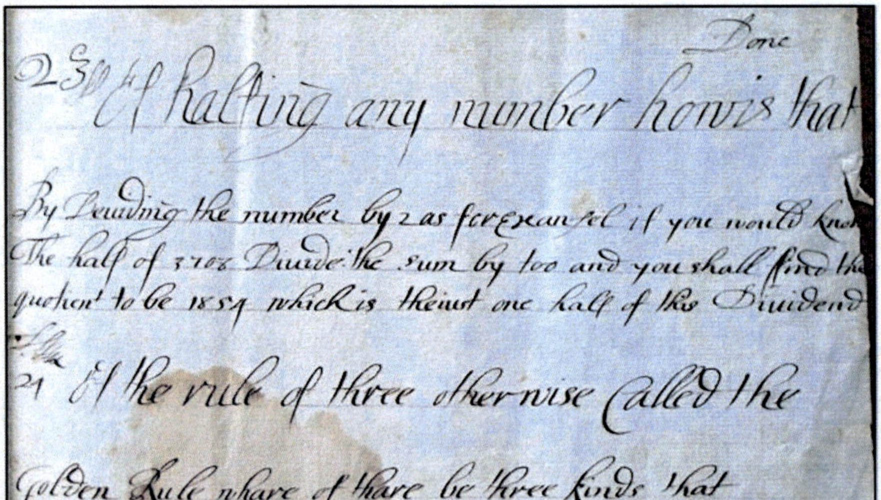
Done
Of halfing any number how is that
By Deuiding the number by 2 as for Exampel if you would know
The half of 3708 Diuide the sum by too and you shall find the
quotient to be 1854 which is theirst one half of the Diuidend
21 Of the rule of three other wise Called the
Golden Rule whare of thare be three kinds that
is the Comond Ruel the Ruel Reuerse and the Double Rule

Figure 3.7. A page from the oldest manuscript (c. 1667) in the E-C cyphering book collection.

Once we had purchased the manuscript, we were able to examine the watermark which had been mentioned in the advertisement. The Gravell Watermark Archive revealed that the paper, which carried Watermark ARMS.047.1, could be dated to 1666, and originated from Holland. So, we conjectured that the paper had been brought to America from Europe, probably in the late 1660s, either by an early settler from Europe or by a commercial exporter.

Although the name of the person who prepared the manuscript, and the place where it was prepared, were not given and, almost certainly, will never be known, we noted the seller's statement that it was purchased at an auction in Camden, Maine, and that it was among other very early American documents. We now assume that the manuscript was prepared in New England in the late 1660s, which would make it the oldest North American cyphering book to have been found. The reader might protest that it has only seven pages, which is less than the minimum that we stated for a cyphering book. Although that is true, it is likely that what remains was part of a larger document. Historically, the document is potentially important to historians of mathematics education because it provides evidence that, quite early, the European cyphering tradition had found its way into colonial North America. Given that nearly all of the white settlers were from Western Europe, that is hardly surprising.

We have commented elsewhere (Ellerton & Clements, 2014, p. 14) that of all the cyphering books that we have seen, this is, mathematically speaking, "the worst"—in the sense that the mathematics is at a very low level and is very roughly presented in the manuscript. However, not all the early cyphering books were of such poor quality. A particularly beautiful early manuscript, by Sarah Cole, dated 1685, is held in the Folger Shakespeare Library in Washington DC, and the standard of arithmetic in that manuscript is at a much higher level (Powell & Dingman, n.d.).

In 2009 and again in 2010 we spent time identifying and analyzing over 200 cyphering books held in the Phillips Library of the Peabody Essex Museum, in Salem, Massachusetts. During our time there we created a finding aid for what is the second largest collection of North American cyphering books. The Phillips Library collection includes a seventeenth century cyphering book which was prepared by an unknown writer between 1692 and 1694 and has 29 pages (with dimensions 12" by 7.5"). The handwritten entries focus on navigation, astronomy, and surveying, with references being made, for example, to the Gunter scale (which had been developed in England in the seventeenth century by Edmund Gunter). Hertel (2016) referred to manuscripts which focused on navigation as "navigation cyphering books."

It is very likely that the seven-page cyphering book (c.1667) is older than any at the Phillips Library. For a detailed analysis of the manuscript see Chapter 2 of Ellerton and Clements (2014). It suffices to say, here, that if indeed the document was prepared in the late 1660s then it should not take a lot of imagination from someone in the twentieth-first century to recognize its significance. Here were settlers, struggling to take their place in a new and dangerous world. Textbooks, teachers, and paper were not readily available to anyone who wanted to engage in an educational enterprise. The labor of all available persons was needed to contribute to the daily struggle of getting enough food and shelter for the family, and to weaving and sewing for the family's clothing. The manuscript reveals that whoever served as teacher did not know much about elementary arithmetic—but that teacher knew *something* and had a desire to pass that something on. Somewhat poetically, one might reflect that in time the small spark of knowledge represented by this 1660s manuscript would symbolize the spread of the cyphering tradition to and within North America—in effect, a precursor to the establishment of strong mathematical roots for the future.

In the 17th and 18th centuries the bustling port of Salem sent many ships to India and to the spice islands, and the Phillips Library has an impressive collection of cyphering books relating to the education of seamen. However, the Library does not claim that its collection is truly representative of cyphering books prepared in North America. From that perspective, the E-C collection, which has more than twice the number of cyphering books than any other collection, has stronger claims, because its manuscripts originated from right across colonial North Americam—in Massachusetts, New Hampshire, Vermont, Maine, Rhode Island, New Jersey, New York, Delaware, Connecticut, Pennsylvania, Virginia, Maryland, North Carolina, South Carolina, Georgia, Alabama, Mississippi, Arkansas, Tennessee, Kentucky, Ohio, Indiana, Missouri, Illinois, and Michigan.

It will be up to future researchers to delve more deeply into the rich data resources provided by cyphering books, and to identify more fully lessons for, and commentary on, issues associated with mathematics curricula and the teaching and learning of mathematics between 1607 and 1865.

The Pine-Chichester (Long Island, New York) Composite Cyphering Book

This 312-page Pine-Chichester hard-cover manuscript (with dimensions 7.75" by 7") is a composite manuscript in the sense that it contains a number of sections obviously prepared by different people in the Chichester/Pine families of Huntington, Long Island, New York. Figure 3.8 shows an undated page which might have been prepared in the seventeenth century.

The manuscript features high-level, multi-color calligraphic headings throughout. It begins with the four elementary operations on whole numbers, and then proceeds to tasks involving measurement of common quantities. Of special interest is the **IRCEE** structure for the coverage of each topic: after a brief introduction to the main ideas, rules and cases are presented, and these are followed by worked examples and exercises. Much of the manuscript was concerned with calculations. Many of the pages in the manuscript were headed with dates from the eighteenth century, but there is an undated 54-page section which could have been prepared in the later years of the seventeenth century. Whenever a new topic was introduced in this section (e.g., "The Single Rule of Three Direct"), the heading is in a very mature hand, and the calligraphy of a high standard.

The Pine (sometimes written as “Pyne”) family originated from England. The first of this family to come to America, James Pine, arrived in Connecticut in the mid-1640s. A few years later he settled in Huntington, near the Dutch settlement of New Amsterdam (renamed New York in 1664). The Pines and their numerous descendants would remain in and around New York for well over a century, and it is likely that the early cyphering sections in this composite manuscript were prepared by children who were among the numerous descendants of the original Pine family settlers. The Chichesters were among the earliest of the settlers on Long Island, and the name “Chichester” became well known—especially after the establishment of the Chichester Tavern in the 1680s. In this manuscript it is recorded that a James

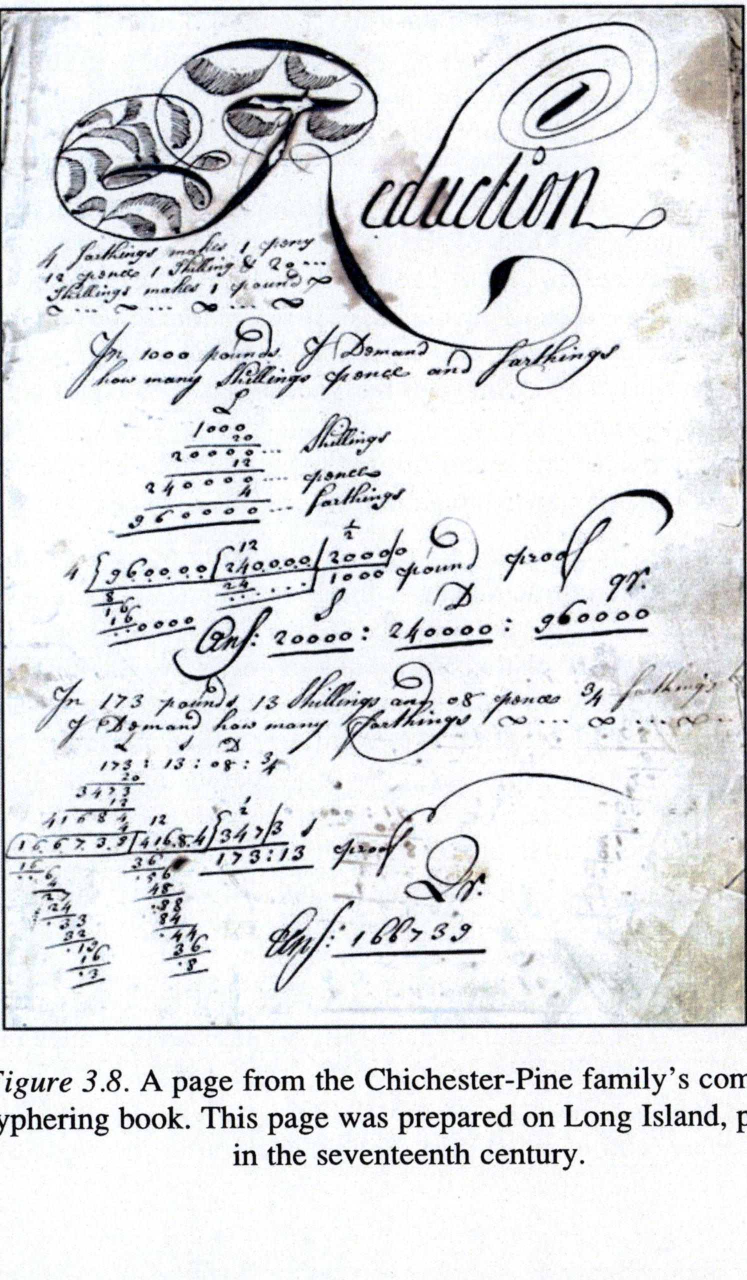

Figure 3.8. A page from the Chichester-Pine family’s composite cyphering book. This page was prepared on Long Island, perhaps in the seventeenth century.

Pine Chichester was born on April 22, 1759, and another James Pine Chichester on June 7, 1779. The name “James P. Chichester” is written on the front cover of this manuscript and it also appears, in handwritten form, on numerous occasions throughout the composite manuscript. Thus, it appears to have been the case that the various cyphering sections within this composite manuscript were prepared by children from two well-known Long Island families.

The latest date recorded in the manuscript is 1872. There is minor foxing throughout, with some noticeable stains—but these do not interfere with the text. Some pages have become detached. Despite all of that, this nationally significant composite document is well preserved and is in good condition. It is now held in the E-C cyphering-book collection within the Library of Congress.

There is a nine-page section (dimensions 13” by 8”), dated either “1701” or “1781”—it is hard to tell which—sewn in at the beginning of this composite manuscript. Although the name of the person who wrote the nine pages is not given, the first part of this large composite manuscript seems to form a cyphering book prepared by Jeremiah Chichester.

The Pine and Chichester families were united in 1758 by the marriage of Mary Pine to Eliphalet Chichester. Most of the pages in the composite cyphering book contain arithmetic exercises (including relatively advanced exercises in business mathematics), or solutions to word problems, or mathematical rules and maxims, or poems (such as, “*As the sum of the several stock, So to the total gain or loss, So is each man's share in stock, To his Share of the gain or loss*”). Several pages contain Pine/Chichester birth records.

There are six features of this manuscript which are of interest from the point of view of the history of mathematics education in the United States:

1. This manuscript was passed on from generation to generation. Signatures by different members of the Pine/Chichester families were added to the existing manuscript, after they had used the book. In one section, there is a list of the birth dates of members of the families. Altogether, there were 13 names listed and, of those, the first was born in 1703 and the last in 1858. Most of the quires which are sewn together to form the composite manuscript were prepared (and dated) in the eighteenth century.
2. There are 312 handwritten pages altogether (the dimensions are mostly 13” by 8.5,” with one of the earliest, if not the earliest, of the quires having dimensions 12.5” by 7.5”). Mary Pine's section, written in 1752, has dimensions 12” by 7.25.”
3. There is plenty of evidence that many of the entries were written by adult teachers. For example, a beautifully-written section attributed to James P. Chichester in 1795 was probably prepared by “A. Kitcham,” for on the bottom of one page we find a very mature signature “James P. Chichester,” and in small print, immediately after the signature we

also find, in the same handwriting, "A Kitcham." On a page dated February 1796 we find "James Pine Chichester, his hand and pen, rote (sic.) at Elias Baylis at scool (sic.) to Abel Kitcham in Sweet holler (sic.) in Huntington." Sweet Hollow was a district within Huntington. It seems that Kitcham wrote the problems, and James Pine Chichester "solved them" (if one can regard scattered masses of calculations, not obviously connected by logic, as "solutions"). The practice of adults preparing calligraphic headings and neatly writing problems in students' cyphering books was very common (Dickens, 1850).

4. The first quire (which appears to be dated 1781, but it could be 1701) was sewn into the composite manuscript. It dealt with numeration and simple operations on measurements (e.g., addition of money, avoirdupois, and troy weight). There are later sections in which more advanced arithmetic—often related to financial matters—was covered. Notes on topics like loss and gain, barter, the various rules of three, and compound interest, are to be found. Jeremiah Chichester prepared one of the sections.
5. All but one of the sections were prepared by boys. The exception was a section prepared in 1752 by Mary Pine Chichester, the mother of James Pine Chichester, but this contained only handwritten admonitions and no arithmetical calculations.
6. There is a page, dated 1778. on which there is a Revolutionary War Oath of Allegiance to King George III (see Figure 3.6).

Mary Pine married Eliphalet Chichester in 1758. The book was then taken over by a succession of young men each named James Pine Chichester. One of them wrote, "*ARITHMETIC/JAMES P. CHICHESTER 1796*" across the front cover.

An Early Composite Cyphering Book, Started in England in 1702 and Completed in New Hampshire Between 1720 and 1722

This manuscript was prepared by two people—Thomas Prust (who sometimes wrote "Priest" for his family name), whose entries were prepared in England between 1702 and 1703, and James Collings, whose entries were prepared in New Hampshire, probably between 1720 and 1722. There are 154 pages altogether. Quires were sewn together, but not necessarily in the order in which they were prepared.

The arithmetic was consistent with the standard *abbaco* sequence—numeration, the four operations, compound operations, rules of three, reduction, fellowship, vulgar fractions, etc. The standard of calligraphy is high, with some of the headings being incredibly elaborate. The cover is detached but present. The name "Thomas Prust" (in superb calligraphy) occupies the front page of the manuscript, and "Thomas Prust his book Amen 1702" also appears on quite a few pages (see, for example, Figure 3.9). Apparently, James Collings was Prust's son-in-law, and on one of the pages James indicated that the book had been given to him by his father-in-law. Although this was a cyphering book, Thomas liked to include some brief poems which, it seems, he himself wrote.

Thomas Prust signed his name on many of the folio-sized pages which remain from the part of the manuscript that he prepared between 1702 and 1703. James Collings also liked to write his name—on three occasions between 1720 and 1722 he indicated in writing, that he now owned the book.

Unfortunately, Prust never indicated *where* he was when he was preparing his book. Genealogical records suggest that he was born in England in 1690 and emigrated to Portsmouth, New Hampshire, in 1720. Toward the end of 1720 he married Sarah Collings in Portsmouth. There is a handwritten note in the cyphering book indicating that it was passed on to James Collings by his "father-in-law." It seems that Prust was James Collings' father-in-law (which, at the time, sometimes meant what "step-father" means today), and that he passed on his highly-valued manuscript to the 13-year-old James, who was his new wife's son by a previous marriage.

If that conjecture is correct, then Thomas prepared his part of the cyphering book as a boy growing up in a well-to-do family in England. However, the book was passed on and used in North America, where it has remained for more than 300 years. It seems reasonable to regard this as an American cyphering book—even though most of it was prepared by Thomas Prust when he was a boy in England.

We purchased the manuscript in 2006 from a collector in Florida who told us that he had acquired it in 1989 at an antiquarian book show in Boston. He explained that he was an artist and that he admired the calligraphy in cyphering books so much that he cut out headings from them and pasted them as special features into his own artwork. With respect to this particular cyphering book, he (or some previous owner (s)), had removed up to 20 of the original pages from the book and, almost certainly, those pages will never be seen again. However, much of what remains of the book is impressive and, we believe, of large historical interest.

The different levels of "maturity" evident in the penmanship in various parts of the cyphering book suggests that Thomas's teacher wrote some of the headings, and also solutions to some of the problems. Thomas wrote other headings, problem statements, and solutions.

This manuscript provides evidence of links between North American and European cyphering traditions. We believe that similar cyphering traditions developed in many Arab and European nations (Cohen, 1993; Denniss, 2012; Stedall, 2012), and that these were translated, as a result of migration, into North America. Over time, the traditions which were established in North America became different from those from which they had emerged (Ellerton & Clements, 2012).

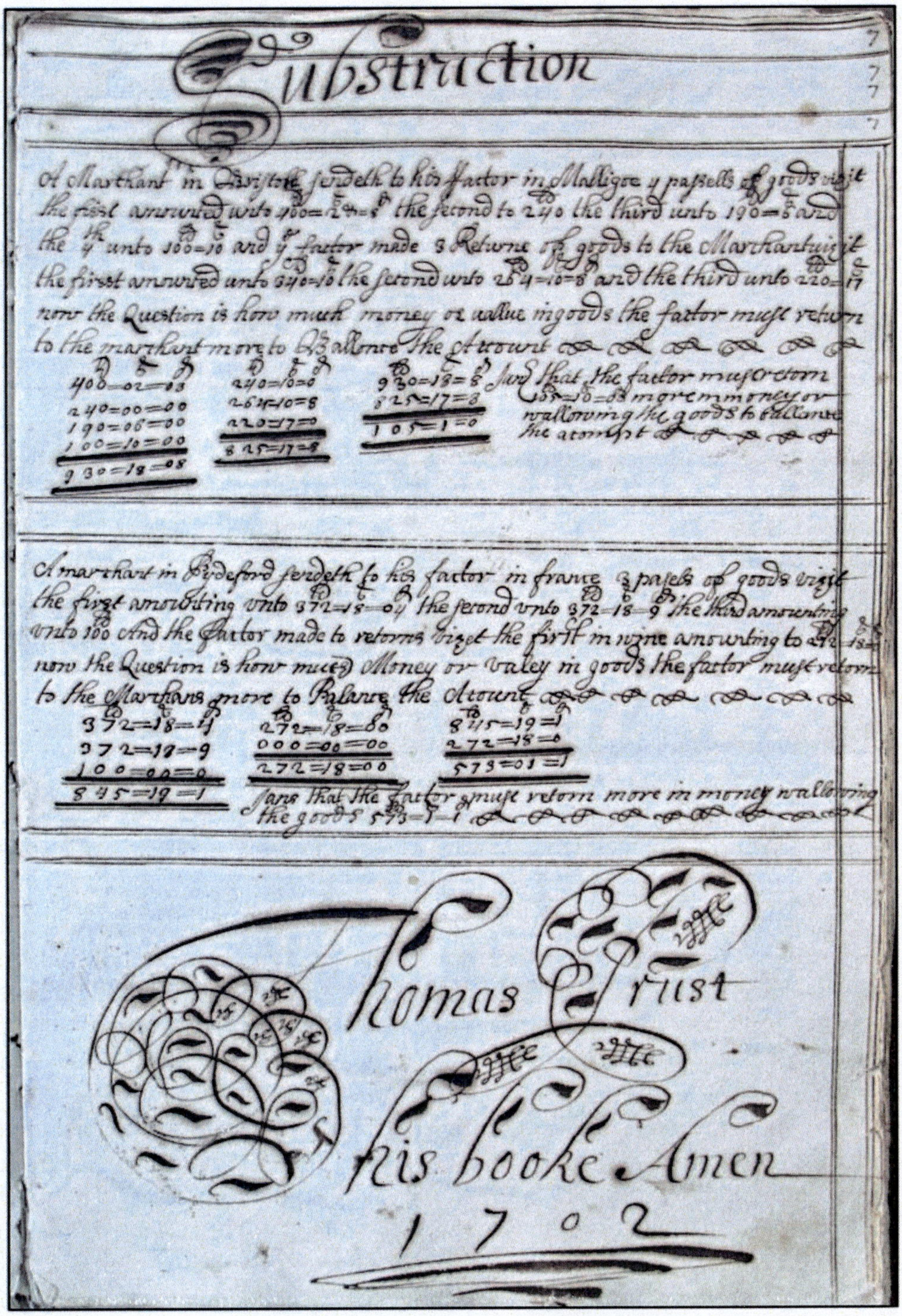

Figure 3.9. "Thomas Prust his booke Amen 1702."

A 1771 Cyphering Book Prepared by a Future Revolutionary War Solder

A manuscript which belonged to a future Revolutionary War soldier, John Grey of Rhode Island, contains more than 100 handwritten pages, each 13" by 8." The contents include mathematical tables, and entries on square roots, computations, the use of money, accounting exercises, weight, mensuration, etc. "John Grey His Book" is found at the bottom of the fifth page and "John Grey His Book February 26, 1771" at the bottom of a page towards the back of the book. The penmanship is of an average to high standard. At the bottom of an early page "L. Little" appended his signature, and on succeeding pages the type of entry changed greatly. Perhaps L. Little was John Grey's tutor.

This cyphering book is typical of those prepared in the colonial era. Much of the content dealt with lower- to middle-level *abbaco* arithmetic. The early pages dealt with measurement, reduction, and practice, but then came single and double fellowship, and various forms of the rule of three. Towards the end of the manuscript, barter and interest were considered. There are some notes (including rules and cases), but most of the manuscript comprises handwritten solutions to model problems. **PCA** genre is clear—state the problem, show some calculations, and write "Answer" next to the last line of the "solution." There are very few explanations in the manuscript.

A 1775–1777 Cyphering Book Prepared by a Revolutionary War Solder

Cornelius Houghtaling, who would become a Revolutionary War soldier, began to prepare his cyphering book in New Paltz (52 miles north of New York City) in 1775. The 250 pages of his colorful, large, full-leather-covered manuscript (dimensions 12" by 7.5") feature two ink colors (black writing and red borders and divisions). This is an attractive example of an extant early American cyphering book. Figures 3.10 and 3.11 reproduce pages from Cornelius's cyphering book. On the inside of the back cover, in beautiful calligraphic penmanship, we find: "If I it loose (sic.) and you it find, Pray give it me, for it is mine." "*Finis Coronat Opus*."

A reasonably standard *abbaco* sequence of arithmetic was followed—from notation and numeration, through elementary whole numbers, common or vulgar fractions, decimal fractions, reduction, simple and compound interest, exchange, rebate and discount, tare and tret, various rules of three, alligation medial and alternate, evolution and evolution, and single and double position. The word "proof" was used throughout to indicate merely a "check." There were numerous practical examples, relating to purchases from a pharmacy, to banking, etc., and there was a section relating to mathematical amusements. A beautifully ornate, and interesting full page included "Decimal Tables," and entries like "shillings reduce to decimals of a pound" (and "1 shilling is .05 of a pound," etc.).

Cornelius was born in 1757, to descendants of early Dutch settlers. He was the sixth of the 11 children of Teunis Houghtaling and Elizabeth Beekman. Their sons Cornelius, Jeremiah, John, Thomas, Wilhelm, Harmon, and Jacob were all soldiers

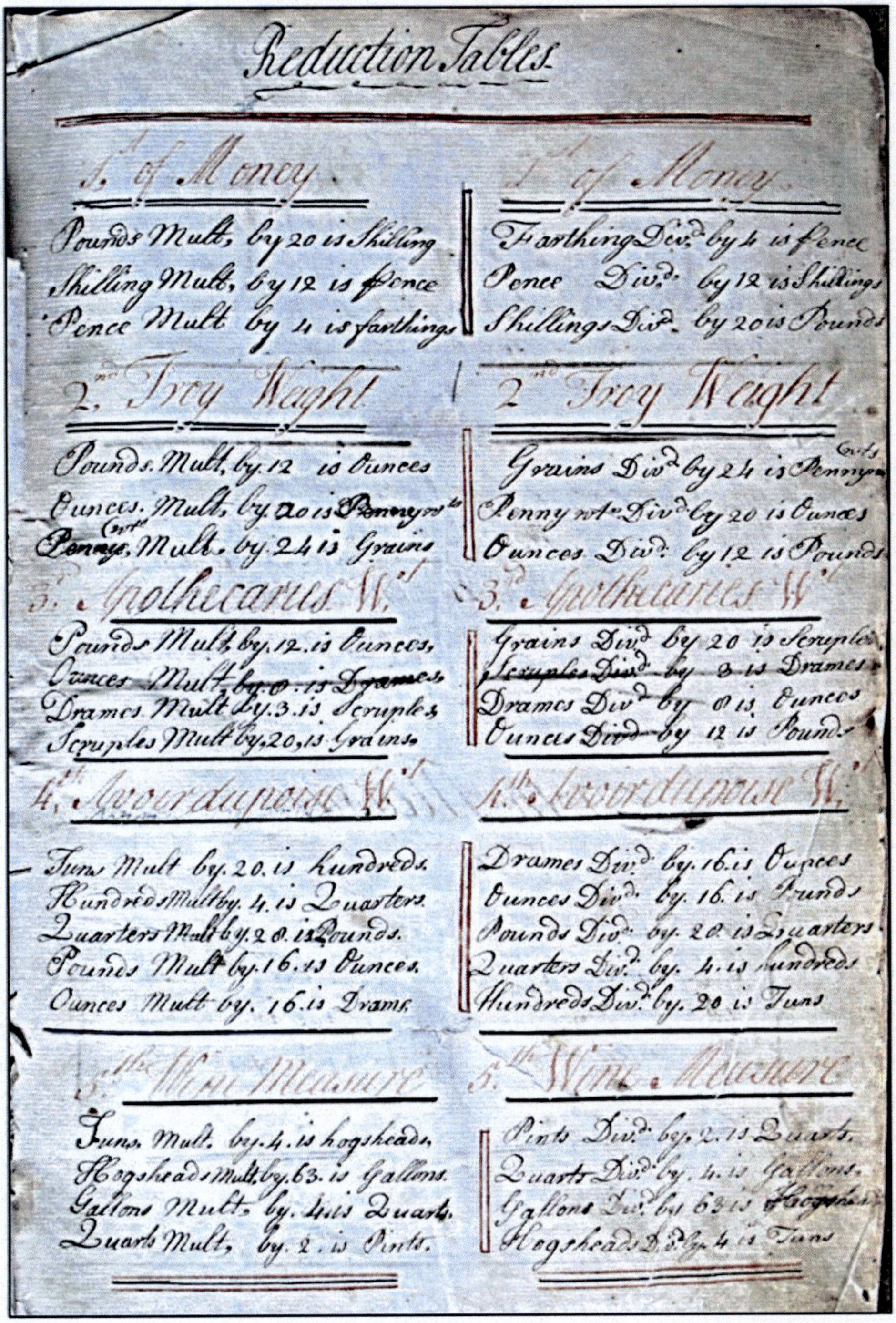

Reduction Tables

1st of Money	1st of Money
Pounds Mult, by 20 is Shilling	Farthing Div.d by 4 is Pence
Shilling Mult, by 12 is Pence	Pence Div.d by 12 is Shillings
Pence Mult by 4 is farthings	Shillings Div.d by 20 is Pounds
2nd Troy Weight	2nd Troy Weight
Pounds. Mult, by. 12 is Ounces	Grains Div.d by 24 is Penny wts
Ounces. Mult, by 20 is ~~Penny~~ wts	Penny wt. Div.d by 20 is Ounces
~~Penny~~ wts. Mult. by. 24 is Grains	Ounces Div.d by 12 is Pounds
3. Apothecaries Wt	3rd Apothecaries Wt
Pounds Mult, by. 12. is Ounces,	Grains Div.d by 20 is Scruples
~~Ounces Mult, by. 8. is Drames~~	~~Scruples Div.d by 3 is Drames~~
Drames Mult by. 3. is Scruples,	Drames Div.d by 8 is Ounces
Scruples Mult by, 20, is Grains.	Ounces ~~Div.d~~ by 12 is Pounds
4th Avoirdupois Wt	4th Avoirdupoise Wt
Tuns Mult by. 20. is hundreds.	Drames Div.d by. 16. is Ounces
Hundreds Mult by. 4. is Quarters.	Ounces Div.d by. 16. is Pounds
Quarters Mult by. 28. is Pounds.	Pounds Div.d by. 28. is Quarters
Pounds Mult by. 16. is Ounces.	Quarters Div.d by. 4. is hundreds
Ounces Mult by. 16. is Drams.	Hundreds Div.d by. 20 is Tuns
5th Wine Measure	5th Wine Measure
Tuns. Mult. by. 4. is hogsheads.	Pints Div.d by. 2. is Quarts.
Hogsheads Mult by. 63. is Gallons.	Quarts Div.d by. 4. is Gallons.
Gallons Mult. by. 4. is Quarts.	Gallons Div.d by 63 is Hogsheads
Quarts Mult, by. 2. is Pints.	Hogsheads Div.d by 4 is Tuns

Figure 3.10. "Reduction" tables in Cornelius Houghtaling's (1775–1777) cyphering book.

Figure 3.11. A double page from Cornelius Houghtaling's (1775–1777) cyphering book.

in the Revolutionary War. The New Paltz Historical Society holds a cyphering book by Jeremiah A. Houghtaling which was prepared between 1825 and 1833, but Cornelius's cyphering book is older. The word problems in Cornelius's book relate to the rule of three, commission, brokerage, insurance, interest, tare and tret, etc. Of interest are poems and phrases found on the final pages of the book, and numerous dated signatures of Jeremiah A. Houghtaling. The influence of **IRCEE** and **PCA** genre expectations is evident throughout.

Apparently "Ordinary" Cyphering Books Which Became Historically Interesting

Some cyphering books which would have appeared "ordinary" when they were being prepared would, with the passage of time, become "special." Thus, for example, when a young backwoods lad in Indiana named Abraham Lincoln was preparing his cyphering book in the early 1820s no-one would have imagined that one day that lad would become perhaps the most admired person in the history of the

United States of America—or that, almost two centuries later, single, loose, pages of that cyphering book would sell for one million dollars (Ellerton & Clements, 2014).

There are several cyphering books in the E-C collection which would later be recognized for their special significance, and here we draw attention to three of them. The first was prepared between 1764 and 1767 in Germantown, Pennsylvania, by Peter Tyson; the second was prepared in Boston, Massachusetts, between 1813 and 1817 by Samuel Fay; the third in New Jersey by Elijah Allen Rockefeller, around 1825.

An Early Pennsylvania (1764–1767) Cyphering Book with Links to William Penn, President Theodore Roosevelt, and President John Tyler

This 106-page manuscript (13" by 8") was prepared between 1764 and 1767 by Peter Tyson (1751–1822). Peter was a grandson of Reynier Tyson, who was mentioned in William Penn's Charter of 1689 as one of the original settlers in Germantown, Pennsylvania. Reynier Tyson, originally a Mennonite, would leave Germantown and settle in Abington township, where he became a leading Quaker.

The manuscript has well-worn vellum covers. It is bound with string, which has become loose. Peter managed to express his individuality in many ways in the cyphering book. It does not have uniformly beautiful penmanship and calligraphy, but it does have other noteworthy features. In particular, there are numerous little poems and personal comments, some of which were:

> "Peter Tyson, his hand and book; and what it cost you may look."
> "Peter Tyson, his hand";
> "Peter Tyson, his cyphering book 1764";
> "Peter Tyson, his book";
> "Independence";
> "Remember me when this you'll C." [That same request appeared three times.]
> "Innumerable anoiances (sic.) and inconveniances (sic.) accompany mankind." [That same comment also appeared three times];
> "Remember me Peter Tyson Philadelphia";
> "Peter Tyson his hand and pen; wrote the day I will not tell you when";
> "Be not proud or onkind (sic.)";
> "Pain wastes the body";
> "Not all the skill that mortals have, can stop the hand of death";
> "The gaudy paint of pride and vanity";
> "Peter Tyson, his arithmetic."

The content is elementary- to middle-level *abbaco* and the influence of **IRCEE** and **PCA** genre expectations is in evidence throughout (see Figure 3.12). The following question was one which could be found (with varying years) in many cyphering books: "How many days, hours and minutes since the birth of our Savior to this present year" (which, for Peter, was 1767—The answer given for the number of minutes was 929371320). At the end of the cyphering book there are three pages of intricate and beautiful compass constructions, all done in ink.

According to a newspaper article (c. 1952) that we found inserted in the manuscript when we purchased it, research by the Pennsylvania Historical Society has revealed that the Tyson family lived in the Philadelphia area (Germantown, Abington), and some were involved in religious activities and in politics. The article

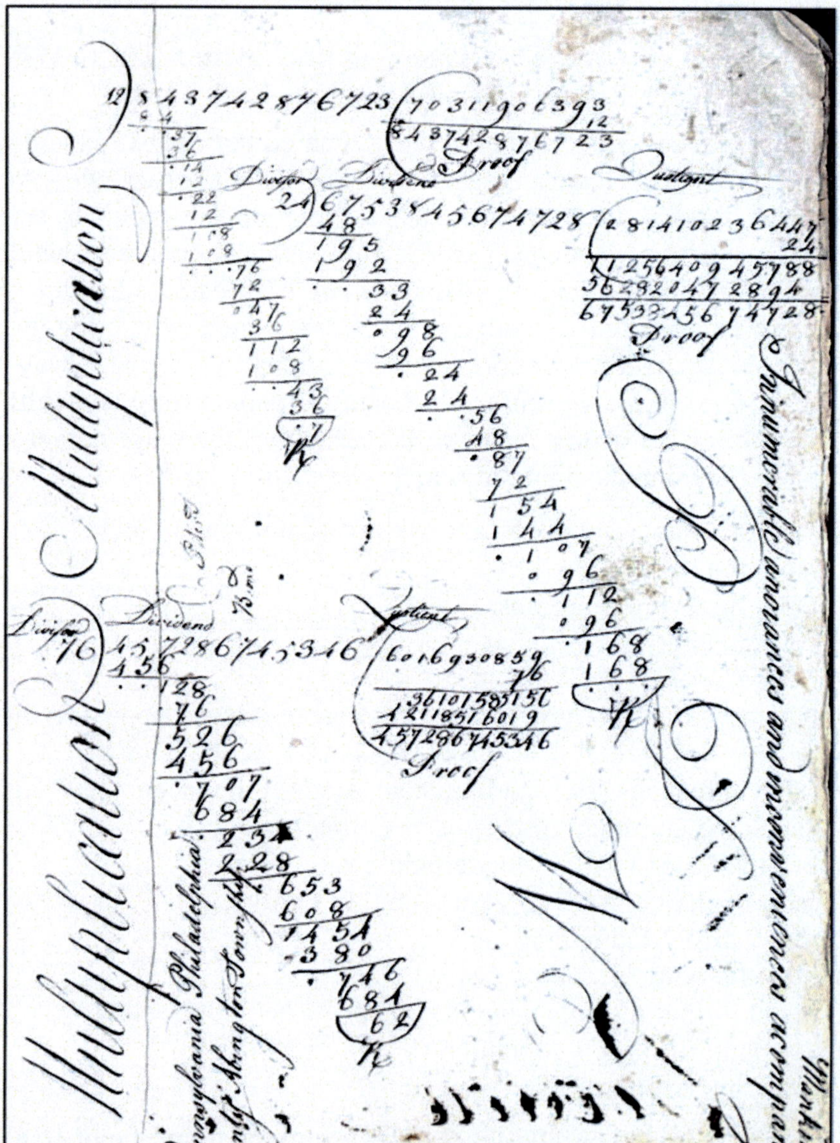

Figure 3.12. A page from Peter Tyson's cyphering book, prepared in Pennsylvania in the 1760s.

indicated that a daughter of Reynier Tyson (Elizabeth) was a third great-grandmother of Theodore Roosevelt and that another descendant of Reynier—Harry Hewlings Tyson—married a grand-daughter, Julia, of President John Tyler, the tenth President of the United States.

On the inside of the front cover and on the inside of the back cover of Peter Tyson's manuscript is a William Bradford newsprint (circa 1766), which helps to add context to the life and times of Peter Tyson. It includes a classified section and an editorial section relating to events of concern to the colonists in the decade before the Revolutionary War. There are references to a major disturbance—which may have been the Tenant-Landlord Conflict in New York, now known as the Great Rebellion of 1766.

The cyphering book is accompanied by Peter Tyson's will and several other documents. There is a list of his personal property, an official British document signed and sealed by one of King George III's justices in Philadelphia which was prepared for a Tyson relative (named Shoemaker), and a newspaper article detailing aspects of Tyson Family history.

Samuel Fay's (1817) Cyphering Book

We bought this manuscript from someone who lived in England. It has 160 pages altogether, of which 87 are covered with English newspaper cuttings—from, for example, the *Times of London*—neatly glued on them. Originally, all pages had formed a cyphering book exclusively devoted to elementary algebra, but now, algebra (in very light, faded ink) can be seen on only 70 of the pages. Some of the algebra provides solutions to problems based on extended stories and poems. Problems are stated, and solutions are given. Some pages have calligraphic headings indicating the topic under consideration, or the date and place of its preparation.

We decided to "unglue" paper covering the front end-paper which had been pasted on the inside of the front cover, and having done so, found on the original inside cover, the name Samuel Fay, and the date October 13th, 1817. We also found a pencil sketch of a three-masted frigate, with the word "Constitution" written below it (see Figure 3.13). Internet research identified that this picture was, almost certainly, a hand-drawn illustration of the *USS Constitution*, a wooden-hulled, three-masted heavy frigate of the United States Navy which became known as "Old Ironsides" during its service in the 1812 War against Great Britain. This frigate, which had been built in Boston, Massachusetts is claimed to have "defeated" five British warships during the 1812 war. It is now in the Charleston Navy Yard and is still regarded as being "in active service."

Further internet research revealed that the "Samuel Fay" named on the inside front cover was Samuel Howard Fay (1804–1847), who lived in Massachusetts, and would have been 13 in 1817. Samuel Howard Fay was the eldest son of Samuel Phillips Prescott Fay (1788–1856), a Harvard graduate, a judge, and at some time a captain in the U.S. army. Young Samuel probably sketched the *Constitution* when

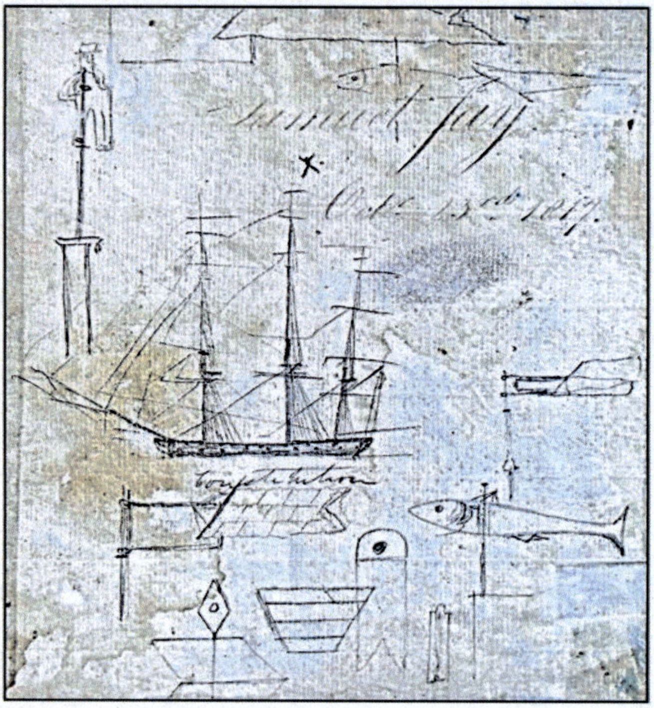

Figure 3.13. "Constitution" on the inside of the front cover of Samuel Fay's cyphering book.

the frigate was celebrating its 20th birthday in the Boston docks in October 1817. One can imagine that he went to the docks and drew a picture of the much-lauded frigate on the inside of the front cover of his new cyphering book. Certainly, the pencil sketch drawing resembles surviving images of the *Constitution*.

It appears that at some later time, the cyphering book was taken to England, where it was used as a scrap book for newspaper cuttings. The name "Eliza Hutchinson" is written in the manuscript, and since Samuel Howard Fay had a younger sister called Eliza it is possible that Samuel handed the cyphering book on to his sister who subsequently took it to England. Samuel Howard Fay would marry Susan Shellman in Georgia in 1825, and although we do not have details of his occupation, he would become treasurer of the Episcopal Institute at Montpelier Springs, Georgia. His daughter, Harriet, would marry the Reverend James Smith Bush, and would become the great, great grandmother of President George Herbert Walker Bush. And therefore, of course, there is also a link with the 43rd President, George Walker Bush. Further genealogical research revealed that there was also a direct link with King Edward 1 of England.

The covers and spine are worn. Internally, hinges are reinforced, and pages have some curvature due to the glue. The manuscript was purchased in 2010 from an online seller in Doncaster, England, who gave no indication that he was aware of its links to two presidents of the United States of America, or to King Edward I.

Figure 3.14 shows a page from Fay's algebra cyphering book. Although it does not give the appearance of being mathematically special, in any way, it needs to be remembered that around 1817 only a tiny percentage of U.S. teenagers ever got to study algebra (see Chapter 5 of this book). Observe the different handwriting in the two sections on the page.

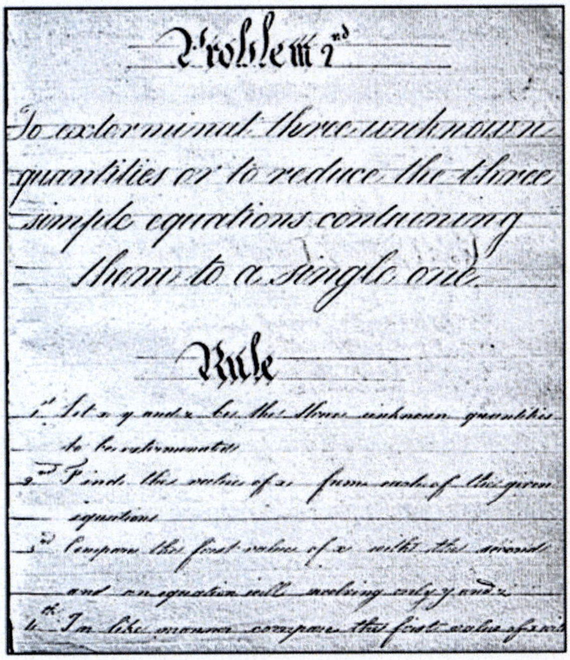
Problem 2nd

To exterminate three unknown quantities or to reduce the three simple equations containing them to a single one.

Rule

1st Let x y and z be the three unknown quantities to be exterminated

2nd Find the value of x from each of the given equations

3rd Compare the first value of x with the second and an equation will arise having only y and z

4th In like manner compare the first value of x

Figure 3.14. A page from Samuel Fay's (1815–1817) algebra cyphering book.

Elijah Allen Rockefeller's (1825–1828) Cyphering Book

The last owner of this manuscript—before we purchased it—lived in Lebanon, New Jersey. The calligraphy with its headings is of reasonable standard, and many of the traditional topics for arithmetic are covered (e.g., inverse proportion, the single rule of three, the double rule of three, practice, simple interest, and rebate). The penmanship, however, is very scrappy, suggesting that Elijah Allen was not much interested in what he was doing. **PCA** genre is evident throughout, with most of the pages showing problem statements and subsequent calculations. There are no explanations. On the few occasions when sentences were written in English used, these were probably written by a tutor, who also made calligraphic headings above them.

Elijah, who was born in Hunterdon County in New Jersey in 1806, seemed to like to write his name "Elijah Rockefeller," for it appeared no less than 19 times throughout the cyphering book. Elijah was not the world's best speller, and on one occasion he wrote: "Elijah Rockefeller his book and doant stealit for the galis will be your end—and shame this be right to notify his friends and enemyes to not dwo it Elijah Allen." This was obviously based on a well-known limerick, which occasionally appeared in cyphering books; Elijah's "galis" should have been "gallows." Another expression to appear in Elijah's handwriting was "Coth measure" (see Figure 3.15). Considering Elijah was about 19 years of age when he began preparing his cyphering book, it appears to have been the case that any strengths he had did not lie in the spelling component of his academic work!

The main reason why this manuscript is of interest is that genealogical research reveals that Elijah Allen Rockefeller was in the direct line of the Rockefeller family, which is considered to be one of the most powerful families in the modern history of the United States of America. It made most of its money during the late 19th and early 20th centuries from petroleum. The family, which would establish control over the Chase Manhattan Bank, originated in Rhineland in Germany, but there are also connections with Scotland and Ireland. Family members moved to the New World early in the 18th century. Unfortunately, it has not been possible to find more details than have been given here of Elijah Allen's life.

A Teacher's Cyphering Book, with an Emphasis on Rules and Cases

Sometimes teachers prepared their own cyphering books in order to use them as models to be followed by students preparing their cyphering books. One would expect that a teacher's cyphering book would display a high standard of calligraphy and penmanship (Thornton, 1996) and that the mathematics itself would be presented in a scholarly way.

Figure 3.16 shows a page from a cyphering book prepared in 1816 by Richard Warner, the proprietor and teacher at Brandywine Boarding School which, according to Internet research, existed in Pennsylvania between 1816 and 1823. The mathematics shown was concerned with finding amounts of fluid which would fill containers whose borders were frustums of paraboloids. Solutions to two problems are elegantly presented, using formulae given in Keith's (1809) *Hawney's Complete Measurer*. At the foot of the page, Warner wrote, using beautiful penmanship: "Brandywine Boarding School, Sept 23rd, 1816. Richard Warner on said day finished this cyphering book. Written by Richard Warner, fortieth year of American Independence." Although on each of the 208 pages Warner maintained the same high level of presentation, the level of creative mathematical thinking on display was never great—the emphasis was on following rules and routines.

Mathematical Errors or Questionable Procedures in Cyphering Books

The emphasis on asking students merely to follow rules had the possible drawback of students recording solutions which were not completely correct but were not recognized as such in recitation sessions. Consider, for example, our brief commentary on the algebra shown in Figure 3.17, which is from an 1838 algebra cyphering book prepared by William Canby in Philadelphia.

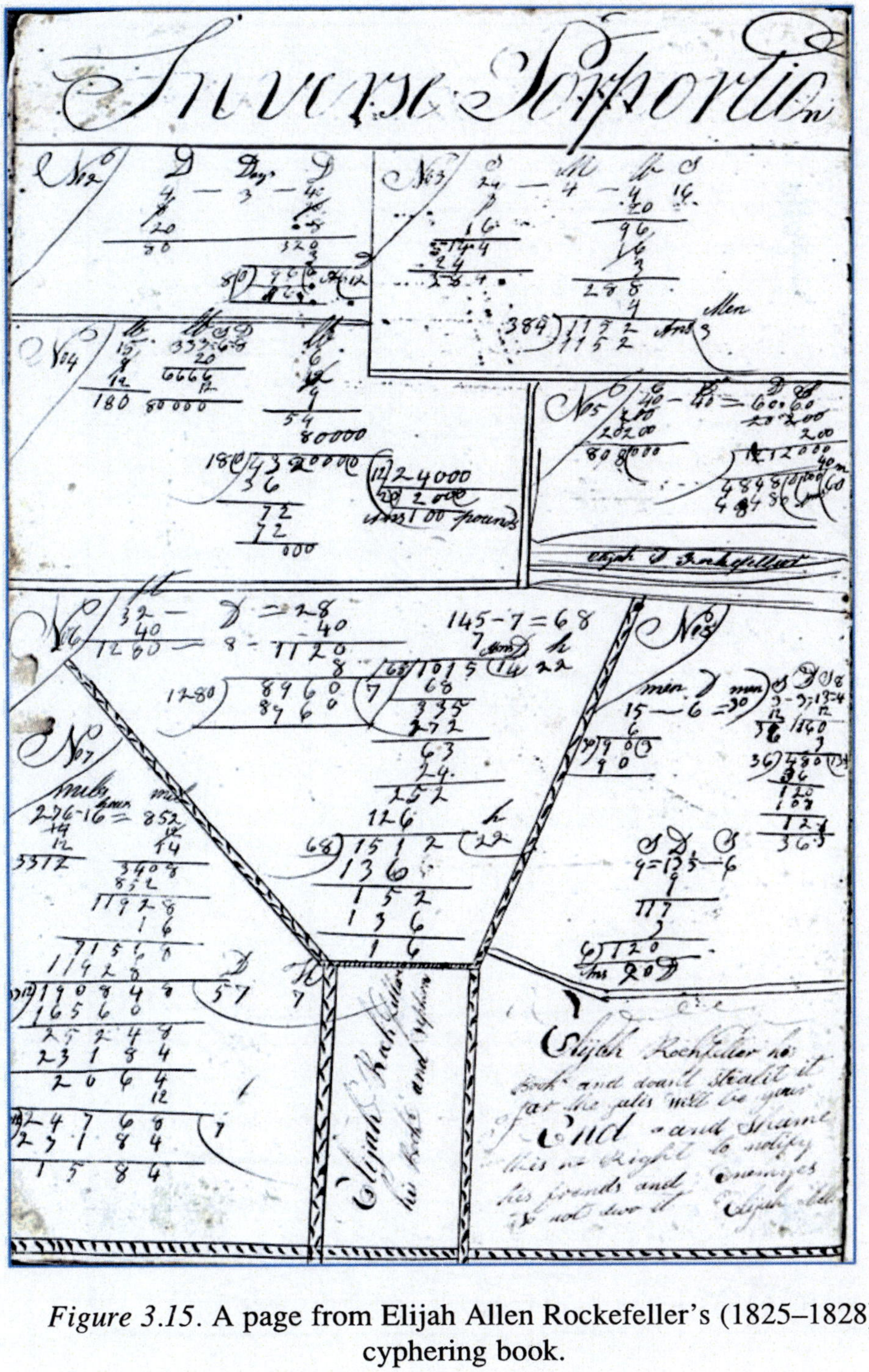

Figure 3.15. A page from Elijah Allen Rockefeller's (1825–1828) cyphering book.

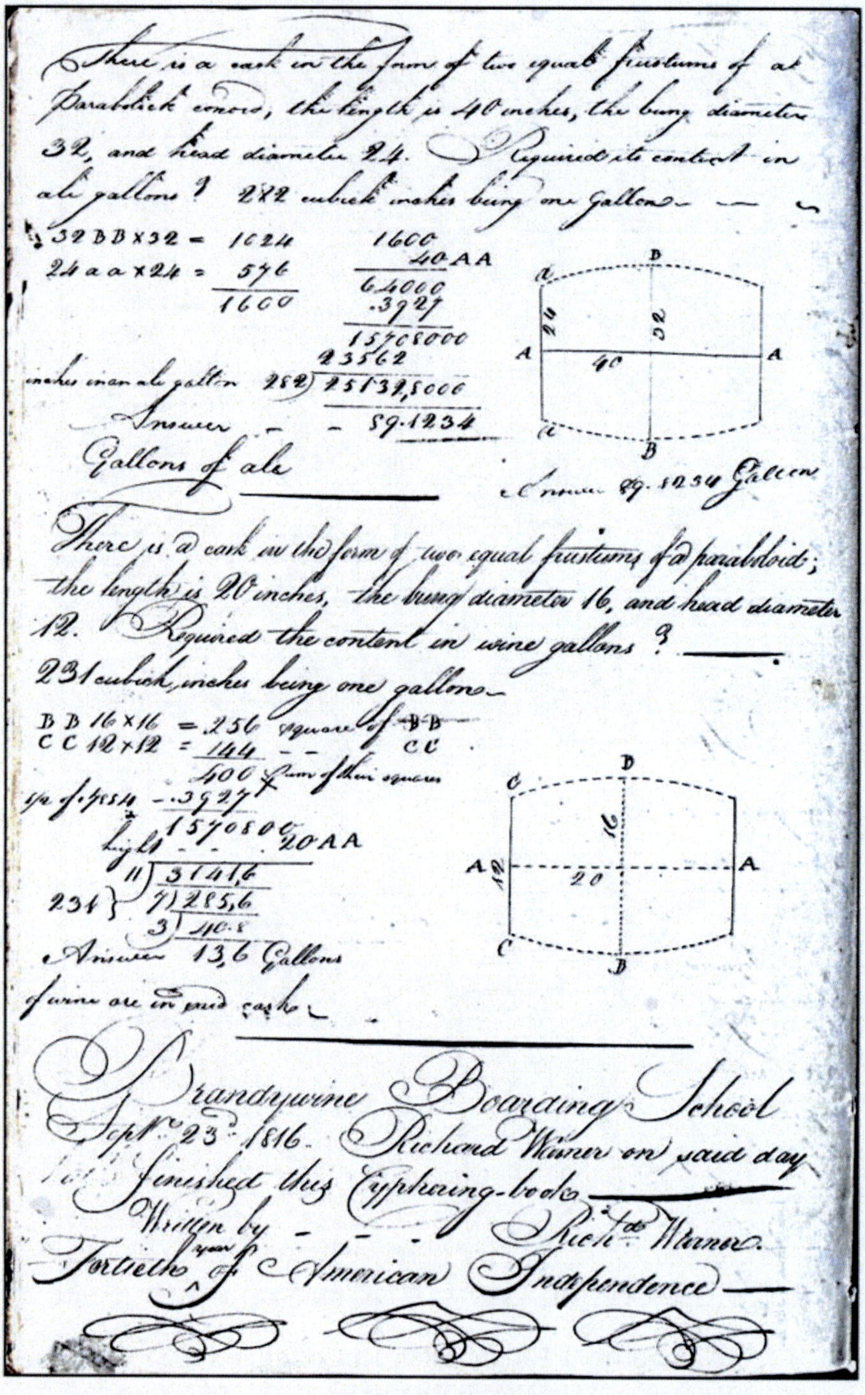
There is a cask in the form of two equal frustums of a Parabolick conoid, the length is 40 inches, the bung diameter 32, and head diameter 24. Required its content in ale gallons? 282 cubick inches being one Gallon.

32 BB × 32 = 1024
24 aa × 24 = 576
1600

1600
40 AA
64000
.3927
15708000
23562

inches in an ale gallon 282) 25132,5000
Answer - - 89.1234
Gallons of ale

Answer 89.1234 Gallons

There is a cask in the form of two equal frustums of a paraboloid; the length is 20 inches, the bung diameter 16, and head diameter 12. Required the content in wine gallons? 231 cubick inches being one gallon.

B B 16 × 16 = 256 square of BB
C C 12 × 12 = 144 - - CC
400 sum of their squares
1/2 of .7854 - .3927
157080
length 20 AA
11) 3141,6
231 { 7) 285,6
3) 40.8
Answer 13,6 Gallons
of wine are in said cask.

Brandywine Boarding School
Septr 23d 1816. Richard Warner on said day
finished this Cyphering-book
Written by - - - Richd Warner
Fortieth year of American Independence

Figure 3.16. A page from a teacher's cyphering book (prepared by Richard Warner in 1816).

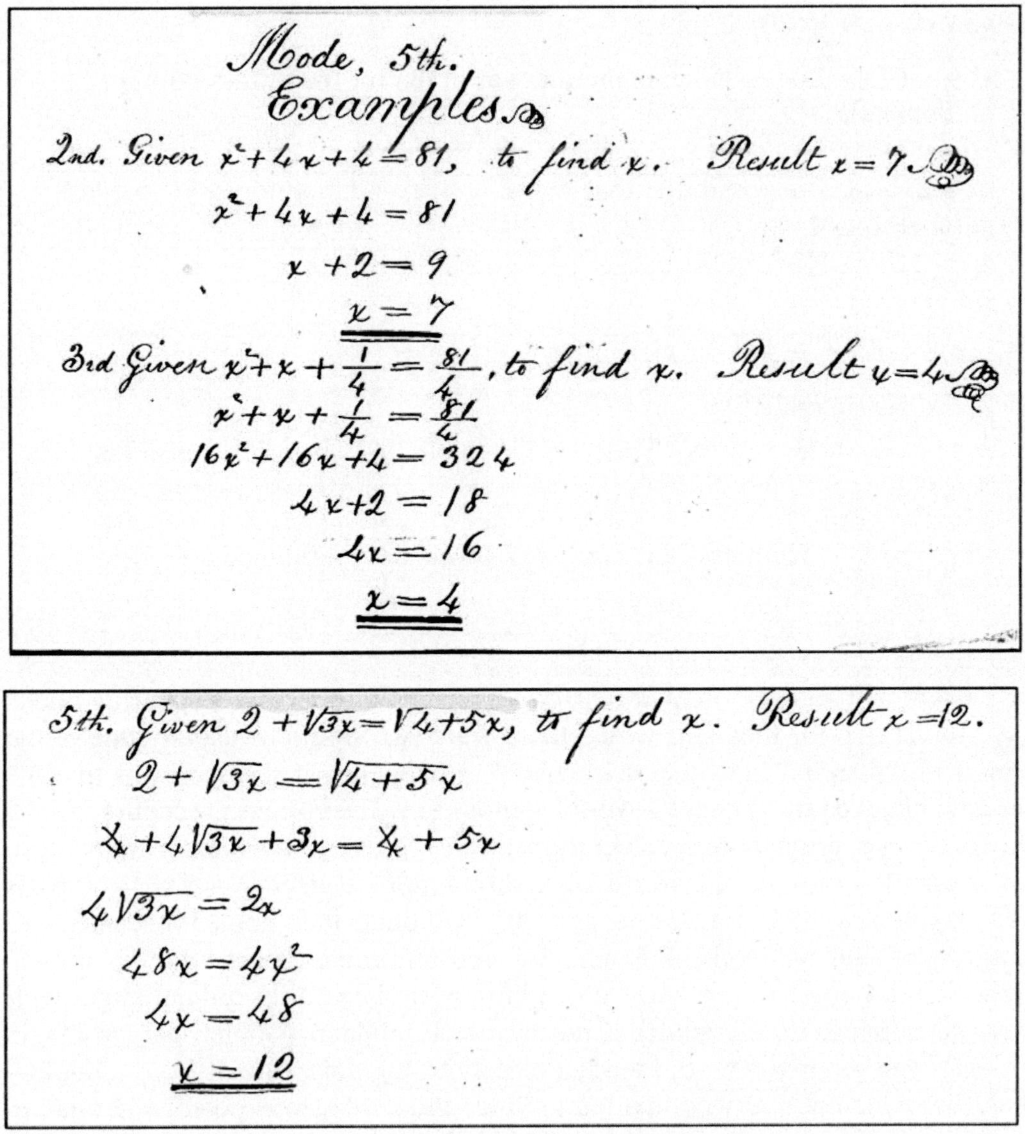

Mode, 5th.

Examples.

2nd. Given $x^2+4x+4=81$, to find x. Result $x=7$.

$x^2+4x+4=81$

$x+2=9$

$x=7$

3rd Given $x^2+x+\frac{1}{4}=\frac{81}{4}$, to find x. Result $x=4$.

$x^2+x+\frac{1}{4}=\frac{81}{4}$

$16x^2+16x+4=324$

$4x+2=18$

$4x=16$

$x=4$

5th. Given $2+\sqrt{3x}=\sqrt{4+5x}$, to find x. Result $x=12$.

$2+\sqrt{3x}=\sqrt{4+5x}$

$4+4\sqrt{3x}+3x=4+5x$

$4\sqrt{3x}=2x$

$48x=4x^2$

$4x=48$

$x=12$

Figure 3.17. Errors with quadratic equations (William Canby's (1838) algebra cyphering book).

Given $x^2 + 4x + 4 = 81$, to find x.

$x^2 + 4x + 4 = 81$

$x + 2 = 9$ Here the possibility that $x + 2 = -9$, and therefore $x = -11$ was not mentioned.

$x = 7$

Given $x^2 + x + = 20$, to find x.

> $x^2 + x + ¼ = 20 + ¼$ No explanation was given for the introduction of ¼ on both sides.
> $16x^2 + 16x + 4 = 324$ One wonders why this was not $4x^2 + 4x + 1 = 81$.
> $4x + 2 = 18$, The possibility that $4x + 2 = -18$, and therefore $x = -5$, was not mentioned.
> $4x = 16$
> $x = 4$

Given $2 + \sqrt{(3x)} = \sqrt{(4 + 5x)}$, to find x.

> $2 + \sqrt{(3x)} = \sqrt{(4 + 5x)}$
> $4 + 4\sqrt{(3x)} + 3x = 4 + 5x$ It appears to be the case that both sides were squared.
> $4\sqrt{(3x)} = 2x$
> $48x = 4x^2$ The possibility that $x = 0$ was not mentioned.
> $4x = 48$
> $x = 12$

We have examined sections on quadratic equations in other algebra cyphering books and have found that for those for which there were two solutions usually one of the solutions was "missed." In every case when that happened it seemed to us that the teacher who checked the student's work did not know the correct procedures.

In many cyphering books calculations were "proved" by merely showing a check. Thus, for example, the result of a subtraction (e.g., 940 − 584 = 356) is "proved" by adding 584 and 356 to get 940. Although it is good for teachers to encourage their students to check results, the constant misuse of the words "prove" and "proof" in schools in the eighteenth and nineteenth centuries did not augur well for the future. Most of the students never met Euclidean geometrical proofs in school, and even the few who did tended merely to memorize them from a book—and therefore were not really compelled to think about what was given, what had to be proved, and what was the most logical sequence of steps needed to achieve the proof. In algebra cyphering books one finds mainly algebraic manipulations, and rarely proofs using algebra. We never found a Cartesian graph in any cyphering book—or, for that matter, in any school algebra textbook published before 1860.

What we are implying in the last few paragraphs is that with nineteenth-century North American mathematics instruction in schools and colleges, neither textbook authors nor teachers prepared their students well for higher-order mathematical studies. In case the reader believes that that was because North American teachers adopted faulty *British* approaches to mathematics education, and that their students would have been better off if they had used "superior" Continental European methods (Hay, 1988), we would hasten to add that we also own a number of cyphering books prepared in France and in Germany and entries in those manuscripts suggest that most French and German school teachers did no better than their English counterparts in preparing future mathematicians. However, many more distinguished mathematicians could be found in European universities than in North American colleges, and this resulted in a greater number of mathematics teachers in secondary schools in Europe having been better trained to "think more mathematically" than in North America. That would have affected the "average" quality of their students' mathematical thinking.

Intended, Implemented and Attained Curricular Considerations

The modern distinction between intended, implemented, and attained curricula (Westbury, 1980) offers a powerful way of thinking about many issues associated with the history of mathematics education. It implies that analyses of old school mathematics textbooks may not tell us much about what students actually studied and what they learned (Littlefield, 1904; Monaghan, 2007). That is to say, although such textbooks provide evidence with respect to author-intended curricula they do not say a great deal about implemented curricula. The latter are much better reflected in handwritten cyphering books.

Certainly, our analysis of topics included in arithmetic textbooks used in North American schools throughout the seventeenth and eighteenth centuries and in the first half of the nineteenth century left us in no doubt that there was an underlying cultural tradition which, among other things, selected and sequenced topics. Elsewhere we have argued that this was a built-in component of the "cyphering tradition" (Ellerton & Clements, 2012, 2014).

The *abbaco* sequence for ordering arithmetic content was an important component of the cyphering tradition. Our analyses of North American cyphering books have indicated that a cyphering book usually began with a student being introduced to the Hindu-Arabic numeration system and its notations. Then would follow topics presented in an order which, in the great majority of cyphering books, conformed to the *abbaco* sequence. Anyone who has studied textbooks only would be likely to get the impression that the author-intended curricula included vulgar fractions and decimal fractions, but our analyses of cyphering books has revealed that often neither vulgar fractions nor decimal fractions were studied (Clements & Ellerton, 2015). That finding raises the question whether textbook authors expected that many students—perhaps a majority—would not go beyond whole-number arithmetic. We examined textbook authors' prefaces to find hints with respect to that issue but did not find many answers. Probably the authors knew that not all students would go beyond whole-number arithmetic—but they still decided to include extra sections on vulgar-fraction arithmetic and decimal-fraction arithmetic for their more advanced students. Probably most of the authors would also have known that any school arithmetic which did not attend to vulgar fractions or decimal fractions would have been regarded as deficient by the educated élite.

The impressive calligraphy and penmanship found in most cyphering books was not something which happened by chance. The students themselves wanted to learn to write well, and parents, teachers, and future employers also wanted students to learn to write well (Littlefield, 1904). Good penmanship was, in fact, an important component of the cyphering tradition. Before he became recognized as an outstanding mathematician, Charles Hutton (1764) described himself as a "writing master" (p. i). This view was entirely consistent with European traditions, and was fully accepted within North America (Karpinski, 1925). Teachers of arithmetic were seen as having more responsibility than teachers of all other subjects for teaching children to write well.

When one examines a large number of students' cyphering books one cannot escape from a strong feeling of authenticity. One is drawn to the thought that this was what students *really* did. But, after studying many cyphering books, we began to question whether that initial very strong feeling of authenticity was always warranted. Often, there was evidence of extensive copying—which could have occurred if students copied from textbooks or from older, "parent," cyphering books. Often there were serious misspellings, suggesting that teachers had dictated notes to students.

The more difficult question arose when it appeared to be the case that the writing on many of the pages of a cyphering book was not done by the students whose names appeared on the covers of the cyphering books (and even on the pages of manuscripts, where assertions like "This is my book" were often written). A beautiful cyphering book prepared between 1776 and 1782 and attributed to sisters Martha and Elisabeth Ryan, of North Carolina, for example, would appear to be a case in point—it seems that often someone else, a tutor, solved the exercises, prepared the calligraphic headings, and wrote the notes (see Ellerton and Clements, 2014, Chapter 4, for a full discussion of the Ryan sisters' cyphering book). We reached the conclusion that no-one should assume that what appeared on the pages of a cyphering book represented the thinking of the student whose name was on the cover. Clearly, more research is needed on this issue.

Despite the above-mentioned difficulties with respect to the concepts of intended, implemented and attained curricula, we still believe that the distinctions are powerful, and should be more often considered by education historians. Textbooks were prepared by authors who were thinking of what was needed in schools, and cyphering books were prepared in schools by students who had differing amounts of assistance from teachers.

So far as the attained curriculum is concerned, no written examination or test data, no interview data, and no recordings of recitation sessions are available from the seventeenth and eighteenth centuries, or from the early nineteenth century. Analysis of students' cyphering books therefore offers the best chance of finding out how much students actually learned. But, given the possibility that some entries were copied, or were the result of dictation, or were even made by teachers, one should not make too many assumptions about how well the students had learned what appeared in their cyphering books.

All that said, when we examined individual cyphering books we usually got a strong sense about how well the students who prepared the books had understood the mathematics written in those books.

Issues associated with attained curricula could also fruitfully be studied by future researchers.

Concluding Comments and Questions

Only about 5 percent of all white European-background males living in North America during the seventeenth century ever got to study mathematics beyond middle levels of *abbaco* arithmetic and during the eighteenth century the corresponding percentage was no more than 15. The corresponding percentages for females and for indigenous American children and children of African-American slaves were much lower. Most of the relatively few students who cyphered to higher-level *abbaco* arithmetic simply copied much of what they wrote, either from "parent" cyphering books or from textbooks.

Any satisfactory history of mathematics or mathematics education in North America between 1607 and the present day calls for much more than merely documenting the achievements of children in certain schools, or the "discoveries" and proofs of famous mathematicians. The period between 1607 and 1865 was one when the place of mathematics within challenging education environments in the New World had to be worked out. Was mathematics something which only academically very capable students should study? What was the best way to prepare youngsters so that they could study, and would want to study, higher levels of mathematics once they left school? Was the definition of a school's intended mathematics curriculum left to individuals within that school? Or to colonial or state officials? Or, after 1776, to the Federal government? How should the attained curriculum of individual students be assessed, and who should be entrusted with that assessment?

We have spent the last 20, or so, years gathering data which would assist us to answer such questions in ways which might be considered objective. One of our most remarkable findings was that before 2005—the year when we decided to link the findings on our earlier research on such issues with the cyphering tradition—nobody had ever researched or written seriously about that tradition. Our first task was to establish appropriate forms of language—for example, we introduced the expression "cyphering tradition," and decided to apply Ian Westbury's (1980) distinctions between intended, implemented, and attained curricula. We were surprised to find that often historians who liked to analyze old textbooks did not seem to recognize that before about 1820 most North American mathematics students did not own a mathematics textbook—but they *did* prepare cyphering books. During the seventeenth and eighteenth centuries, geometry, algebra, and trigonometry were rarely studied in North American pre-college education institutions. Many children *did* study arithmetic, but almost always only a narrow form of that subject—for example, the concept of a set of rational numbers which could be represented on a number line was not taught. Most school students, and many college students, never met decimals, or common fractions, or logarithms. Even within colleges, there was virtually no genuine mathematics research.

We were intrigued by our first encounter with a few cyphering books. They were clearly mathematical, and often included beautiful handwriting. Only after examining more and more cyphering books side-by-side, from different parts of North America, did we begin to recognize common forms of genre (Ellerton & Clements, 2009). Now, after examining about 1500 cyphering books held in major libraries and in the E-C collection, it is clear that during the period from 1607 to about 1820 the cyphering tradition literally *controlled* North American mathematics education.

Toward a More Coordinated Vision of the History of North American Mathematics

Assisting others to develop an understanding of what constituted the cyphering tradition, and how that might affect one's view of the history of mathematics and mathematics education in North America for the period 1607–1865, has been an important goal for us when writing this chapter. Indeed, we established our E-C collection of cyphering books so that we could explore, from primary sources, key aspects of the history of North American mathematics and mathematics education. We wanted to present convincing evidence that a comprehensive history of North American mathematics must include within its ambit a history of North American mathematics education.

When we began our journey toward establishing the E-C cyphering book collection we found that significant, well-resourced archives—such as those at the Houghton Library within Harvard University and the Wilson Library (within the University of North Carolina)—did not have finding aids which readily identified the locations of the cyphering books that they held. That made it difficult for us or, for that matter, for anyone, to locate and examine cyphering books by using the library catalogues. Too often, the cyphering books in their archives were scattered within different collections (e.g., in family and business papers), and could not be considered as representative of cyphering which took place in North American schools and colleges.

Earlier in this chapter, we noted that other historians of mathematics and mathematics education have not conducted in-depth investigations into the history of the cyphering tradition. Perhaps that is because, before now, researchers have not had easy access to a suitably large and representative data set. With the creation of the E-C cyphering book collection and its relocation to the Library of Congress, and the availability of smaller but historically significant collections of cyphering books held in the Phillips Library in the Peabody Essex Museum at Salem, and manuscripts at Harvard University, Yale University, the University of Michigan, the University of North Carolina, the University of Pennsylvania, and Winterthur in Delaware, they can now examine a number of rich data sets. That should make it possible for them to develop a more coordinated vision bringing together important strands of the history of mathematics in North America.

One question which must be asked is this: Did the overwhelming strength of the cyphering tradition in the pre-college education institutions of the 13 colonies and, later, in the United States of America during the period from 1776 through 1865, assist, or hold back, the development of mathematics in North America?

We have no doubt that it greatly assisted that development. It took 10-year-old children who had a very thin preparation in formal mathematics—children who could not do much more than count and read Hindu-Arabic numerals—along a pathway from which some could, and did, progress to college mathematics. The fact is, though, hardly any of the school children were ready to make the most of the *abbaco* sequence which had been translated into the colonies from Europe—we say "Europe," but recognize the fact that the cyphering tradition was the creation of Indian, Arabic *and* Western European nations, and not just Great Britain (Otis, 2017). Given the conditions in the schools, the lack of strong mathematical backgrounds of teachers, and the unwillingness of a majority of parents to allow their children, especially the boys, to attend school other than in winter months, the achievement brought about through the cyphering tradition was commendable.

During the second half of the nineteenth century the strength of the cyphering tradition quickly dissipated (Lancaster, 1805), with the normal schools rejecting its worth. It was referred to as the "old" method, an approach which encouraged children to learn rules and cases by heart, without understanding (Ellerton & Clements, 2019). But in 1902 an elderly distinguished educator, Richard Edwards, who had cyphered at school in Ohio in the 1830s, and later was President of Bridgewater Normal School, in Salem, Massachusetts, and Illinois State Normal University, was moved to write the following recollections of the way arithmetic had been taught in the early schools that he attended:

> It was the custom of the teacher in those ancient schools to devote a certain proportion of every session to private interviews with the pupils concerning their difficulties in arithmetic and other studies. In these interviews a free conversation was carried on, the teacher by questions ascertaining wherein the pupil found himself unequal to the work. I think the usual topic was arithmetic. . . . And may it not be true that if we could, in our own times, modify our rigorous classification of pupils so as to restore something of this old-time method, we should make an improvement in existing conditions? Is there not in our time some danger that the individual shall be submerged in the system? Does not the highest ideal of education involve something of the old-time contact of mind with mind?
> (Edwards, 1902, pp. 396–397)

We have argued, elsewhere (Ellerton & Clements, 2012), that this Richard Edwards—who was an early normal-school graduate and had been taught by Nicholas Tillinghast (who was an USMA (West Point) graduate and the author of a textbook on Euclidean geometry (Tillinghast, 1841))—was one of North America's greatest educators of the nineteenth century.

References

Adams, D. (1848). *Adams's new arithmetic*: *Arithmetic in which the principles of operating by numbers are analytically explained and synthetically explained, thus combining the advantages to be derived both from the inductive and synthetic mode of instructing*. New York, NY: Collins & Brother.

Babcock, T. H. (1829). *The practical arithmetic; in which the principles of operating by numbers are analytically explained and synthetically applied ... Adapted to the use of schools and academies in the United States*. New York, NY: G & C & H. Carvill.

Bowditch, N. I. (1840). *Memoir of Nathaniel Bowditch*. Boston, MA: C. C. Little & Brown.

Brewer, J., & Porter, R. (Eds.). (1994). *Consumption and the world of goods*. New York, NY: Routledge.

Burton, W. (1833). *The district school as it was, by one who went to it*. Boston, MA: Carter, Hendee and Company.

Cajori, F. (1890). *The teaching and history of mathematics in the United States* (Circular of Information No. 3, 1890). Washington, DC: Bureau of Education.

Chessman, R. (1965). *Bound for freedom*. New York, NY: Abelard-Schuman.

Clements, M. A., & Ellerton, N. F. (2015). *Thomas Jefferson and his decimals 1775–1810: Neglected years in the history of U.S. school mathematics*. New York, NY: Springer. https://doi.org/10.1007/978-3-319-02505-6

Cobb, L. (1835). *Cobb's ciphering book, No. 1, containing all the sums and questions for theoretical and practical exercises in Cobb's Explanatory Arithmetic No 1*. Elmira, NY: Birdsall & Huntley.

Cohen, P. C. (1982). *A calculating people: The spread of numeracy in early America*. Chicago, IL: University of Chicago Press.

Cohen, P. C. (1993). Reckoning with commerce. Numeracy in 18th-century America. In J. Brewer, & R. Porter (Eds,). *Consumption and the world of goods* (pp. 320–334). New York, NY: Routledge.

Cohen, P. C. (2003). Numeracy in nineteenth-century America. In G. M. A. Stanic & J. Kilpatrick (Eds.), *A history of school mathematics* (pp. 43–76). Reston, VA: National Council of Teachers of Mathematics.

Colburn, W (1821). *An arithmetic on the plan of Pestalozzi, with some improvements*. Boston, MA: Cummings and Hilliard.

Crackel, T. J., Rickey, V. F., & Silverberg, J. S. (2017). Provenance lost? George Washington's books and papers lost, found, and (on occasion) lost again. *The*

Papers of the Bibliographical Society of America, 111(2), 203–220. https://doi.org/10.1086/691826

Denniss, J. (2012). *Figuring it out: Children's arithmetical manuscripts 1680–1880*. Oxford, England: Huxley Scientific Press.

Dickens, C. (1850). *David Copperfield*. London, England: Bradbury & Evans. https://doi.org/10.1093/oseo/instance.00121331

Doar, A. K. (2006). *Cipher books in the Southern Historical Collection*. Master of Science thesis, Wilson Library, University of North Carolina at Chapel Hill.

Dunlap, L. A. (1959). Lincoln's sum book. *Lincoln Herald*, *61*(1), 6–10.

Edwards, R. (1902). My schools and schoolmasters. *Educational Review, 23*, 385–399.

Ellerton, N. F., Aguirre Holguin, V., & Clements, M. A. (2014). He would be good: Abraham Lincoln's early mathematics, 1819–1826. In N. F. Ellerton & M. A. Clements, *Abraham Lincoln's cyphering book, and ten other extraordinary cyphering books* (pp. 123–186). New York, NY: Springer. https://doi.org/10.1007/978-3-319-02502-5_6

Ellerton, N. F., & Clements, M. A. (2009). Theoretical bases implicit in the *abbaco* and cyphering book traditions. In M. Tzekaki, M. Kaldrimidou, & H. Sakonidis (Eds.), *Proceedings of the 33rd Conference of the International Group for the Psychology of Mathematics Education* (Vol. 3, pp. 9–16). Thessaloniki, Greece: International Group for the Psychology of Mathematics Education

Ellerton, N. F., & Clements, M. A. (2011a, March 13). *Beyond witches: Salem, MA—The cradle of North American mathematics*. Paper presented to the History and Pedagogy of Mathematics (HPM) Americas conference held at the American University, Washington, DC.

Ellerton, N. F., & Clements, M. A. (2011b). Unique mathematics books from a lost tradition. *The Guild of Book Workers' Journal. 11*, 28–39.

Ellerton, N. F., & Clements, M. A. (2012). *Rewriting the history of school mathematics in North America, 1607–1861: The central role of cyphering books*. New York, NY: Springer. https://doi.org/10.1007/978-94-007-2639-0

Ellerton, N. F., & Clements, M. A. (2014). *Abraham Lincoln's cyphering book, and ten other extraordinary cyphering books*. New York, NY: Springer. https://doi.org/10.1007/978-3-319-02502-5

Ellerton, N. F., & Clements, M. A. (2019, September 22). *Major influences on U.S. school mathematics in the nineteenth century*. Paper presented to a meeting of the HPM/AMS Sectional Meeting in Madison, Wisconsin.

Ellerton, N. F., & Clements, M. A. (2021). *Cyphering books prepared in the North American Colonies (but not Canada), or in the United States of America*. Perth, Australia: Meridian Press.

Emerson, F. (1835) *The North American arithmetic: Part third, for advanced scholars*. Boston, MA: Russell, Odiorne & Metcalf.

Gies, J., & Gies, F. (1969). *Leonardo of Pisa and the new mathematics of the Middle Ages*. New York, NY: Thomas Y. Crowell.

Hay, C. (1988). (Ed.). *Mathematics from manuscript to print 1300–1600*. Oxford, England: Clarendon Press.

Heal, A. (1931). *The English writing-masters and their copy-books 1570–1800*. Cambridge, England: Cambridge University Press.

Hertel, J. (2016) Investigating the implemented mathematics curriculum of New England navigation cyphering books. *For the Learning of Mathematics, 36*(3), 4–10.

Hirsch, D., & Van Haften, D. (2010). *Abraham Lincoln and the structure of reason*. New York, NY: Savas Beatie.

Hirsch, D., & Van Haften, D. (2016). *The ultimate guide to the Gettysburg address*. New York, NY: Savas Beatie.

Høyrup, J. (2005). Leonardo Fibonacci and abbaco culture: A proposal to invert the roles. *Revue d'Histoire des Mathématiques, 11*, 23–56.

Høyrup, J. (2008). The tortuous ways toward a new understanding of algebra in the Italian *Abbacus* School (14th–16th centuries). In O. Figueras, J. L Cortina, A. Alatorre, T. Rojano & S. Sepulveda (Eds.), *Proceedings of the joint meeting of PME 32 and PME-NA XXX* (Vol. 1, pp. 1–20). Morelia, Mexico. International Group for the Psychology of Mathematics Education.

Hutton, C. (1764). *The schoolmaster's guide: Or, A complete system of practical arithmetic*. London, England: R. Baldwin.

Jackson, L. L. (1906). *The educational significance of sixteenth century arithmetic from the point of view of the present time*. New York, NY: Columbia Teachers College.

Karpinski, L. C. (1925). *The history of arithmetic*. Chicago, IL: Rand McNally & Company.

Keith, T. (1809). *Hawney's complete measurer; or, the whole art of measuring* (3rd ed.). London, England: Johnson, Rivington & Walker.

Lancaster, J. (1805*). Improvement in education, as it respects the industrious classes of the community*. London, England: Darton & Harvey.

Littlefield, G. E. (1904). *Early schools and school-books of New England*. Boston, MA: The Club of Odd Volumes.

Long, P. O., McGee, D., & Stahl, A. M. (Eds.). (2009). *The book of Michael of Rhodes: A 15th century maritime manuscript*. Cambridge, MA: MIT Press.

Mayo, A. D. (1898). Horace Mann and the American common school. In *Report of the Commissioner of Education for 1896–97* (pp. 715–767). Washington, DC: Education Bureau

Meriwether, C. (1907). *Our colonial curriculum 1607–1776*. Washington, DC: Capital Publishing.

Middlekauff, R. (1963). *Ancients and axioms*. New Haven, CT: Yale University Press.

Monaghan, E. J. (2007). *Learning to read and write in colonial America*. Amhurst, MA: University of Massachusetts Press.

Otis, J. (2017). “Set them to the cyphering schoole”: Reading, writing, and arithmeticale, circa 1540–1700. *Journal of British Studies, 56*(3), 453–482. https://doi.org/10.1017/jbr.2017.59

Parshall, K. H. (2003). Historical contours of the American mathematical research community. In G. M. A. Stanic & J. Kilpatrick (Eds.), *A history of school mathematics* (Vol. 1, pp. 113–158). Reston, VA: National Council of Teachers of Mathematics.

Powell, S., & Dingman, P. (n.d.). Arithmetic is the art of computation: The collation. https://collation.folger.edu/2015/09/arithmetic-is-the-art-of-computation. Folger Shakespeare. This website was viewed for the first time, April 7, 2020.

Stedall, J. (2012). *The history of mathematics: A very short introduction*. Oxford, England: Oxford University Press. https://doi.org/10.1093/actrade/9780199599684.001.0001

Sterry, C., & Sterry, J. (1795). *A complete exercise book in arithmetic, designed for the use of schools in the United States*. Norwich, CT. John Sterry & Co.

Thornton, T. P. (1996). *Handwriting in America: A cultural history*. New Haven, CT: Yale University Press.

Tillinghast, N. (1841). *Elements of plane geometry for the use of schools*. Concord, NH: Luther, Hamilton and Boston.

Van Egmond, W. (1976). *The commercial revolution and the beginnings of Western mathematics in Renaissance Florence, 1300–1500*. PhD dissertation, Indiana University.

Van Egmond, W. (1980). *Practical mathematics in the Italian Renaissance: A catalog of Italian abbacus manuscripts and printed books to 1600*. Firenze, Italy: Istituto E Museo di Storia Della Scienza.

Walkingame, F. (1785). *The tutor's assistant being a compendium of arithmetic and a complete question book* (21st ed.). London, England: J. Scratcherd & I. Whitaker.

Wardhaugh, B. (2012). *Poor Robin's prophecies: A curious almanac, and the everyday mathematics of Georgian Britain*. Oxford, England: Oxford University Press.

Westbury, I. (1980). Change and stability in the curriculum: An overview of the questions. In H. G. Steiner (Ed.), *Comparative studies of mathematics curricula: Change and stability 1960–1980* (pp. 12–36). Bielefeld, Germany: Institut für Didaktik der Mathematik-Universität Bielefeld.

Wickersham, J. P. (1886). *A history of education in Pennsylvania*. Lancaster, PA: Inquirer Publishing Company.

Yeldham, F. A. (1936). *The teaching of arithmetic through four hundred years (1535–1935)*. London, England: Harrap.

Chapter 4
Mathematics Textbooks and the Gradual Decline in the Use of Middle- to Advanced-Level *Abbaco* Arithmetic 1607–1865

Abstract This chapter focuses on the influence of textbooks and textbook authors on the teaching and learning of middle- to more advanced-level abbaco arithmetic in North America during three sub-periods—from 1607 to 1776, from 1776 to 1825, and from 1825 to 1865. During the first sub-period, from 1607 to 1776, there were relatively few students who concentrated on learning any form of mathematics beyond low-level abbaco arithmetic. Those who prepared cyphering books copied statements of rules, cases and model examples from "parent" cyphering books or directly from textbooks. During the second sub-period, from 1776 to 1825, textbooks by North American authors were increasingly used to assist students preparing cyphering books, the most popular authors being Thomas Dilworth, Nicolas Pike, Nathan Daboll, Daniel Adams, Michael Walsh, Stephen Pike, and Warren Colburn. Although algebra and geometry were more studied than in the previous sub-period, any movement away from traditional *abbaco* arithmetic to other forms of mathematics tended to be resisted in the schools. The third sub-period, 1825–1865 witnessed a struggle between those who wanted to revolutionize and expand the teaching and learning of mathematics in the United States of America and those who clung to the content and pedagogical approaches associated with traditional *abbaco* arithmetic intended curricula. In this chapter we concentrate on showing that although initially in school mathematics textbooks were used to complement cyphering, ultimately they came to play a more decisive role.

Keywords *Abbaco* arithmetic • Ciphering • Cyphering • Daniel Adams • Decimals • Edmund Wingate • Edward Cocker • Frederick Emerson • Isaac Greenwood • Joseph Ray • Nathan Daboll • Nicolas Pike • Thomas Dilworth • Warren Colburn

Leading North American Mathematics Textbook Authors Between 1607 and 1865

Between 1607 and 1776 in North America the treatment of *abbaco* topics in arithmetic textbooks and in student cyphering books tended to be similar. Statements of rules and cases, the wording of problems, and the setting out of solutions to problems in cyphering books usually originated from "parent" cyphering books or from textbooks. That said, in most cases neither the students themselves nor their teachers owned a textbook. After 1776, however, more students and teachers gained access to arithmetic textbooks, and by 1865 about 50 percent of students attending

N. F. Ellerton, M. A. (Ken) Clements, *Toward Mathematics for All*, History of Mathematics Education,
https://doi.org/10.1007/978-3-030-85724-0_4

North American schools actually owned an arithmetic textbook (Ellerton & Clements, 2012).

This chapter offers evidence that the quality of arithmetic textbooks used in North American schools between 1607 and 1865 tended to be less than satisfactory, and that that was especially true for the period 1607–1820. Our evidence comes from an examination of statements and problems in more than 20 textbooks published during the period 1607–1865. The authors of the textbooks belonged to one of the following three categories.

- *Authors whose textbooks were mainly concerned with middle- to advanced-level abbaco arithmetic and were initially published in Great Britain between 1607 and 1775*. The authors considered were Robert Record, John Kersey, Edward Cocker, James Hodder, Isaac Greenwood, and Thomas Dilworth.
- *Authors whose textbooks were mainly concerned with middle- to advanced-level abbaco arithmetic and were initially published in the United States of America between 1776 and 1820*. The authors considered were Nicolas Pike, Benjamin Workman, Consider and John Sterry, Erastus Root, Chauncey Lee, Peter Tharp, Nathan Daboll, Daniel Adams, Michael Walsh, and Stephen Pike.
- *Authors whose textbooks were mainly concerned with middle- to advanced-level abbaco arithmetic and were initially published in the United States of America between 1821 and 1865*. The authors considered were Warren Colburn, Charles Davies, Frederick Emerson, Joseph Ray, and Benjamin Greenleaf.

Commentary on Mathematics Textbook Authors Whose Books Were Often Used in North American Schools Between 1607 and 1775

Robert Record(e)

Robert Record (1510–1558), sometimes known as Robert Recorde (with a second "e"), is thought to have been the first person to write an original book on arithmetic in the English language (Roberts & Smith, 2012; Williams, 2011; Yeldham, 1926). He was born in Teby, Wales, and in 1525, when aged about 15 years, entered Oxford University. He graduated with a B.A and, in 1531 was elected a Fellow of All Souls College at Oxford. Later. he studied at Cambridge University—from which he graduated with a Master of Divinity in 1545. *Record's Arithmetick: Or, the Ground of Arts* was first published in London in 1543. It has been claimed that Record was the first person to use the equals sign ("=") (Sanford, 1957).

Record influenced the mathematical thinking of Thomas Harriot (c. 1560–1621)—who has been described as the first significant mathematician to tread North American soil (Lloyd, 2012; Seltman & Goulding, 2007). However, it would be an exaggeration to suggest that other than his introduction of the equals sign, Record had a large influence on the direction of mathematics in North America. *The Ground of the Arts* concentrated almost entirely on topics in commercially-oriented *abbaco* arithmetic—from numeration, the four operations using Hindu-Arabic numerals, measurement, and reduction, through to more advanced topics like tare and tret, simple and compound interest, alligation, fellowship, and false position. There can be little doubt that these would have been studied in North America even if Record had never lived. They were part of the long-established *abbaco* sequence and the cyphering tradition which were translated into North American settings and became the cornerstone of intended and implemented curricula for school mathematics in the New World. Readers should note, though, that cyphering books came before textbooks.

A 535-page "Robert Record" text (see Figure 4.1) is held in the E-C textbook collection. It was published in London in 1658 and, because it includes lengthy introductions by John Dee and John Mellis, is one of the best-known editions of Robert Record's book.

Figure 4.1. The copy of Robert Record's (1658) *Record's Arithmetick: Or, the Ground of the Arts* held in the Ellerton-Clements textbook collection.

Edmund Wingate (and John Kersey)

Late in the sixteenth century both François Viète (1579), of France, and Simon Stevin (1585), of Holland, developed ways of writing "decimal fractions" using the familiar Hindu-Arabic numerals. The algorithms they developed for carrying out the four operations with decimals were extensions of well-established algorithms for adding, subtracting, multiplying, and dividing natural numbers expressed in Hindu-Arabic form. Soon after that, the concept of a logarithm was introduced (Briggs, 1617; Napier, 1614, 1619), and early in the seventeenth century some authors of arithmetic textbooks for schools began to show how both decimal fractions and logarithms could be useful when solving problems arising in real-life contexts.

Edmund Wingate (1596–1656), an English mathematician who was temporarily based in Paris in the 1620s, was one of the first to advocate the idea that a combination of decimal fractions and common logarithms—logarithms to the base 10—could not only simplify the daily work of practitioners such as surveyors, navigators, and builders, but could also be taught in schools and colleges (Wingate, 1624; Yeldham, 1926). On returning to England, Wingate (1630) courageously used the "decimal point" in his *Arithmetique Made Easie* (Glaisher, 1873), which was aimed at schools. He distinguished between "natural or common arithmetick" and "artificial arithmetick"—the latter referring to an arithmetic employing decimals and logarithms.

In John Kersey's (1689) revision of Wingate's *Arithmetique Made Easie* there was a return to an emphasis on the traditional *abbaco* arithmetic sequence—with numeration, the four operations, compound operations on money and weights and measures, reduction, practice, the rules of three, alligation, fellowship, and false position being developed through whole numbers before there was any mention, in Chapter 22, of decimals or vulgar (or "common") fractions (Ellerton & Clements, 2012). Kersey (1616–1677) claimed that before Wingate died, he asked him to revise *Arithmetique Made Easie* and that in particular he wanted Kersey to delete much of what he had included in relation to decimal fractions. Kersey (1689) maintained that decimals should not be used for problems which could be "resolved with much more facility by vulgar arithmetic" (p. 168).

It is apposite to provide an example of Kersey's description of how simple interest problems were best solved. In his Appendix to the ninth edition of Wingate's *Arithmetick* Kersey introduced his solution to a problem on interest with the following statement:

> When the gain of (or allowance for) 100 integers consists of some number of pounds not exceeding 10, the gain of as many like integers and known parts of an integer as one will, may be found very briefly by the following method. (Kersey, in Wingate, 1689, p. 335)

What Kersey meant by this statement was certainly not clear from the wording itself—which is interesting because Kersey achieved fame in his time as an English-language purist (Wallis, 2004). In any case, the model problem which Kersey provided was: "If 100 *l* gain 3 *l*, what is the gain of 246 *l*, 18 *s*, 10 *d*?" (Kersey, 1689, pp. 335–336). Note that the *l* represents a pound, *s* a shilling, and *d*, a penny. Before giving his model solution, Kersey gave his answer: 7 *l*, 8*s*, 1 98/100 *d*.

Figure 4.2 shows how Kersey set out his model solution to the problem.

336 *Rules of Practice* *Appendix.*

gain of 246 *l.* 18 *s.* 10 *d.*) *Anſwer* 7 *l.* 8 *s.* 1 $\frac{98}{100}$ *d.* Firſt, I multiply 246*l.* 18 *s.* 10 *d.* by 3(the ſecond term) after the manner delivered in the 17 Rule of this Chapter,and write down the product which is 740 *l.* 16 *s.* 6 *d.* Then I divide the ſaid product by 100 (the firſt term in this *Rule of Three*) in this manner, *viz.* I divide 740 pounds by 100,which is performed by cutting off towards the right hand

l. . *l.* *l.* *s.* *d.*
100 . . 3 . . 246 : 18 : 10
3
l. 7|40 : 16 : 06
|20
s. 8|16
|12
d. 1|98

the two laſt places of 740, ſo the quotient gives 7 pounds,and there will be a remainder of 40 pounds, which 40 pounds I reduce into ſhillings,ſo there will ariſe 800 *s.* to which adding the 16 *s.* which ſtand in the place of ſhillings,the ſum will be 816ſhillings; theſe are alſo to be divided by 100 (by cutting off two places as before,) ſo the quotient will give 8 ſhillings, and there will remain 16 ſhillings, which being reduced to pence, and unto them 6 pence being added (to wit the 6 pence which ſtands in the place of pence) there will ariſe 198 pence; theſe alſo are to be divided by 100 (by cutting off two places to the right hand as before,) ſo the quotient gives

Figure 4.2. Simple interest according to Edmund Wingate and John Kersey (1689, p. 336).

We (Ellerton and Clements) wonder how well you (the reader) were able to follow the mathematics in the model solution. We invite you to think about readers around 1690—most of whom would have had little formal training in mathematics. Although students at the Universities of Cambridge and Oxford (or at Harvard College) would probably have understood what Kersey wrote, they might have had to devote some time to interpreting each line. And, this was simple interest, with a principal of 246 pounds, 18 shillings and 10 pence, and an interest rate of 3 percent. Kersey had not yet got to more complicated problems in simple interest (with a fractional interest rate, or fractional parts of a year, for example), or to compound interest. There was no "simple interest formula" to be found in textbooks of the time, and often students reached the topic "simple interest" before they had studied either common or decimal fractions, or percentage. They used the "direct rule of three." In the 1820s, a young Abraham Lincoln attending school in Pigeon Creek, Indiana, would do just that (Ellerton, Aguirre-Holguin & Clements, 2014; Roberts, 2019).

There is evidence that Wingate's arithmetics were used in colonial North America (see, e.g., Sarjeant, 1788, p. 94). But the method which Wingate used in his "model examples" for simple interest calculations was exactly the same as that used by authors of arithmetics in other nations (see, for example, the setting out by the French author, Barrème, 1744, p. 167). Figure 4.2 shows how one was expected to "do" simple interest problems when they arose in the *abbaco* sequence. The method shown had been used for centuries in Europe and was used in the first half of the nineteenth century in the United States of America (see, for example, Daboll, 1818, p. 120).

Writing in 1890, Florian Cajori maintained that the best teachers of mathematics in North America were "college students or college graduates who engaged in teaching as a stepping stone to something better" (Cajori, 1890. p. 9). He was probably right—the college students had been *forced* to learn and to remember the methods, but very few others would have tried to master such "remote" reasoning.

This chapter tells the story of how mathematics in colonial North America, and in the United States during the period from 1776 to 1865 was held back because most students who entered colleges lacked an understanding of basic mathematics. When in school most of them had studied arithmetic according to *abbaco* traditions, and had been forced to remember rules and cases, and to copy model examples into cyphering books. During recitation sessions they had been expected to regurgitate rules and cases which they had been required to remember.

Edward Cocker

John Kersey effectively reversed James Wingate's *intention* that teachers should make sure that decimal fractions were integrated fully into implemented arithmetic curricula. Edward Cocker's (1677) *Arithmetic*, published by John Hawkins, was another textbook which avoided decimal fractions, but in the preface to Cocker's (1685) *Decimal Arithmetick* it was argued that Kersey had correctly noted that teachers who did not want their students to learn about decimals directly could nevertheless achieve that by using Cocker's (1677, 1678) "non-decimal" book. It is not surprising that the non-decimal version proved to be much more popular than the decimal version, both in Great Britain, where it was much used in schools, and in North America (where it was used, but in relatively few schools) (Clements & Ellerton, 2015; Yeldham, 1926).

Ironically, the 1685 first edition of *Cocker's Decimal Arithmetick* provided an excellent summary of how the introduction of decimals might modernize learners' approaches to problem solving in arithmetic. It is not clear who wrote this decimal version attributed to Cocker (or, for that matter, who wrote the original non-decimal version—see Wallis, 1997) but, whoever it was took pains to show how multiplication of decimals could be used to find measures of areas and volumes—although, in fact, specific general measurement terms like "area," "volume," and "capacity" were rarely used. For example, Cocker (1685), having asked his readers to find the "content" of a table whose length was 18.75 feet and breadth 3.5 feet, commented:

> Here by the way, take notice, that although amongst artificers the two foot rule is generally divided, each foot into 12 inches, &c., yet for him that at any time is employ'd in the practice of measuring, it would be most necessary for him to have his two foot rule, each foot divided into 10 equal parts, and each of those parts divided again into 10 other equal parts: so would the whole foot be divided into 100 equal parts, and thereby would it be made fit to take the dimensions of anything whatsoever, in feet and decimal parts of a foot; and thereby the content of anything may be found exactly, if not more exactly and near, than if the foot were divided into inches, quarters, and half quarters. (p. 45)

Cocker proceeded to demonstrate how cumbersome the calculation of the "content" of the table-top would be if the "old" method were to be used—18 feet 9 inches would be converted to 225 inches and the 3 feet 6 inches to 42 inches. Then, after multiplying 225 by 42 to get 9450, this product would be divided by 144 to get 65 5/8, and that mixed fraction would need to be interpreted in relation to the original problem. Cocker (1685) described that approach as "tedious" in comparison with "the decimal way" (p. 47).

We cannot be sure why Kersey resisted and reversed Wingate's decimal approach to that used in the traditional *abbaco* sequence—perhaps it was nothing more than a commercial decision to stick to the well-known *abbaco* approach because Kersey and Hawkins thought that that was likely to result in more books being sold. Parents, teachers, merchants, textbooks authors, and even many mathematicians preferred to continue with time-honored algorithmic approaches. Whatever was the reason, revised editions of the non-decimal book would be published in England for the next 150 years (see, for example, Cocker, 1720), but there is no evidence that *Cocker's Decimal Arithmetick* was often used outside of Great Britain (Karpinski, 1980).

Cocker also adopted time-honored *abbaco* approaches when dealing with other standard topics in arithmetic. Take for example this question, which appeared in his section on "the single rule of three inverse," which was a standard *abbaco* topic. The question was:

> How many yards of 3 quarters broad are required to double, or be equal in measure to 30 yards, that are 5 quarters broad? (Cocker, 1697, p. 126)

Cocker immediately gave the answer as "50 yards," and explained his working to that problem (see Figure 4.3) in the following way:

> For, say, if 5 quarters wide require 30 yards long, what length will 3 quarters broad require? Here I consider that 3 quarters broad will require more yards than 30, for the narrower the cloath is, the more in length will go to make equal measure with a broader piece. (Cocker, 1697, p. 126)

Arithmetically-speaking, the question is quite simple—provided the student was able to comprehend its meaning. It seems that the part of the question which mentioned doubling should not be there.

The most difficult part of following Cocker's working (shown in Figure 4.3) was to work out *why* he did what he did. The arithmetic itself—multiplying by 5 and then dividing by 3—should have been easy. Obviously, a well-qualified and experienced teacher could easily have explained the method to a student, but having access to such a teacher was not guaranteed. For a school student who was preparing a cyphering book there was often a daily recitation period, but when a student was attempting an exercise for which no solution was shown, that student had to work out which numbers needed to be multiplied or divided, and in which order, and why. There is no hint that anyone was aware of the most common difficulties that students had with word problems of this kind. That same issue still arises in the twenty-first century.

Queſt. 4. How many yards of 3 quarters broad are required to double, or be equal in meaſure to 30 yards, that are 5 quarters broad? *Anſwer* 50 yards, For ſay, if 5 quarters wide require 30 yards long, what length will three quarters broad require? Here I conſider that 3 quarters broad will require more yards than 30, for the narrower The Cloath is, the more in length will go to make equal meaſure with a broader piece.

qrs.——*long*——*qrs*
5 30 3
5
———
3) 150 (50 *yds*
15
———
(0)

Figure 4.3. Edward Cocker's (1697) method for a standard inverse-rule-of-three question.

Karpinski (1940) described the extent to which Cocker's traditional arithmetic was accepted in North America in the following way:

> *Cocker's Arithmetick* of 1678 attained a record of approximately one hundred English editions in something over one hundred years. Though it was often imported the work appears never to have been the basis of an American publication. Diligent search in practically all great American collections has not revealed a copy. Despite the fact that Evans (*American Bibliography*) mentions publication in Philadelphia in 1779; he does not locate a copy nor indicate the source of his information.
>
> (Karpinski, 1940, pp. 4–5)

Benjamin Franklin, himself, with a background in printing and publishing (Franklin, 1964), probably could have arranged for Cocker's book to be published. But, as we shall see later in this chapter, and in Chapter 5, the experiences of Isaac Greenwood (1729) and Pieter Venema (1730) suggested that publishing mathematics textbooks in North America was not likely to be a profitable commercial venture.

Some students did not use a textbook at all, and others might have used a textbook but avoided the chapter(s) dealing with the topic. In other words, the implemented curriculum may not necessarily have been the same as the author-intended curriculum (Clements & Ellerton, 2015).

James Hodder

One of the most popular arithmetic textbooks in the early 1700s in British colonial North America was authored by James Hodder. This was very widely used in England, and the 25th edition of it was reprinted in Boston in 1719 by Benjamin Franklin's elder brother, James (Karpinski, 1940, see pages 39–41). Figure 4.4 shows Hodder's model solution to a direct rule-of-three problem involving common fractions. The problem was: "If 2/5 of an ell cost 2/3 of a *l*, what cost 4/5?"

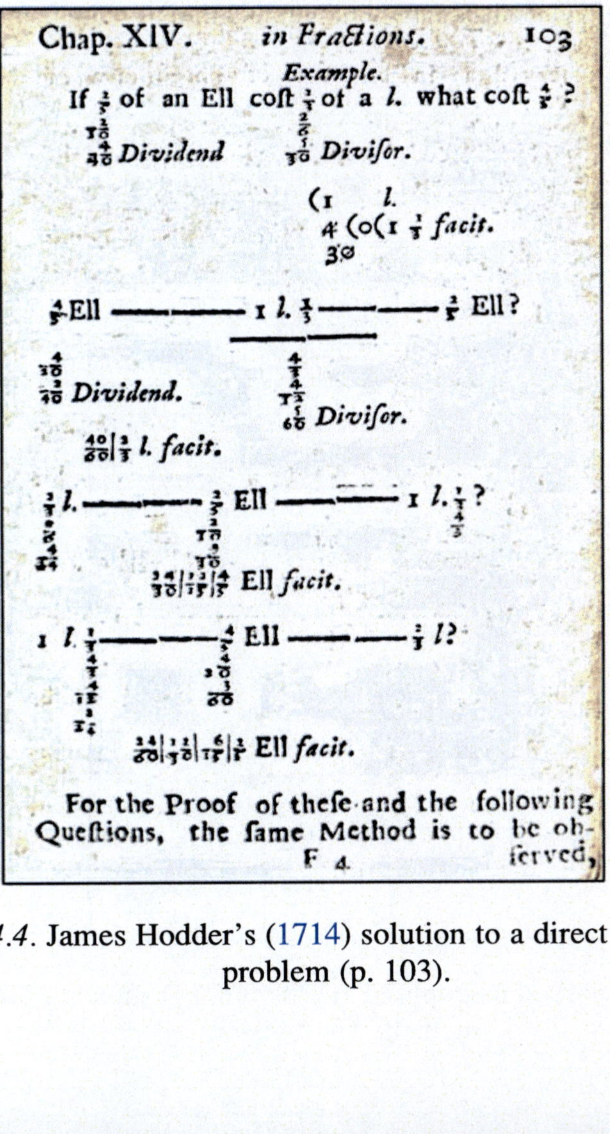
Chap. XIV. *in Fractions.* 103

Example.

If $\frac{2}{5}$ of an Ell coſt $\frac{2}{3}$ of a *l.* what coſt $\frac{4}{5}$?

Dividend *Diviſor.*

facit.

$\frac{4}{5}$ Ell ———— 1 *l.* $\frac{1}{3}$ ———— $\frac{2}{5}$ Ell?

Dividend. *Diviſor.*

l. facit.

$\frac{2}{3}$ *l.* ———— $\frac{2}{5}$ Ell ———— 1 *l.* $\frac{1}{3}$?

Ell *facit.*

1 *l.* $\frac{1}{3}$ ———— $\frac{4}{5}$ Ell ———— $\frac{2}{3}$ *l.*?

Ell *facit.*

For the Proof of theſe and the following Queſtions, the ſame Method is to be obſerved,

F 4

Figure 4.4. James Hodder's (1714) solution to a direct rule-of-three problem (p. 103).

One would expect that, once again, the student's main problem would have been to work out why Hodder did what he did in the working shown in Figure 4.4.

Hodder was the only author of a popular arithmetic textbook used in North American schools who, around 1700, still used and recommended what was known as the "scratch" (or "galley") method for division (see Ellerton & Clements, 2014, pp. 193–197 for details related to the method). The algorithm for this method took Hodder 12 pages to explain (pp. 35–46).

Hodder introduced Figure 4.5 in the following way:

> I shall not (I hope) need to trouble myself, or . . . to shew the working of this sum, or any other, having now (as I suppose) sufficiently treated of division; but will leave it to the censure of the most experience'd to judge, whether this manner of dividing be not plain, lineal, and to be wrought with fewer figures than any which is commonly taught. (Hodder, 1714, p. 55)

Figure 4.5 also shows a "proof" for the division. Can you work out what is being divided by what? What is the answer? And what is the meaning of the word "Proof" on the bottom line?

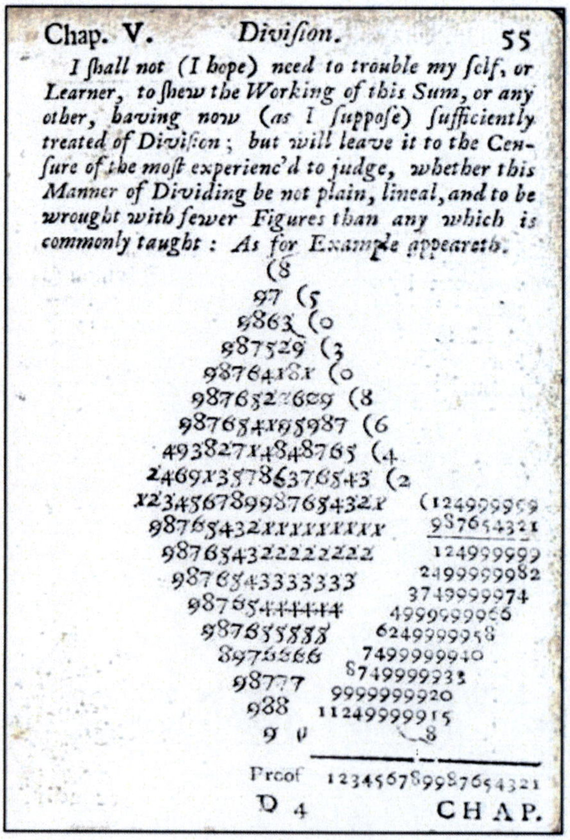

Chap. V. *Division.* 55

I shall not (I hope) need to trouble my self, or Learner, to shew the Working of this Sum, or any other, having now (as I suppose) sufficiently treated of Division; but will leave it to the Censure of the most experienc'd to judge, whether this Manner of Dividing be not plain, lineal, and to be wrought with fewer Figures than any which is commonly taught: As for Example appeareth.

(8
97 (5
9863 (0
987529 (3
98764181 (0
9876527609 (8
987654195987 (6
49382714848765 (4
2469135578637654 3 (2
1234567899876543 21 (124999999
987654321111111111 987654321
9876543222222222 124999999
98765433333333 2499999982
987654444444 3749999974
9876555555 4999999966
89756666 6249999958
98777 7499999940
988 8749999933
9 9999999920
11249999915
8

Proof 12345678998765432 1

D 4 CHAP.

Figure 4.5. A worked example in Hodder (1714) showing the "scratch method" for division.

Sometimes it is argued that the mathematics textbooks written and published on the Continent were superior to those written for students of corresponding ages in Great Britain (see, e.g., Cajori, 1890). We examined books in the Ellerton-Clements textbook collection to see whether data support that contention. Without going into detail, we do not think they do. The Continental textbooks tended to have more complete verbal explanations of rules, but generally speaking those explanations were no less confusing than those in textbooks prepared in Great Britain. Consider, for example, the text in Figure 4.6, which is from page 35 of an arithmetic by Guillaume Prévost (1677) which was published in France. The question gave the cost of 174 books and asked for the cost of 2. The working shown included a French form of the scratch method for division.

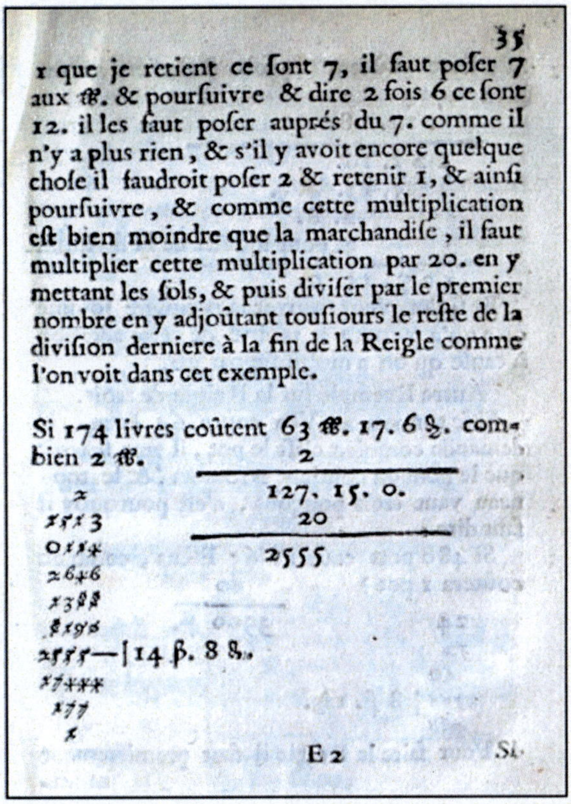
35

r que je retient ce ſont 7, il faut poſer 7 aux ₶. & pourſuivre & dire 2 fois 6 ce ſont 12. il les faut poſer auprés du 7. comme il n'y a plus rien, & s'il y avoit encore quelque choſe il faudroit poſer 2 & retenir 1, & ainſi pourſuivre, & comme cette multiplication eſt bien moindre que la marchandiſe, il faut multiplier cette multiplication par 20. en y mettant les ſols, & puis diviſer par le premier nombre en y adjoûtant touſiours le reſte de la diviſion derniere à la fin de la Reigle comme l'on voit dans cet exemple.

Si 174 livres coûtent 63 ₶. 17. 6 ₰. combien 2 ₶.

2

127. 15. 0.

20

2555

2555—|14 ß. 8 ₰.

E 2 Si

Figure 4.6. Scratch division in a French arithmetic textbook (Prévost, 1677).

Isaac Greenwood

Between 1727 and 1737 Isaac Greenwood was Harvard's first Hollis Professor of Mathematics and Natural and Experimental Philosophy. He became the first North American author to write an English-language mathematics textbook specifically aimed at North American college and school students. Greenwood's (1729)

162-page textbook did not achieve a wide circulation and, except for one thing, was not much different from existing British arithmetic textbooks aimed at schools. The one difference was that Greenwood provided space for readers to carry out calculations and to write working for questions on the actual pages of the textbook as they occurred in the book. There can be little doubt that Greenwood's (1729) text had an overriding commercial, *abbaco* thrust, and mimicked the kind of emphasis which had emerged from European reckoning schools in which elementary *abbaco* arithmetic computations and their applications to tasks involving money, and weights and measures, were of central importance.

Although the Ellerton-Clements textbook collection does not hold an original copy of Greenwood's book—there are very few extant copies—it does have a good facsimile copy. Figure 4.7 shows page 39 from that copy—it is concerned with the reduction of fractions "to their lowest or least denomination" (p. 38). The responses of the original owner of the textbook to two exercises are shown. The rule at the top of the page is presented in highly complex language, and one wonders whether a more simply-stated rule could have been presented in order that the fraction 72/108 could have more convincingly been shown to equal 2/3. In an "advertisement" at the front of the book Greenwood stated that "the language and manner of writing is such, as the author hopes, will be easily apprehended by those that have not been very much conversant with books."

Greenwood's (1729) textbook did not sell well, and there was no second edition (Karpinski, 1940).

Thomas Dilworth

The most widely-used arithmetic textbook in North America up to 1820 was Thomas Dilworth's *The Schoolmaster's Assistant, Being a Compendium of Arithmetic both Practical and Theoretical,* (Karpinski, 1940). Dilworth, who died in 1780, was an English cleric who authored numerous textbooks, the most popular being his *New Guide to the English Tongue*. He also authored textbooks on arithmetic, bookkeeping and geography (Clements & Ellerton, 2015). Figure 4.8—taken from Dilworth (1806)—reproduces a rarely-shown image of Dilworth. The first British edition of his *Schoolmaster's Assistant* (for arithmetic) was published in 1740, and in 1773 the 17th British edition was republished—by two different publishers—as the First North American edition. The two publishers were John Dunlap (of Philadelphia and who, as official printer to the Continental Congress in 1776, would print the Declaration of Independence during that same year), and Joseph Crukshank, also of Philadelphia (Karpinski, 1940, p. 73). Karpinski (1980) listed the Crukshank reprinting but not the Dunlap reprinting. As far as we know, the copy of the 1773 John Dunlap printing of Dilworth's *Schoolmaster's Assistant* in the E-C Textbook Collection is the only extant copy. Dauben and Parshall (2014, p. 178) mistakenly claimed that the first American edition of Dilworth was published in 1781.

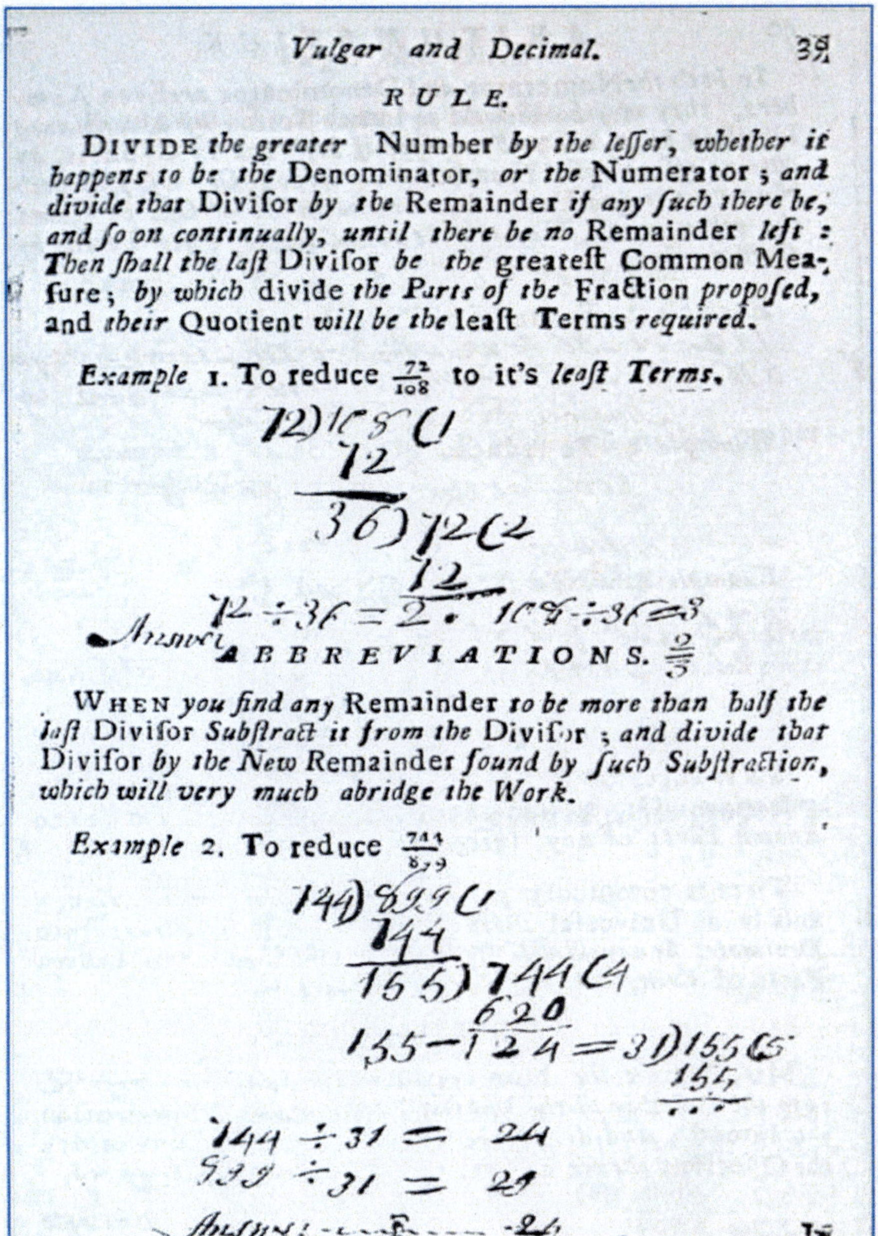

Vulgar and Decimal. 39

RULE.

DIVIDE the greater Number by the lesser, whether it happens to be the Denominator, or the Numerator; and divide that Divisor by the Remainder if any such there be, and so on continually, until there be no Remainder left: Then shall the last Divisor be the greatest Common Measure; by which divide the Parts of the Fraction proposed, and their Quotient will be the least Terms required.

Example 1. To reduce $\frac{72}{108}$ to it's *least Terms*.

72)108(1
72
36)72(2
72
72 ÷ 36 = 2. 108 ÷ 36 = 3

Answer $\frac{2}{3}$

ABBREVIATIONS.

WHEN you find any Remainder to be more than half the last Divisor Subtract it from the Divisor; and divide that Divisor by the New Remainder found by such Subtraction, which will very much abridge the Work.

Example 2. To reduce $\frac{744}{899}$

144)899(1
744
155)744(4
620
155 − 124 = 31)155(5
155
144 ÷ 31 = 24
899 ÷ 31 = 29

Answer $\frac{24}{29}$.

Figure 4.7. Page 39 from Greenwood's (1729) *Arithmetick Vulgar and Decimal,* showing writing by a student who used the textbook.

Figure 4.8. Image of Thomas Dilworth, fshown in Dilworth (1806).

Dilworth often adopted a different method from other authors when explaining how to solve problems. He liked to state rules verbally. Consider, for example, his explanation for "extracting roots of all powers." Dilworth (1773a) considered that there were 11 parts to the rule:

1. Prepare the given number for extraction by pointing off from the unity place, as the root required directs.
2. Find the first figure in the root by your own judgment, or by inspection to the table of powers.
3. Subtract it from the given number.
4. Augment the remainder by the next figure in the given number, that is, by the first figure in the next point, and call this your dividend.
5. Involve the whole root last found, into the next inferior power to that which is given.
6. Multiply it by the index of the given power and call this your divisor.
7. Find a quotient figure by common division, and annex it to the root.
8. Involve all the roots thus found, into the given power.
9. Subtract this power (always) from as many points of the given power as you have brought down, beginning at the lowest place.

10. To the remainder bring down the first figure of the next point for the new dividend.
11. Find a new divisor as before, and in like manner proceed till the work is ended (p. 141).

Immediately after these instructions—which Dilworth stated he had "received" from his "worthy friend William Mountain Esq. F.R.S." (p. 142)—there was a model example: "What is the cube root of 115501303?" One wonders how many school students could have followed these instructions, or even the steps in Dilworth's model example. We confess that although we spent time studying the solution Dilworth set out as a model example, we could not even identify Dilworth's answer.

This, like many other sections of Dilworth's text, would have succeeded in giving students the impression that the mathematics that they were being asked to learn had all been worked out by remote but expert outsiders—their only task as students was simply to copy the rules, remember them, and apply them until they could get correct answers.

In case the reader might think that this is an unfair evaluation of Dilworth's text, one based on a difficult section, consider the following passage from "Practice," which was an early topic in the *abbaco* sequence:

OF PRACTICE

Q. What is practice?
A. It is a short way of finding any quantity of goods by the given price of one integer.
Q. How do you prove questions in practice?
A. By the single rule of three direct. Or practice may be proved by itself, by varying the parts.

Then followed:

CASE 1

Q. What must be done with the price of an integer when it is less than a penny?
A. Find the aliquot parts of that price contained in a penny, which must be divisors to the given sum; that is, if the price be a farthing, say a farthing is a fourth of a penny, and let it thus, |1/4 |1/4|. If the price be a half-penny, then say a half-penny is the half, thus |1/2|1/2|. If it be three farthings, then say, a half-penny is the half of a penny, and a farthing is a fourth of a penny, thus |1/2 |1/2|
|1/4 |1/4|.
Q. What do you observe concerning these columns?
A. The first column contains the money and the other the parts.

Although the tasks which were solved in the section on "practice" were elementary, when reading the passage we found it difficult to work out what Dilworth meant. Because of language complexity it was not easy to work out the line-to-line reasoning provided in the model examples.

Nevertheless, between 1773 and 1800, 27 editions of Dilworth's *The Schoolmaster's Assistant, Being a Compendium of Arithmetic both Practical and Theoretical* were printed in North America (Karpinski, 1940). In the prefaces to all 27 editions Dilworth dedicated his books to the "revered and worthy schoolmasters in Great Britain and Ireland" (see, e.g., Dilworth, 1797, pp. vii–x), and there was never a mention of the United States. Even well into the nineteenth century, in sections on currency in North American editions of Dilworth's arithmetic there were no references to American dollars and cents and, except in the section on "Exchange," all money calculations were based on sterling pounds, shillings, pence and farthings.

One might expect that because Dilworth's publishers made no effort to simplify the text, or to "Americanize" the editions published in North America, there would not have been much demand for them in North America, especially after 1776. But, in fact, Dilworth's *Schoolmaster's Assistant* was easily the most popular mathematics textbook used in the United States of America between 1776 and 1800, and between 1773 and 1820 there were more editions of Dilworth's textbook printed in the United States than of any other mathematics textbook (Karpinski, 1940).

The popularity of Dilworth's *Schoolmaster's Assistant* in the United States during the period 1773–1800, and beyond that, was probably due to the fact that the book was totally consistent with, and indeed gradually helped to define, the cyphering approach to school mathematics in England and North America (Ellerton & Clements, 2012). As we examined manuscripts in cyphering-book collections it became obvious to us that many students simply copied Dilworth's questions and answers, and his model examples, into their cyphering books.

Authors from the Federal Period, 1787–1801

Nicolas Pike

As stated above, before the Revolutionary War most of the arithmetic textbooks used by students and teachers in the North American British colonies were written by British authors who never set foot in North America (e.g., Cocker, 1685, 1719; Dilworth, 1773a, b). It is hardly surprising that after the first U.S. Constitution had been approved by Congress in 1787 there was a surge of activity in North American publishing for schools and colleges (Monroe, 1917) and that, in particular, there was a sharp increase in the number of arithmetics written by U.S. citizens. The sales of arithmetics by British authors correspondingly declined.

The first major school arithmetic textbook by a North American author was Nicolas Pike's (1788) 512-page *A New and Complete System of Arithmetic Composed for Use of Citizens of the United States* (see Figure 4.9). Pike (1743–1819), a

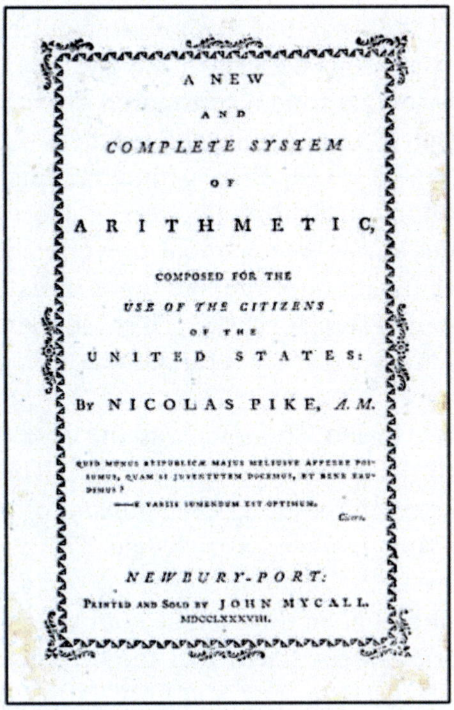

A NEW
AND
COMPLETE SYSTEM
OF
ARITHMETIC,
COMPOSED FOR THE
USE OF THE CITIZENS
OF THE
UNITED STATES:
BY NICOLAS PIKE, A.M.

QUID MUNUS REIPUBLICÆ MAJUS MELIUSVE AFFERRE POSSUMUS, QUAM SI JUVENTUTEM DOCEMUS, ET BENE ERUDIMUS?
—E VARIIS SUMENDUM EST OPTIMUM.
Cicero.

NEWBURY-PORT:
PRINTED AND SOLD BY JOHN MYCALL.
MDCCLXXXVIII.

Figure 4.9. Title page of Nicolas Pike's (1788) *A New and Complete System of Arithmetic Composed for Use of the Citizens of the United States.*

native of New Hampshire, had graduated from Harvard College in 1766, and had then taught mathematics in academies for about 20 years. His confidence in the value of his pioneering work was evidenced by his registering himself as an author in Pennsylvania, South Carolina, Massachusetts, and New York, such registration serving as copyright notice. He also corresponded with George Washington, seeking the first President's endorsement of the book. Washington politely replied, saying he valued greatly what Pike had done—but he did not explicitly offer the recommendation which Pike had sought (Clements & Ellerton, 2015; George Washington to Nicolas Pike, June 20, 1788).

Although numerous editions of Pike's arithmetics would appear throughout the period 1788–1843, all editions published after the first 1788 edition were not fully prepared by Pike himself. That was because around 1790 Pike left the profession of teaching and became a judge, and at that time he sold his rights to royalties from future editions of his textbook—even though the title pages of the future editions would indicate that he was still the author. A sequence of publishers purchased the right to amend and publish previous versions of Pike's texts and by 1820 books attributed to Pike differed in important ways from the 1788 text. The E-C textbook collection includes many copies of all but one of the arithmetics attributed to Nicolas Pike.

"Old Pike," as Pike's (1788) original *Arithmetic* would ultimately come to be known, would go through six editions between 1788 and 1843 (Karpinksi, 1980). In 1788 it sold for about $2.50—at that time a substantial price for a textbook, and one which placed it out of the reach of most pupils and teachers (Monroe, 1917). Besides arithmetic, it introduced sections on algebra, geometry, logarithms, trigonometry, and conic sections. Applications of arithmetic were made to problems associated with mechanics, gravity, pendulums, mechanical powers, and problems in astronomy requiring calculations of the moon's age, the times of its phases, and the date of Easter. This represented a major step forward in thinking about what mathematics might be most needed in the "new age." As it turned out, though, this was a step into an educational unknown.

A well-known publisher, Isaiah Thomas, was the first to acquire the right to amend and publish Nicolas Pike's writings, and in 1793 Thomas released a book, purportedly authored by Pike, with the title, *Abridgement of the New and Complete System of Arithmetick*. The *Abridgement*, which had 371 pages, was significantly smaller than the original *New and Complete System of Arithmetic*. It was aimed at schools, whereas the original had been directed at both schools and colleges. Isaiah Thomas's preface to the *Abridgement* stated that the original *Arithmetic* was "now used as a classical book in all the New England universities," and excelled "everything of the kind on this content." The author had "reason to hope that this abridgement will not be less esteemed as a schoolbook" (Pike, 1793, p. ix). The 1793 *Abridgment* dropped those sections on algebra, trigonometry, plane geometry, conic sections, and regular solids which had appeared in Pike's (1788) text. The "dropped" sections had probably been deemed to be irrelevant by most schoolmasters and school students. The result was the 1793 *Abridgement,* and later editions, became strikingly similar to the European arithmetics that Pike had hoped to supplant.

In his choice and ordering of content and in his approaches to the various topics, Pike leant heavily on the content of school arithmetics in textbooks written and published in England—especially those by John Bonnycastle, Edward Cocker, Thomas Dilworth, and John Ward. Like the authors of those arithmetics, Pike claimed that his book was "practical." For example, they all had rules for the "reduction of coin." Naturally, those written in England assumed that English "pounds, shillings, pence" would be solely used in the schools. This same assumption was translated into the American colonies. Thus, although Pike's (1788) *Arithmetic* devoted 28 pages (pages 96–123) to currency conversion, only three of those (pages 96–98) were concerned with the new federal currency which had been approved by Congress. The remaining pages stated and illustrated rules for transferring between the various "legal" currencies which had been used in the old British colonies. Pike included sections on converting New Hampshire, Massachusetts, Rhode Island, Connecticut, and Virginia currencies to New York and North Carolina currencies, and on converting Pennsylvania, New Jersey, Delaware, and Maryland currencies, to Irish money, to Canadian and Nova Scotia currencies, to *livres tournois*, and to Spanish milled dollars. He gave specific rules for reducing federal money to New England and Virginia currencies.

Other than in the brief section on federal money, Pike's (1788) text reflected a viewpoint that much of the time spent in schools on mathematics—usually, just arithmetic—should be dedicated to rules and calculations related to conversion of units. However, he expected students to follow rules on monetary conversions suited to old-society currencies and measures and offered hardly any examples of how the new federal currency could be applied in farming, trade, and business transactions. In fairness to Pike, it is important to note that although in 1788 a new decimalized Federal currency had already been approved by Congress, the U.S. Mint had not yet been established, and no new "federal" coins were available.

There were some mathematically strange things in Pike's (1788) textbook. For example, when introducing the topic "duodecimals" in *The New and Complete System of Arithmetic, Composed for the Use of the Citizens of the United States,* Pike asserted that "pounds multiplied by pounds are pounds; pounds multiplied by shillings, are shillings, &c., shillings multiplied by shillings are twentieths of a shilling; shillings multiplied by pence, are twentieths of a penny; pence multiplied by pence, are 240ths of a penny, &c" (p. 123). Pike then showed how 9 f. 8' 6" could be multiplied by 7 f. 9' 3" to get 75 f. 5' 3" 7''' 6''''. Of course, present-day mathematicians would tend to question the legitimacy of such a calculation, even though it gave the right answer (if one doesn't think too hard about units of area).

In another example, Pike multiplied 3 pounds 6 shillings and 8 pence by 2 pounds 5 shillings and 7 pence and got 7 pounds 11 shillings and 11 pence (p. 115). A century later, Florian Cajori (1890), the noted U.S. historian of mathematics and mathematics education, described that kind of arithmetic as "absurd," and in opposition to the fundamental ideas of multiplication in arithmetic. He blamed "the English" (p. 17) for its presence in school mathematics in North America.

Cajori's criticism was unfair to the "English." Certainly, all editions of Dilworth's arithmetic included a section on duodecimal "cross multiplication"—offering numerous examples showing how lengths (measured in feet, inches, seconds, thirds, etc.) could be multiplied by lengths. But, the same kind of thing appeared in French textbooks—for example, Barrème (1744), when considering the problem "Un mur a 22 toises 3 pieds de long sur 6 toises a pieds de haut, savoir combien il y a des toiles quarries," multiplied "22 toiles 3 pieds de long" by "6 toiles 2 pieds de haut," and got "142 toiles and 3 pieds" (pp. 187–188). Furthermore, between 1810 and 1825, John Farrar, Hollis Professor of Mathematics and Natural Sciences at Harvard College—someone who had chosen to specialize in translating Continental mathematics texts for use in the United States—included sections on duodecimal cross multiplication in three editions of his "translations" of Sylvester Lacroix's *Elementary Treatise on Arithmetic* (see, e.g., Lacroix, 1818). It would make little sense to blame "the English" for that. And, Charles Davies, a key player in a strong Continental European mathematics push at the West Point U.S. Military Academy in the first half of the 19th century, included sections on duodecimal cross multiplication in his school arithmetics.

Benjamin Workman

Although Nicolas Pike clearly became the best-known of "local" authors during the period 1788–1800, there were others who became known. We shall draw special attention to books authored by Benjamin Workman, Consider and John Sterry, Zachariah Jess, Erastus Root, Chauncey Lee, and Peter Tharp.

In 1788, a 370-page book attributed to John Gough, of Ireland, was published in Philadelphia. The title of this book was *A Treatise of Arithmetic in Theory and Practice Containing Everything Important in the Study of Abstract and Applicant Numbers, Adapted to the Commerce of Great Britain and Ireland*. Despite the fact that Benjamin Workman, a U.S. teacher, claimed, in a preface to Gough's book, that he had added "many valuable amendments more particularly fitting to the work for the improvement of the American youth," the book was hardly different from Gough's text written for Ireland. Except for a section on "Exchange," all of the questions involving money referred to pounds, shillings, pence, and farthings. There was no mention of the fact that Congress had approved a system of decimal arithmetic for the United States of America and, indeed, none of the word problems in the text had been revised so that they would be set in North American contexts (Gough, 1788).

This failed attempt motivated Benjamin Workman (1789) to cause to have published—also in Philadelphia, and through the same publisher who had published Gough's text in 1788—a 224-page textbook entitled *The American Accountant or Schoolmasters' New Assistant*. In his preface, Workman said that he was dropping the theoretical components of Gough's book, because he wanted to "furnish the scholar, at a cheap rate, with a complete system of practical arithmetic" (p. iii). This book was more successful, with second and third editions appearing in 1793 and 1796, but even in those later editions there were hardly any references to the new federal currency. Workman (1789) informed his readers that "in England, Ireland, and America, accounts are kept in pounds, shillings and pence" (p. 33), and almost all money examples were consistent with that backward-looking statement. Workman (1789) did include a brief section on "decimal fractions" (pp. 93–100), but nowhere in that section was there any mention of the new federal currency. In later editions, Workman referred, briefly, to "federal money," which he said had been approved by "Acts of Congress in 1792 and 1793" (see, e.g., Workman, 1793, p. 34). He also stated that "10 mills make a cent, 10 cents make a dime, 10 dimes make a dollar [for which he used the symbol *D*] and 10 dollars make an eagle" (p. 34). But that was all he had to say about federal money—no examples were given, and no exercises were set in relation to the new currency.

Consider Sterry and John Sterry

In 1790 two brothers, Consider Sterry and John Sterry, caused to have published, in Providence, Rhode Island, a 388-page book titled *The American Youth: Being a New and Complete Course of Introductory Mathematics, Designed for the Use of Private Students*. The most notable feature of this book was that, like Nicolas Pike's (1788) book, it included a section on algebra—indeed, 147 pages were devoted to algebra (whereas Pike had had 39 pages on algebra). The Sterrys were private teachers, outside of college circles, and so it was not to be expected that they would include such an extensive, and mathematically ambitious, section on algebra (Simons, 1924). The extent of the algebra covered was such that the text provided the widest coverage on algebra of any textbook written by North American authors in the eighteenth century.

In their Preface, Sterry and Sterry (1790) maintained that existing mathematics textbooks were "not adapted to the capacity of young and tender minds" (p. v), mainly because the authors had paid too much attention to "close and refined reasoning" and not enough to "simplicity, plainness and brevity" (p. v). The Sterrys probably had Nicolas Pike's book in mind when they commented that some books were "so prolix and voluminous, as even to discourage a learner at the sight of their works" (p. v).

The Sterrys paid considerable attention to the new federal currency, showing how operations on sums of money could be carried out by decimal operations. Curiously, though, they did that before they reached the 32-page section on decimal fractions.

The Sterry's book did not go to a second edition, and so it is reasonable to assume that not many teachers or students ever used it. Probably, the section on algebra was off-putting for many. Furthermore, it was published in 1790, two years before the United States Mint was established—at that time there were no decimal coins in circulation, and nearly everyone was still using sterling currency, or Spanish dollars. It is likely that neither the teachers nor their students wanted to spend their time learning to calculate with a system of currency which was not even being used at the time.

So far as weights and measures were concerned, the Sterrys offered a traditional coverage of all the different kinds of measures. In the absence of any progress in Congress on decimalizing weights and measures, who could blame the Sterrys for that?

A few years later, in 1795, the Sterry brothers made another attempt to publish a successful school mathematics textbook. This was a much smaller text, with only 121 pages. There was no algebra, and a larger treatment of decimal currency was provided than in the earlier book—which was appropriate, given that by 1795 the U.S. Mint had been open for three years, and people were now expected to learn to deal with the new coinage. Like the 1790 textbook, the Sterrys' (1795) book did not go to a second edition.

In their Preface to their new publication the Sterrys (1795) emphasized that this smaller book was aimed at schools. The section on compound operations offered specific instructions with respect to decimal currency. Relevant excerpts were:

> *[For addition of federal money]* In addition of money of the United States, add the numbers as in simple addition, and separate with a point, as many figures on the right hand as are equal to the greatest number in the inferior denominations given in the question; then decimate those on the right hand, beginning at the point, and call the first figure dimes, the second cents, third mills, etc. (p. 20)
> *[For subtraction of federal money]* In the money of the United States, point and decimate as in addition. (p. 25)
> *[For multiplication]* If one of the factors is dollars, cents, &c, multiply and separate on the right hand of the product as many figures as there are cents, mills, &c, and decimate as before in addition. (p. 28)
> *[For division]* In division of money of the United States, divide as in simple division, and separate with a point as many figures on the right of the quotient as there are contained in the inferior denominations in the dividend; then decimate as before taught. (p. 31)
> *[For reduction]* In reducing dollars, dimes, cents, and mills, to mills, the given numbers wrote as one in a line will be reduced as required. 46 dol 2 d 6 c 3 m reduced to mills is 46263 mills. (p. 35)

The language used in these instructions would have been formidable for many students. Furthermore, the rules were given well before the section in the book on decimal fractions—which occupied pages 49 through 58.

The Sterrys (1795) began their second book with the following "definitions":

> Arithmetic is the art of composition by numeral figures, called *digits* [original emphasis]. Which are considered either integral or fractional, and therefore vulgar or decimal.
>
> *Vulgar* arithmetic contemplates these digits integrally or dividedly.
>
> *Decimal* arithmetic considers those divide digits in a decimal ratio of those parts to unity.
>
> The digits made use of are these, 1, 2, 3, 4, 5, 6, 7, 8, 9, and for convenience in computation is added the cypher, 0.
>
> All arithmetical operations are performed by addition, subtraction, multiplication, and division. (p. 5)

The Sterrys (1795) were obviously attempting to sum up the achievements of those who had developed the Hindu-Arabic numeration system, and of those who had extended it to include vulgar and decimal fractions. This introduction to numbers should already have been known by students before they began to prepare a cyphering book, but the Sterrys' introduction went further than most when it explicitly referred to vulgar and decimal fractions (Clements & Ellerton, 2015).

In keeping with this early mention of vulgar and decimal fractions, Sterry and Sterry (1795) provided more than 12 pages on vulgar fractions quite early in their book (pages 37–49). They immediately followed their section on vulgar fractions with a 10-page section on decimal fractions (pages 49–58). Thus, 22 pages of their 120-page book were dedicated to vulgar and decimal fractions, and these pages came early in the book. That approach was unusual for the 1790s.

Furthermore, the 22 pages were placed in the book before more advanced *abbaco* topics were introduced—in fact, the Sterrys followed their introduction to decimal fractions with the rules of three, the rules of practice, tare and tret, simple and compound interest, commission, brokerage, rebate or discount, barter, loss and gain, fellowship, and alligation. That ordering of those topics was consistent with the *abbaco* tradition (Ellerton & Clements, 2012), but in this book vulgar and decimal fractions came earlier than usual. Thus, the Sterrys (1795) were able to ask students to tackle exercises such as:

- A goldsmith sold a tankard for 29 dol. 97 cts. It weighed 270 oz. What is that per ounce? (p. 60)
- If 20 bu. of grain at 50 cts. per bushel will pay a debt, how much at 2 d 60 ct will pay the same? (p. 63)
- Two partners, *A* and *B*, constitute a joint stock of 300 dollars, whereof *A* had 200 dollars and *B* 100. They gain 150 dollars in trade. What is each person's share of the gain? (p. 87)

In model examples, the Sterrys showed how such tasks could be tackled using decimal fractions. They made a determined effort to link the new decimal currency to the formal study of decimal fractions. Although such an approach seemed to be demanded by the times, it was unusual, and perhaps that was why Sterry and Sterry's (1795) book, like their (1790) book, was never published beyond the first edition (Karpinski, 1980).

Zachariah Jess and "Sundry Teachers in and Near Philadelphia"

According to Louis Karpinski (1980), a textbook prepared by John Todd, Zachariah Jess, William Waring and Jeremiah Paul with the assistance of "sundry teachers in and near Philadelphia" (p. 97) was published in 1791. Although, there are no extant copies of the 1791 edition, there are extant copies of later editions which were published in 1794, 1796, 1797, 1799, and 1800 as well as of 15 other editions which appeared well into the nineteenth century. What makes the books interesting, historically, is the claim, on the title pages, that each was a "practical arithmetic prepared by practicing teachers" (Todd, Jess, Waring, & Paul, 1800, p. i). Various editions of a related textbook authored by Zachariah Jess himself were also published.

The earliest edition in the E-C Collection of the text by "Sundry Teachers in and Near Philadelphia" was published in 1800, and of particular interest were the sections on vulgar fractions, decimal fractions, and federal money. Analysis of these sections revealed that aside from a summary of the relative values of the new federal coins, and some additions, subtractions, multiplications and divisions of federal money, the book was essentially no different from the kind of school arithmetic found in standard British arithmetic textbooks like those written by Thomas Dilworth and Francis Walkingame. Almost all of the practical money problems were expressed in sterling pounds, shillings, pence, and farthings. Simple and compound interest tasks involved sums of money expressed in sterling rather than federal money. That remained true of later editions of the book.

The chapters on vulgar fractions appeared after chapters on equation of payments, barter, loss and gain, and fellowship—and the chapters on decimal fractions appeared after the chapters on vulgar fractions. Cyphering-book data have revealed that most students who used this book never got to study formally vulgar or decimal fractions and were only briefly introduced to federal money. Almost all word problems requiring money calculations were expressed in terms of pounds, shillings, pence, and farthings.

The role of Zachariah Jess is unclear because he seemed to be involved as an author with two very similar textbooks. One cannot deny, though, that Jess's books remained popular for 40 years. Indeed, questions which Abraham Lincoln answered in his cyphering book in the 1820s could be found in Jess's textbooks (Ellerton et al., 2014).

Erastus Root

The title of Erastus Root's (1795) text, *An Introduction to Arithmetic for the Use of Common Schools*, would have sounded attractive to some teachers, as would the following statement in Root's preface:

> To be candid, fellow citizens, the object of this publication is to furnish common schools, with an easy, accurate and cheap volume, containing all the arithmetical knowledge necessary for the farmer or the mechanic. The *manner* may be new if the *matter* is not. Several very excellent treatises on arithmetic have lately been published; yet none of them seem to be exactly calculated for common schools. The size and consequent dearness of some, forbid their general use, while the deficiency and unnecessary learning of others, ought to exclude them. Transatlantic authors will no longer do for independent America. We have coins and denominations of money peculiar to ourselves—In these our youth ought to be instructed and familiarized. The simplicity alone, of this our federal money, is its sufficient recommendation. Its denominations are the simplest possible—being purely decimal. Almost two centuries have elapsed since the invention of decimal arithmetic; yet never, till lately, has it been applied to the

> weights, measures, or monies of any nation. But it remained for the United States to make the beginning. (pp. v–vi)

The message seemed to be—the other arithmetics are too expensive, and too difficult, and this arithmetic will give you the kind of arithmetic that you, as an American, will need to know.

Root's first edition had only 105 small-sized pages, but early in the book there were nine successive pages totally devoted to "Federal Money" (pp. 20–28). Later pages also focused on the arithmetic of federal money (e.g., page 49 and page 64, were both concerned with reduction). The writing was unusually clear, and perhaps that is why the book went to 10 editions, the last being issued in 1814. And, as he indicated in his preface, Root (1795) was prepared to compromise on the currency question: "I have given many of the examples in pounds, shillings and pence—supposing it necessary to instruct our youth in the *old way* [original emphasis] for some time yet to come. The customs of a great nation cannot be wholly changed in a month, nor a year" (p. vi). So far as both coinage and weights and measures were concerned, this statement was prophetic.

Root was also prepared to ask students to answer questions like (a) "Divide 66 dollars and 66 cents by 2 dollars and 5 dimes" and (b) "Divide 74 dollars by 4 dollars 75 cents. The answers he expected (see Figure 4.10) were, for (a) 17 dollars 8 dimes 6 cents and 4 mills, and, for (b) 15 dollars 5 dimes 7 cents and 8+ mills. This illustrates part of the problem with an author like Erastus Root. He was not well-trained in mathematics, and this often showed. That said, at that time college professors of mathematics were *not* among the names of early U.S. authors of popular school mathematics textbooks. The historian of mathematics education should ask, "Why was that?"

As an introduction to mathematics, Root's little book was both inadequate and deficient. From a historical perspective, the most serious weakness was connected to Root not being aware of his mathematical deficiencies. His solutions to the questions shown in Figure 4.10 were revealing. He did not include any work on vulgar fractions because they were "not absolutely necessary" (p. vi). He did not seem to recognize that, from a mathematical perspective, fractions are not only an important component of the number system, but they can also be usefully applied in many daily situations. Later in this chapter we will see that he was certainly not the last author to maintain that fractions should not be part of the common-school arithmetic curriculum. But, all attempts to rid schools of the bogey of fractions would fail, probably because, in fact, they *are* mathematically important and because they *can be* useful.

The 1795 copy of Root's book held in the E-C Collection has wooden covers. Such books were known in the book trade as "scabbards." There are 40 scabbards in the E-C Collection.

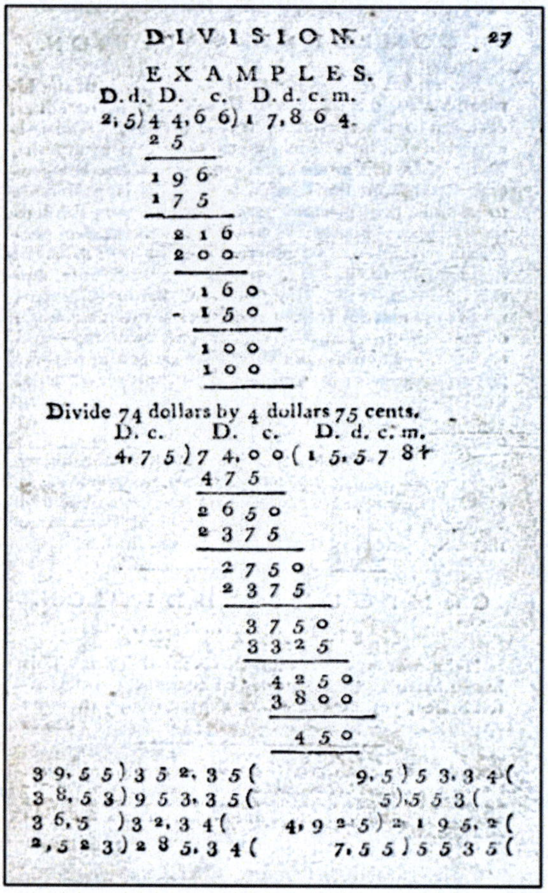
DIVISION. 27

EXAMPLES.

D.d. D. c. D. d. c. m.
2,5)4 4,6 6)1 7,8 6 4
2 5
1 9 6
1 7 5
2 1 6
2 0 0
1 6 0
1 5 0
1 0 0
1 0 0

Divide 74 dollars by 4 dollars 75 cents.
D. c. D. c. D. d. c. m.
4,7 5)7 4,0 0(1 5,5 7 8+
4 7 5
2 6 5 0
2 3 7 5
2 7 5 0
2 3 7 5
3 7 5 0
3 3 2 5
4 2 5 0
3 8 0 0
4 5 0

3 9,5 5)3 5 2,3 5(9,5)5 3,3 4(
3 8,5 3)9 5 3,3 5(5),5 5 3(
3 6,5)3 2,3 4(4,9 2 5)2 1 9 5,2(
2,5 2 3)2 8 5,3 4(7,5 5)5 5 3 5(

Figure 4.10. Division with federal currency (in Root, 1795, p. 27).

Chauncey Lee

Chauncey Lee (1763–1842) offered a particularly radical outlook on the arithmetic curriculum. After graduating from Yale College, and after practicing Law for some years, Lee turned to Theology and in 1789 he was licensed to preach. But he then took up school teaching and, in 1796, became Principal of Lansingburgh Academy, a new school in Troy, New York.

In 1797, Lee authored a 300-page textbook with the title *The American Accomptant being a Plain, Practical and Systematic Compendium of Federal Arithmetic; in Three Parts; Designed for the Use of Schools, and Specially Calculated for the Commercial Meridian of the United States of America*. Although the publication was financed by a long list of subscribers, only one edition of the book ever appeared (Karpinski, 1980). It has been well remembered by historians, however, because it has been claimed that in it Chauncey Lee became the first author to use a symbol which resembled the now familiar dollar sign ($)—see Figure 4.11. But that claim has been challenged (see Fanning & Newman, 2011).

CASE 5.*

Q. *How do you multiply any whole number by a ſum in Federal Money, in order to find the price or value of the whole?*

A. 1. Set down the factors as before; for the ſake of convenience, placing that for the Multiplier which has the fewer figures.

2. Multiply the factors together as if they were both whole numbers, according to the general rule in Leſſon XIX.

3. Point off, in the total product, as many right hand figures, as there were figures in the factor or price below dollars, for cents and mills—the reſt are Dollars.

EXAMPLES.

Q. *What is the amount of 3257 yds. of Velvet, at* $ × 3.55 7 *per yard?*

```
   22799
  16285
 16285
9771
------------
Anſwer, 11585.14 9
```

Figure 4.11. Chauncey Lee's (1797) use of a sign which resembled the dollar sign when multiplying $3.55 7 by 3257 (p. 87).

If one reads Lee's lengthy (38-page) introduction to *The American Accomptant* one can get a good idea why the publication did not go beyond its first edition. The longer version of the title of the book inferred that its author was concerned to write an arithmetic suited to the needs of North American school students, and in his introduction he made it clear that he intended his textbook to live up to its title. He did not name Nicolas Pike in the introduction, but it was obvious from what he wrote that he felt that existing arithmetic texts were inappropriate for school children, and Pike's arithmetics "fitted" the criticisms that Lee made. Nevertheless, from a historical perspective, Lee's introduction is, perhaps, the most important extant statement on the weaknesses inherent in standard approaches to 18th-century arithmetic education.

Lee (1797) was prepared to tread where previous authors of arithmetic texts had rarely trod. Thus, for example, he recommended the use of the "decomposition" method of subtraction, rather than the then-popular "equal additions" algorithm (Ellerton & Clements, 2012). Most importantly, not only did he recommend that a whole new decimalized system of weights and measures be introduced, he also provided details for a possible system that might be used.

However, his criticisms sometimes missed their mark because, like Erastus Root, he did not know enough mathematics. Like Root, he took a stand against the use of vulgar fractions—on page *ix* he argued that "these absurd, untoward fractional numbers" needed to be "banished from practice and the several denominations in all commercial tables of mixed quantities conformed to our federal money and established upon a decimal scale." He then pointed out that "to accomplish all this is a task too great for any individual in a republican government" (p. *ix*). What was needed, he wrote, was "the arm of Congress to effect it," and it was "equally to be hoped and expected, that their wisdom and patriotism will not be inattentive to so important an object of legislation" (p. *xix*).

Lee then put forward his plan for a new system of units (pages *xx* to *xxvi*). After pointing out that "an unnecessary multiplication of the tables of compound quantities will not facilitate the study or practice of arithmetic, but have a contrary effect" (pp. *xxvii*), he warmed to his theme:

> And, let me ask, what real necessity can there be of having such a diversity of weights? What even imaginary necessity, abstract from the current of arbitrary custom and habit? What benefit from it to society in general, or to the tuition of schools in particular? What good purposes are answered by it in the transaction of any kind of business, or in the operation of any arithmetical calculation whatever, which would not be as well, and on the whole much better answered, by reducing them all to practice to a single standard; and ascertaining the gravity of gold, iron, medicines, and all kinds of substances, now classed under three different sorts of weights, by one common table of weights, distinguished and dignified by the name of *American weight*? (p. *ix*)

He then urged Congress to introduce his scheme for a decimalized system of weights and measures, arguing that that should be possible because Congress had shown foresight by putting into place a decimalized form of currency, and that what was needed now was a "matching" system of weights and measures which was consistent with the new currency.

Lee (1797) next proposed that "federal avoirdupois" be based on the following relationships: "10 drams make 1 ounce; 10 ounces make 1 pound; 100 pounds make 1 hundred weight; and 10 hundreds make one thousand" (p. *xxi*). So far as "federal troy weight" was concerned, he pointed out that by a 1793 Act of Congress the weight of the American dollar was called a "pennyweight," one-tenth of that, a "cent," and one-tenth of a cent, a "mill." He then proposed that "10 cents should be 1 grain, 10 grains 1 pennyweight, 10 pennyweights 1 ounce, and 10 ounces 1 pound" (p. *xxii*). He also put forward a "Federal Apothecary Weight," by which "10 grains equaled 1 scruple, 10 scruples would be 1 dram, 10 drams 1 ounce, and 10 ounces 1 pound" (p. *xxv*), and showed decimalized tables for liquid measure, dry measure, long measure, and cloth measure.

But Lee (1797) realized that, ultimately, it was not really his task to be putting forward such a radical proposal. He wrote:

> I need not be reminded that it becomes not a private individual, in a great Republic, to dictate rules and reforms of this kind: I am not so weak as to aspire to it; but only to exercise the republican private privilege of *proposing* [original emphasis] what the more enlightened public may judge of, and candor will not reject without reason. (p. *xxix*)

In sections on weights and measures, Lee included tables of traditional measures and also tables of his own proposed measures. This would have been confusing for some students and teachers. He also showed how the arithmetic of decimal fractions could greatly simplify calculations for weights and measures, and even used his dollar sign ($) when doing it (see, Figure 4.11). Notice that in Figure 4.11 the price of a yard of velvet is given as 3 dollars, 55 cents, and 7 mills, which was recorded as $3.55 7.

Although Lee's thinking about decimalization was ahead of his time, it was politically and educationally naïve. Unlike Thomas Jefferson, Lee failed to recognize that the time had passed when Congress would accept legislation creating a decimalized federal system of weights and measures (Clements & Ellerton, 2015). Attempts had been made to bring the matter to a vote in Congress at various times between 1790 and 1792, and again in 1795 and 1796, but it had never formally been committed to a vote. Indeed, according to Boyd (1961), the debate in 1796 had seen opponents to the idea treating the proposal "with levity" (p. 617). There was no way the nation would accept Lee's proposals for a unified system, and therefore there was little chance that Lee's textbook, which assumed that his proposals would be in use, would be supported by the public.

Thomas Jefferson continued to believe strongly in the desirability of a unified system but when he was President he recognized that, politically speaking, he had lost his opportunity. In 1801, when he first became President, Jefferson seemed resigned to the likelihood that the political forces which would line up in Congress against any attempt to introduce such a national system would prove to be too strong. The sun of the new era which had dawned in 1775, had reached its high point in the 1780s with the agreement to introduce a decimalized coinage, but was now on its way to a quiet sunset.

Still, a question remains about Jefferson's influence on the French, because he spent five or six years in France in the 1780s before actions were put in place there to establish formally a metric system, and it is known that while in France Jefferson had serious discussions with influential mathematicians about the sense of such a system (Clements & Ellerton, 2015). By the mid-1790s Jefferson had recognized that his battle to achieve a coordinated system of weights and measures had been lost, but Chauncey Lee was prepared to fight on.

Peter Tharp

Tharp's (1798) small-sized 120-page book was another which was not published beyond the first edition. Tharp designated himself, on the title page, as "Math," which was presumably meant to imply that he taught mathematics on a private basis. He lived in the town of Marlborough, in the State of New York. In his preface he commented that he had "laid down a very plain and concise rule for reducing the currency of each state into federal money, and the contrary" (p. iii).

Tharp was another writer who chose not to include a section on vulgar fractions. However, he gave a strong place to both decimal fractions and to federal currency. Understandably, though, because Congress had decided against introducing a federal system of weights and measures, all of the compound operations tasks that he included were based on traditional units.

Tharp's presentation of material often left much to be desired. Consider, for example, how he solved the problem: "What is the interest of 27.5 dol. for five months at 7 per cent? Tharp expected his readers to work out that the .07 on the second line corresponded to the 7 per cent interest, and that the .4166 on the fourth line corresponded to the five months (or 5/12 of a year); in the long-multiplication of 1.925 by .4166, both of the 11553s on the fifth and sixth lines should be 11550s, unless some account was taken of the fact that 5/12 equals .41667 to five decimal places, on the seventh line 1952 should be 1925; and it was not clear what the 10000, on the last line, meant.

$$\begin{array}{l} 27.5 \\ \underline{.07} \\ 1.925 \\ \underline{.4166} \\ 11553 \\ 11553 \\ 1952 \\ \underline{7700} \\ .8019553 \text{ dol.} \\ \underline{10000} \text{ (Tharp, 1798, p. 59)} \end{array}$$

From this example, one can begin to understand why only one edition of Tharp's book was ever published (Karpinski, 1980, p. 122).

North American Authors from the Period 1800–1820

The early years of the nineteenth century witnessed the appearance of numerous mathematics textbooks which were written by American authors. But six authors—Nicolas Pike, Nathan Daboll, Daniel Adams, Michael Walsh, Stephen Pike (who was not related to Nicolas), and Jeremiah Day—became better known than the others. The works of Jeremiah Day were especially important with respect to algebra education and will be dealt with in the next chapter. Here we comment on the work and influence of the other five who, we argue, were responsible for the standard *abbaco* sequence continuing to be emphasized in U.S. schools.

In 1797 the second edition of Nicolas Pike's *The New and Complete System of Arithmetic, Composed for the Use of the Citizens of the United States* appeared, and four more editions were published in the nineteenth century (in 1808, 1820, 1832 and 1843) (Karpinski, 1980). However, even the abridged editions of Pike's book—those prepared especially for use by schoolchildren—tended to be very difficult for ordinary learners in schools, and that presented an opportunity for Adams, Daboll, and Walsh, whose arithmetics were first published in 1800 and 1801. Their books would increasingly be accepted in schools, particularly in the cities and larger towns in New England, in New York, and in Philadelphia (Clements & Ellerton, 2015). Stephen Pike, who was based in Philadelphia, had the first edition of his *The Teacher's Assistant or a System of Practical Arithmetic* published in Philadelphia in 1811, and over the next 40 years many new editions and *Keys* would be published and widely used in schools in Pennsylvania and Virginia (Karpinski, 1940).

Nathan Daboll

Nathan Daboll Senior (c. 1750–1818) was born in Groton, Connecticut. He had little formal education but mastered mathematics quickly while earning a living as a cooper. He then moved into the world of writing and publishing and, during the 1770s and 1780s, he achieved fame as an early U.S. teacher of mathematics and navigation, and as a publisher of almanacs. He edited the *Practical Navigator* (Daboll, 1820), and it was claimed, that he instructed as many as 1500 persons in navigation. In the late 1790s Daboll developed a method for "dead reckoning" which, he claimed, was preferable to existing methods for determining longitude (Daboll, 1820). In 1787, 13 years before his own arithmetic textbook would be first published, Daboll, when "signing" a recommendation for one of the textbooks authored by Consider and John Sterry (1790), described himself as a "teacher of Mathematics and Astronomy, in the Academic School in Plainfield."

Daboll's most famous textbook was his *Schoolmaster's Assistant: Being a Plain Practical System of Arithmetic Adapted to the United States*. This was first published in 1800 and was the first in a series of arithmetics carrying the name "Nathan Daboll" which would appear over the next 60 years—although after 1818 (the year when Daboll Senior died), Nathan Daboll Junior (1782–1863) became the driving force behind the revised versions. Textbooks authored by a Daboll would be so widely used that the expression "according to Daboll" became commonplace in arguments about or discussions on arithmetic.

Both Nathan Daboll Senior and Nathan Daboll Junior claimed that so far as fractions were concerned, they had "taken an entirely new method" (Daboll, 1818, p. v). Like Daniel Adams (1801) and Erastus Root (1795), they argued that decimal fractions were much easier than vulgar fractions, and because they were "more simple, useful, and necessary, and soonest wanted in more useful branches of arithmetic, they ought to be learned first, and vulgar fractions omitted until further progress in the science shall make them necessary" (Daboll, 1818, p. vi). Accordingly, in their textbooks (e.g., Daboll, 1813) they delayed their treatment of vulgar fractions later than pure mathematical logic would have demanded. Vulgar fractions were briefly dealt with early in their textbooks and then a detailed treatment of decimal fractions was given. Much later in their books and then in a later section vulgar fractions received more systematic attention.

Phillip Jones and Arthur Coxford (1970), editors of NCTM's *A History of Mathematics Education in United States and Canada* (pp. 11–23), maintained that Nathan Daboll Senior's text was originally published in England, and that it resembled the arithmetic by his "countryman," Thomas Dilworth (p. 15). The distinguished British historian, Geoffrey Howson (1982), repeated this error in his book, *History of Mathematics Education in Great Britain.* In the 1830s, in the United States of America, Daboll's arithmetics were probably more used in the United States of America than all other arithmetics (Karpinski, 1980).

False position problems in Daboll (1804). The content in Daboll's (1804) textbook was strictly in line with the *abbaco* sequence. "Double false position" was a topic that came toward the end of that sequence, and it will be worth looking at an example given by Daboll. Before the example, Daboll had described "position" as a method which "by false or imperfect numbers, taken at pleasure, discovers the true ones required" (p. 196). Of course, such a description was sufficiently vague to be almost meaningless. Daboll gave model examples in order to show the types of situations for which false position was relevant, and also the rules by which one needed to proceed. He started with "single position," and then proceeded to "double position," which he described, once again vaguely, as teaching "to resolve questions by making two suppositions of false numbers" (p. 198). Having made that statement, he immediately offered five rules:

> 1. Take any two convenient numbers and proceed with each according to the conditions of the question.
> 2. Find how much the results are different from the result in the question.
> 3. Multiply the first position by the last error, and the last position by the first error.
> 4. If the errors are alike, divide the difference of the products by the difference of the errors, and the quotient will be the answer.
> 5. If the errors are unlike, divide the sum of the products by the sum of the errors, and the quotient will be the answer.
>
> *NOTE—The errors are said to be alike, when they are both too great or both too small; and, unlike, when one is too great and the other is too small.* [original emphasis]
>
> (Daboll, 1804, p. 198)

An uninitiated reader would be struggling to make sense of what Daboll had written. He then offered the following model example:

> A purse of 100 dollars is to be divided among 4 men, *A*, *B*, *C*, and *D*, so that *B* may have 4 dollars more than *A*, and *C*, 8 dollars more than *B*, and *D* twice as much as *C*, what is each one's share of the money? (Daboll, 1804, p. 199)

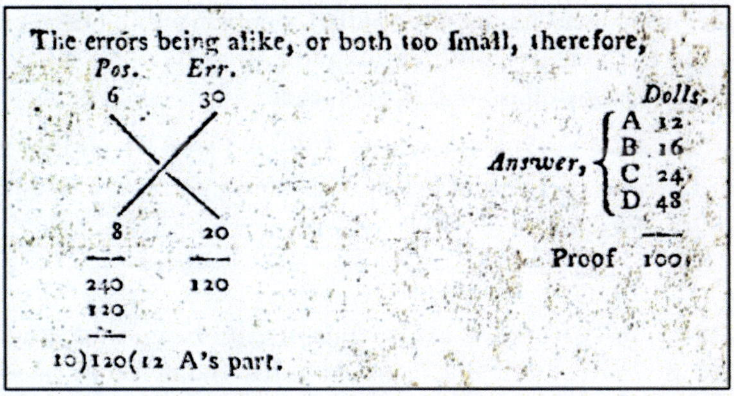

Figure 4.12. Daboll's (1804) solution to a "double false position" task (p. 199).

Daboll immediately gave the solution shown in Figure 4.12.

Double false position tasks like this can be found in most of the early arithmetic textbooks, and the explanations given were no better or worse than that offered by Daboll. From a mathematics education perspective, at least four points arise:

1. In real-life, would a student who had learned double false position in this way be able to recognize when it was appropriate to use the method?
2. Readers were not informed which contexts were likely to generate problems for which the method of double false position would be appropriate.
3. Readers were not taught *why* the rule worked; they were merely given a rule which would, if accurately applied, help them "get the right answer."
4. Readers were not informed whether there were any other ways of solving problems like the one shown—and, if there were, whether these other ways might be better than the one shown.

This was the standard approach to advanced *abbaco* arithmetic, in Western Europe and, later, in the Americas, throughout the period 1200–1850. Students who had advanced to the point where double false position arose in the *abbaco* sequence read the notes, studied the model example, and then attempted to find solutions to the exercises set. There were no written examinations, and a student's solutions to exercises were only allowed to be entered into his or her cyphering book after they had been checked by the teacher during a recitation session. A problem arose, of course, if the teacher did not know how to solve the problems. It was relevant, though, that only a small proportion of students ever reached the stage where they would be asked to consider double false-position problems.

Most pre-college teachers had never studied algebra, and therefore the following algebraic solution to the above problem would hardly have been a possibility: if x represented the number of dollars that A gets, then:

> $x + (x + 4) + (x + 4 + 8) + 2(x + 4 + 8) = 100$, and so $5x + 40 = 100$, and therefore $x = 12$. Hence, A would get \$12, B would get \$16, C would \$24, and D would get \$48.

Daniel Adams

Adams was born in Townsend, Massachusetts, in 1773, and graduated from Dartmouth College in 1797. In 1799, he received a Bachelor of Medicine from Dartmouth, and in 1822 he was awarded the Doctor of Medicine degree by Dartmouth. For the rest of his life he would manage, somehow, to combine his interests in writing, publishing, education, medicine, theology, and politics.

After graduating, Adams settled in Leominster, Massachusetts, to practice medicine, and it was there that he began a career in publishing. As a medical graduate he received recognition within his local society—for example, in February 1800, he was chosen to read a eulogy to George Washington, and he subsequently published an oration on the life of Washington. In 1801, he caused to have published his first school arithmetic text from a press that he, himself, had established in Leominister. He moved to Boston in 1806, where he opened a "select school." After settling in Mount Vernon in New Hampshire in 1813, he resumed his medical practice and, in 1827, published a revised version of his arithmetic, titled *Adams's New Arithmetic*. From 1838 until 1840 he served as a state senator, and he was for some time president of the New Hampshire Bible Society. In 1846, he settled in Keene, New Hampshire, where he spent the remainder of his life. He was the author of many school textbooks, principally—but not only—on mathematics. From 1838 till 1840 he served as a state senator, and he was for some time president of the New Hampshire Bible Society. He was also president of the New Hampshire Medical Society. He died in Keene in 1864, aged 90.

Adams's *Scholar's Arithmetic*, and his *New* and *Revised* versions of that text, were extensively used for many years (see, e.g., Adams, 1848). Like Nathan Daboll, Michael Walsh, and Stephen Pike, three other very successful U.S. arithmetic textbook authors of the time, Adams adopted a very pragmatic stance so far as curriculum content was concerned—their school arithmetics were based on the traditional *abbaco* sequence of arithmetic topics and were especially concerned with commercial aspects of everyday life (Ellerton & Clements, 2012).

In a "preface dedicated to schoolmasters," Adams (1801) thanked the schoolmasters for their "kind and very ready exceptance (sic.) of the first edition of his *Scholar's Arithmetic*." He added:

> The testimony of many respected teachers has inspired a confidence to believe that this work, where it has been introduced into schools, has proved a kind assistant towards a more speedy and thorough improvement

> of scholars in numbers, and at the same time, has relieved masters of a heavy burden of writing out rules and questions, under which they have so long labored, to the manifest neglect of other parts of their schools. A most flattering proof of their approbation is that the first edition has met with an entire sale, within ten months after its publication. (p. iv)

In the same preface, Adams went on to say that the blank spaces which had been left after each exercise in his arithmetic text were "designed for the operation by the scholar, which being wrought upon a slate, or waste paper, he may afterwards transcribe into his book" (p. v). Clearly, Adams expected teachers and students to use his textbook in such a way that elements of the longstanding cyphering tradition would need to be modified (Ellerton & Clements, 2012).

In Section II of his widely used *Scholar's Arithmetic or Federal Accountant*, Adams (1817) claimed that an understanding of the following 10 aspects of arithmetic was necessary for every person who would subsequently engage in business transactions:

> Reduction, fractions, federal money, exchange, interest, compound multiplication, compound division, single rule of three, double rule of three, and practice. A thorough knowledge of these rules is sufficient for every ordinary occurrence in life. Short of this, a person in any kind of business will be liable to repeated embarrassments. It is the extreme usefulness of these rules which commends them to the attention of every scholar. (p. 50)

This was a very conservative view of curriculum sequencing for arithmetic—although one might wonder why compound addition and compound subtraction, and vulgar and decimal fractions, were not mentioned. Adams added that in his book, fractions were taken no further than was necessary to show their significance, and to illustrate the principles of federal money.

No less than 55 copies of *The Scholar's Arithmetic—Or Federal Accountant* are held in the Ellerton-Clements textbook collection. On the title page of each copy it was stated, in small print, that the book's contents could be summarized in the following way:

- Common arithmetic, the rules and illustrations;
- Examples and answers with blank spaces, sufficient for their operation by the scholar;
- To each rule a supplement, comprehending questions on the nature of the rule, its use and the manner of its operations, and exercises;
- Federal money, with rules for all the various operations in it, to reduce federal to old lawful and old lawful to federal money;
- Interest cast in federal money, with compound multiplication, compound division, and practice, wrought in old lawful and in federal money; the same questions being put in separate columns on the same page in each kind of money, these two modes of account become

contrasted, and the great advantage gained by reckoning in federal money easily discerned;
- Demonstrations by engravings of the reason and nature of the various steps in the extraction of the square and cube roots, not to be found in any other treatise on arithmetic;
- Forms of notes, deeds, bonds, and other instruments of writing.

Adams (1817) then wrote: "The whole is a form and method altogether new, for the sake of the master and greater progress of the scholar" (p. *i*).

Included at the front of an undated (c. 1848) edition of *The New Scholar's Arithmetic* (pp. v–vi) were comments by "W B. B." on the ways "recitation" occurred in many school arithmetic classes. W. B. B. began by severely criticizing how arithmetic was often taught. After referring to "unqualified," and "ignorant" teachers, he called for "understanding" to be the main aim of arithmetic instruction and added that much work needed to be done to bring teachers to a higher standard. According to W. B. B., that was more important than improving textbooks.

Adams gave scant attention to "vulgar fractions." On page 75 of his 1802 second edition, Adams offered a general definition of a fraction and then stated:

> The arithmetic of vulgar fractions is tedious and even intricate to beginners. Besides, they are not of necessary use. We shall not, therefore, enter into any further consideration of them here. This difficulty arises chiefly from the variety of denominators; for when numbers are divided into different kinds, or parts, they cannot easily be compared. This consideration gives rise to the invention of decimal fractions. (pp. 75–76)

Then followed, immediately, eight pages on decimal fractions.

As was common with the *abbaco* tradition, rules were given, and students were expected to learn how to apply them by examining model examples. The most iconic topic in the *abbaco* sequence was the "direct rule of three," and Adams described his approach to solving problems using the following rule:

1. State the question by making that number which asks that question the third term, or putting it in the third place; that which is of the same name or quality as the demand, the first term, and that which is of the same name or quality with the answer required, the second term.
2. Multiply the second and third terms together, divide by the first, and the quotient will be the answer to the question, which (as also the remainder) will be in the same denomination in which you left the second term, and may be brought into any other denomination required. (Adams, 1802, p. 119)

Adams seemed to assume that this wording of his two-step rule would not be troublesome for readers. As was normal with *abbaco* arithmetic, no reason for the rule was given. *Abbaco* arithmetic was originally devised for apprentices who would

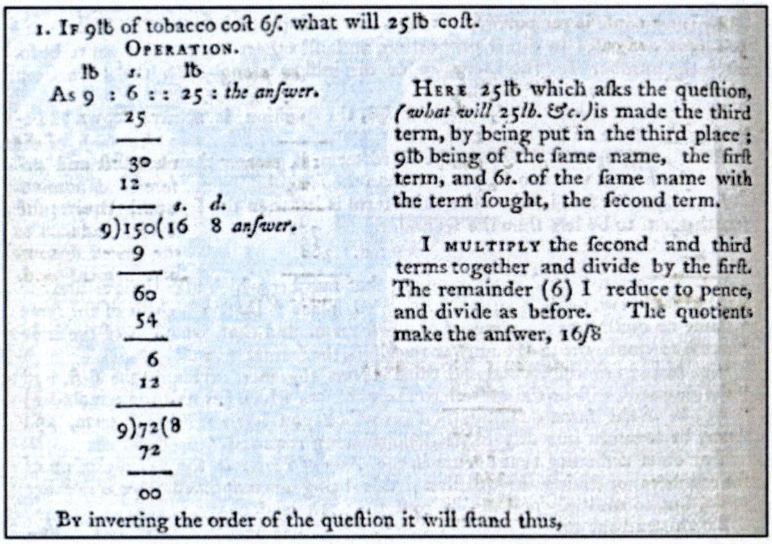

1. If 9lb of tobacco coſt 6ſ. what will 25lb coſt.

Operation.

lb s. lb

As 9 : 6 :: 25 : *the anſwer.*

25

30

12

s. d.

9)150(16 8 *anſwer.*

9

60

54

6

12

9)72(8

72

00

Here 25lb which aſks the queſtion, (*what will 25lb. &c.*) is made the third term, by being put in the third place; 9lb being of the ſame name, the firſt term, and 6s. of the ſame name with the term ſought, the ſecond term.

I multiply the ſecond and third terms together and divide by the firſt. The remainder (6) I reduce to pence, and divide as before. The quotients make the anſwer, 16/8

By inverting the order of the queſtion it will ſtand thus,

Figure 4.13. Rule-of-three solution to a model problem in Adams (1802, p. 120).

be involved in commercial activities, and many teachers were not interested in whether the arithmetic made any sense to anyone. They just wanted the apprentices to be able to get correct answers.

Figure 4.13 shows Adams's model solution to the problem: "If 9 lb of tobacco costs 6 *s*, what will 25 lb cost?" Even though Adams gave two explanatory notes (on the right in Figure 4.13), there would have been many middle-level *abbaco* arithmetic students puzzled by what was written.

That said, Adams's explanation was clearer than that found for the "rule of three" in many other textbooks. The "direct rule of three" was regarded as the "golden rule," the most important rule in *abbaco* arithmetic, the rule by which many elementary problems in commercial practice were solved—for example, the rule by which simple and compound interest problems were solved. The direct rule of three could also be invoked with heights and distances problems in mensuration, and in many other contexts. What mattered most, teachers and students were taught to think, was that students got the idea that one multiplied the second by the third term, and then divided by the first. Teachers wanted their students to get correct answers as soon as possible.

In the early 1820s a young Abraham Lincoln learned how to apply the direct rule of three. For the question: "If 3 lb of ginger cost 3 *s* what cost 26 lb?" Abraham multiplied 3 by 26 and got 78; he then divided the 78 by 3 and got 26. His solution, in his cyphering book, did not suggest that he had thought that since 1 lb of ginger would cost 1 *s*, then 26 lb would cost 26 *s* (Ellerton et al., 2014, p. 158).

Michael Walsh

Michael Walsh was born in Ireland in 1763 and came to Massachusetts, around 1785. In 1786 he became a naturalized citizen of the United States of America, and in 1792 he took up a position as teacher at Marblehead Academy, in Massachusetts. In 1794, he was appointed master of the grammar school at Newbury-Port Town, in Massachusetts. The school initially had 38 students and, in addition to his work at the grammar school, he conducted a private school at which he taught writing, arithmetic, and accountancy. In 1798, girls were admitted to that school for the first time. In 1803, Harvard College awarded Walsh the degree of A.M. and in 1805 he moved to Salisbury Point, Massachusetts. He died in Massachusetts in 1840 (Karpinski, 1980).

Between 1801 and 1806, an influential Newbury-Port printer, Edmund M. Blunt, published three editions of Walsh's *A New System of Mercantile Arithmetic Adapted to the Commerce of the United States* (see, e.g., Walsh, 1801). The book was aimed directly at parents who wanted their boys to be well prepared for commercial positions. Its front pages reproduced numerous supporting letters from groups of businessmen. Typical of these letters was the following from leading merchants of the bustling seaport of Salem, Massachusetts. Nathaniel Bowditch was named as one of the merchants to be associated with the recommendation.

> We the subscribers, merchants of Salem, convinced of the necessity of rendering the forms of business, the value of coins, and the nature of commerce more familiar to the United States as a commercial people, do approve of the *Mercantile Arithmetic* of Mr. Walsh, and recommend it as calculated to subserve in the best manner the instruction of our youth, and the purposes of a well-informed merchant. (Walsh, 1804, p. 5)

Among other recommendations for Walsh's (1804) book were statements similar to the one just quoted made by leading merchants of numerous towns other than Newbury-Port and Boston.

Like most commercially successful arithmetic textbooks of the period, Walsh's arithmetic followed the traditional *abbaco* sequence of topics and adopted standard **IRCEE** (**I**ntroduction, **R**ules, **C**ases, **E**xamples, **E**xercises) and **PCA** (**P**roblem. **C**alculation, **A**nswer) genres expected of students preparing cyphering books (Ellerton & Clements, 2012). Many thousands of copies of Walsh's book were sold over several decades. According to Karpinski (1980, p. 140), the last edition of Walsh's arithmetic appeared in 1832.

"Equation of payments" was one of the middle-level *abbaco* topics. Walsh (1804) introduced it by stating that the aim was "to find a mean time for the payment of several sums due at different times" (p. 132). He then stated the rule for solving such problems:

> Multiply each sum by its time and divide the sum of the products by the whole debt; the quotient is accounted the mean time. (Walsh, 1804, p. 132)

Walsh then showed the solution to a model problem: "*A* owes *B* 200 dollars, whereas 40 dollars is to be paid in 3 months, 60 dollars in 5 months, and the remainder in 10 months. At what time may the whole be paid without any injustice to either? Walsh set out his solution in this way:

Dols mo,

40 × 3	= 120	
60 × 5	= 300	
100 × 10	= 1000	
200	200/ 1420,	
		7 months and 3 days

The arithmetic was not difficult to follow, but some explanation of the meaning of the expression "without any injustice to either" might have been helpful for learners. Also, the inclusion of a clear statement of why the rule worked might have given the impression that learning arithmetic should be something more than merely applying rules in order to get correct answers. Also, a comment on how the "3 days" in the given answer was obtained would have been helpful for many readers.

Stephen Pike

Not much is known about Stephen Pike, except that after his only textbook—*The Teachers Assistant or a System of Practical Arithmetic; Wherein the Several Rules of that Useful Science are Illustrated by a Variety of Examples, a Large Proportion of Which are in Federal Money*—was first published in Philadelphia in 1811, there would be 22 later editions of the book, all published between 1813 and 1850 (see, e.g., Pike, 1822). Also, 12 *Keys* for the book would appear between 1813 and 1850 (Karpinski, 1980, pp. 185–187). Genealogical research suggests that Pike was a Quaker who, when the book was first published, lived in Philadelphia. Reports from education officials indicated that the book achieved massive support from schools in Virginia and Pennsylvania—according to an annual report of country-school commissioners in Virginia in 1836, Pike's book was "used more than all others together," and five years later, in 1842, "Pike continued in widest use" (quoted in Karpinski, 1980, p. 598).

Our examination of an 1813 edition of Pike's textbook suggested to us that there was not much to distinguish it from other arithmetics of the time. The content was sequenced in an order which conformed totally with the traditional *abbaco* sequence, with the sections on vulgar fractions and decimal fractions both coming well after sections on federal currency, the various rules of three, simple interest, compound interest, discount, commission, and loss and gain. There does not seem to be any logical reason why the book was so popular, except for the fact that teachers were familiar with the traditional *abbaco* sequence and felt comfortable with Pike's standard approaches to standard topics. Quite a few of the exercises which, sometime around 1825, occupied the mind of a young Abraham Lincoln, could be found in Stephen Pike's *Arithmetic*, and Lincoln showed how exercises on simple interest, for

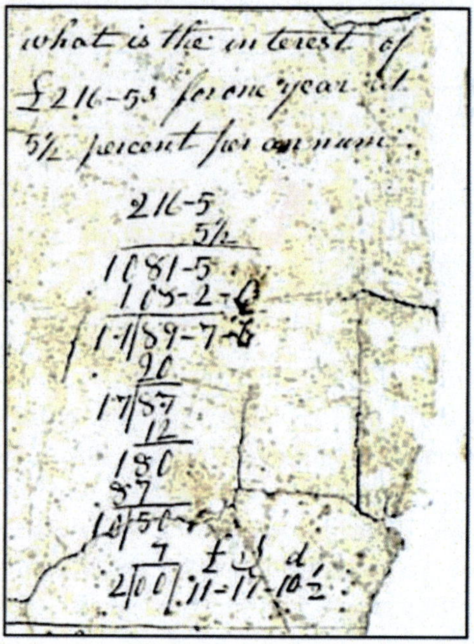

Figure 4.14. Abraham Lincoln's solution to a problem similar to one set by Stephen Pike (1811, p. 101). From image courtesy Lilly Library, Indiana University, Bloomington, Indiana.

example, could be solved using a rule-of-three method which had been developed in Pike's book (Ellerton et al., 2014). Figure 4.14 shows Lincoln's solution to the simple interest task: "What is the interest of £216-5*s* for one year at 5½ percent per annum?" The question is very similar to questions found in Pike (1811, p. 101), and the method used by Lincoln was identical to the method that Pike demonstrated in a model example.

Abraham obtained the correct answer, but had he studied decimals he might have obtained the answer more felicitously by the decimal multiplication, 216.25 × 0.055. However, he had not studied decimals (and probably did not study them while he was at school) (Ellerton et al., 2014). The section on decimals in Stephen Pike's textbook came after the section on simple interest and that probably explains why the method that Pike used in his model example did not make use of decimals.

Overview of Middle- to Advanced-Level *Abbaco* Arithmetic Textbooks, 1776–1820

The best-selling middle- to advanced-level *abbaco* arithmetic textbooks in the United States of America between 1776 and 1820 were those written by Thomas Dilworth, Nicolas Pike, Nathan Daboll, Daniel Adams, Michael Walsh ,and Stephen Pike. There are many cyphering books in the E-C cyphering book collection which include word problems obviously taken from one or more of those textbooks. That did not mean, of course, that most students who prepared cyphering books actually owned any of the books—often the problems were copied from "parent" cyphering books or were dictated to students by teachers (Ellerton & Clements, 2012).

Apart from the fact that five of the six authors just named—Thomas Dilworth was the exception—were all residents of the recently-created United States of America when they wrote their books, the arithmetics were not very different from those used in Great Britain at the time. The books featured *abbaco* topics which were mostly sequenced according to the traditional *abbaco* sequence. The only exception so far as sequence was concerned was the placement of sections on vulgar fractions and decimal fractions. American authors did not seem to know where to place them. Vulgar fractions were notoriously difficult for learners, and many authors (e.g., Daniel Adams) chose to hold back their introductions to them, in their textbooks, as long as possible. Some authors (e.g., Erastus Root, Chauncey Lee, Peter Tharp, and Nathan Daboll) went so far as to say that it was not necessary for students to learn about vulgar fractions. The authors of the most widely used arithmetics during this period did not seem to recognize the mathematical importance of vulgar fractions.

So far as decimal fractions were concerned, they were not introduced into school curricula anywhere in the world until the seventeenth century, and there was doubt in many minds where they should be placed within an *abbaco* curriculum sequence which had prevailed since the twelfth century CE, or even before that (Clements & Ellerton, 2015). Anyone who knew about the education of navigators was aware that decimals and logarithms were important (Ellerton & Clements, 2017)—but most teachers of arithmetic in schools did not know anything about navigation education.

From the perspective of the history of mathematics in North America, those who taught *abbaco* arithmetic during the period 1607–1820 did not feel pressured to change what they taught, or how they taught, to any great extent. Cyphering books were deemed to be important indicators of the quality of teaching, and of learning, and so teachers of *abbaco* arithmetic pressured their students to prepare impressive-looking manuscripts. Colleges did only rough checks on how much their prospective students knew about arithmetic, and so the teachers merely continued to do what was expected of them—get students to prepare attractive cyphering books which went as far as the rules of three. Anything further than that was rarely deemed to be necessary. Educationally, this was not a healthy situation for a nation which might want to become recognized as a leader in the applications of mathematics. In Chapter 6 of this book we will draw attention to some national leaders who worked toward improving the situation.

The Move Away from the *Abbaco* Sequence, 1820–1850

During the period 1820–1850 the *abbaco* sequence and the associated cyphering tradition lost ground in North America. In Chapter 3 a move away from the cyphering tradition after 1830 was documented and this present section offers related documentation of our thesis that the power of the *abbaco* sequence began to diminish between 1820 and 1850. In the next two chapters (Chapters 5 and 6) the growth in popularity of an emerging curriculum featuring a combination of arithmetic, algebra, geometry, trigonometry, surveying, and navigation will be described. In Chapters 7 and 8 the gradual development of a genuine, albeit embryonic, mathematics research culture in the colleges will be a focus.

Warren Colburn's Fundamental Challenge to Teachers of Mathematics

Toward the end of Chapter 2, Warren Colburn's (1822) ringing challenge to those responsible for the education of young children was summarized. Colburn maintained that *all* young children—male or female, from rich or from poor families, irrespective of race, or whether they lived in cities or in remote rural settings—should learn mathematics, and that learning should be the result of getting students to think about situations which were familiar and of interest to them. Here we consider what Colburn had to say about mathematics teaching and learning for older children. He wrote a book (Colburn, 1822), *Arithmetic upon the Inductive Method of Instruction: Being a Sequel to Intellectual Arithmetic,* which appeared shortly after his first book (Colburn 1821). In his preface to this *Sequel* he made the following highly provocative, comment:

> One general maxim to be observed with pupils of every age is never to tell them directly how to perform any example. If a pupil is unable to perform an example, it is generally because he does not fully comprehend the object of it. The object should be explained, and some questions asked, which will have a tendency to recall the principles necessary. If this does not succeed his mind is not prepared for it. After he has been told, he is satisfied and will be no better prepared for another case of the same kind than he was before. When the pupil knows that he is not to be told, he learns to depend on himself; and when he once contracts the habit of understanding what he does, he will not easily be prevailed upon to do anything which he does not understand. (p. vi)

What did Colburn think this would mean in practice?

We answer that question with an example from Colburn's *Sequel*. In the second half of the *Sequel* Colburn tackled a number of what, traditionally, had been regarded as advanced-level *abbaco* topics—one such topic was "alligation alternate," which was concerned with arithmetic related to the mixing of quantities. In all previous arithmetic textbooks which dealt with this topic (e.g., Pike, 1788, pp. 328–334), the **IRCEE** genre was evident—That was something inherited from

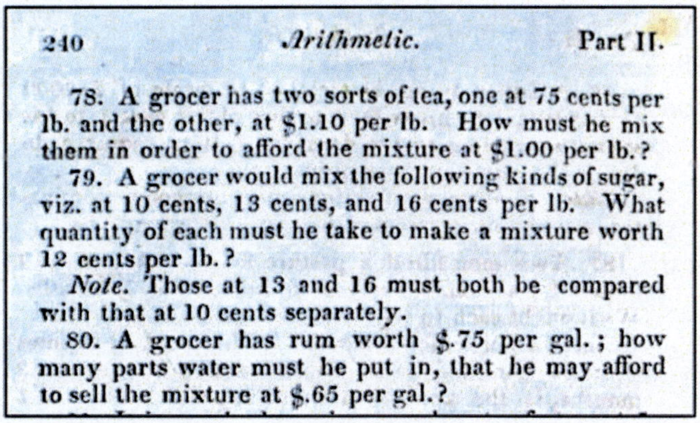
240 *Arithmetic.* Part II.

78. A grocer has two sorts of tea, one at 75 cents per lb. and the other, at $1.10 per lb. How must he mix them in order to afford the mixture at $1.00 per lb.?

79. A grocer would mix the following kinds of sugar, viz. at 10 cents, 13 cents, and 16 cents per lb. What quantity of each must he take to make a mixture worth 12 cents per lb.?

Note. Those at 13 and 16 must both be compared with that at 10 cents separately.

80. A grocer has rum worth $.75 per gal.; how many parts water must he put in, that he may afford to sell the mixture at $.65 per gal.?

Figure 4.15. Alligation alternate in Colburn's (1827) *Sequel.*

British authors (Stedall, 2012)—after an introduction, rules and cases were stated, model examples were shown, and exercises were set. Almost invariably, the model examples illustrated an ingenious approach which had been developed by algorists for solving "alligation alternate" problems, and students were expected to use that approach to obtain solutions to the exercises.

Figure 4.15 shows Colburn's (1827) approach, in his *Sequel*, to assisting students to make progress learning and applying key ideas for the topic "alligation alternate." Colburn did not give a heading "Alligation Alternate"; he made no introductory comment on what alligation alternate was all about; there were no statements of rules or cases; and there was not even a model example. Question 79 was simply posed, and students were asked to tackle it:

> *Question 79:* A grocer would mix the following kinds of sugar, viz. at 10 cents, 13 cents, and 16 cents per lb. What quantity of each must he take to make a mixture worth 12 cents a lb.? (p. 240)

The reader is invited to solve this problem, and to reflect on her or his thinking while doing so.

Immediately after stating the problem, Colburn offered a comment: "*Note*: Those at 13 or 16 must both be compared with those at 10 cents separately" (p. 240). What did that mean? Do you think all of the students who arrived at a correct answer to the problem would have got the same answer? What do you think would be the most common method used? Would it be a good idea to ask different groups of two or three students to work together on a problem like this?

Basically, Colburn was calling for a revolution in the teaching and learning of mathematics. Colburn would also write an algebra textbook—which will be considered in the next chapter. But how were his ideas with respect to middle- to advanced-level *abbaco* arithmetic received at the time? The fact is, although 53 editions of the *Sequel* were published between 1822 and 1849 (including 6 in Hawaii), the *Sequel* did not sell nearly as well as Colburn's first book (Karpinski, 1940). That is not surprising, because the *abbaco* sequence, the individualized methods associated with the cyphering tradition, and the assumptions concerning who should, and should not, study mathematics, had been developing over a period of at least 600 years, and teachers would have hesitated before departing from time-honored, "trusted and proved" approaches which incorporated **IRCEE** and **PCA** genres. But, the publication of 53 editions of the *Sequel* points to the likelihood that Colburn's *approach* did gain a good measure of support.

Charles Davies's Vision of a National Textbook Series

Charles Davies (1798–1876) (see Figure 4.16) was born in Connecticut in 1798 but received his school education in common schools in New York. He entered the United States Military Academy (USMA), at West Point, at the age of 13 and was commissioned to the Army when aged just 15. He became an assistant professor in mathematics at West Point in 1816, and became full professor in 1823 (Zitarelli, 2019). Despite the fact that his own teaching positions were always in college mathematics he was not a great mathematician. That said, he did try to be a great educator.

Davies' main contribution to the history of U.S. school mathematics was that he recognized the possibility of writing a coordinated series of textbooks, with separate books for arithmetic, algebra, geometry, trigonometry, surveying, and calculus. He wanted to define an intended curriculum for all the years of school education, from elementary school to college. Although he no longer thought of pre-college mathematics comprising solely, or mostly, the *abbaco* sequence, he still adhered to that sequence for middle- to more advanced levels of arithmetic.

From 1823 to 1837, as professor of mathematics at USMA, Davies quickly gained recognition as a significant scholar—in 1824, he was awarded an A.M. degree from the College of New Jersey, Princeton, New Jersey, and in 1825

Figure 4.16. Charles Davies (image from Wilson, Fiske, & Klos, 1889).

another A.M. degree, this time from Williams College, Williamstown, Massachusetts. Between 1826 and 1839 he authored 10 different books (Karpinski, 1940; Monroe, 1917), some of which were aimed directly at schools (e.g., Davies, 1833, 1838).

Between 1839 and 1841 Davies was professor of mathematics at Trinity College in Hartford, Connecticut, and between 1841 and 1845 he was back at West Point, as paymaster and treasurer. He left West Point in September 1848 to become professor of mathematics at the University of New York, and from 1857 to 1865 he was professor of higher mathematics at Columbia College, New York.

Edward Deering Mansfield, a student at West Point between 1815 and 1819, provided the following description of Davies as a teacher and textbook author:

> With the exception of two or three intervals of civil and military service, he was practically a teacher; and whether at West Point for many years or in civil institutions, whether in the instruction of a class, or writer of textbooks, or the author of various essays and treatises, he has made his mark on the educational system of this country probably quite as much, if not more, than any man in his generation. It was not merely the class teaching of 32 years to thousands of young men who have gone forth to instruct again the millions of their countrymen, but it was also the producing of the best textbooks on the exact sciences, which have gone into the schools, academies, and colleges of our country, directing the studies and enlightening the minds of millions of our rising youth. The books and writings of Professor Davies were not those of a brilliant genius. Neither the character of his mind nor the subjects upon which he wrote admitted that; but, with two or three exceptions, they were those simple, familiar textbooks which concentrate and crystallize the light of science. (Quoted in Callum, 1891, p. 152)

Colonel Sylvanus Thayer, the superintendent of USMA during much of Davies' tenure there, brought to West Point the influence of the French *École Polytechnique*. Under Davies and Thayer, the mathematics curriculum grew out of the shadows of Charles Hutton's textbooks through to Monge's descriptive geometry and to the calculus.

Later, however, Davies felt compelled to admit that the French influence, which initially he had embraced, had not been well received in many schools—teachers regarded the writings of Bourdon, etc., as too abstract. Although the high-achieving students at USMA tended to like the French textbooks, by 1839 all the mathematics textbooks used at the Academy had been authored by Davies.

In the following passage, Edward Mansfield discussed the transition which occurred because of Davies' insights:

> When we old cadets came to the higher branches, the application of mathematics, such as mechanical philosophy and engineering, we were

> completely at sea; no textbook of any sort existed. Professor Crozet, my professor, taught us descriptive geometry and engineering with nothing but a blackboard and a piece of chalk. It was in this state of things that Professor Davies conceived the idea of preparing textbooks. In the meanwhile, he had been promoted to be professor of mathematics, in which office he served 14 years. In that period, he not only aided in placing the Military Academy on that better footing and perfect classification it now has but began that series of textbooks he was many years in completing, which stands and will stand a great and noble monument to his name and usefulness. (Quoted in Callum, 1891, p. 153)

Davies' first textbook was on descriptive geometry, the "new" subject that Claudius Crozet, a French import to USMA, had been teaching in engineering. On this, he was heavily influenced by the writings of the important French scholar, Gaspard Monge, the "father of descriptive geometry." Part of Crozet's problem was that most of his West Point students had not been well prepared, before coming to West Point, to study such a difficult theoretical subject. But Sylvanus Thayer, USMA superintendent, believed that French mathematics was superior to British mathematics and he assembled in the USMA library numerous French books on geometry (Roberts, 2019). Davies chose to translate and improve the textbook on geometry by Legendre. He followed that up by translating Bourdon's *Algebra*. He seemed to think that the West Point students liked the books he had written, and he decided to write mathematics textbooks for schools within the United States.

In May 1837, Davies was forced to resign because of illness, possibly brought on by the effort he had expended on writing mathematical textbooks. After spending two years studying and traveling, he subsequently settled in Hartford, Connecticut, and once again devoted himself to textbook writing and publication. For the next 38 years, he was one of the most prolific and successful authors of mathematics textbooks in North America. He wrote mathematics textbooks for common schools, grammar schools, high schools, and colleges. In all, he prepared about 40 books with different titles and 350 different editions (including 177 up to and including 1850) (Karpinski, 1940). One commentator maintained that, for Davies, textbook writing became "a life of labor, of duty, of usefulness, and of success seldom equaled, scarcely ever surpassed" (Callum, 1891, p. 154). But, because of his initial, and not-necessarily-successful, flirtation with French approaches to school mathematics, and his tendency toward plagiarism, other commentators have not always evaluated Davies' textbooks in such a positive light. There can no denying, though, that his influence on U.S. school mathematics in the nineteenth century was large.

Davies was not shy of plagiarizing the writings of others. In 1834, Frederick Emerson, the author of a well-known series of school arithmetics, "entered suit against Davies, alleging the too generous use of an elementary arithmetic written by Emerson" (Karpinski, 1980, p. 11). The case was finally settled "out of court." There is little doubt that Davies used the Carlyle/Brewster translation of Legendre's

Geometry liberally, and the extent of his wrongdoing on this was emphasized by the fact that later in his life he did not even acknowledge the level to which he had used Carlyle/Brewster's translation in his own versions of Legendre's works. He also leant heavily on Edward C. Ross's English translations of Bourdon's *Elements of Algebra* (Davies, 1837; Karpinski, 1980, p. 11). In relation to his *Common School Arithmetic* of 1833, Davies admitted that "some of the examples in the rule of three, and most of those at the end of the book" were selected from Hutton's and Walkingame's arithmetics, and that he had adopted Ferdinand Hasler's term "denominate numbers." He also admitted taking exercises from John Bonnycastle's *Algebra* and from an *Arithmetic* by the English mathematician, Thomas Keith. For his *Elements of the Differential and Integral Calculus* (Davies, 1836), he admitted that he had relied on "the works of Bouchariat and Lacroix" (p. iii).

In the preface to Davies' (1840) textbook on arithmetic for young children Davies stated that the book was designed for beginners, starting with counting and advancing step by step through all the simple combinations of numbers. In order that pupils might be impressed with the idea that a number expresses a collection of units, Davies represented numbers by illustrated "stars." At the end of the oral arithmetic section Davies added a supplement involving exercises using slates. Davies said that that section was intended to cater for the needs of more gifted children.

The books that Davies prepared for young children were totally Colburn-like, with the texts set out entirely in lessons within sections. Each lesson was intended to be implemented via an oral exchange between teachers and pupils. Some of the linguistic constructions in the text were more difficult than Davies probably realized. For example, he asked "Five are how many times one?" (Davies, 1840, p. 48), and "How many fourths are there in eight? In eight and one fourth? In eight and one-half? In eight and three fourths?" (Davies, 1840, p. 79). For mathematicians such questions seem to be totally trivial, but for very young children they are linguistically complex and therefore challenging.

Although Davies himself did not have great experience teaching young children he tried to take account of teachers' needs. Thus, for example, in a preface to a "Key" that he prepared for one of his arithmetics he stated: "It was not intended as a means of aiding the teachers in working the examples, but to assist them in teaching the subject of arithmetic," for "every competent teacher is of course able to work any example in an arithmetic, but has not always time to do so in the school-room" (Davies, 1847, p. 3). Every example in the *Arithmetic* was fully worked out in this *Key*, and Davies commented: "If, therefore, a pupil makes a mistake in working an example, the teacher, by comparing his slate with the *Key*, can at once detect the error without the trouble of working the example from the beginning" (p. 3). In a later series of arithmetics, Davies (1852) gave answers to all the exercises toward the end of each book—which, at that time, was an innovation.

In his treatment of middle-level and more advanced arithmetic, Davies adhered to the time-honored *abbaco* sequence—ratio and proportion, common fractions, decimal fractions, analysis, duodecimals, applications to business, involution, evolution, arithmetical and geometrical progressions, promiscuous questions, mensuration, and gauging (see, e.g., Davies, 1844). At all times he seemed to have his eye on the market, and he recognized that many teachers would still like to teach the kind of arithmetic that they themselves had entered into their own cyphering books when they were school pupils, and that textbooks which retained the traditional order of topics would be most likely to appeal to teachers and their students, and therefore to sell well.

He liked to include, in his prefaces, statements which suggested that he had developed strong positions relating to education theory, especially as that related to the learning of mathematics. Consider, for example, the following, which appeared in the preface to his *School Arithmetic, Analytical and Practical* (Davies, 1852):

> 1. The unit 1 is regarded as the base of every number and the consideration of it as the first step in the analysis of every question relating to numbers.
> 2. Every number is treated as a collection of units, or as made up of sets of such collections, each collection having its own base, which is either 1, or some number derived from 1.
> 3. The numbers expressing the relations between the different units of a number are called the scale; and the employment of this term enables us to generalize the laws which regulate the formation of numbers.
> 4. By employing the term "fractional units," the same principles are made applicable to fractional numbers; for, all fractions are but collections of fractions units, these units having a known relation to 1. (p. iii)

Each of these four points might have *sounded* as if was important, but what it actually meant was not easy to determine. With the first point, for example, is it true that consideration of the idea of the unit 1 was the first step in the analysis of every question relating to numbers? Davies continued by stating that he had two objects in mind when he prepared the book: first, to make it educational; and second, to make it practical. Presumably, most authors of school arithmetic would have had the same two objectives.

All that said, there can be no doubt that Davies' vision of defining a national curriculum which would cover all the years of schooling contributed to an emerging trend toward acceptance of the idea that a national school mathematics curriculum was needed (Ellerton & Clements, 2012).

Frederick Emerson (1787–1865) and the Concept of Ability in Mathematics

Certainly, Warren Colburn should be credited with making clear to U.S. educators that young children acquired mathematical ideas as they grew up, and that it was the responsibility of the school, and society, to assist the development in the minds of young learners, of strong, well-structured, concepts—and that statement applied to all children, not just boys, and not just those from well-to-do European-background families. Colburn also emphasized the importance of developing good teaching methodologies—and that, in particular, teachers needed to assist groups of children to develop important numerical and other mathematical concepts by asking age-appropriate sequences of questions which prompted thinking toward generalization.

After 1823, Charles Davies regularly authored books and papers on mathematics, but it was not until 1840 that his *First Lessons in Arithmetic, Designed for Beginners* was published. Although that book was Colburn-like in its approach to teaching young children, it was even more like an arithmetic authored by Frederick Emerson, the Principal of the Department of Writing and Arithmetic at Boylston School, in Boston.

The name Frederick Emerson is less well-known than either ofWarren Colburn 1575 or Charles Davies so far as the history of North American mathematics is concerned. Emerson had had his small, 48-page *The North American Arithmetic, Part First, Containing Elementary Lessons,* published in 1829. It was a heavily illustrated little book which offered "elementary lessons" for teachers to use with children from five to eight years of age (Emerson, 1829). The plan of the book was entirely original, but like Colburn's *First Lessons* it required teachers to ask sequences of leading questions. As previously mentioned, in 1845 Emerson took Davies to court alleging plagiarism, and although the case was settled out of court, the judge's written comments made it clear that if he had had to rule then Emerson would have been the winner. Davies had argued that nobody owned elementary mathematics, but Emerson countered by maintaining that although nobody owned the mathematics in his book, *he* owned his approach to organizing it (Karpinski, 1940). The ramifications of the judge's written comments would be massive so far as the future of mathematics in North America was concerned.

But Emerson's "victory" over Davies was not the only way Emerson influenced thinking about practices related to mathematics. His *Second Part*—which first appeared in 1832—contained within itself a complete system of mental and written arithmetic sufficiently extensive for all common purposes regarding school mathematics and was designed as the final standard book for common schools. Emerson's *Third Part*—which first appeared in 1834—was designed for *advanced* scholars. This *Third Part* stressed the need to cater for the special needs of "advanced" scholars, and in time that would be interpreted as preparing texts specifically for students who had "high ability" in mathematics. Recognition of the need to prepare special mathematics programs for children with disabilities of various kinds would come later.

The last section of *The North American Arithmetic, Part Third* comprised 137 miscellaneous problems. Problem 137, that is to say, the very last problem in a textbook designed for advanced learners, could be expected to be difficult, and it was. The problem is stated in Figure 4.17.

If 12 oxen eat up 3½ acres of grass in 4 weeks, and 21 oxen eat up 10 acres in 9 weeks, how many oxen will eat up 24 acres in 18 weeks; the grass being at first equal in every acre, and growing uniformly? (From Emerson, 1834, p. 286)

Figure 4.17. The pasturage problem (from Emerson, 1834, p. 286).

Although a quick reading of this problem might suggest that its solution could be reached without much ingenuity, we have found that the problem, as it was stated, is well beyond most students in U.S. secondary schools in the twenty-first century. In fact, after the publication of *Part Third* (Emerson, 1834), the problem quickly achieved a notoriety within educated classes in the North-Eastern States of the United States of America (Clements & Ellerton, 2006).

According to Emerson (1838), in June 1835, a premium of $50 was offered, publicly, for the most "lucid analytical solution" to Question 137. Subsequently, a committee, chaired by a Mr P. Mackintosh, was appointed to examine the solutions presented, and the committee reported that of the 112 solutions submitted, only 48 had given the correct answer—despite the fact that almost all submissions had come from mathematicians or teachers of mathematics. After excluding all submissions with incorrect answers, the committee reduced the remaining number by excluding those which were "algebraical and, also, those which were performed by *position* or by *proportion*; retaining for the comparative examination, such only as were strictly analytical" (p. 110). The committee awarded the prize to a certain James Robinson, Principal of the Department of Arithmetic at Bowdoin School, in Boston. Cajori (1890) commented:

> Neither Mr. Emerson, nor the committee, nor Mr. Robinson, nor Mr. Bigger [then Registrar of the Treasury], nor the national Teachers' Association, nor the *Mathematical Monthly*, alludes to the fact that the question is taken from the *Arithmetica Universalis* of Sir Isaac Newton, published in 1704, which contains a "lucid analytical solution." Mr. Emerson's statement of the problem differs from that of Newton in this, that, owing to a misprint the fraction ½ instead of 1/3 is given by the former in the number of acres contained in the first pasture, which mistake produces the absurd result of 37 113/175, instead of 36. (p. 110)

An educational issue arising in relation to the pasturage problem was whether it is unwise to include very difficult problems in mathematics textbooks written for school students.

Cajori (1890) criticized Emerson for including the problem. But the task was the last problem in a textbook specifically written for *advanced* scholars and, from that perspective, we think Emerson's decision to include it was educationally sound. The fact that the question could be linked to no less an intellectual giant than Isaac Newton (Evans, 1876) made it all the more appropriate. The story of the pasturage problem suggests that in the 1830s in North America there was much interest in problems which were difficult but "do-able," or had elegant solutions. The challenge was to devise ways and means by which that interest could be equitably translated into school mathematics across the nation.

At the beginning of Emerson's *Third Part* there were four recommendations, the first of which was provided by Benjamin Peirce, Professor of Mathematics and Natural Philosophy at Harvard University, Peirce wrote:

> To the publishers of Emerson's *Arithmetic—Gentlemen*—I have examined the *Third Part* of Mr. Emerson's *Arithmetic*, with a great pleasure. The perspicuity of the arrangement, and the clearness and brevity of its explanations, combined with its happy adaptation to the purposes of practical business are its great recommendations. I hope it will soon be introduced into all our schools and take the place of the ill-digested treatises to which our instructors have hitherto been compelled to resort. (Statement by Benjamin Peirce, reproduced in Emerson, 1834, p. 2)

That was quite a recommendation for an author of a school mathematics textbook to have printed at the front of his book!

Joseph Ray (1807–1855), a Best-Selling Author

It is common to read that Joseph Ray's school mathematics textbooks topped the North American best-seller list in the nineteenth century so far as mathematics textbooks were concerned (Roberts, 2019)—and, perhaps, the best-seller list for mathematics textbooks written by any single author and sold in North America at any time since 1607.

Many first-edition and later-edition books carrying Ray's name were written during his 24 years at Woodward High School (later Woodward College). His best-selling books were *Primary Arithmetic, Intellectual Arithmetic, Practical Arithmetic, Higher Arithmetic, Elementary Algebra,* and *Higher Algebra.* After his death many "new" and "revised" editions were issued under his name. Later titles on geometry, trigonometry, analytic geometry, surveying, astronomy, and calculus, almost all of which were totally authored by persons other than Ray himself, were listed as belonging to "Ray's Mathematical Series." It has been claimed that even during the first decade of the twentieth century, average yearly sales of Ray's textbooks were 250,000. During Ray's lifetime, his books were officially adopted by Ohio, New York, and the New England states, and they were used almost universally in West Virginia (Ray, 1985). Total sales of his arithmetic books alone have been estimated at 120 million (U.S. Department of Education, 1985). Such was his reputation that the Smithsonian Institution created a web-based exhibit on "Teaching Mathematics in America" which showed an image of the cover of an edition of *Ray's Practical Arithmetic*.

Ray (see Figure 4.18) was born to Quaker parents in West Virginia, in November 1807. He grew up on a farm and attended local district schools during

Figure 4.18. Joseph Ray (c.1850)—A best-selling author (Kullman, 1998)

winter months each year. He excelled in reading, writing and arithmetic and at the age of 15 was sent to an academy at West Alexander, Pennsylvania. The next year he began teaching in rural schools near his home (Kullman, 1998).

In April 1825, Ray enrolled in Franklin College in New Athens, Ohio, but his money ran out and he withdrew without a degree—although he was listed in the graduating class of 1828. He then based himself in Cincinnati, Ohio, and enrolled in a medical course, taking teaching positions during vacations. Ray completed his M.D. degree from the medical college in 1831, but a lack of money led him to return to teaching. In November 1831, he took up a teaching position at the recently established Woodward High School in Cincinnati, Ohio (Kullman, 1998).

A Cincinnati businessman, William Woodward, set up a trust in 1826 for "better educating the poor children of Cincinnati," and in 1830 Woodward was granted land to establish Woodward High School. In November 1831, Ray was appointed as a teacher in the Preparatory Department. Initially, he was not involved in teaching "collegiate" students who were enrolled in the "classical" course which required the study of Latin and Greek, along with algebra, geometry, trigonometry, conic sections, logarithms, surveying and navigation, fluxions (calculus), and astronomy. Nearly 150 pupils attended the school during its first year of operation.

Woodward High School thrived, and in January 1836 the Ohio legislature permitted the Woodward College of Cincinnati to be established. A third story was added to the high-school building, and teachers in the high school were immediately appointed as professors in the College. By this time, Joseph Ray had moved to the collegiate department, in which he became professor of mathematics. Thomas J. Matthews, formerly a professor of mathematics at Transylvania College in Kentucky, had been elected as Woodward High School's first President in September 1832, and Joseph Ray and William Holmes McGuffey—another Woodward College professor, and someone who became the author of the most famous set of readers in the history of the United States—soon became active in the Western Literary Institute and College of Professional Teachers. This was the first professional organization for the advancement of education in Ohio and the West. Most meetings were held in Cincinnati, and Ray served as a director for the Ohio section during 1837 and 1838, and as recording secretary during 1839 and 1840.

Ray was appointed to various committees charged with preparing reports on matters such as the teaching of English composition, the science of arithmetic, the use of blackboards, and the utility of cabinets in natural science education. He advocated the grading of schools and called for the appointment of a state superintendent of instruction. He also wrote articles for educational journals and led many professional-development courses for teachers. He would continue working at Woodward College until his death in 1855. In 1851, Ray, then Principal of Woodward High School, delivered an address on "The Qualifications of Teachers" at the fourth annual meeting of the Ohio State Teachers Association. He emphasized the importance of knowing *what* to teach, *how* to teach it, and the *ability* to teach it well. Such was his standing among teachers that he was elected President of the Ohio State Teachers Association in 1853, and in 1854 he became associate editor of *The Ohio Journal of Education*—for which he created a "mathematical department," which discussed, and solved, interesting mathematics problems.

Charles Matthews, one of his pupils and someone who completed a book on advanced arithmetic after Ray died, wrote of Ray: "In every line of duty he was conspicuous for unremitting industry, and in all his relations of life, his first desire was to be of service to others." Ray was on the board of directors of the Cincinnati House of Refuge and was an elder in the Disciple Church. In 1849, during a cholera epidemic, he was much weakened. He never fully recovered, and in 1855, at the age of 48, he died from tuberculosis.

As early as 1830, Cincinnati had become the center of the Western book trade, and it was soon the country's fourth largest publishing center. In 1834 Truman, Smith and Co., one of Ohio's most successful publishers, published Ray's *The Little Arithmetic* (Karpinski, 1980; Ray, 1834), and that would become the foundation stone for a series of arithmetic and algebra textbooks which, according to Louis Karpinski (1980) were "surpassed in popularity by no other arithmetical series in America" (p. 366).

In 1821 Warren Colburn had dared to assert that all children, both males and females, should study school arithmetic, from the age of six, and that idea steadily gained traction. Like Colburn, Ray emphasized that the teacher should be expected to play a decisive and central role in small-group or whole-class learning, and should consciously direct the thinking of students, via carefully framed questions, toward generalizing. Although this was the inductive approach, a careful reading of Ray's (1838) *Eclectic Arithmetic* reveals that Ray's own text was often in standard **IRCEE** genre form. Ray sought to combine the best of both analysis *and* induction—just as, from 1827 onwards, Daniel Adams had tried to do with his "new arithmetics."

Like Colburn, Ray advocated the importance of linking mental arithmetic with written arithmetic and argued that it was "desirable that *all* pupils, and especially those who are young, should have gone through a course of exercises in intellectual arithmetic" (Ray, 1838, p. 8). For that purpose, he recommended that all students should read *Ray's Little Arithmetic,* which had been published in 1834.

There has been much modern-day romanticizing about what should be the place of Ray's arithmetics in the history of U.S. school mathematics. According to Hughes

(1932), the explanation of their phenomenal popularity lay in the way they contrasted with earlier texts by other authors—which, so the story goes, had been little more than statements of laws, principles, theories, and hypotheses which entirely lacked any human interest. Ray's texts, Hughes claimed, were practical, insofar as they dealt with buying and selling such articles as sugar, tea, coffee, bacon, butter, and beer. In his preface to *Ray's Algebra—Part First*, Ray (1838) stated that his aim was "to combine the clear explanatory methods of the French mathematicians with the practical exercises of the English and German, so that the pupil should acquire both a practical and theoretical knowledge of the subject" (p. iv). Our analyses of Ray's books suggest that although they were well written, they were not very different in structure and genre from books written by authors such as Benjamin Greenleaf (1850) and Roswell Smith (1850). In Ray's texts there were many rules, cases, model examples, and exercises; indeed, the **IRCEE** ("Introduction-Rules-Cases-Examples-Exercises") genre, so common in the cyphering era (Ellerton & Clements, 2014), could be found on almost every page of Rays' textbooks. A special feature of Ray's texts, however, was his willingness to explain, through worked examples, why calculations were being made. In that way, he moved beyond the cyphering era's **PCA** ("Problem-Calculation-Answer) genre.

The title page of one of Ray's most popular arithmetics is reproduced in Figure 4.19. On the cover it was claimed that this was the 1000th edition. Unless one allows the publisher to have had an idiosyncratic definition of the word "edition," that claim was almost certainly false.

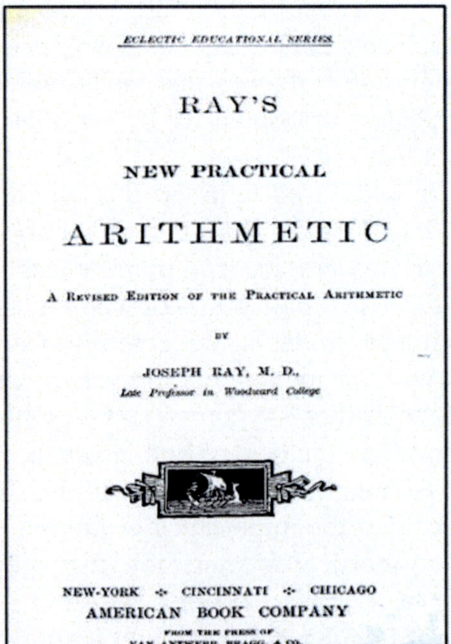

ECLECTIC EDUCATIONAL SERIES.

RAY'S

NEW PRACTICAL

ARITHMETIC

A Revised Edition of the Practical Arithmetic

BY

JOSEPH RAY, M. D.,

Late Professor in Woodward College

NEW-YORK · CINCINNATI · CHICAGO

AMERICAN BOOK COMPANY

FROM THE PRESS OF

VAN ANTWERP, BRAGG, & CO.

Figure 4.19. This shows the cover of an 1857 textbook attributed to Joseph Ray. It was claimed to be the "Thousandth Edition."

Ray's first arithmetic, *The Little Arithmetic,* was published in 1834, three years before McGuffey published his *First Reader. The Little Arithmetic* sold for six cents and emphasized drill. Motivated by the successes of his early books, Ray authored book after book, and his *Ray's Arithmetics, Part One, Part Two*, and *Part Three* became best sellers. Some of his books were re-published in 1903, when his *Practical Arithmetic*, first published in 1879, was re-titled *Ray's Higher Arithmetic*. When, in the 1980s, Mott Media decided to republish Ray's arithmetics—for the use, mainly, of home-schoolers—it made the decision that nothing should be changed. Even the original monetary prices in word problems were left unchanged!

Ray often linked school mathematics with real-life situations. A typical word problem was:

> A farmer has a flock of 30 sheep, of which 10 are worth $3 each, 12 worth $4 each, and the remainder worth $9 each. What is the average value? (Ray, 1838, p. 261)

Ray did not seem to give much thought to the difference between what he wrote in his books, and what teachers had the time to do with their students—in modern terminology, the difference between *intended* and *implemented* curricula. Thus, for example, he did not deal with common (or vulgar) fractions, or with decimals, until after he had gone through all the early *abbaco* arithmetic topics (numeration, simple and compound operations, reduction).

Analyses of data from cyphering book research (Clements & Ellerton, 2015) have shown that many students never got to study common fractions, or decimals, at all. Ray (1838) actually commented on the matter in the following way: "While federal money may be considered in connection with decimals, yet it is truly a species of compound numbers, and is so regarded in all the ordinary computations of business. Hence, the propriety of assigning it the place which it occupies in this work" (p. 6). In other words, Ray was saying, "I'll expect all students to learn about federal money, but only a small proportion of them need to learn about decimals."

The extent to which Ray (1838) adopted the **IRCEE** genre is illustrated in Figure 4.19, which shows the page following an introduction to the topic "arithmetical progressions." In the text for that Figure, Ray informed his readers:

> Any rank or series of numbers, increasing or decreasing by a common difference, is called an *arithmetical series* or progression. Thus, 1, 2, 3, 4, 5 and 11, 9, 7, 5, 3, 1 are both a series of numbers in arithmetical progression. When the series increases it is called an *ascending series*; and when it decreases it is called a *descending series*. ... The numbers forming the series are called the *terms;* the first and last terms of a series are called the extremes, the other terms are called the means. (p. 225)

Then came page 226, which is shown in Figure 4.20. The font size used in his book was very small—so small, in fact, that almost anyone would find it hard to read. There were no diagrams. Rules, cases, and model examples are shown, in the same

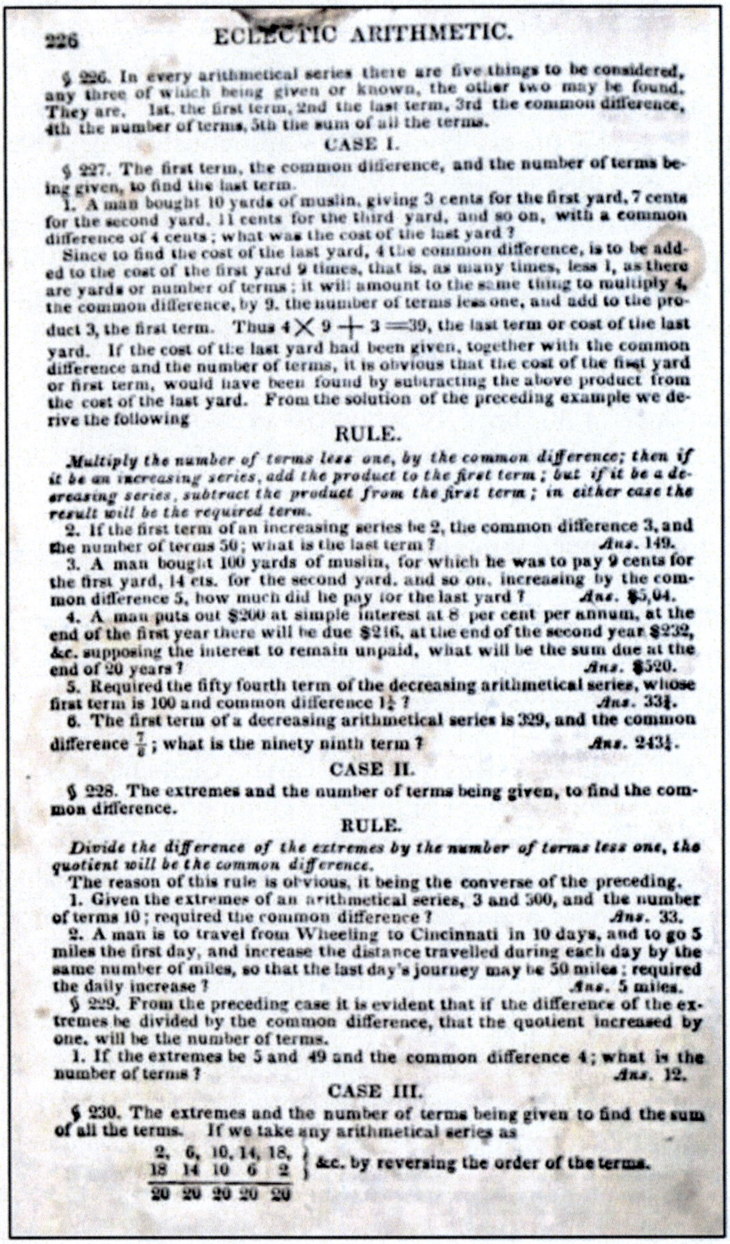

226 ECLECTIC ARITHMETIC.

§ 226. In every arithmetical series there are five things to be considered, any three of which being given or known, the other two may be found. They are, 1st, the first term, 2nd the last term, 3rd the common difference, 4th the number of terms, 5th the sum of all the terms.

CASE I.

§ 227. The first term, the common difference, and the number of terms being given, to find the last term.

1. A man bought 10 yards of muslin, giving 3 cents for the first yard, 7 cents for the second yard, 11 cents for the third yard, and so on, with a common difference of 4 cents; what was the cost of the last yard?

Since to find the cost of the last yard, 4 the common difference, is to be added to the cost of the first yard 9 times, that is, as many times, less 1, as there are yards or number of terms; it will amount to the same thing to multiply 4, the common difference, by 9, the number of terms less one, and add to the product 3, the first term. Thus $4 \times 9 + 3 = 39$, the last term or cost of the last yard. If the cost of the last yard had been given, together with the common difference and the number of terms, it is obvious that the cost of the first yard or first term, would have been found by subtracting the above product from the cost of the last yard. From the solution of the preceding example we derive the following

RULE.

Multiply the number of terms less one, by the common difference; then if it be an increasing series, add the product to the first term; but if it be a decreasing series, subtract the product from the first term; in either case the result will be the required term.

2. If the first term of an increasing series be 2, the common difference 3, and the number of terms 50; what is the last term? *Ans.* 149.

3. A man bought 100 yards of muslin, for which he was to pay 9 cents for the first yard, 14 cts. for the second yard, and so on, increasing by the common difference 5, how much did he pay for the last yard? *Ans.* $5,04.

4. A man puts out $200 at simple interest at 8 per cent per annum, at the end of the first year there will be due $216, at the end of the second year $232, &c. supposing the interest to remain unpaid, what will be the sum due at the end of 20 years? *Ans.* $520.

5. Required the fifty fourth term of the decreasing arithmetical series, whose first term is 100 and common difference $1\frac{1}{4}$? *Ans.* $33\frac{3}{4}$.

6. The first term of a decreasing arithmetical series is 329, and the common difference $\frac{7}{8}$; what is the ninety ninth term? *Ans.* $243\frac{1}{4}$.

CASE II.

§ 228. The extremes and the number of terms being given, to find the common difference.

RULE.

Divide the difference of the extremes by the number of terms less one, the quotient will be the common difference.

The reason of this rule is obvious, it being the converse of the preceding.

1. Given the extremes of an arithmetical series, 3 and 300, and the number of terms 10; required the common difference? *Ans.* 33.

2. A man is to travel from Wheeling to Cincinnati in 10 days, and to go 5 miles the first day, and increase the distance travelled during each day by the same number of miles, so that the last day's journey may be 50 miles; required the daily increase? *Ans.* 5 miles.

§ 229. From the preceding case it is evident that if the difference of the extremes be divided by the common difference, that the quotient increased by one, will be the number of terms.

1. If the extremes be 5 and 49 and the common difference 4; what is the number of terms? *Ans.* 12.

CASE III.

§ 230. The extremes and the number of terms being given to find the sum of all the terms. If we take any arithmetical series as

$$\left.\begin{array}{ccccc} 2, & 6, & 10, & 14, & 18, \\ 18 & 14 & 10 & 6 & 2 \\ \hline 20 & 20 & 20 & 20 & 20 \end{array}\right\} \text{\&c. by reversing the order of the terms.}$$

Figure 4.20. A page, on arithmetical progressions, from a textbook by Joseph Ray (1838).

manner as they had been recorded for all topics in cyphering books for centuries, according to the cyphering tradition.

The text for arithmetical progression continued to the next page, where the topic "geometrical progression, or continuing proportions" was introduced, and then dealt with in a way entirely consistent with the **IRCEE** genre. There was no diagram in the text for either arithmetical or geometrical progression. Only one page was devoted to geometrical progressions, and the treatment included cases when the "number of terms [was] infinite" (Ray, 1838, p. 228). The text stated that to find the sum of all the terms of a geometric series one could apply the rule "multiply the greatest term by the ratio, and from the product subtract the least term. Divide the remainder by the ratio less one, and the quotient will be the sum of the series" (p. 228). Readers were told that "when the series is decreasing and the number of terms infinite, the sum may be found by this rule, the least term evidently being nothing" (p. 228).

Ray's expectation that a student would understand how to obtain the sum of the terms of an infinite series after such a brief (and poorly worded) introduction, without any suggestion of a reason for the rule, was problematic—but there are many other parts of his book where a similar comment would be appropriate. The fact is, however, the sales of Ray's books went from strength to strength. Ray's (1838) book was written in the decade before Horace Mann was responsible for introducing written examinations as the main way by which mathematical learning would be evaluated, and so at that time there was no strong, and simply administered, way of evaluating the attained curriculum. In the 1830s the perceived quality of a textbook was still greatly linked to the reputation of the author. By the time written examinations were introduced into North American education systems, in the late 1840s, Joseph Ray's reputation was well established.

Benjamin Greenleaf Works Toward a National Curriculum

Benjamin Greenleaf (1786–1864) was born in Haverhill, Massachusetts, and died in Bradford, Massachusetts. His mathematics textbooks were clearly aimed at a national audience, and there can be no doubt he achieved his aim. Like Charles Davies, and Joseph Ray, Greenleaf wished to establish a wide range of textbooks, covering various aspects of mathematics. Like Joseph Ray, many of the books carrying his name were, in fact, written by other anonymous persons after his death in 1864.

Greenleaf was a seventh-generation descendant of the Greenleafs from England who settled in Newbury-Port, Massachusetts, in the 1630s. He attended a college in Atkinson, New Hampshire, and then taught in Plaistow, Atkinson, Haverhill and Marblehead. He graduated from Dartmouth in 1813 and, after teaching in grammar schools in Haverhill, became the proprietor of Bradford Academy—located about 32 miles north of Boston. He held that position between 1814 and 1836. While at the Academy he received an annual salary of $400 and one-half of the surplus for all pupils over 30 in number. In 1837, he became Bradford's representative in the State legislature, a position he held until 1839. During that period, he took the opportunity to urge for the establishment of normal schools. Between 1839 and 1848 he was director of Bradford Teachers' Seminary (Dartmouth College Library, n.d.).

Greenleaf was a successful teacher. When he became principal of Bradford Academy in 1814 it had only 10 students, but between 1814 and 1817 he succeeded in increasing its enrollment to 147. He helped found the Essex County Teachers' Association, and was regarded as a strict disciplinarian, interested not only in developing student character but also in insisting upon academic rigor. Ironically, he became well known for his ability to teach mathematics without referring to textbooks. His first book, *National Arithmetic*, was published in Boston in 1835. Its title attested to his idea of establishing a national school mathematics curriculum. His textbooks would quickly become well known, and it would be claimed, in 1864, that more than a million copies had been sold (Dartmouth College Library, n.d.).

A two-page advertisement in the August 1859 *Vermont School Journal and Family Visitor* described the books in Greenleaf's "popular series of mathematics." Professor Perry, formerly of Dartmouth College, was quoted as saying the books were "standard and imperishable works of their kind; the richest and most comprehensive as a series, that had ever appeared in the nineteenth century" (pp. 130–131). Three textbooks for "district schools" were advertised (*New Primary Arithmetic, Intellectual Arithmetic,* and *Common School Arithmetic*), as were three textbooks for "schools and academies" (*National Arithmetic, Treatise of Algebra*, and *Elements of Geometry*). The advertisement pointed out that in December 1858 the Vermont Board of Education had adopted the entire Greenleaf series for Vermont's district schools, and that that decision was "authoritative and binding upon the Board of Education superintendents and teachers until January 1, 1864" (p. 131). The advertisement also claimed that although "liberal inducements" were offered to interested parties to displace the Greenleaf texts from the schools, a trial of the other works was "generally sufficient to prove the superiority of Greenleaf's arithmetic" (p. 131).

Other boards of education adopted the Greenleaf series, including those for New Hampshire and New York City. It would be claimed—in *Appleton's Cyclopedia of American Biography* (Wilson, Fiske, & Klos, 1889)—that Greenleaf's *Common School Arithmetic* was used "in upwards of 270 cities and towns in Massachusetts, in nearly every city and town in Maine, Connecticut, and Rhode Island, and very extensively in nearly every state in the union" (p. 398). But, although he was named as the author of numerous texts on arithmetic, algebra, and trigonometry, it is likely that he did not write many of the books—others did, but Greenleaf's name appeared on the covers.

In an advertisement included toward the end of Greenleaf's *The National Arithmetic, on the Inductive System, Combining the Analytic and Synthetic Methods; Forming a Complete Course of Higher Arithmetic*, published by Robert S. Davis and Co. in Boston in 1870, six years after Greenleaf's death, there appeared the following remarkable statement:

> Greenleaf's arithmetics and algebra, are approved text-books for normal schools and commercial colleges in all parts of the country. Greenleaf's system is now used, with great acceptance, in the public schools of upwards of 1000 cities and towns in New England, and in the public schools of New York City, Philadelphia and New Orleans, and is generally

> introduced in the middle states, and has a growing popularity in all the Western states. Greenleaf's is the only mathematical series for which the demand steadily increases despite unparalleled competition. (Greenleaf, 1850, p. 453)

One wonders whether more copies of school mathematics texts authored by, or attributed to, Benjamin Greenleaf were sold than the mathematics texts of any other nineteenth-century American author, including Colburn, Davies, and Ray.

A careful examination of texts written by Greenleaf raises the same kinds of questions that were asked with respect to Joseph Ray's textbooks. Figure 4.21 shows the introduction to the topic "Reduction of Circulating Decimals" which appeared on page 153 of Greenleaf's (1850) *National Arithmetic*. There can be no denying that the topic is important, but Greenleaf's introduction failed to get students to consider whether 0.99999... (with "an infinite sequence of 9s") is less than 1, or is equal to 1, or is greater than 1. Intriguing questions such as "what is the number closest in value to 1 but less than 1?" or "what is the next decimal fraction after 1?" could have been asked.

Rather, Greenleaf missed the moment, and proceeded with the following uninspiring, but nevertheless mathematically accurate, note:

> If there be integral figures in the circulate, as many ciphers must be annexed to the numerator as the highest place of the repetend is distant from the decimal point. (p. 153)

REDUCTION OF CIRCULATING DECIMALS.

CASE I.

To reduce a simple repetend to its equivalent vulgar fraction.

If a unit with ciphers annexed to it be divided by 9 *ad infinitum*, the quotient will be one continually; that is, if $\frac{1}{9}$ be reduced to a decimal, it will produce the circulate $.\dot{1}$; and since $.\dot{1}$ is the decimal equivalent to $\frac{1}{9}$, $.\dot{2}$ will be equivalent to $\frac{2}{9}$, $.\dot{3}$ to $\frac{3}{9}$, and so on, till $.\dot{9}$ is equal to $\frac{9}{9}$ or 1. Therefore every single repetend is equal to a vulgar fraction, whose numerator is the repeating figure, and denominator 9. Again, $\frac{1}{99}$, or $\frac{1}{999}$, being reduced to decimals, makes .01010101, and .001001001 *ad infinitum*, $= .\dot{0}\dot{1}$ and $.\dot{0}0\dot{1}$; that is, $\frac{1}{99} = .\dot{0}\dot{1}$, and $\frac{1}{999} = .\dot{0}0\dot{1}$; consequently $\frac{2}{99} = .\dot{0}\dot{2}$, and $\frac{2}{999} = .\dot{0}0\dot{2}$; and, as the same will hold universally, we deduce the following

RULE. — *Make the given decimal the numerator, and let the denominator be a number consisting of as many nines as there are recurring places in the repetend.*

If there be integral figures in the circulate, as many ciphers must be annexed to the numerator as the highest place of the repetend is distant from the decimal point.

Figure 4.21. Benjamin Greenleaf (1850) on "reduction of circulating decimals" (p. 153).

The *Abbaco* Sequence Loses Popularity

Toward the end of Chapter 2 in this book, and also in this chapter, the radical ideas of Warren Colburn with respect to the mathematics education of young children were summarized. In fact, in the 1820s Colburn wanted to revolutionize thinking about all of mathematics and mathematics education in the United States of America—not just the mathematics education for young children (Kilpatrick, 2015).

Colburn wanted to change intended and implemented school mathematics curricula in the United States of America. He wanted to do away with the acceptance of the commercially-oriented *abbaco* sequence of arithmetical topics which had been translated from Western Europe to North America in the seventeenth century. He rejected the assumption that young children, especially girls, would not benefit from learning arithmetic beyond counting, and reading and writing the Hindu-Arabic numerals. He also wanted to change the implemented curriculum—he thought that the 600-year-old cyphering tradition had finally run its race, and that teachers should now strive to help students to create their own mathematical knowledge as a result of careful questioning based on arithmetic related to the children's everyday lives and interests.

One might have thought that that kind of thinking had little to no chance of gaining acceptance in the United States of America in the 1820s. After all, the opportunistic work of Nicolas Pike, Zachariah Jess, Nathan Daboll, Daniel Adams, Michael Walsh, and Stephen Pike had created a way of thinking among teachers and parents that the United States had responded well to the challenge presented by the longstanding success of Thomas Dilworth. By 1820, Dilworth's books were no longer the leader in sales in the United States of arithmetic textbooks (Karpinski, 1940). But the success of the U.S. authors raised a problem which mitigated against the likelihood of significant curriculum change being achieved quickly. Each of the authors just named held strongly to the importance of the *abbaco* sequence, and their books were used by those preparing cyphering books (Ellerton & Clements, 2012). Their publishers were unlikely to support any strong movement for change.

The 1820s witnessed a succession of authors who tried to demonstrate that they were even more in line with Pestalozzi's theories than Colburn had been (see, e.g., Beecher, 1828; Davis, 1826; Emerson, 1822; Fowle, 1826; Merchant, 1824; Ruter, 1827; Smith, 1826). In 1822 Colburn had his *Sequel* published in which his inductive teaching ideas were applied to middle- and higher-level *abbaco* topics. However, during the period 1830–1860, the influential, but more conservative, works of Frederick Ray, and Benjamin Greenleaf served to restore the balance. In the 1840s, the recently created normal schools helped accelerate a decline in the power of the cyphering tradition. It seemed to be possible that a radical transformation might be achieved so far as school mathematics was concerned, and by 1850 that possibility was quickly becoming a reality. Two reasons for that were (a) the emerging influence of the high schools, and (b) the controlling influence of high-school and college-entrance examinations. We will hear more of those factors in the next chapter.

References

Adams, D. (1801). *The scholar's arithmetic—Or Federal accountant*. Leominster, MA: Adams and Wilder.

Adams, D. (1802). *The scholar's arithmetic—Or Federal accountant* (2nd ed.). Leominster, MA: Adams & Wilder.

Adams, D. (1817). *Scholar's arithmetic or federal accountant: Containing, 1. Common arithmetic, the rules and illustrations, II. Examples and answers with blank spaces, sufficient for their operation by the scholar, III. To each rule a supplement, comprehending, 1. Questions in the nature of the rule, its use and manner of its operations, 2 Exercises. IV. Federal money, with rules for all the various operations in it to reduce Federal to the Old Lawful, and the Old Lawful to Federal money. V. Interest cast in Federal money, with compound multiplication, compound division and practice, wrought in Old Lawful and Federal money; the same questions being put in separate columns on the same page in each kind of money, these two modes of account being contrasted, and the great advantage gained by reckoning in federal money easily discerned. VI. Demonstrations by engravers of the reason and nature of the various steps in the extraction of the square and cube roots, not to be found in any other treatise on arithmetic. VII. Forms of notes, deeds, bonds and other instruments of writing. The whole in a form and method altogether new, for the ease of the master and the greater progress of the scholar* (9th ed.). Keene, NH: John Prentiss

Adams, D. (1848). *Adams's new arithmetic: Arithmetic in which the principles of operating by numbers are analytically explained and synthetically explained, thus combining the advantages to be derived both from the inductive and synthetic mode of instructing*. New York, NY: Collins & Brother.

Barrème, N. (1744). *L'arithmétique du Sr Barrème ou le livre facile*. Paris, France: Gandouin.

[Beecher, C. E.]. (1828). *Arithmetic, explained and illustrated, for the use of the Hartford Female Seminary*. Harford, CT: P. Canfield.

Boyd, J. P. (1961). Report on weights and measures: Editorial note. In J. Boyd (Ed.), *The papers of Thomas Jefferson 16, November 1789 to July 1790* (pp. 602–617). Princeton, NJ: Princeton University Press. https://doi.org/10.1515/9780691185224-014

Briggs, H. (1617). *Logarithmorum chilias prima*. London, England: Author.

Cajori, F. (1890). *The teaching and history of mathematics in the United States* (Circular of Information No. 3, 1890). Washington, DC: Bureau of Education.

Callum, G. W. (1891). *Biographical register of the officers and graduates of the U.S. Military Academy at West Point, NY: From its establishment, in 1802, to 1890, with the early history of the United States Military Academy*. Boston, MA: Houghton Mifflin.

Clements, M. A., & Ellerton, N. F. (2006). Historical perspectives on mathematical elegance—To what extent is mathematical beauty in the eye of the beholder? In P. Grootenboer, R. Zevenbergen & M. Chinnappan (Eds.), *Identities, cultures and learning spaces* (pp. 147–154). Adelaide, Australia: Mathematics Education Research Group of Australasia.

Clements, M. A., & Ellerton, N. F. (2015). *Thomas Jefferson and his decimals 1775–1810: Neglected years in the history of U.S. school mathematics*. New York, NY: Springer. https://doi.org/10.1007/978-3-319-02505-6

Cocker, E. (1677). *Cocker's arithmetick: Being a plain and familiar method suitable to the meanest capacity for the full understanding of that incomparable art, as it is now taught by the ablest school-masters in city and country*. London, England: John Hawkins.

Cocker, E. (1678). *Cocker's arithmetic: Being a plain and familiar method suitable to the meanest capacity* ... London, England: H. Tracey.

Cocker, E. (1685). *Cocker's decimal arithmetick,* ... London, England: J. Richardson.

Cocker, E. (1697). *Cocker's arithmetic: Being a plain and familiar method suitable to the meanest capacity* ... London, England: Eben Tracey.

Cocker, E. (1719). *Cocker's arithmetic: Being a plain and familiar method suitable to the meanest capacity* ... London, England: H. Tracey.

Cocker, E. (1720). *Decimal arithmetic, wherein is shewed the nature and use of decimal fractions in the usual rules of arithmetic, ...* (5th ed.). London, England: J. Darby for M. Wellington.

Colburn, W (1821). *An arithmetic on the plan of Pestalozzi, with some improvements*. Boston, MA: Cummings and Hilliard.

Colburn, W. (1822). *Arithmetic upon the inductive method of instruction: Being a sequel to Intellectual Arithmetic*. Boston, MA: Cummings & Hilliard.

Colburn, W. (1827). *Arithmetic upon the inductive method of instruction: Being a sequel to Intellectual Arithmetic* (3rd ed.). Boston, MA: Hilliard , Ray, Little and Wilkins.

Colburn, W. (1830/1970). Teaching of arithmetic. In J. K. Bidwell & R. G. Clason (Eds.), *Readings in the history of mathematics education* (pp. 24–37). Washington, DC: National Council of Teachers of Mathematics.

Daboll, N. (1804). *Daboll's schoolmaster's assistant: Being a plain, practical system of arithmetic; adapted to the United States* (3rd ed.). New London, CT: Samuel Green.

Daboll, N. (1813). *Daboll's schoolmaster's assistant: Improved and enlarged being a plain practical system of arithmetic; adapted to the United States* (7th ed.). New London, CT: Samuel Green

Daboll, N. (1818). *Daboll's schoolmaster's assistant: Improved and enlarged being a plain practical system of arithmetic; adapted to the United States* (10th ed.). New London, CT: Samuel Green

Daboll, N. (1820). *Daboll's practical navigator: Being a concise, easy, and comprehensive system of navigation; calculated for the daily use of seamen, and also for an assistant to the teacher: containing plane, traverse, parallel, middle latitude, and Mercator's sailing; with all the necessary tables: Concise rules are given, with a variety of examples in every part of common navigation; also, a new, scientific, and very short method of correcting the dead reckoning; with rules for keeping a complete reckoning at sea, applied to practice, and exemplified in three separate journals, in which may be seen all the varieties which can probably happen in a ship's reckoning*. New London, CT: Samuel Green.

Dartmouth College Library (n.d.). *Guide to the papers of Benjamin Greenleaf, 1807–1865*. Manuscript MS-1108). Hanover, NH: Author.

Dauben, J. W., & K. H. Parshall (2014). Mathematics education in North America to 1800. In A. Karp & G. Schubring (Eds), *Handbook on the history of mathematics education* (pp. 175–185). New York, NY: Springer.

Davies, C. (1833). *The common school arithmetic, prepared for the use of academies and common schools in the United States, and also for the use of the young gentleman who may be preparing to enter the Military Academy at West Point*. New York, NY: N. & J. White.

Davies, C. (1836). *Elements of the differential and integral calculus*. New York, NY: Wiley and Long.

Davies, C. (1837). *Elements of algebra: Translated from the French of M. Bourdon*. New York, NY: Wiley & Long

Davies, C. (1838). *Mental and practical arithmetic designed for the use of academies and schools*. Hartford, CT: A. S. Barnes & Co.

Davies, C. (1840). *First lessons in arithmetic*. Hartford, CT: A. S. Barnes & Co.

Davies, C. (1844). *Arithmetic, designed for academies and schools, uniting the reasoning of the French with the practical methods of the English with full illustrations of the method of cancellations*. New York, NY: A. S. Barnes and N. L. Burr.

Davies, C. (1847). *Key to Davies' arithmetic*. New York, NY: A. S. Barnes and Burr.

Davies, C. (1852). *School arithmetic, analytical and practical*. New York, NY: A. S. Barnes & Co.

Davis, S. (1826). *The pupil's arithmetick*. Boston, MA: Lincoln and Edmands.

Dilworth, T. (1773a). *The schoolmasters assistant: Being a compendium of arithmetic, both practical and theoretical* (17th ed.). Philadelphia, PA: John Dunlap.

Dilworth, T. (1773b). *The schoolmasters assistant: Being a compendium of arithmetic, both practical and theoretical* (17th ed.). Philadelphia, PA: Joseph Crukshank.

Dilworth, T. (1797). *The schoolmasters assistant: Being a compendium of arithmetic, both practical and theoretical: The latest edition*. New London, CT: S. Green.

Dilworth, T. (1806). *The schoolmasters assistant: Being a compendium of arithme-2109 tic, both practical and theoretical.* New York, NY: McFarllane & Long, for Evert Duyckinck.

Ellerton, N. F., Aguirre Holguín, V., & Clements, M. A (2014). He would be good: Abraham Lincoln's early mathematics, 1819–1826. In N. F. Ellerton & M. A. Clements, *Abraham Lincoln's cyphering book and ten other extraordinary cyphering books* (pp. 123–186). New York, NY: Springer. https://doi.org/10.1007/978-3-319-02502-5_6

Ellerton, N. F., & Clements, M. A. (2012). *Rewriting the history of school mathematics in North America 1607–1861. New* York, NY: Springer.

Ellerton, N. F., & Clements, M. A. (2014). *Abraham Lincoln's cyphering book and ten other extraordinary cyphering books.* New York, NY: Springer. https://doi.org/10.1007/978-3-319-02502-5

Ellerton, N. F., & Clements, M. A. (2017). *Samuel Pepys, Isaac Newton, James Hodgson and the beginnings of secondary school Mathematics: A history of the Royal Mathematical School at Christ's Hospital 1673–1868.* New York, NY: Springer. https://doi.org/10.1007/978-3-319-46657-6

Emerson, F. (1822). *Female education ... To which is added, the little reckoner, consisting principally of arithmetical questions for infant minds.* Boston, MA: Samuel T. Armstrong and Crocker and Brewster.

Emerson, F. (1829). *The North American arithmetic: Part First, containing elementary lessons.* Boston, MA: Lincoln & Edmands.

Emerson, F. (1832). *The North American arithmetic: Part Second, uniting oral and written exercises in corresponding chapters.* Boston, MA: Lincoln & Edmands.

Emerson, F. (1834). *The North American arithmetic: Part Third, for advanced scholars.* Boston, MA: Russell, Odiorne, & Metcalf.

Emerson, F. (1838). *Key to the North American arithmetic Part Second and Part Third for the use of teachers.* Boston, MA: G. W. Palmer & Co.

Evans, A. (1876). The problem of the pasturage. *The Analyst, 3*(3), 75–78. https://doi.org/10.2307/2635526

Fanning, D. F., & Newman, E. P. (2011, July 24). More on the *American Accomptant* and the first printed dollar sign. *The E-Sylum, 14,* Numosmatic Bibliomania Society.

Fowle, W. B. (1826). *The child's arithmetick, or the elements of calculation, in the spirit of Pestalozzi's method, for the use of children between the ages of three and seven years.* Boston, MA: Thomas Wells.

Franklin, B. (1964). *The autobiography of Benjamin Franklin* (2nd ed.). New Haven, CT: Yale University.

George Washington to Nicolas Pike, June 20, 1788 (in Electronic Text Center, University of Virginia Library). Retrieved December 7, 2006, from https://etext.virginia.edu/etcbin/toccernew2?id=WasFi30.xml&ima

Glaisher, J. W. L. (1873). On the introduction of the decimal point into arithmetic. *Report of the Meeting of the British Association for the Advancement of Science, 43*, 13–17.

Gough, J. (1788). *A treatise of arithmetic in theory and practice containing everything important in the study of abstract and applicant numbers, adapted to the commerce of Great Britain and Ireland*. Philadelphia, PA: B. Workman.

Greenleaf, B. (1850). *The national arithmetic, on the inductive system, combining the analytic and synthetic methods. In which the principles of arithmetic are explained*. Boston, MA: R. S. Davis.

Greenleaf, B. (1870). Greenleaf's new intellectual arithmetic. Boston, MA: Robert S, Davis & Co.

Greenwood, I. (1729). *Arithmetick, vulgar and decimal, with the application thereof to a variety of cases in trade and commerce*. Boston, MA: Kneeland & Green.

Hodder, J. (1714). *Arithmetick, or that necessary art made most easie* (520th ed.). London, England: N. & M. Baddington.

Howson, G. (1982). *A history of mathematics education in England*. Cambridge, England: Cambridge University Press. https://doi.org/https://doi.org/10.1057/9780230305090

Hughes, R. G. (1932, February). *Joseph Ray*, the mathematician, and the man. *West Virginia Review*.

Jones, P. S., & Coxford, A. F. (1970). From discovery to an awakened concern for pedagogy: 1492–1821. In National Council of Teachers (Ed.), *Thirty-second handbook* (pp. 11–23). Washington, DC: National Council of Teachers of Mathematics.

Karpinski, L. C. (1940). *Bibliography of mathematical works printed in America through 1850*. Ann Arbor, MI. The University of Michigan Press.

Karpinski, L. C. (1980). *Bibliography of mathematical works printed in America through 1850*. (2nd ed.). New York, NY: Arno Press.

Kersey, J. (1689). An appendix containing choice knowledge in arithmetick, both practical and theoretical. In E. Wingate, Mr. Wingate's arithmetick containing a plain and familiar method for attaining the knowledge and practice of common arithmetick (9th ed., pp. 303–544). London, England: James Wingate.

Kilpatrick, J. (2013). Warren Colburn and the inductions of reason. In K. Bjarnadóttir, F. Furinghetti, J. Prytz, & G. Schubring (Eds.), *"Dig where you stand" 3. Proceedings of the Third International Conference on the History of Mathematics Education* (pp. 219–232). Uppsala, Sweden: Uppsala University.

Kullman, D. E. (1998). Joseph Ray: The McGuffey of mathematics. *Ohio Journal of School Mathematics, 38*, 5–10.

Lacroix, S. F. (1818). *An elementary treatise of arithmetic, and an introduction to the elements of algebra* ... Cambridge, MA: Hilliard and Metcalf.

Lee, C. (1797). *The American accountant; being a plain, practical and systematic compendium of federal arithmetic ...* Lansingburgh, NY: William W. Wands.

Lloyd, H. A. (2012). Commonwealth and empire: Robert Recorde in Tudor England. In G. Roberts & F. Smith (Eds.), *The life and times of a Tudor mathematician* (pp. 145–164). Cardiff, Wales. University of Wales Press

Merchant, A. M. (1824). *The first lines of arithmetic: Made easy and adapted to the capacity of junior learners.* New York, NY: John C. Totten.

Monroe, W. S. (1917). *Development of arithmetic as a school subject.* Washington, DC: Government Printing Office.

Napier, J. (1614). *Mirifici logarithmorum canonis descriptio.* Edinburgh, Scotland: Andrew Hart.

Napier, J. (1619). *The wonderful canon of logarithms.* Edinburgh, Scotland: William Home Lizars, 1857 [English translation by Herschell Filipowski].

Pike, N. (1788). *The new and complete system of arithmetic, composed for the use of the citizens of the United States.* Newbury-Port, MA: John Mycall.

Pike, N. (1793). *Abridgement of the new and complete system of arithmetick composed for the use, and adapted to the commerce of the citizens of the United States.* Newbury-Port, MA: John Mycall, Isaiah Thomas.

Pike, S. (1811). *The teacher's assistant; or a system of practical arithmetic; wherein the several rules of that useful science, are illustrated by a variety of examples, a large proportion of which are in federal money. The whole is designed to abridge the labour of teachers, and to facilitate the instruction of youth.* Philadelphia, PA: Johnstonn (sic.) and Warner.

Pike, S. (1822). *The teacher's assistant or a system of practical arithmetic; wherein the several rules of that useful science, are illustrated by a variety of examples, a large proportion of which are in federal money. The whole is designed to abridge the labour of teachers, and to facilitate the instruction of youth.* Philadelphia, PA: Benjamin Warner.

Prévost, G. (1677). *Briefve méthode et instruction pour apprendre l'arithmétique.* Tournay, France: Jacques Coulon.

Ray, J. (1834). *The little arithmetic.* Cincinnati, OH: Truman, Smith and Co.

Ray, J. (1838). *Ray's eclectic arithmetic on the inductive and analytic methods of instruction* (4th ed.). Cincinnati, OH: Truman and Smith.

Ray, J. (1985). *Ray's new practical arithmetic.* Milford, MI: Mott Media, Inc.

Record, R. (1658). *Grounde of the arts:* Teaching the worke and practise, of arithmeticke. London, England: R. Wolff.

Roberts, D. L. (2019). *Republic of numbers: Unexpected stories of mathematical Americans through history.* Baltimore, MD: Johns Hopkins University Press.

Roberts, G., & Smith, F. (2012). *Robert Recorde: The life and times of a Tudor mathematician.* Cardiff, Wales: University of Wales Press.

Root, E. (1795). *An introduction to arithmetic for the use of common schools.* Norwich, CT: Thomas Hubbard.

Ruter, M. (1827). *The juvenile arithmetick, and scholar's guide*. Cincinnati, OH: N and G. Guilford.

Sanford, V. (1957). Robert Recorde's "Whetstone of witte," 1557. *The Mathematics Teacher*, 50(4), 258–266. https://doi.org/https://doi.org/10.5951/MT.50.4.0258

Sarjeant, T. (1788). *Elementary principles of arithmetic with their application to the trade and commerce of the United States of America*. Philadelphia, PA: Dobson and Lang.

Seltman, M., & Goulding, R. (Eds.). (2007). *Thomas Harriot's Artis Analyticae Praxis*. London, England: Springer.

Simons, L. G. (1924). *Introduction algebra into American schools in the 18th century*. Washington, DC: Department of the Interior Bureau of Education.

Smith, R. C. (1826). *Practical and mental arithmetic on a new plan*. Boston, MA: S. G. Goodrich and Richardson and Lord.

Smith, R. C. (1850). *Practical and mental arithmetic on a new plan*. New York, NY: Cady & Burgess.

Sterry, C., & Sterry, J. (1790). *The American youth: Being a new and complete course of introductory mathematics, designed for the use of private students*. Providence, RI: Authors.

Sterry, C., & Sterry, J. (1795). *A complete exercise book in arithmetic, designed for the use of schools in the United States*. Norwich, CT. John Sterry & Co.

Stevin, S. (1585). *De Thiende*. Leyden, The Netherlands: The University of Leyden.

Tharp, P. (1798). *A new and complete system of federal arithmetic*. Newburgh, NY: D. Denniston.

Todd, J., Jess, Z., Waring, W., & Paul, J. (1800). *The American tutor's assistant, or a compendium system of practical arithmetic. ...* Philadelphia, PA: Zachariah Poulson.

U.S. Department of Education. (1985). *Early American textbooks*. Washington, DC: Author.

Venema, P. (1730). *Arithmetica of Cyffer-Konst, volgens de Munten Maten en Gewigten te Nieu-York, gebruykelyk als mede een kort Ontwerp van de Algebra*. New York, NY: Jacob Goelet.

Viète, F. (1579). *Canon mathematicus seu ad triangula cum appendicibus*. Paris, France: Jean Mettayer.

Wallis, R. (1997). Edward Cocker (1632?–1676) and his arithmetick: De Morgan demolished, *Annals of Science, 54*, 507–522. https://doi.org/10.1080/00033799700200471

Wallis, R. (2004). Kersey, John, the Elder. *Oxford dictionary of national biography*. Oxford, England: Oxford University (online).

Walsh, M. (1801). *A new system of mercantile arithmetic: Adapted to the commerce of the United States, in its domestic and foreign relations; with forms of accounts, and other writings usually occurring trade*. Newbury-Port, MA: Edmund M. Blunt.

Walsh, M. (1804). *A new system of mercantile arithmetic: Adapted to the commerce of the United States, in its domestic and foreign relations; with forms of accounts, and other writings usually occurring trade* (4th ed.). Newbury-Port,-MA: Edmund M. Blunt.

Williams, J. J. W. (2011). *Robert Recorde: Tudor polymath, expositor, and practitioner of computation*. London, England: Springer-Verlag. https://doi.org/10.1007/978-0-85729-862-1

Wilson, J., Fiske, J., & Klos, S. L. (1889), Benjamin Greenleaf. In J. Wilson & J. Fiske (Eds.), *Appleton's Cyclopedia of American Biography* (p. 399). New York, NY: D. Appleton & Co.

Wingate, E. (1624). *L'usage de la règle de proportion en arithmétique*. Paris, France: Author.

Wingate, E. (1630). *Arithmétique made easie*. London, England: Stephens and Meredith.

Wingate, E. (1689). *Mr. Wingate's arithmetick containing a plain and familiar method for attaining the knowledge and practice of common arithmetick* (9th ed.). London, England: Author.

Workman, B. (1789). *The American accountant or schoolmaster's new assistant*. Philadelphia, PA: John M'Culloch.

Workman, B. (1793). *The American accountant or schoolmaster's new assistant ... Revised and corrected by Robert Patterson*. Philadelphia, PA: W. Young.

Yeldham, F. A. (1926). *The teaching of arithmetic through four hundred years (1535–1935)*. London, England: George A. Harrap.

Zitarelli, D. A. (2019). *A history of mathematics in the United States and Canada. Volume 1: 1492–1900*. Washington, DC: Mathematical Association of America. https://doi.org/10.1090/spec/094

Chapter 5
The Struggle for Algebra

Abstract This chapter focuses on the emergence of algebra in the intended and implemented curricula of U.S. schools between 1607 and 1865. Before 1776 only a tiny proportion of school-age children, in what is now the mainland part of the United States, studied algebra. The chapter begins by providing evidence that until about 1820 the study of mathematics other than abbaco arithmetic was not something seriously engaged in by most young people in North America. Very few textbooks on any branch of mathematics other than arithmetic were suitable for school children, and relatively few cyphering books which focused on mathematics other than arithmetic were prepared. That changed in the early 1820s, after the first public high schools were opened, and after colleges began to require prospective students to demonstrate a knowledge of algebra. Nevertheless, even in the 1850s less than 10% of school-age North American children studied any of algebra, geometry, trigonometry, surveying, navigation, or calculus. The cyphering tradition was strongly linked to both abbaco arithmetic and algebra, but algebra was much less studied. In 1730 a Dutch-language textbook, by Pieter Venema, on arithmetic and algebra, was published in New York, but at that time there was little demand for it and a second edition never appeared. Documentary evidence—never before available to historians—from a "precursor" document prepared by Venema in New York in 1725, is discussed and analyzed. In that document Venema demonstrated how might could be used to prove and to generalize. Venema was ahead of his time and offered North American mathematics education an opportunity which it failed to grasp. Venema was ahead of his time and offered North American mathematics education an opportunity which it failed to grasp.

Keywords Assessment of mathematical learning • Cyphering books • Graphs • History of algebra education • History of U.S. high schools • Horace Mann • John Bonnycastle • John Farrar • Nicolas Pike • Normal schools • Pieter Venema • U.S. college entrance requirements • U.S. Military Academy (West Point) • Warren Colburn • Written examinations

The Need for Algebra

In this chapter evidence will be presented showing that very little algebra was studied in North American schools before 1820. This was the case despite the fact that important advances in algebra featuring scholars like François Viète, Simon Stevin, René Descartes, John Napier, Henry Briggs, Blaise Pascal, Isaac Newton, Gottfried Leibniz, Leonhard Euler, and the Bernoulli brothers, occurred in Europe

N. F. Ellerton, M. A. (Ken) Clements, *Toward Mathematics for All*, History of Mathematics Education,
https://doi.org/10.1007/978-3-030-85724-0_5

during the period 1500–1750 (Thomas & Kempis Kloyda, 1937). Comparable advances in algebra did not occur in North America during the same period.

In 1820 Harvard College made knowledge of elementary algebra a pre-requisite for entry, and at various times during the next 30 years Columbia, Yale and Princeton followed suit (Kilpatrick & Izsák, 2008). Before 1820, however, there were hardly any teachers in schools who knew elementary algebra well enough to teach it, and it is even doubtful whether many *college* teachers of mathematics knew a sufficient amount of algebra to consider teaching beyond what was written in the early chapters of elementary algebra textbooks published in Europe.

There *were* a few who did choose to study algebra. The Houghton Library, at Harvard University, holds a magnificent manuscript—filed as fMs Typ245 and prepared in 1797 by Phoebe Folger (1771–1857). Phoebe paid much attention to forms of algebra, dealing with simple equations, quadratic equations, extraction of square roots, cube roots, and "promiscuous problems" in arithmetic and algebra. The manuscript features beautiful penmanship and calligraphic headings, and superb water-color paintings. Phoebe was a distant cousin of Benjamin Franklin and her brother, Walter, was a mathematician and famous clockmaker. It seems that Phoebe, unlike most students, had the opportunity, desire, and talent, to pay special attention to algebra.

Another person who studied algebra at a young age was Nathaniel Bowditch (Bowditch, 1840), and he would become possibly the greatest of early U.S. mathematicians (see Chapter 8).

We begin by looking at the main body of evidence relating to the extent of algebra being studied in colonial schools between 1607 and 1865. None of the several recent histories of algebra education (e.g., da Ponte & Guimarães, 2014; Kanbir, Clements & Ellerton, 2017; Kilpatrick, 2014a, 2014b), have focused on algebra in North America (excluding Canada) during the period 1607–1865. The most important evidence is found in reports on early American textbooks on algebra and in other early documents on algebra education examined by Lao Genevra Simons (1924, 1931, 1936).

Mathematics Beyond the *Abbaco* Sequence: Evidence from Cyphering Books

There is evidence indicating the extent of interest in education related to any or all of algebra, geometry, trigonometry, mensuration, surveying, navigation, and calculus in North America during the period 1607–1865. For all but the last 20 years of that period there were no public examination papers, or associated statistics, relating to school curricula because public examinations of the written kind were not held. As a result, it is very difficult, and probably impossible, to provide details of subjects which were being studied by students in most of the academies and other pre-college schools in North America at that time. The evidence that we offer in this chapter comes mostly from our analyses of cyphering books held within the Ellerton-Clements (E-C) cyphering book collection, which is now held by the Library of Congress, in Washington, DC.

As mentioned in Chapter 3, at the time of writing there are 549 manuscripts in the E-C cyphering book collection, and these were prepared in many colonies and states—but especially in Massachusetts, New York, and Pennsylvania. Of these, 536 cyphering books were prepared between 1607 and 1865, with the remaining 13 "transition" cyphering books being prepared between 1866 and 1907. The mathematics covered in most of the manuscripts was solely concerned with *abbaco* arithmetic, which had a strong commercial orientation. However, in a few of the cyphering books attention was given to algebra or geometry or trigonometry or surveying or navigation but none to calculus. Occasionally, after topics in the *abbaco* arithmetic sequence had been dealt with in a cyphering book there were a few pages on which other branches of mathematics were considered. The best measure we can offer on the issue of "what parts of mathematics did students study the most?" would be to see how many of the cyphering books in the E-C collection dealt with the various branches.

Table 5.1 reports data with respect to the 536 cyphering books (CBs) in the E-C collection prepared during the period 1607–1865—and, in particular, CBs to those which dealt with at least one of arithmetic, algebra, geometry, trigonometry, surveying or navigation.

The sharp decline during the period 1840–1865, in the number of cyphering books in the E-C collection corresponds to what we have called the "demise of the cyphering tradition" (Ellerton & Clements, 2012). Although that tradition—which emphasized individual learning—had controlled school mathematics in Europe for about six centuries, it was quickly replaced in North America between 1840 and 1855, by a small-group teaching approach in which cyphering books were much less prepared than previously see Ellerton & Clements (2012) for discussion of reasons for the change.

Note that with respect to Table 5.1:

- Between 1607 and 1829, only 6 of 339 CBs (i.e., 2%) included entries on algebra, 28 included entries on geometry (8%), 21 included entries on trigonometry (6%), 20 included entries on surveying (i.e., 6%) and 8 included entries on navigation (2%). Of the 339 CBs, 314 (94%) included entries dedicated to arithmetic only.
- Between 1830 and 1865, 19 CBs (i.e., 10%) included entries on algebra; 14 included entries on geometry (7%); 12 included entries on trigonometry (6%); 11 included entries on surveying (6%) ; and 3 included entries on navigation (2%).
- Some of the 536 CBs included entries dedicated to more than one of arithmetic, algebra, geometry, surveying and navigation. For example, most CBs which had entries on surveying also had entries on geometry and trigonometry.

- Over the whole period (1607–1865), arithmetic was easily the dominant form of mathematics studied in the schools by children who were at least 10 years of age.

Table 5.1
Number of Cyphering Books (CBs) in the E-C Cyphering Book Collection Dealing with Different Components of Mathematics During the Period 1607–1865 (n = 536 CBs)

Period	Total CBs In Period	Total CBs including Arithmetic	Total CBs including Algebra	Total CBs including Geometry	Total CBs including Trigonometry	Total CBs including Surveying	Total CBs including Navigation
1607–1799	67	62	1	4	4	5	2
1800–1809	57	55	1	0	2	2	3
1810–1819	94	87	3	9	7	7	1
1820–1829	121	113	1	9	8	6	2
1830–1839	94	90	5	6	4	4	1
1840–1849	50	43	9	6	5	3	2
1850–1859	42	37	5	2	3	3	0
1860–1865	11	11	0	0	0	1	0
Totals	536	498	25	36	33	31	11

During the 258-year period from 1607 to 1865 over 90 percent of persons who prepared cyphering books (and, therefore, were likely to have been at least 10 years old) did not make entries on any aspect of mathematics other than *abbaco* arithmetic. About 5% attended to algebra, 7% to geometry, 6% to trigonometry, 6% to surveying, and 2% to navigation. Those percentages sum to more than 100 because some students made entries on two or more of *abbaco* arithmetic, algebra, geometry, trigonometry, surveying and navigation in the same cyphering book—notice, however, that sections on algebra (or geometry, etc.) occupied far fewer entries than sections on *abbaco* arithmetic. Unlike data gained from analyzing the contents of textbooks, cyphering book data attest to implemented rather than intended curricula.

As indicated in Table 5.1, of the 339 CBs in the Ellerton-Clements collection of cyphering books prepared between 1607 and 1829, only 2% included any entries on algebra. Such a finding raises doubts about the validity of Simons' (1924) claim that "algebra was an important part of . . . American education of the eighteenth century" (p. 74). During the period between 1607 and 1799 there were only five textbook authors in that part of North America which excludes Canada who included sections on algebra in their published works—they were Pieter Venema, Nicolas Pike, Consider Sterry and John Sterry, and John Gough—and in fact the textbooks of all of those authors were mainly concerned with arithmetic. Hugh Jones, Professor of

Mathematics at William and Mary College between 1717 and 1722, apparently authored a textbook *Accidence to the Mathematick and all its Parts and Applications, Algebra, Geometry, Surveying of Land, and Navigation* (Cajori, 1890, p. 33), but no extant copy remains, and the book was not listed by Louis Karpinski (1940).

That said, during the eighteenth century, algebra was studied in some North American grammar schools or colleges, and by some apprentices and private students in evening classes—see Seybolt (1921) and Simons (1924) for summaries of who studied algebra, and where, during the eighteenth century. Nevertheless, it was still true that at the beginning of the nineteenth century very little algebra was being studied in North American pre-college education institutions. That was not surprising given that no college required algebra as a pre-requisite for admission (Kilpatrick & Iszák, 2008; Simons, 1924). Yale College required its first-year students to take an algebra course (Simons, 1924) and students in most other colleges were required to take an elementary algebra course at some stage of their degree programs. But there was no real pressure on those administering pre-college education institutions to include algebra in their curricula. In summary, we can say that analysis of cyphering book data reveals that much *abbaco* arithmetic, but not much algebra, was studied in North American schools during the eighteenth century.

Given the data reported in Table 5.1 it is easy to see why textbook publishers in North America were reluctant to publish algebra or geometry textbooks, and authors were reluctant to prepare textbooks on algebra or geometry. The fact that so few algebra or geometry textbooks were published helps to explain why so few students prepared cyphering books which included entries on algebra or geometry. Notice that only 38 of the 536 CBs (i.e., 7%) did not include any section on arithmetic.

With the introduction of public high schools after 1821, and with an increasing number of colleges demanding knowledge of algebra or geometry as pre-requisites for entry, one might have expected more attention would be given to algebra and geometry in schools than would appear to have been the case. That said, 10% of cyphering books prepared between 1830 and 1859 included entries on algebra, and 9% included entries on geometry, and each of those percentages represented an increase on what had gone before. That was only to be expected, given that the first public high schools opened in 1821 and after that time the best-known colleges gradually made an elementary knowledge of algebra a pre-requisite for admission (Kilpatrick & Iszák, 2008).

Although it is clear that not much algebra was studied in schools until after 1830 it will be useful to report and interpret data pertaining to what constituted "algebra" and when it was studied. But first we need to discuss a very early text, prepared in the 1720s, which has escaped the notice of previous writers on the history of mathematics education.

A Dutch-Language Textbook Published in New York in 1730, Which Included Algebra

Pieter Venema

Sometime around 1725, a certain Pieter Venema migrated to New York from Holland. From a mathematics perspective he was an unusual migrant because just 11 years earlier, in 1714, his book *Een Korte en Klare Onderwysinge in de Beginselen van de Algebra ofte Stel-Konst* had been published in Holland. Danny Beckers (2006) has described that book as "the most important Dutch algebra textbook of the 18th century" (p. 938).

It is thought that Venema left the Netherlands for New York because of religious conflict in his home city of Groningen (Pelletreau, 1907; Simons, 1923, 1924). Upon his arrival in New York he began to prepare a printed text combining arithmetic and algebra for Dutch-speaking students in New York. It seems, though, that he was undecided whether the book should be written in the English or in the Dutch language. Although many of the Dutch-background citizens of New York City at that time communicated with each other in Dutch, it was not obvious whether publication of a Dutch-language mathematics textbook would be commercially viable. No school mathematics textbook, in any language, had ever been prepared by an author who was living in North America and it was not clear, therefore, whether there would be a market for a textbook written in Dutch which combined arithmetic and algebra. That consideration was especially relevant in "a small town" which was such that "above Wall Street there were only a few outlying dwellings surrounded by vegetable gardens and sown pastures" (Baldwin, 1908, p. 175). New York's total population at the time was about 15000 with only a relatively small percentage speaking Dutch. Why should a mathematics textbook which included algebra, written in the Dutch language, be successful?

As it turned out, Venema's 120-page *Arithmetica of Cyffer-Konst, Volgens de Munten Maten en Gewigten, te Niew-York, Gebruykelyk Als Mede Een Kort Ontwerp van de Algebra* was published in Dutch in New York in 1730. Its 120 pages comprised 75 pages of arithmetic and 45 of algebra. The publisher was J. Peter Zenger—who, around 1734, would become famous for his leading role in the first major "freedom-of-the-press" dispute in North America. An English translation of a short version of the title of Venema's (1730) book was "Arithmetic, According to the Coins, Measures and Weights Used in New York." Figure 5.1 shows the title page of Venema's (1730) book—copied from Karpinski (1980). An English version of the book was never published, and there were no further editions published in English or in the Dutch language (Pelletreau, 1907).

Venema's and Zenger's decision to use the Dutch language testified to a level of mistaken confidence by Venema and his financial backers within New York that the Dutch sub-population of the city was sufficiently large, and mathematically-inclined, to support a Dutch-language mathematics textbook. It is likely that they chose Dutch because that had been the language of instruction in some New York educational institutions for about 100 years (Kilpatrick, 1912), and in the Netherlands there had been a strong curricular emphasis on algebra for more than 100 years (Thomas & Kempis Kloyda, 1937).

It is not surprising that from a purely commercial point of view the book failed. Venema's book was pitched at a high level for the time, and it is difficult to believe that it could ever have been imagined that its publication would generate a profit for Zenger and Venema (Howsam & Raven, 2011). Simons (1924) pointed out that it would be another 58 years before another textbook with a section on algebra, and written by a North American resident, was published. That next book was Nicolas Pike's (1788) *The New and Complete System of Arithmetic, Composed for the Use of the Citizens of the United States*. It was mostly concerned with *abbaco* arithmetic, with Pike devoting 39 of the 512 pages to algebra.

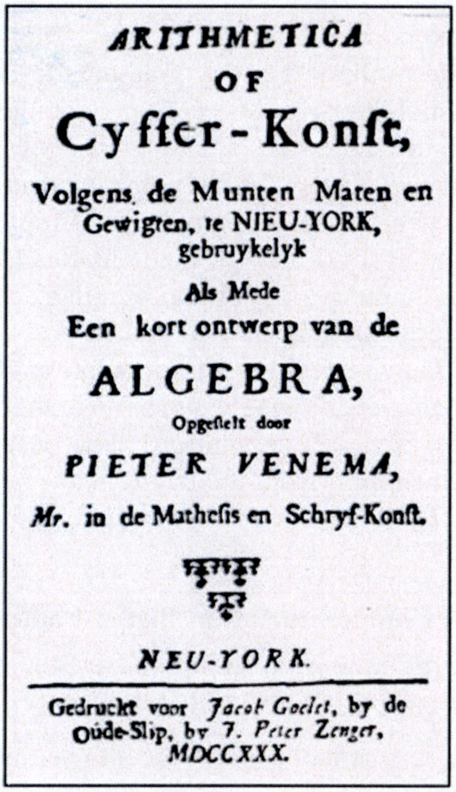
ARITHMETICA
OF
Cyffer-Konst,
Volgens de Munten Maten en
Gewigten, te NIEU-YORK,
gebruykelyk
Als Mede
Een kort ontwerp van de
ALGEBRA,
Opgestelt door
PIETER VENEMA,
Mr. in de Mathesis en Schryf-Konst.

NEU-YORK.

Gedruckt voor Jacob Goelet, by de
Oude-Slip, by J. Peter Zenger,
MDCCXXX.

Figure 5.1. The title page of Pieter Venema's (1730) textbook (reproduced from Karpinski, 1980, p. 46).

The Ellerton-Clements textbook collection does not include an original copy of Venema's (1730) Dutch-language textbook. Karpinski (1980) reported that he knew of only two such copies in existence. Despite Venema's book not being a commercial success, he remained in North America for the rest of his life. Indeed,

Karpinski (1940, p. 576) documented the fact that a mathematical problem published by Peter Zenger in his *New York Weekly Journal* of August 23, 1742, was solved by Pieter Venema, with Venema's solution being published in the *Journal* on November 12, 1742. Venema was buried in the "Dutch Church" in New York in 1748 (Bradley, 1949; Karpinski, 1940; Stocker, 1922).

In New York in 1730, there did not seem to be much demand for a mathematics textbook with a substantial section on algebra, especially one written in the Dutch language. Venema's (1730) book was only the second mathematics textbook written by an American-based author to be published in North America—the first was that by Isaac Greenwood (1729), Harvard's Hollis Professor of Mathematics and Natural and Experimental Philosophy. Greenwood's book was mainly concerned with arithmetic.

We make two conjectures with respect to consequences arising from Venema's (1730) textbook failing to be commercially successful. The first conjecture is that publishers would have begun to think that any mathematics textbook aimed at the North American market was not likely to generate a profit unless it was written in the English language. Despite the fact that a small number of persons living in New York in the eighteenth century studied algebra in evening classes—most of those were apprentices (Seybolt, 1921)—there were simply not enough Dutch-background students in New York to make a mathematics textbook written in the Dutch language, or in any language other than English, a commercially viable proposition. The second proposition is that it was likely to be true not only in New York, with its strong Dutch-speaking sub-population (Pelletreau, 1907), but also in Philadelphia and other parts of Pennsylvania where there were strong German and Pennsylvania-Dutch-speaking sub-populations. The only kind of mathematics textbook likely to gain much support from the public was one which was written in English and dealing with *abbaco* arithmetic—and not containing a large section on algebra.

Historical Errors Made in Commentaries on Pieter Venema

We now provide an extra dimension to the Pieter Venema story. Of Venema, Florian Cajori (1890) (see Figure 5.2), wrote:

> In New York the Dutch teachers of the seventeenth century imported from Holland an arithmetic called the "Coffer Konst," written by Pieter Venema, a Dutch school-master who died about 1612. So popular was the book that an English translation of it was published in New York in 1730. (Cajori, 1890, p. 13)

There were two details in that statement that Cajori got wrong. The first error was the statement that the original text, published in Holland in 1714, had been authored by someone who died around 1612. In fact, the author was the same Pieter Venema who arrived in New York around 1725. The second error was the statement that the book published in New York in 1730 was written in the English language. In fact, it was in the Dutch language (Pelletreau, 1907). Zitarelli (2019) also stated that the 1730 book was written in English. Smith and Ginsburg (1934) mistakenly believed that Venema's (1730) book contained no algebra.

It is easy to understand how the errors were made because it has never been easy to locate and examine a copy of Venema's (1730) book. As already pointed out, many years ago Louis Karpinski (1940) found only two extant copies of Venema's (1730) 120-page *Arithmetica of Cyffer-Konst, Volgens de Munten Maten en Gewigten, te Niew-York, Gebruykelyk Als Mede Een Kort Ontwerp van de Algebra.*

The appearance of Venema's (1730) book did raise the important question—one which is still being asked today—whether school students in North America who do not have English as their first language should be expected to study subjects like arithmetic and algebra in English only, or whether they should be allowed to study from textbooks written in their first languages.

Figure 5.2. Florian Cajori (c. 1890) (Wikipedia contributors (2021, May 9).

During his time in North America, Pieter Venema was recognized as both a competent mathematician and an effective, if somewhat controversial, Moravian evangelist (Bradley, 1949; Goodfriend, 2017; Stocker, 1922). Simons (1931) described Venema's (1730) textbook as "the first book to be printed in what is now known as the United States of America which bears in part of its contents the subject of algebra" (p. 6).

Venema's (1725) Precursor Manuscript

What we are addressing here is a 276-page unpublished manuscript, dealing with both algebra and arithmetic, which was prepared by Pieter Venema in or before 1725. We purchased this manuscript in 2018 from a dealer in Pennsylvania who had alerted us to its existence. It is the jewel in the crown of the E-C collection of early North American mathematics textbooks (we agree though, that technically, it is not a textbook and it is possible that Venema prepared it before he arrived in New York).

About 40 percent of the pages in this "precursor" manuscript are in Venema's handwriting—sometimes he wrote in the Dutch language, sometimes in English).

The other pages are commercially-printed (and always in the Dutch language). This manuscript pre-dates all mathematics textbooks published in that part of North America which would become the United States of America and written by a person living in North America.

Although there are at least two extant copies of Venema's (1730) book, our "part-handwritten, part-printed," 276-page bound precursor text is the *only* extant copy of what Venema wrote in 1725. As previously argued, it seems that when Venema arrived in New York he was torn between whether the text of his proposed book should be in English or in Dutch. Venema's (1725) "precursor," was signed and dated on several pages by "Pieter Venema" himself (one of the six pages reproduced in Figures 5.3a, 5.3b, 5.3c was signed).

The Pieter Venema story as we have told it, and the cyphering book data analysis summarized by the entries in Table 5.1, illustrate how, too often, articles and books on the history of mathematics in North America (and, for that matter, elsewhere) have not been based on data from primary sources. In a domain of research in which there has been tension between historians of mathematics, historians of mathematics education, historians of science, and general historians of education, it has not been easy to obtain primary sources other than textbooks, some of which contain inaccurate information, and we think we have shown, above, that even competent modern scholars have made serious errors in their descriptions of historical figures and associated textbooks. And, once an error is made that same error is likely to be repeated, not only later in the book but also by later writers.

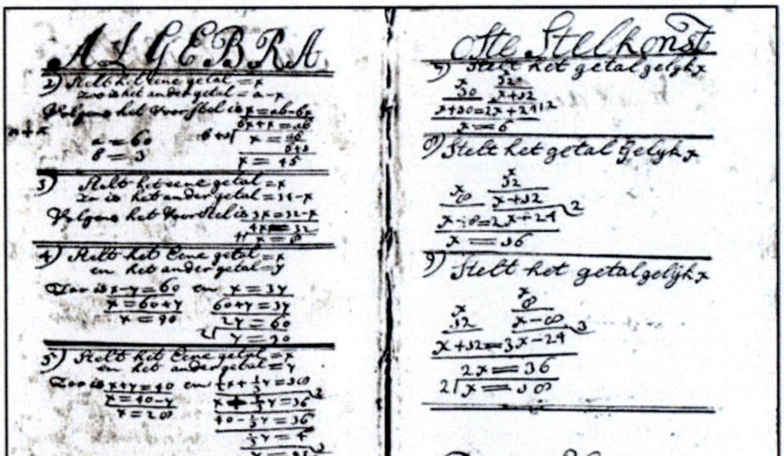

Figure 5.3a. Pages from Venema's (1725) precursor manuscript (including one signed page).

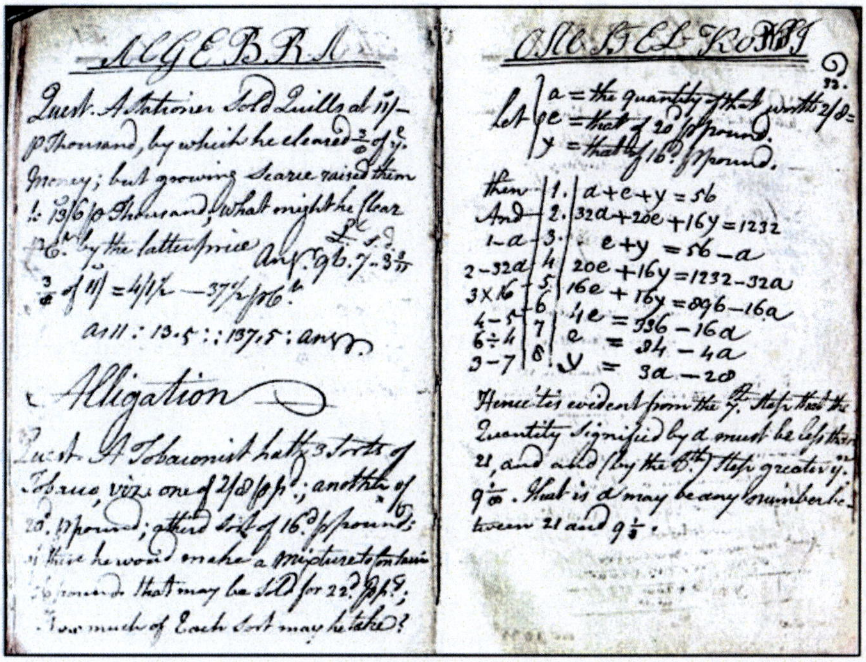
ALGEBRA

Quest. A Stationer sold Quills at 11/- p Thousand, by which he cleared $\frac{3}{8}$ of ye Money; but growing scarce raised them to 13/6 p Thousand, what might he clear p Ct. by the latter price. Ansr. 96.7-$3\frac{3}{11}$

$\frac{3}{8}$ of 11/ = 4/1½ — 37/½ p 96

as 11 : 13.5 : : 137.5 : ansr.

Alligation

Quest. A Tobacconist hath 3 sorts of Tobacco, viz. one of 2/8 p lb.; another of 20d p pound; a third sort of 16d p pound; of these he would make a mixture to contain 56 pounds that may be sold for 22d p lb; How much of each sort may he take?

ALLIGATION

Let a = the quantity of that worth 2/8; e = that of 20d p pound; y = that of 16d p pound.

then	1.	$a+e+y=56$
And	2.	$32a+20e+16y=1232$
1−a	3.	$e+y=56-a$
2−32a	4.	$20e+16y=1232-32a$
3×16	5.	$16e+16y=896-16a$
4−5	6.	$4e=336-16a$
6÷4	7.	$e=84-4a$
3−7	8.	$y=3a-28$

Hence 'tis evident from the 7th step that the Quantity signified by a must be less than 21, and also (by the 8th) step greater than $9\frac{1}{3}$. That is a may be any number between 21 and $9\frac{1}{3}$.

Figure 5.3b. Pages from Venema's (1725) precursor manuscript (on alligation).

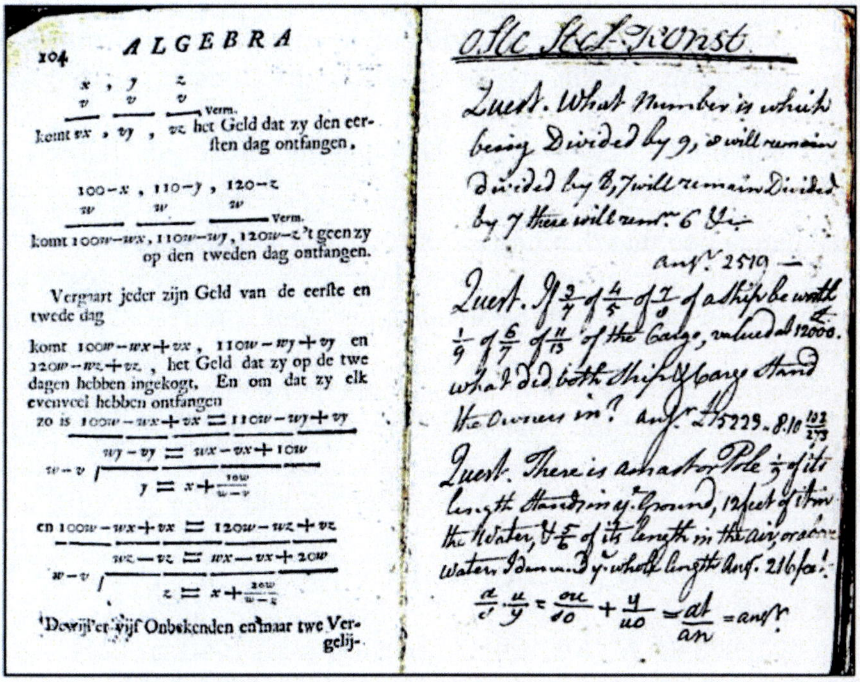
104 ALGEBRA

komt vx, vy, vz het Geld dat zy den eersten dag ontfangen.

komt $100w-wx$, $110w-wy$, $120w-z$ 't geen zy op den tweden dag ontfangen.

Vergaart jeder zijn Geld van de eerste en twede dag

komt $100w-wx+vx$, $110w-wy+vy$ en $120w-wz+vz$, het Geld dat zy op de twe dagen hebben ingekogt. En om dat zy elk evenveel hebben ontfangen

zo is $100w-wx+vx = 110w-wy+vy$

$wy-vy = wx-vx+10w$

$y = x+\frac{10w}{w-v}$

en $100w-wx+vx = 120w-wz+vz$

$wz-vz = wx-vx+20w$

$z = x+\frac{20w}{w-v}$

Dewijl'er vijf Onbekenden en maar twe Vergelij-

ALLE SLET-KONST.

Quest. What Number is which being Divided by 9, 8 will remain, Divided by 8, 7 will remain, Divided by 7 there will remr. 6 &c. Ansr. 2519

Quest. If $\frac{3}{7}$ of $\frac{4}{5}$ of $\frac{7}{8}$ of a ship be worth $\frac{1}{9}$ of $\frac{6}{7}$ of $\frac{11}{13}$ of the Cargo, valued at £12000. What did both Ship & Cargo stand the Owners in? Ansr. £15229.8.10

Quest. There is a water Pole $\frac{1}{5}$ of its length standing in ye Ground, 12 feet of it in the Water, & $\frac{5}{6}$ of its length in the air, or above water. I demand ye whole length. Ans. 216 feet.

Figure 5.3c. More pages from Venema's (1725) precursor manuscript.

Consider Sterry's and John Sterry's (1790) 147-Page Text on School Algebra

Around 1790 the Sterry brothers, Consider and John, were private teachers working outside of college circles in Rhode Island. Their 388-page textbook *The American Youth: Being a New and Complete Course of Introductory Mathematics, Designed for the Use of Private Students,* published in Providence, Rhode Island, devoted 147 pages to algebra (Simons, 1924). As mentioned in Chapter 4, this book provided the widest coverage on algebra of any textbook written by persons living in North America and published in the New World in the eighteenth century (Simons, 1924). By contrast, in Nicolas Pike's (1788) *New and Complete System of Arithmetic* only 39 of the 512 pages in the book were devoted to algebra. Like Pike, the Sterrys communicated with George Washington, seeking his support for their book, but unlike Pike, the Sterrys gave much attention to the new federal decimal currency. A second edition of the Sterrys' (1790) book did not appear, so it is reasonable to assume that not many teachers or students purchased the first edition. In her *Introduction of Algebra into American Schools in the Eighteenth Century,* Simons (1924) stated that the Sterrys' (1790) textbook "deserved more popularity than extant evidence shows it to have attained" (p. 65). According to Simons, the Sterrys were engaged entirely in work with private pupils.

In their book, the Sterrys (1790) dealt with algebra topics such as:

> Infinite series, the binomial theorem, proportion or analogy algebraically considered, arithmetical, geometrical and harmonical, genesis of equations in general, concerning the transformation of equations, and exterminating their immediate terms, resolution of equations by divisors, finding the roots of numerical equations in general by the method of approximation, concerning unlimited problems and Diophantine equations. (Listed in Simons, 1924, pp. 65–66)

Simons (1924) maintained that the Sterrys' (1790) work on algebra represented "an ambitious course in algebra set forth ... at a time when students in some colleges were still dependent on taking mathematical notes from lectures and setting them down in notebooks" (p. 66).

Simons' statement that "only three books containing algebra appeared in print in the American colonies and the young American Republic during the eighteenth century" (p. 66) was slightly inaccurate—five different authors wrote texts including sections on algebra (Venema, Pike, the two Sterrys, and John Gough), and four books were prepared. Each of the four books contained both arithmetic and algebra, with *abbaco* arithmetic being allocated far more pages than algebra. As Simons (1936, p. 9) pointed out, John Gough's (1788) textbook had an appendix dealing with algebra written by "the late W. Atkinson" (see Karpinski, 1940, p. 121).

The evidence is overwhelming that before 1800 most schools did not make algebra an important component of their implemented curricula. This was made clear in 1793 when an "abridged" version, aimed at schools, of Pike's original 1788 textbook was published. Unlike the larger 1788 book, the 1793 version did not contain a section dedicated to algebra.

Influence of British Authors

Charles Hutton, Samuel Webber and Robert Adrain

In 1764, Charles Hutton's *The Schoolmasters Guide, or a Complete System of Practical Arithmetic* was published in England, and in 1773 Hutton became Professor of Mathematics at the Royal Military Academy at Woolwich, England. He remained in that position until his retirement in 1807. In 1809, his original 1764 publication was revised and republished in the United States, with the title *A Complete System of Practical Arithmetic and Bookkeeping; Corrected, Enlarged, and Adapted to the Use of Schools and Men of Business in the United States by D. P. Adams*. A second edition of this book was published in 1810. The Ellerton-Clements North American cyphering book collection includes a manuscript, prepared by a certain Sharpless Green in 1839, titled "Book Keeping by Single Entry—Extracted from the Works of Charles Hutton, LL.D., F.R.S."

Hutton's most important influence on mathematics education in North America came through the adaptation—perhaps "plagiarizing" would be a better word—of his writings by two highly-placed American scholars, Samuel Webber and Robert Adrain (see, e.g., Hutton, 1812, 1831; Simons, 1936; Webber, 1801, 1808). Between 1789 and 1806, Webber was Hollis Professor of Mathematics and Natural Philosophy at Harvard College and between 1806 and 1810 he was President of the College. Karpinski (1980) maintained that Webber took "from Charles Hutton about 600 pages to make Webber's *Mathematics*" (p. 11) and added that there was "nothing approaching originality in Webber's work" (p. 11). An advertisement at the front of Webber's *Mathematics Compiled from the Best Authors and Intended to be the Textbook of the Course of Private Lectures on These Sciences in the University at Cambridge* [Massachusetts] stated that "the authors of the principal part of most of the branches are Dr. Hutton and Mr. Bonnycastle."

In 1798, Hutton's large, and influential, *A Course of Mathematics*, was published in London, and seven editions of this, with "additions by Robert Adrain," were published in America between 1812 and 1831 (Karpinski, 1980, pp. 192–193) (see Figure 5.4). The book presented a full, traditional coverage of *abbaco* arithmetic, to permutations and combinations. It also included sections on logarithms, algebra, and geometry (including conic sections). Adrain claimed that he had completely rearranged much of the English edition to suit the needs of American students.

The Influence of John Bonnycastle on North American Mathematics Education

Between 1782 and 1821, John Bonnycastle, the mathematical master at the Royal Military Academy in Woolwich, England, was a prolific writer of mathematics textbooks—many of which were reprinted and sold in North America. Three of his books went through many editions, both in Great Britain and in North America (see Karpinski, 1980, pp. 86–87 for publication details in North America for one of Bonnycastle's arithmetics; see pp. 160–162, for publication details related to an algebra textbook; and pp. 189–190, for publication details related to a textbook on mensuration and practical geometry).

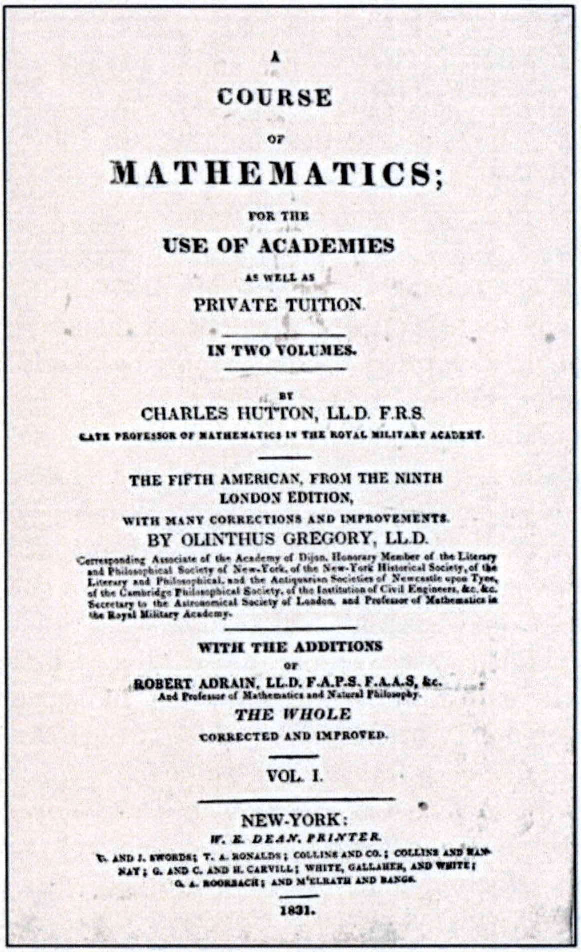
A

COURSE

OF

MATHEMATICS;

FOR THE

USE OF ACADEMIES

AS WELL AS

PRIVATE TUITION.

IN TWO VOLUMES.

BY

CHARLES HUTTON, LL.D. F.R.S.

LATE PROFESSOR OF MATHEMATICS IN THE ROYAL MILITARY ACADEMY.

THE FIFTH AMERICAN, FROM THE NINTH LONDON EDITION,

WITH MANY CORRECTIONS AND IMPROVEMENTS.

BY OLINTHUS GREGORY, LL.D.

Corresponding Associate of the Academy of Dijon, Honorary Member of the Literary and Philosophical Society of New-York, of the New-York Historical Society, of the Literary and Philosophical, and the Antiquarian Societies of Newcastle upon Tyne, of the Cambridge Philosophical Society, of the Institution of Civil Engineers, &c. &c. Secretary to the Astronomical Society of London, and Professor of Mathematics in the Royal Military Academy.

WITH THE ADDITIONS

OF

ROBERT ADRAIN, LL.D. F.A.P.S. F.A.A.S, &c.

And Professor of Mathematics and Natural Philosophy.

THE WHOLE

CORRECTED AND IMPROVED.

VOL. I.

NEW-YORK:

W. E. DEAN, PRINTER.

T. AND J. SWORDS; T. A. RONALDS; COLLINS AND CO.; COLLINS AND HANNAY; G. AND C. AND H. CARVILL; WHITE, GALLAHER, AND WHITE; G. A. ROORBACH; AND M'ELRATH AND BANGS.

1831.

Figure 5.4. Title page of a North American edition of Charles Hutton's (1831) *A Course of Mathematics*. (Notice the reference to Robert Adrain.)

Given the antipathy in the United States of America toward many things British—arising, quite naturally, from the Revolutionary War and the 1812 War—the widespread acceptance in the United States of school textbooks authored by John Bonnycastle, Thomas Dilworth and Charles Hutton, during the period 1780–1850 is puzzling— suggesting that assertions by Amy Ackerberg-Hastings (2010), A. J. Angulo (2012), Florian Cajori (1890), Stanley Guralnick (1975), Edward Hogan (1981), Karen Hunger Parshall (2003), Lao Genevra Simons (1931), and J. R. Young

(1833), that Continental European authors, especially French authors, had an enlivening influence on mathematics education in the United States in the antebellum period should be subjected to careful examination. Certainly, there can be no doubt that John Farrar, at Harvard, and Charles Davies, at the West Point Military Academy, were strongly influenced by French thinking about mathematics, but the question arises whether that influence generated positive educational outcomes in the United States of America. Davies (1835), for example, later stated that the French approaches proved to be too "scientific" for U.S. school students and that therefore in his own writing he decided to adopt "the practical methods of the English school" (p. iii). He even admitted that many of the examples he had used in his *Elements of Algebra* were selected "from the *Algebra* of Bonnycastle" (p. iv). Another to have second thoughts on the matter was J. R. Young (1833), who like Davies, taught calculus to American students in the 1820s and 1830s.

Bonnycastle's *Algebra* textbook, titled *An Introduction to Algebra, with Notes and Observations: Designed for the Use of Schools and Places of Public Education*, was even more popular in North America than his *Arithmetic* textbook. It was published in Philadelphia in 1806 by Joseph Crukshank, as a reprint of Bonnycastle's *Algebra* which had first been published in London in 1788. It was much used in North America, with 15 editions, and 5 keys being released by North American publishers between 1806 and 1842. Karpinski (1980) described Bonnycastle's *Algebra* as "the first widely popular American algebra reprint of an English text" (p. 160).

Although the title page of Bonnycastle's (1806) *Algebra* textbook announced that this was the first American edition, there was no reference in the book to anything American, either in the preface or in the text itself. Whenever money was mentioned in word problems, sterling currency was assumed to be the basis for calculations. There was no attempt to hide the fact that, really, this was a reprint of a book that had been aimed at students in British schools. Bonnycastle stated that all the elementary texts he had seen were "extremely defective," were "unfit for the purpose of teaching," and were "generally calculated to vitiate the taste and mislead the judgment" (p. vi). He added that "there is a certain taste and elegance which is only to be obtained from the best authors, and a judicious use of their instructions" (p. vii). In his preface, Bonnycastle (1806) stated that in particular, he admired the writings on algebra of "Newton, Maclaurin, Sanderson, Simpson, and Emerson." He drew special attention to what he called their "patterns of elegance and perfection" and maintained that his book was "formed entirely upon the model of those writers" and was "intended as a useful and necessary introduction to them" (p. vi). No pains had "been spared to make the whole as easy and as intelligible as possible" (p. vi).

Figure 5.5 shows nine model examples given for "Case III of Addition," on page 11 in an 1822 New York edition of Bonnycastle's *An Introduction to Algebra*. The preface could not have made clearer the acceptance of the mental-discipline theoretical position: "The powers of the mind, like those of the body, are increased by frequent exertion; application and industry supply the place of genius and invention; and even the creative faculty itself may be strengthened by use and perseverance" (Bonnycastle, 1822, p. iii). From the outset, the **IRCEE** ("Introduction-Rule-Case-Example-Exercise) genre was adopted. "Like quantities" were defined as "those which consist of the same letters or combination of letters" (p. 4), Taken literally, this implied that ab and $a + b$ were "like quantities" which, of course, most people who know algebra would reject.

Bonnycastle might have known algebra well but all of the American editions of his book demonstrated that knowing algebra well and teaching it well were two different things. On page 11 of the 1822 edition, for example, he began a section on operations on algebraic quantities with the rule for "Case III" on addition." This was stated as: "When the quantities are unlike, or some like and others unlike . . . collect all the like quantities together by taking their sums or differences." Without any further discussion, he stated the rule: "Collect all the like quantities together by the last rule, and let down those that are unlike, one after another, with their proper sign" (p. 11). He then showed the nine model examples, in Figure 5.5. The seventh was $3a^2y + -2xy^2 + -3y^2x + -8x^2y + 2xy^2$ and the answer immediately shown was $3a^2y - 3y^2x - 8x^2y$. Anyone who has taught algebra to beginners would know that for

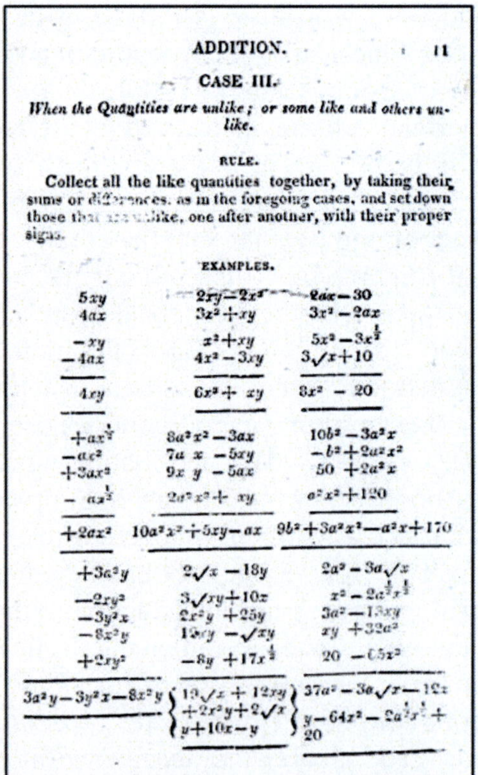
ADDITION. 11

CASE III.

When the Quantities are unlike; or some like and others unlike.

RULE.

Collect all the like quantities together, by taking their sums or differences, as in the foregoing cases, and set down those that are unlike, one after another, with their proper signs.

EXAMPLES.

$5xy$	$2xy-2x^2$	$2ax-30$
$4ax$	$3x^2+xy$	$3x^2-2ax$
$-xy$	x^2+xy	$5x^2-3x^{\frac{1}{2}}$
$-4ax$	$4x^2-3xy$	$3\sqrt{x}+10$
$4xy$	$6x^2+xy$	$8x^2-20$

$+ax^{\frac{1}{2}}$	$8a^2x^2-3ax$	$10b^2-3a^2x$
$-ac^2$	$7ax-5xy$	$-b^2+2a^2x^2$
$+3ax^2$	$9xy-5ax$	$50+2a^2x$
$-ax^{\frac{1}{2}}$	$2a^2x^2+xy$	a^2x^2+120
$+2ax^2$	$10a^2x^2+5xy-ax$	$9b^2+3a^2x^2-a^2x+170$

$+3a^2y$	$2\sqrt{x}-18y$	$2a^2-3a\sqrt{x}$
$-2xy^2$	$3\sqrt{xy}+10x$	$x^2-2a^{\frac{1}{2}}x^{\frac{1}{2}}$
$-3y^2x$	$2x^2y+25y$	$3a^2-13xy$
$-8x^2y$	$13xy-\sqrt{xy}$	$xy+32a^2$
$+2xy^2$	$-8y+17x^{\frac{1}{2}}$	$20-65x^2$
$3a^2y-3y^2x-8x^2y$	$13\sqrt{x}+12xy+2x^2y+2\sqrt{xy}+10x-y$	$37a^2-3a\sqrt{x}-12xy-64x^2-2a^{\frac{1}{2}}x^{\frac{1}{2}}+20$

Figure 5.5. Page 11 from Bonnycastle's (1822) New York edition of *An Introduction to Algebra.*

persons being introduced to algebra that task is much more difficult than it might appear to be. One has to wonder how an introductory textbook to algebra, aimed at school children, which started like that could ever have gone to 15 North American editions.

Although the title of Bonnycastle's book indicated that the text was aimed at schools and other places of public education, its content quickly got difficult. After 15 pages of introductory discussion and exercises, Bonnycastle chose to address algebraic fractions, and then followed involution and evolution, surds, infinite series, arithmetical and geometrical proportions, simple and quadratic equations, cubic equations, biquadratic equations, approximations to roots, exponential equations, Diophantine problems, summation of infinite series, and logarithms. This represented a standard sequence of algebra topics. Although the approach adopted was uncompromisingly academic, Bonnycastle did not include a Cartesian graph anywhere in his book.

The Sterry brothers admitted to having been influenced by Bonnycastle's *Scholar's Guide to Arithmetic* but said that there was "too much superfluous and unpopular material in that book to make it suitable for schools" (Sterry & Sterry, 1790, p. iv). Our analysis of Bonnycastle's *Introduction to Algebra* led us to a similar conclusion with respect to that book.

Jeremiah Day's (1814) *Algebra*—The First Dedicated Algebra Textbook Published in North America and Written by a U.S. Citizen

Soon after Jeremiah Day (see Figure 5.6) had graduated from Yale College with high honors in 1795, he served as headmaster of Greenfield School, and in 1797, he became a tutor at Williams College, in Connecticut, where he remained until 1798. Then, he accepted an offer to be a tutor at Yale College. Around that time, he became a candidate for the ministry, and in 1801, on the same day that he was ordained, he was appointed Professor of Mathematics and Natural Philosophy at Yale College. Later, he would be President of Yale for 29 years—between 1817 and 1846.

The son of a clergyman, Day was described by an admirer as "a wise disciplinarian, a judicious governor, a thorough and accurate scholar, a valuable teacher, and a man of intelligent and penetrative mind" (Dwight, 1903, p. 42). According to Dwight, he combined serenity, self-control, modesty, and unselfishness to such a degree that all of the students who came under his influence at Yale would have unquestionably declared him the best man they had ever known. He was born in New Preston, Connecticut, in 1773 and died in New Haven, Connecticut, in August 1867.

Day would become the first American-born person to have a book published which was devoted solely to algebra (Simons, 1936). That text, *An Introduction to Algebra, Being the First Part of a Course of Mathematics, Adapted to the Method of Instruction in the American Colleges,* first appeared in 1814 and quickly became a classic, being used in colleges across the United States. In addition to his *Algebra,* he also authored numerous other mathematical textbooks, all of which were prescribed for students at Yale College.

Figure 5.6. Jeremiah Day (c. 1820) (Wikipedia contributors, 2021, April 8).

The references to other texts in many editions of Day's *Algebra* reveal the wide range of mathematical texts that Day consulted. Among books mentioned were Dugall Stewart's *Philosophy of the Mind,* Isaac Barrow's *Mathematical Lectures,* Isaac Newton's work on fluxions and on *Universal Arithmetic*, William Emerson's *Method of Increments*, Leonhard Euler's *Analysis of Infinity*, algebras by John Bonnycastle, John Fenn, Sylvestre François Lacroix, Nicholas Saunderson, and William Wallace, treatises on *Fluxions* by Samuel Vince, Colin Maclaurin, and Edward Waring, Abraham Rees' *Cyclopaedia*, Robert Woodhouse's *Analytical Calculations,* Sterling's *Summation of Series*, Thomas Simpson's *Essays and Dissertations*, and Joseph Louis Lagrange's *Theory of Analytical Functions*. He also often referred to the writings of Abraham De Moivre.

Karpinski (1980) indicated that Yale University published 72 editions of Day's *Algebra*, between 1814 and 1864. Each was stated to be a "new edition," and each had 404 pages. A *Key* was published in 1853 by Durrie and Peck of New Haven. For each "new edition," 2000 copies were printed. According to Karpinski (1980), "no other American mathematical work to 1850 had so long a series of consecutively numbered editions," and only Daboll's and Dilworth's arithmetics had as many different editions (p. 202).

Jeremiah Day's (1814) *Algebra* was written for college-level students, and not school students. At the beginning of the 1840s, Day asked one of his former Yale students, James Bates Thomson, to help him write an algebra textbook which would be suitable for school students. In 1843 Thomson's 252-page *Elements of Algebra, Being an Abridgment of Day's Algebra, Adapted to the Capacities of the Young, and the Method of Instruction, in Schools and Academies* was published, and by 1865 about 20 editions of that book had appeared (Karpinski, 1940).

The preface to Thomson's (1843) *Elements of Algebra* began with some high words:

> Public opinion has pronounced the study of algebra to be a desirable and important branch of popular education. This decision is one of the clearest proofs of an onward and substantial progress in the cause of intellectual improvement in our country. A knowledge of algebra may not indeed be regarded as strictly necessary to the discharge of the common duties of life; nevertheless, no young person at the present day is considered as having a "finished education" without an acquaintance with its rudiments. The question with parents is, not "how little learning and discipline their children can get through the world with"; but "how much does their highest usefulness require"; and "what are the best means to secure this end?" (p. iii).

According to the preface, it had long been recognized that an abridgement of Day's *Algebra*, adapted to the wants of schools and academies, would "facilitate the object" (p. iii). Accordingly, Thomson (1843) argued that a smaller and cheaper work than Day's *Algebra* was needed for schools, a work which combined "the simplicity of language and the unrivalled clearness with which the principles of the sciences are there stated," so that the subject might be brought "within the means of the humblest child in the land" (p. iii). Thus, the aim of the book was "to furnish an easy and lucid transition from the study of arithmetic to the higher branches of algebra and mathematics" (p. iv). The contents of the book provided an elementary introduction to algebra (four operations, fractions, powers—up to the binomial theorem—roots, surds, equations in one and two unknowns, ratio and proportion, progressions, and applications of algebra to geometry).

Thomson's abridgement resembled Day's original, larger *Algebra*. It proceeded as far as geometrical proportion and geometrical progressions, and evolution and involution. There was a 15-page section on the application of algebra to geometry. The treatment was quite formal with **IRCEE** genre being evident throughout. One reason why the book was commercially successful was that teachers knew that the book provided a strong preparation for prospective college students, especially those hoping to go to Yale.

Colburn's, Bailey's and Ray's "Inductive" Algebras, Written Specifically for Schools

Warren Colburn's (1825) *An Introduction to Algebra*

Colburn's (1825) sole volume on algebra was written when he was no longer working as a teacher in a school. The topics ranged from elementary definitions, operations and simplifications through to the binomial theorem and logarithms. Like other school algebra textbooks of the time, no Cartesian graph was shown or called for. Also, proofs of theorems using algebra were not shown—although the inductive method employed by Colburn did encourage students to think about the basic structural properties surrounding "advanced *abbaco*" topics like alligation, fellowship, and false position.

The preface started with the statement that the first object of the author had been "to make the transition from arithmetic to algebra as gradual as possible" (p. 3). Colburn asserted, naively, that in his book all of the fundamental aspects of algebra had "been applied so simply that scarcely anyone can mistake them, if left entirely to himself" (p. 3). He added that "the learner is expected to derive most of his knowledge by solving the examples himself," and "therefore care has been taken to make the explanations as few and as brief as is consistent with giving an idea of what is required" (p. 3). He added that showing how to do problems through model examples was "apt to embarrass than aid the learner, because he is apt to trust too much to them, and neglect to employ his own powers; and because the explanation is frequently not made in the way that would naturally suggest itself to him, if he were left to examine the subject by himself" (p. 3). Colburn's (1825) "induction" thesis was explicitly stated in the following way:

> The best mode, therefore, seems to be, to give examples so simple as to require little or no explanation, and let the learner reason for himself, taking care to make them more difficult as he proceeds. This method, besides giving the learner confidence in making him rely on his own powers, is much more interesting to him, because he seems to himself to be constantly making new discoveries. Indeed, an apt scholar will frequently make original explanations much more simple than would have been given by the author. (p. 3)

This was "discovery learning," 1820s style. The question arises: Was Colburn's text in his *Introduction to Algebra* consistent with his claims? After a brief recapitulation, at the beginning of Chapter 1, of the meanings of certain "signs" used in arithmetic (specifically: $=$, $+$, $-$, $\times$, and $\div$), and a statement that "signs generally used to express the unknown quantities . . . are some of the last letters of the alphabet, as x, y, z. &c" (p. 10), Colburn launched straight into the following two examples:

> Two men A and B trade in a company, and gain 267 dollars, of which B has twice as much as A. What is the share of each?
>
> Four men, A, B, C, and D, found a purse of money containing \$325, but not agreeing about the division of it, each took as much as he could get; A got a certain sum, B got 5 times as much, C 7 times as much, and D as much as B and C both. How many dollars did each get? (Colburn 1825, p. 11).

Colburn gave model solutions to both problems—his solution for the second problem was:

Let x represent the number of dollars that A got; then B got $5x$, C $7x$, and D $(5x + 7x = 12x)$. These, added together, must make \$325, the whole number to be divided.

$x + 5x + 7x + 12x = 325.$

Putting all the x's together, $25x = 325$

$x = 13 = A$'s share.
$5x = 65 = B$'s share.
$7x = 91 = C$'s share.
$12x = 156 = D$'s share.

Colburn (1825) then told readers that the correctness of the answer could be "proved" by "adding the four shares and noting that the sum was equal to 325" (p. 11).

What Colburn did not say was that students who were steeped in *abbaco* arithmetic would have recognized this as a "double false position" task and would have applied an algorithm to get the answer. But Colburn, almost at the start of his book on algebra, led students to use the much more direct algebraic method. Then, immediately, students were asked to solve seven more tasks for which an algebraic approach was desirable. There were no side headings, no introductions to topics, and from the start students were invited to think about the logic behind each line. That was the approach Colburn used throughout his *Algebra*.

We do not have data on whether students benefited from the approach to algebra which Colburn adopted. Of one thing, though, the reader can be sure—Colburn's approach to algebra for school students was radically different from methods adopted by Pike, Bonnycastle, and other textbook authors who had written about algebra before him. Karpinski (1940) has informed us that 20 editions of Colburn's algebra appeared between 1825 and 1848, and also 6 keys were published giving answers to questions asked (see pp. 262–263).

Later, in Colburn's *Algebra,* when the material got more difficult, Colburn started to give explanations. He closed his preface with the comment that some critics might find his *Algebra* too easy (as indeed, he wrote, some critics had found his *Arithmetic*). He summarily dismissed that thought as unimportant by saying that with this book the learner "must give a great many explanations which he does not find in the book" (p. 5).

By 1846 the publishers Jordan and Wiley had picked up publishing rights for Colburn's *Sequel* and Colburn's *Algebra* and on page 279 of their version of the *Algebra* the publishers stated that "the study of algebra is singularly adapted to discipline the mind, and gives direct and simple modes of reasoning, and is usually regarded as one of the most pleasing studies in which the mind can be engaged." The question remained, however—when should the formal study of algebra begin?

That question would be addressed by the next major contributor to discussions on algebra education—Ebenezer Bailey.

Ebenezer Bailey's (1833) *An Introduction to Algebra*

Warren Colburn was not the only author to challenge the traditional top-down approach to algebra which was evident in texts like those by John Bonnycastle and Jeremiah Day (1814). The major partner—"challenger" might be a better word—to Colburn on the matter was Ebenezer Bailey (1795–1839), whose *First Lessons in Algebra, Being an Easy Introduction to that Science Designed for the Use of Academies and Common Schools* first appeared in 1833. The title of Bailey's book spoke for itself—clearly the author was aiming his *Algebra* at pre-college students.

On the title page of his *Algebra*, Bailey (see Figure 5.7) was described as "Principal of the Young Ladies' High School, Boston," and author of *Young Ladies Class Book.* As early as 1828, Boston's Mayor, Josiah Quincy Sr., had declared that Bailey's period as head of the Boston High School for Girls had been a failure, and upon that pronouncement Bailey lost his position. He then became head of a private school for young ladies and was active in the establishment of the American Institute for Education.

Although Bailey was a Yale graduate, he was much connected to ordinary people. Besides being a schoolteacher, he was at various times a member of the city council of Boston, director of the home of reform, president of the city lyceum, and director of the Boston mechanics' institute. He made frequent contributions to the *Boston Courier* and to other periodicals. Furthermore, Bailey was known to be actively involved with the education of girls, and therefore the publication of his book represented the idea that algebra should be for girls as well as for boys.

Figure 5.7. Ebenezer Bailey (c. 1839) (from *The Kouroo Contexture*, 2021).

On a preliminary page opposite the preface in Bailey's *First Lessons in Algebra,* there was a statement: "At a meeting of the school committee of the city of Boston, March 11, 1834, it was voted that "Bailey's *Algebra*" be used in the writing schools in which algebra could be taught." A note, dated January 1834, and placed opposite the book's table of contents, indicated that the first edition of this book, consisting of 2000 copies, had sold out in a few months, and as a result, it was decided to put it into stereotype form, so that it would become more permanent.

Bailey's book, which would have 18 editions and 7 associated "keys," was intended for those who were beginning algebra. In the preface, Bailey stated that he had "long wished that algebra might be introduced into common schools, as a standard branch of education," and there seemed "to be no good reason why the study of this most interesting and useful science should be confined to the higher seminaries" (p. 3). He wrote that there was nothing much new in his *Algebra*. "If there be any peculiarity in this work, it is its simplicity" (p. 3). That was reminiscent of what Warren Colburn wrote in his preface to his algebra text. Bailey added that there was "little danger that the student will find the beginning of any art or science too easy," and "in algebra, he is required to learn a peculiar language, to determine new principles, and to accustom himself to an abstract mode of reasoning with which he has been little acquainted" (p. 4). That was different from the "it's easy" theme in Colburn's preface. After dealing with definitions, notations, and the four operations, Bailey's text proceeded to cover fractions, powers, equations, exercises in generalization, evolution, and equations of the second degree. But there were no graphs.

The most historically significant section of Bailey's (1833) *Algebra*, perhaps, was Chapter 9, which was titled "Exercises in Generalization." In that chapter Bailey attempted to show that algebra enabled one "to deduce general truths from particular instances, and thus to form rules for conducting numeral calculations" (p. 172). He began the discussion in that chapter with a numerical example which suggested that "if we subtract the difference of two quantities from their sum, the remainder is equal to twice the smaller quantity." He then used algebra to *prove* that property of numbers. This chapter had a much more progressive ring about it than chapters appearing in most other algebra texts of the period.

Henry Barnard (1859), a famous educator, described Bailey's *Algebra* as "the first work on the science that pretended to be adapted to the wants of beginners" (p. 440). Bailey's *Algebra* was more a bottom-up production than a top-down directive. Numerous new editions of Bailey's *First Lessons in Algebra* would appear between 1833 and 1850 (see Karpinski, 1980, pp. 345–346) and during that period it would become clear that many thought that even Bailey's text was too difficult for many high-school students. Others thought it was not difficult enough.

As Colburn had done, early in his *Algebra*, Bailey showed how a number of "sharing" problems, which traditionally had been solved by "false-position" algorithms, could more easily be solved by algebra. One such problem was:

> A gentleman gave a purse, containing a certain sum of money, to his three children, to be divided among them in such a manner that Mary should have twice as much as Ellen, and John should have as much as both his sisters. What was the share of each? (Bailey, 1833, pp. 16–17)

Bailey's model solution is shown in Figure 5.8.

> As the sum contained in the purse is not named, we will call it a.
>
> Let x denote Ellen's share; then Mary's share is twice as much, or $2x$; and John's, x and $2x$, that is, $3x$; and the sum of their shares is x and $2x$ and $3x$, or $6x$, which must be equal to a, the sum to be divided, whatever the value of a may be. And if $6x$ is equal to a, x is equal to one sixth part of a, which is Ellen's share.
>
> If the purse contained 18 dollars, Ellen's share was 3, Mary's 6, and John's 9 dollars.
>
> If the purse contained 24 dollars, Ellen's share was 4, Mary's 8, and John's 12 dollars.
>
> In this manner the share of each may be determined, whatever be the sum indicated by a.

Figure 5.8. Algebraic solution to a false-position task (from Bailey, 1833, p. 17).

The gender relations suggested by the problem itself reflected the thinking of the time—which applied even though the writer was principal of a ladies college. Aside from the fact that this "false-position" task was now being treated as an algebra problem, it was also interesting that the actual amount of the sum of money being shared was not specified in the problem statement. Bailey overcame this by treating the sum as a variable, which he represented by *a*—the first letter of the alphabet. Then, the amounts to be received by the three children had to be expressed as fractions of *a*. Bailey commented: "In this manner the share of each may be determined whatever be the sum indicated by *a*" (p. 17). After introducing *a*, Bailey also introduced *x* (a letter toward the end of the alphabet) as representing how much Mary should get. The ideas being presented were subtle and sophisticated for a beginner in algebra. Bailey followed the question with a section on "algebraic signs" in which he summarized methods for combining algebraic terms using the four operations. Bailey's approach would have been easily understood by persons with a background in algebra, but for beginners it would have been conceptually challenging.

In his preface Bailey stated that "the upper classes, at least, in common schools might be profitably instructed in its elements without neglecting any of those branches to which they usually attend" (p. 3). Bailey admitted that the mathematics he was presenting in his book was not new—there were "no new discoveries" (p. 3). What was new was his organization of the material in order that it might be brought within the reach of younger students.

Aguably, Colburn and Bailey were the foundation fathers of an "algebra-for-all" movement which continues into the twenty-first century (Clements, Keitel, Bishop, Kilpatrick, & Leung, 2013). Algebra had been made part of the high-school curriculum when the first public high school was established in Boston in the early 1820s, and now Bailey's quest was to get algebra into the upper grades of common schools.

Joseph Ray's (1848) *Elements of Algebra*

As shown in Chapter 3, Joseph Ray's arithmetic textbooks proved to be very popular in the late 1830s and throughout the 1840s and, therefore, given the growing number of high schools in which algebra was compulsorily studied, it was not surprising that Ray prepared an elementary text on algebra. In his preface to *Elements of Algebra*, Ray (1848) presented a disciplines-of-the-mind justification for making algebra an important part of high-school curricula. He argued that "the science of algebra, properly taught, stands among the first of those studies essential to both the great objects of education—learning to reason correctly, and exercising in all relations, the energies of a cultivated and disciplined mind" (p. v). He also asserted that the study of algebra naturally followed the study of arithmetic "and should be taught immediately after it" (p. v).

Ray's 240-page book began with a lengthy section called "intellectual exercises" which were to be done "in the head." This section was divided into 14 lessons, and each lesson included about 20 fairly simple questions, such as "the sum of the ages of a father and his son is equal to 35 years, and the age of the father is six times that of the son; what is the age of each?" (p. 11). The aim appeared to be to get a student to "think algebraically," but it is likely that this father-son problem would have been solved using a trial-and-error method that was essentially arithmetical.

Like both Colburn and Bailey, Ray made considerable use, in this first section, of problems which would previously have been solved by false position. Following the first section came a section on "definitions and notation," and then the "four operations". Then came sections giving formal introductions to elementary algebra topics—as far as quadratic equations and arithmetical and geometrical progressions. As the book progressed, more and more rules and cases were given, as well as model examples. This was consistent with a move during the late 1840s away from Colburn's inductive approach, and back to a more traditional "analytic" approach (Michalowicz & Howard, 2003).

One of the main forces which contributed to this reversion was the belief among some persons teaching mathematics in colleges that algebra textbooks written by Continental writers were educationally and mathematically superior to those written by British mathematicians. Another influence was the increasing popularity of algebra textbooks authored by Charles Davies, the Professor of Mathematics at the United States Military Academy at West Point. Davies was well regarded by teachers, and when he decided to write an algebra text, *Elements of Algebra, Translated from the French of M. Bourdon*, which was aimed at both college and school students, some teachers immediately decided to adopt this new text in their schools. Further comment on Davies' influence will be offered shortly.

The Influence of French Approaches on Algebra Education in North America

Florian Cajori (1890), in *The Teaching and History of Mathematics in the United States*, devoted almost 200 pages to a section titled "Influx of French Mathematics." He began that section with the following commentary on the quality of French mathematics during the nineteenth century:

> In 1794 was opened in Paris the Polytechnic School and in the following year the Schools of Application. The Polytechnic School gained a world-wide celebrity. The professors of this institution were men whose names are household words wherever science has a votary. Lagrange, Lacroix, and Poisson laid the basis to its course in analytic mathematics; Laplace, Ampère, and others to that of analytical mechanics and astronomy. Descriptive geometry and its applications had for their first teachers the founder of this science the illustrious Monge and his celebrated pupils, Hachette and Arago.
>
> The success of the Polytechnic School was phenomenal. It was the nurse of giants. Among its pupils were Arago, Biot, Bourdon, Cauchy, Chasles, Duhamel, Gay-Lussac, Le Verrier, Poncelet, Regnault. The Polytechnic School is of special interest to those who live in America because the U.S. Military Academy at West Point mimicked its algebra program on it.
>
> Compared with the French mathematicians who flourished at the beginning of this century the contemporary American professors were mere Lilliputians. The masterpieces of French scholars were unknown in America. What little mathematical knowledge existed here came to us through English channels. ... There was a great dearth in original thinking on mathematics among us. The genius of our people was exercised in different fields, and so the little success we had was borrowed from others.
>
> But the time came when French writers were at last beginning to make their influence among us. We recognized their superiority over the English and profited by it. Mathematical studies received new impetus. (p. 99)

Cajori (1890) then extended his thesis to cover mathematics education:

> The improvements in mathematical text-books and reforms in mathematical instruction were due to French influences. French authors displaced the English in many of our best institutions. It is somewhat of a misfortune, however, that we failed to gather in the full fruits of the French intellect; ... many of the works which were adopted were beginning to be "behind the times," when introduced in America. We used works of Bézout, Lacroix, and Bourdon. But Bézout flourished before the French Revolution and Lacroix wrote most, if not all, of his books before the beginning of the century. (p. 99)

Cajori (1890) claimed that in North America "mathematical teaching has been bad" (p. 100) and added that he was applying that judgment to teachers in "preparatory schools" (p. 101) as well as to college teachers.

There are two questions which arise immediately. First, was Cajori's claim correct that around 1800 the research of top French mathematicians was decidedly superior to that of the best British mathematicians? And second, were French textbooks for school mathematics better than the British and North American textbooks for school mathematics?

So far as the first question is concerned, although one does not know, definitely, whether the answer should be "Yes" or No," it is generally held that the answer is "Yes" (Ackerberg-Hastings, 2010; Cajori, 1890; Parshall, 2003). David Zitarelli's (2019) view that "after 1800 France had become the undisputed leader under the direction of the 'three L's—Lagrange, Laplace, and Legendre" (p. 111), summarizes the most commonly-held position.

But, with respect to the second question, we would assert that there is no strong evidence that French textbooks for school mathematics were better than British or North American textbooks for school mathematics—such as those by Hutton or Bonnycastle or Colburn—which were often used in North American schools. Our own position is that if the main criterion is assisting children to learn significant mathematics well, then *both* the French and British textbooks were poor; and, if anything, the British books were slightly better than the French books. We have examined carefully most of the algebra textbooks by the most highly-regarded French authors which were translated into English and used in North America during the period 1815–1830, and have not been impressed. They were very discursive and would have been too difficult for most pre-college students in North America. We shall now offer data to support our contention, concentrating on algebra textbooks. Since formal written tests were not administered in North America in the eighteenth and early nineteenth centuries, there are no achievement data available from tests. But there are other data.

The first data were presented in the 1930 doctoral dissertation by Amy Olive Chateauneuf, of the University of Pennsylvania. The title of the dissertation was "Changes in the Content of Elementary Algebra Since the Beginning of the High School Movement as Revealed by the Textbooks of the Period." Chateauneuf analyzed the contents of 257 textbooks, authored by 158 different persons over the period from 1818 to 1928. What Chateauneuf (1930) found was that for most of that period the balance of topics in the textbooks remained constant from decade to decade. Definitions, four operations, fractions, factors, proportion, word problems, linear equations, quadratic equations, the binomial theorem, and progressions were dealt with in all elementary- to middle-level algebra textbooks, with different amounts of emphasis being given to each topic. But, from decade to decade, the proportion of pages devoted to any particular topic remained virtually unchanged.

From the time of Descartes, many French mathematicians had emphasized the importance of functions and graphs (Minto, 1788), but in our analyses of more than 100 British and 10 French-background school algebra textbooks in the Ellerton-Clements North American mathematics textbook collection covering the period 1780–1865 we found only two which provided a definition of a function and displayed Cartesian graphs, and they were both authored by Benjamin Peirce (1837, 1841) and mainly used in colleges, not schools (see Chapters 7 and 8 in this book). We compared the treatments of topics in British-background and French-background algebra textbooks and concluded that the French books were more discursive and more difficult than the British books. Chateauneuf's (1930) analysis showed graphs and functions slowly appearing, but only after 1880. In other words, the intended curriculum for algebra did not change much during the period 1780–1865. That would appear to contradict any hypothesis that the influx of French thinking changed dramatically what was happening with respect to algebra in North American schools.

There are other data which point to the same conclusion. Ackerberg-Hastings's (2010) analysis of the situation at Harvard after the "Cambridge Course of Mathematics" was instituted by Professor John Farrar provided some information on this. Farrar was Hollis Professor of Mathematics and Natural Philosophy at Harvard College between 1807 and 1836, and the first of the textbooks he selected for his "Cambridge course" was Leonhard Euler's (1818) *An Introduction to the Elements of Algebra: Designed for the Use of Those who are Acquainted only with the First Principles of Arithmetic*. Euler, of course, was not a Frenchman, but the book had been translated from French into English in 1797 (see Euler, 1797), and so it was relatively easy to get an English "translation" for Farrar's Cambridge course. Euler included some elementary "practical" problems—like, for example, "A mule and an ass were carrying burdens amounting to some hundred weight. The ass complained of his, and said to the mule, I need only one hundred weight of your load, to make mine twice as heavy as yours. The mule answered, Yes, but if you gave me a hundred weight of yours, I should be loaded three times as much as you would be. How many hundred weight did each carry?" Such pseudo-reality problems were common in most early U.S. textbooks on mathematics. But Euler's book got difficult quickly and because of that it quickly became unsuited to U.S. schools below the college level. That said, four editions of the book—printed in 1818, 1821, 1828, and 1836—would be used by Harvard students (Karpinski, 1940, pp. 215–216).

Most of the textbooks used by the Harvard students were books translated from textbooks originally prepared in France by leading French mathematicians, including Lacroix, Bézout and Bourdon. Harvard students did not respond well to these books. In fact, students formally complained about the new textbooks and the methods employed in teaching mathematics at Harvard (Ackerberg-Hastings, 2010, p. 23). Farrar left Harvard in 1836, and not one of the French-background mathematics textbooks which he introduced at Cambridge continued to be widely used outside of Harvard—and, in fact, even at Harvard they were soon abandoned (Karpinski, 1940). Farrar's translation of Bourdon's *Elements of Algebra* did not go beyond its first edition, despite a glowing review of it appearing in an 1832 edition of *The American Monthly Review*, and despite the claim that the book was suitable for both colleges and schools. Farrar's translation of *Lacroix's (1818) *Elements of Algebra* was not published for Harvard students after 1837 (Karpinski, 1980, p. 220].

Depending on one's educational orientations, one might conclude that the algebra textbooks prepared by North American authors Jeremiah Day, Warren Colburn, Ebenezer Bailey and Joseph Ray were more user-friendly than any of the French—or, for that matter, British—textbooks. They were certainly more successful from a commercial perspective, and we believe that those by Colburn, Bailey and Ray would have promoted better learning of mathematics than the translations of the French textbooks. Lao Genevra Simons (1936), referring to a lecture "On Teaching the Elements of Mathematics," delivered by Thomas Sherwin to the American Institute of Instruction in Boston, in August 1834, quoted Sherwin as follows:

> A young gentleman of fine talents with a mind somewhat matured, at Harvard University, asked me by what means he could make himself well acquainted with algebra. I directed him to study Colburn's work on that subject. At the expiration of six months, he assured me that he had obtained much more knowledge of the science from that treatise than from the less inductive ones of Euler and Lacroix, which he had previously studied. Until I perused this book, he said, I knew nothing about the subject.
>
> (Quoted in Simons, 1936, p. 22)

Later in his talk, Sherwin added: "No man among us has contributed so much to a correct method of studying mathematics as the lamented Colburn" (quoted in Simons, 1936, p. 22).

Charles Davies and School Algebra, 1818–1865

The influence of textbooks with French origins on the quality of U.S. mathematics during the period 1820–1865 is a matter for debate, as is the extent to which the French positively influenced mathematics teaching methods in the United States during the same period. Peter Molloy (1975), in his Brown University doctoral dissertation on "Technical Education and the Young Republic: West Point as America's École Polytechnique, 1802–1833," argued that the introduction of blackboards into the United States came as a result of the United States Military Academy's adoption of blackboards after observing the introduction of that new technology into French colleges—see Phillips (2015).

Charles Davies' *Elements of Algebra: Translated from the French of M. Bourdon* was first published in 1835 and would prove to be highly successful from a commercial point of view, with 17 editions appearing by 1857 (Karpinski,

1980). In 1837, two years after the appearance of his *Elements of Algebra*, and after teaching mathematics at USMA for 19 years, Davies resigned from USMA to take up a position at Trinity College in Hartford, Connecticut, where he remained for three years before returning to West Point. Davies' *Elements of Algebra* was the first of his published books to deal directly with algebra.

Davies (1837) admitted, in his preface, that his *Elements of Algebra*, which had 353 pages, was a simplified and reduced version of *Bourdon's Algebra* (the original French version of which contained 673 pages), and relied heavily on the English translation, by fellow USMA mathematician, Lt. Edward C. Ross, which had been used at West Point since 1831. He also admitted that "the work here presented to the public is an abridgement of Bourdon, from the translation of Lt. Ross with such modifications as experience in teaching it, and a very careful comparison with other standard works, have suggested" (pp. iii–iv). He added that "many of the examples have been selected from the Algebra of Bonnycastle" (p. iv). It seems that this was Charles Davies at his most opportunistic and plagiarizing best.

With respect to the French influence—which, it was well known, had been strongly pushed at West Point—Davies (1837) stated in his preface:

> It has been the intention to unite in this work, the scientific discussions of the French with the practical methods of the English school, that theory and practice, science and art, may mutually aid and illustrate each other. (p. iv)

In other words, it seemed that Lt. Ross's lengthy translation of Bourdon's *Algebra* had not been entirely satisfactory for USMA students—it was too long and too theoretical. Davies did not say, directly, that he expected that his *Elements of Algebra* should be suitable for use in high schools, but he did say he was presenting it "to the public" (p. iii). Cyphering book evidence proves that it would be adopted in some high schools as well as in colleges. Thus, for example, in 1849 John Haskins Winfree, while attending the Episcopal School in Fairfax, Virginia, prepared a 150-page algebra cyphering book which was entirely based on Davies' *Elements of Algebra*. Winfree's cyphering book is held in the Ellerton-Clements cyphering book collection.

Our analysis of Davies' (1835, 1837) *Elements of Algebra* revealed that the sequencing of topics was traditional and quickly reached a level of difficulty which would have been far too great for most U.S. high-school students at that time. That is hardly surprising given that in France Bourdon's algebra was mainly used by secondary-school students preparing to go to the *École Polytechnique* in Paris (da Ponte & Guimarães, 2014). Consider, for example, the text shown in Figure 5.9, which is from a section titled "Of Algebraic Fractions" in Davies (1835).

The text immediately before what is shown in Figure 5.9 asked readers to find "the greatest common divisor between the two polynomials $a^4 + 3a^3b + 4a^2b^2 - 6ab^2 + 2b^4$ and $4a^2b + 2ab^2 - 2b^3$," and then there was the comment "or simply $2a^2 + ab - b^2$, since the factor $2b$ can be suppressed, being a factor of the second polynomial and not of the first" (p. 52). This page came early in a 353-page book. Any person who has had experience teaching algebra to high-school students would know that the text shown in Figure 5.9 would be almost impossible for most young high-school students to follow. Yet, at the top of the next page came the assertion:

> These examples are sufficient to point out the course the beginner is to pursue, in finding the greatest common divisor of two polynomials, which may be expressed by the following general rule.
>
> I. Take the first polynomial and suppress all the monomial factors common to each of its terms. Do the same with the second polynomial, and if the factors so suppressed have a common factor, set it aside as forming a part of the common divisor sought.
>
> II. Having done this, prepare the dividend in such a manner that its first term shall be divisible by the divisor; then perform the division, which gives a remainder of a degree less than that of the divisor, in which suppress all the factors that are common to the co-efficients of the different powers of the principal letter. Then take this remainder as a divisor, and the second polynomial as a dividend, and continue the operation with these polynomials, in the same manner as with the preceding.
>
> III. Continue this series of operations until a remainder is obtained which will exactly divide the preceding divisor; this last divisor will be the greatest common divisor; but if a remainder is obtained which is independent of the principal letter, and which will not divide the co-efficients of each of the proposed polynomials, it shows that the proposed polynomials are prime with respect to each other, or that they have not a common factor.
>
> (Davies, 1835, p. 54)

Every experienced teacher of high-school algebra knows that when students are learning to add or subtract algebraic fractions it can be important for them to find the least common multiple of the denominators. But the above rule for finding the greatest common divisor of two polynomials, as given by Davies, would have been very difficult for most high-school students to comprehend—the language is opaque and, in any case, it is not clear whether the statement of the rule was necessary.

Many other similarly far-fetched, just-as-opaque statements of rules and their applications, for other topics, could have been chosen from Davies (1835) to illustrate the main point being made. When many young learners were being introduced to algebra they must have felt discouraged when trying to untangle meanings and procedures. For them, the complexity of the language used would have made it almost impossible to get a good grasp of the subject. For youngsters who were required to use the textbook it is likely that algebra would have come to be

OF FRACTIONS. 53

First Operation.

$$8a^4+24a^3b+32a^2b^2-48ab^3+16b^4 \;\|\; 2a^2+ab-b^2$$
$$+20a^3b+36a^2b^2-48ab^3+16b^4 \;|\; 4a^2+10ab+13b^2$$
$$+26a^2b^2-38ab^3+16b^4$$

1st. Rem. . . . $-51ab^3+29b^4$

or, . . . $-b^3(51a-29b)$.

Second Operation.

Multiply by 2601, the square of 51.

$$5202a^2+2601ab-2601b^2 \;\|\; 51a-29b$$
$$5202a^2-2958ab \;|\; 102a+109b$$

1st. Rem. . $+5559ab-2601b^2$

$5559ab-3161b^2$

2d. Rem. . . . $+560b^2$.

The exponent of the letter a in the dividend, exceeding that of the same letter in the divisor by *two* units, we multiply the whole dividend by the cube of 2, or 8. This done, we perform three consecutive divisions, and obtain for the first principal remainder,

$$-51ab^3+29b^4.$$

Suppressing b^3 in this remainder, it becomes $-51a+29b$ for a new divisor, or, changing the signs, which is permitted, $51a-29b$: the new dividend is $2a^2+ab-b^2$.

Multiplying this dividend by the square of 51, or 2601, then effecting the division, we obtain for the second principal remainder, $+560b^2$, which proves that the two proposed polynomials are *prime with respect to each other*, that is, they have not a common factor. In fact it results from the second principle (Art. 67), that the greatest common divisor must be a factor of the remainder of each operation; therefore it should divide the remainder $560b^2$; but this remainder is *independent* of the principal letter a; hence, if the two polynomials have a common divisor, it must be *independent* of a, and will consequently be found as a factor in the co-efficients of the different powers of this letter, in each of the proposed polynomials; but it is

5*

Figure 5.9. Page 53 from Davies (1835), in a section titled "Of Algebraic Fractions."

seen as some remote, extremely difficult subject, something almost impossible for them to learn.

The Emergence of Normal Schools, and Its Effects on School Algebra

During the late eighteenth century and throughout all of the nineteenth century, "normal schools," specifically created for the purpose of teacher education, became increasingly popular in Europe (Harper, 1935, 1939). That was also true in North America after 1839—when the first public normal school in the United States was established, in Massachusetts, by Horace Mann. From a mathematics education perspective, the normal schools would teach generations of prospective and practicing teachers that cyphering approaches to teaching and learning were antiquated and that the inductive methods advocated by Colburn, 1821; Pestalozzi and Colburn were what was needed in schools (Barnard, 1851, 1856, 1859; Colburn, 1821; Monroe, 1969).

Those teaching in the normal schools were expected to believe that from a quality-of-learning perspective, the cyphering tradition had produced unsatisfactory results (Colburn, 1830/1870; Harper, 1935, 1939; Henry, 1843; Page, 1877; Reisner, 1930; Wayland, 1842). It was also expected that those studying algebra in normal schools would acquire a sufficiently strong knowledge of mathematics that they would be well positioned to use whole-class methods effectively when teaching mathematics.

From the outset, algebra became part of mathematics curricula in U.S. normal schools, and it was expected that a new generation of teachers who had not only studied algebra but had also studied how to teach mathematics would be forthcoming. Graduates of normal schools who were assessed as having strong academic knowledge and teaching abilities were appointed to high schools, and by 1865 there were a few teachers in each high school who not only had a reasonable knowledge of algebra but were also determined to teach it using Pestalozzian/Colburn teaching methodologies. That resulted in a rapid reduction in the number of cyphering books which included pages on algebra being prepared—Entries in Table 5.1 (earlier in this chapter) reveal that none of the cyphering books in the Ellerton-Clements collection prepared after 1859 dealt with algebra.

The views on cyphering of Nicholas Tillinghast (1804–1855), a principal in the early normal school movement, are especially worthy of consideration. Tillinghast, who was chosen in 1841 by Horace Mann to be the foundation Principal of Bridgewater Normal School, in Massachusetts, held that position until his death in 1857. He had been trained in mathematics at the West Point Military Academy, and from the beginning, at Bridgewater, he used his *Elements of Plane Geometry for the Use of Schools* (Tillinghast, 1844). He was imbued with the spirit of Pestalozzi and believed that, for all branches of mathematics, teachers should have a vibrant but rigorous presence in the classroom. For him, the chief aim was for students to *understand* what they learned. Richard Edwards (1857), one of Tillinghast's Bridgewater students, said of Tillinghast:

> There was a thoroughness in his teaching, but there was also another element, which if we could coin a word we might call "logicalness"—an arranging of the subject taught according to the character and wants of the mind to be instructed. In every operation, there was not only thorough knowledge, but also thorough reasoning. Every point was not only to be thoroughly understood, but it was to be understood rationally not only by itself, but also in its relations. The pupil was himself required to discover if possible, or at least to appreciate, the connection between one part of the subject and another, to see how much of one statement could be inferred from a previous one. Mere thoroughness in the knowledge of facts, or of principles, learned and remembered, is a very different matter from the

> thoroughness that characterized the teaching of Mr. Tillinghast. The one can be accomplished by the industry of the pupil; the other requires, in addition, careful thought and ready skill on the part of the teacher. (p. 14)

In 1847, David Page, a highly-regarded Principal of the State Normal School at Albany, in New York (see Page, 1877), referred to the cyphering approach as "the old plan" (p. 53).

With leaders like Tillinghast (1844), Page (1877), Edwards (1857), Henry Barnard (1851, 1856), Edward Brooks (1879), Horace Mann (in 1852 Mann became President of Antioch College, a normal school in Ohio), and Cyrus Peirce (Harper, 1939), the normal schools responded to the challenge of improving the intended, implemented and attained algebra curricula of North American schools, at all levels (Barnard, 1859). Normal school students were taught that successful mathematics teaching and assessment not only required careful verbal questioning of what students knew, but also of how and why they knew it, and how and why it might be useful (Brooks, 1879; Edwards, 1857; Executive Committee of the State Normal School, New York, 1846). And that was something which, most of the normal school leaders believed, had *never* been achieved through the cyphering approach (State of Massachusetts, 1855).

Many faculty and graduates of normal schools wrote mathematics textbooks. On that score, the lengthy honor roll included Robert F. Anderson, M. A. Bailey, Howard Griffith Burdge, Edward Brooks, Dana P. Colburn, John W. Cook, Charles Davies, James B. Dodd, David Felmley, S. A. Felter, Benjamin Greenleaf, Daniel B. Hagar, W. D. Henkel, Alfred Holbrook, George W. Hull, Edwin C. Hewett, Malcolm MacVicar, Horace Mann, Charles A. McMurry, Frank M. McMurry, William J. Milne, George Perkins, Albert N. Raub, Martha H. Rodgers, John Herbert Sangster, David M. Sensenig, G. C. Shutts, David Eugene Smith, L. M. Sniff, John F. Stoddard, Nicholas Tillinghast, Electa N. Walton, and George A. Walton (Ellerton & Clements, 2012).

Whether all of these educators were well qualified to write mathematics textbooks is a moot point. But, on key curriculum, teaching, and assessment issues, their voices were heard, and from a mathematics education perspective they represented a new era. One result was a fillip in algebra education. By 1865, the idea that all high-school students should learn algebra well, as a result of being taught it by knowledgable teachers who emphasized understanding rather than mere memorization, had been spread abroad. But what would be the effects? Did the normal-school teachers and graduates know their algebra sufficiently well to have a positive effect? Might the normal-school thrust prove to be counter-productive?

The Effect of the Introduction of Written Examinations on Algebra Education

Throughout the eighteenth century and the early nineteenth century the cyphering tradition dictated the quality of both student learning and teacher instruction. At the end of each semester a "school committee" (made up of locally-respected persons like physicians, church officials, lawyers) would conduct a public examination of a school's work. Parents would be present for the occasion and committee members would ask questions of individual pupils. On these occasions, cyphering books and "sewing samplers" would be displayed for inspection by the school committee and by parents. It was understood that a teacher's future at the school partly depended on the attractiveness of the cyphering books—and hence there was an emphasis on excellent penmanship and calligraphy (Ellerton & Clements, 2012).

From the 1840s onwards, however, individual assessment by public committees was gradually replaced by assessment based on written examinations. This trend had begun in Europe in the late 1700s but was not introduced into the United States until around 1840 (Henry, 1843; Roach, 1971; Rotherham, 1852; Watson & Kandel, 1911). It challenged the importance of the cyphering tradition in U.S. education, and its effect on algebra education is worthy of comment.

Horace Mann, Secretary to the Massachusetts Board of Education, roundly criticized the old committee system which had long been used for student and teacher evaluation (State of Massachusetts, 1848). He advocated the use of externally-set, written examinations, arguing that it should be possible for all students in a particular grade at different schools to take the same written examination at approximately the same time. Mann (1845/1925) argued that that would make assessment of the quality of student learning and of the effectiveness of teachers more objective. In 1845, he arranged for members of his Board of Education to test senior pupils in Boston public schools using written tests which had been prepared by the Board but had not been seen by the teachers before the examinations were conducted. After the examinations had been administered, and students' scripts assessed, Mann maintained that students had performed poorly and argued that the results showed that the cyphering approach did not help students to learn well and that the committee system did not provide valid assessment of student knowledge, and therefore of teacher efficiency.

Although those who supported the introduction of externally-set written examinations did not anticipate some of the weaknesses of that system of assessment (Katz, 1968; White 1886), such was the influence of those who supported the approach that it was soon adopted in many states (Kilpatrick, 1992; Landis, 1854; Reisner, 1930; S. H. M., 1856). Under the new system, the assessment of a teacher's worth was no longer intimately related to the quality of students' cyphering books, and that hastened the demise of cyphering books in schools—see Table 5.1. Teachers began to think that they could not afford to allow their students to spend time preparing cyphering books, because they needed as much time as possible to prepare students for the forthcoming high-stakes written examinations. So, teachers began to "teach to the test."

The growth of algebra education in U.S. schools during the second half of the nineteenth century was linked to the increasing power of externally-set examinations—how and why that occurred is outside the scope of this book. It is important to note, here, though, that the easiest written questions to create in relation to school algebra were those which assessed knowledge of fundamental operations (e.g., "Simplify as much as possible: $3a - 4 - 5a + 3a^2 - 7$"). It was obviously easy to determine whether answers given by students to such questions should be assessed as right or wrong. As a result, the rapid move toward written examinations in the 1840s and 1850s quickly led to much emphasis being given in school algebra classes to "simplifications," and "equation solving." Algebra quickly became a subject which emphasized symbol manipulation and getting correct answers. Despite calls for "teaching for understanding," teachers determined that it was difficult to test for understanding if assessment was by externally-set examinations. And, so, the die was cast.

Who Should Be Given the Most Credit for Improving School Algebra in North America?

As Kilpatrick and Izsák (2008) have pointed out, "in the United States and Canada before 1700, algebra was absent not only from the school curriculum but also from the curriculum of the early colleges and seminaries" (p. 3). In this chapter, cyphering-book evidence has been presented showing that so far as schools were concerned, the same was true for most schools as late as 1820.

It seemed that any movement with respect to the inclusion of algebra within school curricula had to be initiated by colleges because at that time hardly anyone in the schools knew much about algebra. With the advent of public high schools from 1821 onward, and decisions by colleges to make knowledge of elementary algebra a pre-requisite for entry, it was inevitable that algebra would increasingly become part of intended and implemented curricula of post-elementary schools, and especially of public high schools and academies. But who would be in control of this development—the schools, the colleges, or local or state education authorities?

Because algebra, even of the most elementary kind, was not well known by many teachers of mathematics in the United States of America, and because algebra was becoming a pre-requisite for entry to colleges, there arose a perceived need, from both teachers and students, for good algebra textbooks. Back in 1730 Pieter Venema had had a textbook published for the use of Dutch-speaking students in New York, but that had failed to be a commercially-viable venture, and it was not until 1788 that another textbook, this one written by Nicolas Pike in English, introduced elementary algebra to U.S. students. The section on algebra in Pike's (1788) book was not well received, however, and although Pike's (1793) "Abridgement" was aimed at schools, no section on algebra was included in that book. In 1790, a mathematics textbook authored by Consider and John Sterry, brothers from Rhode Island, was published and it included a substantial section on algebra. But it was not published beyond the first edition.

In fact, John Bonnycastle's (1806) *An Introduction to Algebra*, which was a reprint of an English textbook, was the first commercially-successful algebra textbook for schools published in the United States, with 17 editions appearing between 1806 and 1847. The brief analysis of Bonnycastle's text in this chapter suggested that it left much to be desired so far as getting students to think about what was being presented. The same was true of Charles Davies' (1835) *Elements of Algebra*, which was largely based on a translation of Louis Bourdon's *Elements of Algebra*. It was less true of Jeremiah Day's (1814) *An Introduction to Algebra*, but that book was specifically prepared with Yale College students in mind and its chapters were, often, too difficult for school students. To his credit, Day recognized that fact, and persuaded one of his former students, James B. Thompson, to prepare *Elements of Algebra* as part of the "Day and Thomson series." The title of their main book, *Elements of Algebra, Being an Abridgment of Day's Algebra, Adapted to the Capacities of the Young, and the Method of Instruction, in Schools and Academies* carried a message to schools, and between 1843 and 1850 no less than 14 editions were published (Karpinski, 1940). For a moment, it seemed that college-professor/school-teacher partnerships in the preparation of school mathematics textbooks were likely to become the order of the day.

Our analysis in this chapter identified the algebras written by Warren Colburn, Ebenezer Bailey, James B. Thomson, and Joseph Ray as the best of the early U.S. algebras for use in schools—at least the forms of language used by those authors were more appropriate for beginning algebra students than the forms of language used by the other authors. Colburn, Bailey, and Ray adopted problem-based approaches as they attempted to get students to reflect on and apply what they were asked to read.

This chapter has revealed the origins of a conflict between mathematicians and mathematics educators which has continued into the twenty-first century. The mathematicians knew their mathematics well but had difficulty communicating what they knew to school children; the mathematics educators did not know their mathematics so well, but students found it much easier to comprehend what they wrote. Other algebras aimed at high-school and grammar-school students soon appeared (see, e.g., Bridge, 1832; Green, 1839; Harney, 1840; Perkins, 1845; Sherwin, 1842, 1845) and answers to questions like the following began to be considered:

- What content should there be in an algebra textbook for schools?
- How should the topics which constituted school algebra be sequenced?
- How should school algebra be taught, and what forms of pre-service and in-service education should be made available to teachers?
- How should the quality of algebra learning and algebra teaching be assessed?
- What relationships between algebra, arithmetic, geometry and trigonometry should be introduced into schools?

Finally, it could be argued that what was missing from the algebra cyphering books, and even from the algebra textbooks, was how algebra could and should be used for "proving." The concept of proof was not well understood in the schools, with the word "proof" almost always being regarded as synonymous with "check." Given the absence of geometry in the curricula of most schools one might have thought that, with older children, algebra might have been used to establish proofs—for, after all, the concept of proof, and methods of proving, have always been regarded as fundamentally important by mathematicians. But, in 1865 a level of success with the difficult work of devising methods for enabling school learners to prove using algebra was still a century away. That is why the following "Epilogue" to this chapter is important for this book.

Epilogue to Chapter 5

From a history-of-mathematics perspective there is one part of a story told in this chapter which needs elaboration. That part refers to the section on Pieter Venema in New York, in the mid-1720s. It bears repeating that, somehow, as if by fate, Pieter Venema, a person who had authored a highly successful algebra textbook in The Netherlands (Venema, 1714)—one of the world's powerhouse nations at that time—had come to live in the still small distant outpost of New York (which had previously been known as "New Amsterdam"). The contents of Venema's book on algebra indicated that its author was someone who could elevate the tone of school mathematics in the New World. And, yes, soon after his arrival in New York, he made it clear that he was willing to write a mathematics textbook which included algebra and would be used by education institutions and families in New York. A key question was: "Would the language he used in the textbook, be Dutch, or would it be English?" In 1725, soon after his arrival in New York, Venema prepared a draft textbook, which, in this chapter, we have called his "precursor." Printed pages were written in Dutch most of the handwritten pages in English. Ultimately, he and his advisors decided that the whole book should appear in the Dutch language. It was published in New York in 1730, and a copy is held today in the Plimpton collection in the Butler Library, at Columbia University; another copy is held by the New York Historical Society (Karpinski, 1980). But the only copy of the (1725) precursor is held by us. This chapter represents the first time anyone has ever referred to the precursor.

The reader is referred to the set of reproductions of six pages from the precursor shown as Figure 5.3 earlier in this chapter. Two of the six pages show Venema's English-language handwritten solution to the following "alligation" problem.

An Alligation Question from Venema's (c. 1725) *Precursor*

> A tobacconist hath 3 sorts of tobacco, viz of 2/8d per pound; another of 20d per pound; a third sort of 16d per pound; of these he would make a mixture containing 56 pounds that may be sold for 22d per pound. How much of each sort must he take?

Venema's solution, shown in Figure 5.3, went something like this (see Figure 5.10):

Let: a = of that worth 2/8d per pound;
e = that of 20d per pound;
y = that of 16d per pound.

Then	**1.**	$a + e + y = 56$
And	**2.**	$32a + 20e + 16y = 1232$
$1 - a$	**3.**	$e + y = 56 - a$
$2 - 32a$	**4.**	$20e + 16y = 1232 - 32a$
3 ⊠16	**5.**	$16e + 16y = 896 - 16a$
$4 - 5$	**6.**	$4e = 336 - 16a$
6 ⊠ 4	**7.**	$e = 84 - 4a$
$3 - 7$	**8.**	$y = 3a - 28$

> Note that the first column in this text shows operations to be performed on an earlier line. Thus, in the third line, "a" is to be subtracted from each side of the first line, and in the sixth line, entries in Line 5 are to be subtracted from entries in Line 4. Note, also, that the second column shows (in bold print) the number of the step.

Figure 5.10. Venema's algebraic solution to an alligation ("mixture") problem.

Venema concluded that it was evident from the 7th step that the quantity signified by a must be less than 21 and, from the 8th step, greater than 9 1/3. That is to say, a could represent any number between 9 1/3 and 21. It was interesting that Venema was clearly indicating that, provided the tobacconist had enough of each kind—and assuming fractional amounts in pounds were permitted, there were many solutions to the problem. For example, a could be 10, and then e would be 44 (from step 7), and y would be 2 (from step 8); or, a could be 19.5, and then e would be 6, and y would be 30.5. There was an infinite set of solutions to the problem. Etc. The algebra itself was not difficult, but the thinking behind what was done was important.

Most of the cyphering books in the E-C collection deal mainly with *abbaco* arithmetic, and "alligation" (i.e., the arithmetic of "mixing quantities") was an important middle- to high-level topic in the *abbaco* sequence. Thus, many cyphering books show solutions to problems not unlike the one Venema considered (above). In every case a standard (and extremely "clever") algorithm was used to arrive at an arithmetic solution quickly. Any "reason for the rule" was not mentioned, The "rule" was simply stated and then assumed to be true. It could be found in textbooks (see, e.g., Pike, 1788, pp. 333–338). In most of the entries on alligation in cyphering books and textbooks only a single solution for the mixture was shown (see, e.g., Stephen Pike, 1822, p. 165), and there was rarely any discussion of whether that solution was unique—although both Nicolas Pike (1788) and John Ward (1719) did, in fact, indicate that more than one mixture might be possible, and how other solutions could be obtained.

Venema used algebra to show that there could be an infinite set of solutions (i.e., possible mixtures). Unfortunately, he did not comment on what he did, other than what is shown in Figure 5.10. But the reasoning behind the mathematics presented in Figure 5.10 was profound—the algebra shown, was essentially, a *proof*. It is clear that Venema had much to offer so far as the development of mathematics in his newly adopted country was concerned. But, as previously stated, the opportunity was missed, and it would be almost 60 years before another book written by someone in North America would appear with a substantial section on algebra.

A typical cyphering book entry on alligation is shown in Figure 5.11. It was made by David Townsend in a manuscript prepared in New York in the early 1770s. That cyphering book is now held in the E-C cyphering book collection. The task was: "If I have 4 sorts of tea – one at 9 pence per oz., another at 12 pence, another at 24 pence and another at 30 pence—how much of each sort must I take to make the compound worth 20 pence per oz.?"

David did not state his solution clearly—but his answer was 10 ounces of the 9 pence variety, 4 ounces of the 12 pence variety, 8 ounces of the 24 pence variety, and 11 ounces of the 30 pence variety. David used a standard "alligation alternate" algorithm, to arrive, magically, at that single solution to the problem. Readers of this book are invited to use algebra to solve the same problem, and also to decide whether, in fact, there were solutions other than the one given by David.

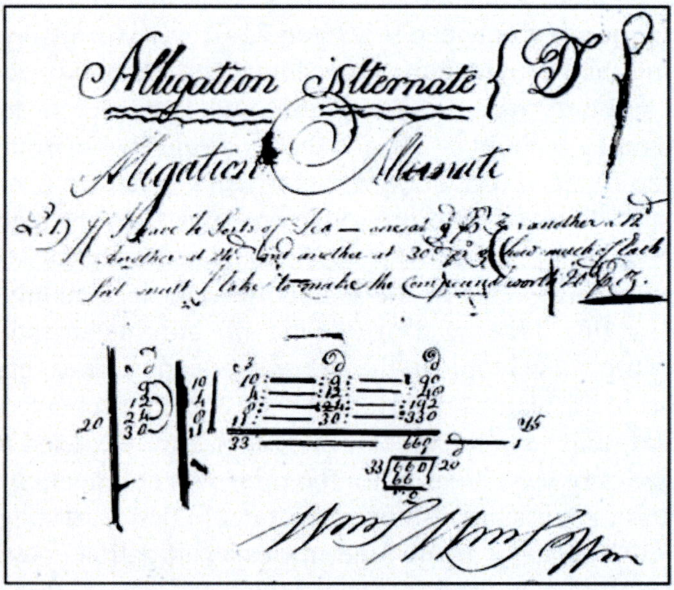

Figure 5.11. David Townsend's cyphering book solution to a standard alligation task.

References

Ackerberg-Hastings, A. (2010). John Farrar and curricular transitions in mathematics education. *International Journal for the History of Mathematics Education, 5*(2), 17–30.

Angulo, A. J. (2012). The polytechnic comes to America: How French approaches to science instruction influenced mid-nineteenth century American higher education. *History of Science, 50*(168), 315–338. https://doi.org/10.1177/007327531205000304

Baldwin, J. (1908). *Barnes's elementary history of the United States*. New York, NY: American Book Company.

Bailey, E. (1833). *First lessons in algebra, being an easy introduction to that science; designed for the use of academics and common schools*. Boston, MA: Carter, Hendee and Co.

Barnard, H. (1851). *Normal schools, and other institutions, agencies and means designed for the professional education of teachers*. Hartford, CT: Case, Tiffany and Company.

Barnard, H. (1856). Graduation of public schools with special reference to cities and large villages. *American Journal of Education, 2*, 455–464.

Barnard, H. (Ed.). (1859). *Life, educational principles, and methods of John Henry Pestalozzi*. New York, NY: F. C. Brownell.

Beckers, D. (2006). Elementary mathematics education in the Netherlands ca. 1800: New challenges, changing goals. *Bulletin of the Belgian Mathematical Society, 13*, 937–940. https://doi.org/10.36045/bbms/1170347816

Bonnycastle, J. (1806). *An introduction to algebra: With notes and observations*. Philadelphia, PA: Joseph Crukshank.

Bonnycastle, J. (1822). *An introduction to algebra: With notes and observations*. New York, NY: Evert Duyckinck and George Long.

Bourdon, L. P. M. (1831). *Elements of algebra by Bourdon: Translated from the French for colleges and schools*. Boston, MA: Hilliard, Gray, Little & Wilkins.

Bowditch, N. I. (1840). *Memoir of Nathaniel Bowditch*. Boston, MA: C. C. Little & Brown.

Bradley, A. D. (1949). Pieter Venema, teacher, textbook author and free thinker. *Scripta Mathematica, 15*(1), 13–16.

Bridge, B. (1832). *A treatise on the elements of algebra*. Philadelphia, PA: Key, Mielke and Biddle.

Brooks, E. (1879). *Normal methods of teaching containing a brief statement of the principles and methods of the science and art of teaching*. Philadelphia, PA: Normal Publishers.

Cajori, F. (1890). *The teaching and history of mathematics in the United States* (Circular of Information No. 3, 1890). Washington, DC: Bureau of Education.

Chateauneuf, A. O. (1930). *Changes in the content of elementary algebra since the beginning of the high school movement as revealed by the textbooks of the period*. PhD dissertation, The University of Pennsylvania.

Clements, M. A., Keitel, C., Bishop, A. J., Kilpatrick, J., & Leung, F. (2013). From the few to the many: Historical perspectives on who should learn mathematics. In M. A. Clements, A. J. Bishop, C. Keitel, J. Kilpatrick, & F. Leung (Eds.), *Third international handbook of mathematics education* (pp. 7–40). New York, NY: Springer. https://doi.org/10.1007/978-1-4614-4684-2_1

Colburn, W (1821). *An arithmetic on the plan of Pestalozzi, with some improvements*. Boston, MA: Cummings and Hilliard.

Colburn, W. (1822). *Arithmetic upon the inductive method of instruction being a sequel to intellectual arithmetic*. Boston:, MA: Cummings and Hilliard.

Colburn, W. (1825). *An introduction to algebra upon the inductive method of instruction*. Boston, MA: Cummings, Hilliard, and Co.

Colburn, W. (1830/1970). Teaching of arithmetic. In J. K. Bidwell & R. G. Clason (Eds.), *Readings in the history of mathematics education* (pp. 24–37). Washington, DC: National Council of Teachers of Mathematics.

da Ponte, J. P., & Guimarães, H. M. (2014). Notes for a history of the teaching of algebra. In A. Karp & G. Schubring (Eds.), *Handbook on the history of mathematics education* (pp. 323–334). New York, NY: Springer.

Davies, C. (1835). *Elements of algebra: Translated from the French of M. Bourdon*. New York, NY: Wiley & Long.

Davies, C. (1837). *Elements of algebra: Translated from the French of M. Bourdon*. New York, NY: Wiley & Long.

Day, J. (1814). *An introduction to algebra, being the first part of a course of mathematics, adapted to the method of instruction in the American colleges*. New Haven, CT: Howe & Deforest.

Dwight, T. (1903). *Memories of Yale life and men*. New Haven, CT: Dodd, Mean and Co.

Edwards, R. (1857). *Memoir of Nicholas Tillinghast, first Principal of the State Normal School at Bridgewater, Massachusetts*. Boston, MA: James Robinson & Co.

Ellerton, N. F., & Clements, M. A. (2012). *Rewriting the history of mathematics education in North America 1607–1861*. New York, NY: Springer. https://doi.org/10.1007/978-94-007-2639-0

Euler, L. (1797). *Elements of algebra*. London, England: J. Johnson.

Euler, L. (1818). *An introduction to the elements of algebra designed for the use of those who are acquainted only with the first principles of arithmetic, selected from the algebra of Euclid*. Boston, MA: Hilliard, Gay, Little & Wilkins.

Executive Committee of the State Normal School, New York. (1846). *Annual report*. Albany, NY: Author.

Goodfriend, J. D. (2017). *Who should rule at home? Confronting the elite in British New York City*. Ithaca, NY: Cornell University Press. https://doi.org/10.7591/cornell/9780801451270.001.0001b

Gough, J. (1788). *Practical arithmetic in four books . . . with an appendix of algebra by the late W. Atkinson of Belfast*. Wilmington, DE: Peter Brynberg.

Green, R. W. (1839). *Gradations in algebra, in which the first principles of algebra are inductively explained*. Philadelphia, PA: Thomas, Cowperthwait and Co.

Greenwood, I. (1729). *Arithmetick, vulgar and decimal, with the application thereof to a variety of cases in trade and commerce*. Boston, MA: Kneeland & Green.

Guralnick, S. (1975). *Science and the antebellum American college*. Philadelphia, PA: American Philosophical Society.

Harney, J. H. (1840). *An algebra upon the inductive method of instruction*. Louisville, KY: Morton and Griswold.

Harper, C. (1935). *Development of the teachers college in the United States with special reference to Illinois State University*. Bloomington, IL: McKnight & McKnight.

Harper, C. (1939). *A century of public teacher education: The story of the state teachers colleges as they evolved from the normal schools*. Washington, DC: American Association of Teachers Colleges.

Henry, J. (1843). An address upon education and common schools. *The Common School Journal, 6*(2), 26–29.

Hogan, E. R. (1981). Theodore Strong and ante-bellum American mathematics. *Historia Mathematics, 8*, 439–455. https://doi.org/10.1016/0315-0860(81)90052-5

Howsam, L., & Raven, J. (Eds.). (2011). *Books between Europe and the Americas: Connections and communities, 1620–1860*. Basingstoke, England: Palgrave Macmillan. https://doi.org/10.1057/9780230305090

Hutton, C. (1764). The schoolmaster's *guide: Or, A complete system of practical arithmetic*. London, England: R. Baldwin.

Hutton, C. (1812). *A course of mathematics for the use of academies*, as well as private tuition. New York, NY: Samuel Campbell

Hutton, C. (1831). *A course of mathematics for the use of academies*, as well as private tuition. New York, NY: W. E. Dean.

Kanbir, S., Clements, & Ellerton, N. F. (2017). *Using design research and history to tackle a fundamental problem with school algebra*. New York, NY: Springer. https://doi.org/10.1007/978-3-319-59204-6

Karpinski, L. C. (1940). *Bibliography of mathematical works printed in America through 1850*. Ann Arbor, MI: University of Michigan Press.

Karpinski, L. C. (1980). *Bibliography of mathematical works printed in America through 1850* (2nd ed.). New York, NY: Arno Press.

Katz, M. B. (1968). *The irony of early school reform: Educational innovation in mid-19th century Massachusetts*. Cambridge, MA: Harvard University Press.

Kilpatrick, J. (1992). A history of research in mathematics education. In D. A. Grouws (Ed.), *Handbook of research on mathematics teaching and learning* (pp. 3–38). New York, NY: Macmillan Publishing Company.

Kilpatrick, J. (2014a). Warren Colburn and the inductions of reason. In K. Bjarnadóttir, F. Furinghetti, J. Prytz, & G. Schubring (Eds.). *"Dig where you stand" 3: Proceedings of the Third International Conference on the History of Mathematics Education* (pp. 219–232). Uppsala, Sweden: Uppsala University Department of Education.

Kilpatrick, J. (2014b). Mathematics education in the United States and Canada. In A. Karp & G. Schubring (Eds.), *Handbook on the history of mathematics education* (pp. 323–334). New York, NY: Springer. https://doi.org/10.1007/978-1-4614-9155-2_16

Kilpatrick, J., & Iszák, A. (2008). Historical perspectives on algebra in the curriculum. In C. E. Greenes & R. Rubinstein (Eds.), *Algebra and algebraic thinking in school mathematics: Seventieth yearbook* (pp. 3–18). Reston, VA: National Council of Teachers of Mathematics.

Kilpatrick, W. H. (1912). *The Dutch schools of New Netherland and colonial New York*. Washington DC: US Bureau of Education.

Lacroix, S. F. (1818a). *An elementary treatise on arithmetic, taken principally from the arithmetic of S. F. Lacroix and translated into English with such alterations and additions as were found necessary in order to adapt it to the use of the American student*. Cambridge, New England: Hilliard and Metcalf, at the University Press.

Lacroix, S. F. (1818b). *Elements of algebra, translated from the French … Cambridge,* New England: Hilliard and Metcalf, at the University Press.

Landis, J. H. (1854). Examination of teachers' certificates. *Pennsylvania School Journal, 2*(1), 11–12.

Mann, H. (1845/1925). Boston grammar and writing schools. *The Common School Journal*. In O. W. Caldwell & S. A. Courtis, *Then and now in education 1845–1923* (pp. 237–272): New York, NY: World Book Company.

Michalowicz, K. D., & Howard, A. C. (2003). Pedagogy in text: An analysis of mathematics texts from the nineteenth century. In G. M. A. Stanic & J. Kilpatrick (Eds.). *A history of school mathematics* (Vol. 1, pp. 77–112). Reston, VA: National Council of Teachers of Mathematics.

Minto, W. (1788). *An inaugural oration, on the progress and importance of the mathematical sciences*. Manuscript, held in the Clements Library, The University of Michigan.

Molloy, P. M. (1975). *Technical education and the Young Republic: West Point as America's Ecole Polytechnique, 1802–1833*. PhD Dissertation, Brown University.

Monroe, W. (Will) S. (1969). *History of the Pestalozzian movement in the United States*. New York, NY: Arno Press & the New York Times.

Page, D. P. (1877). *Theory and practice of teaching: The motives and methods of good school-keeping* (90th ed.). New York, NY: A. S. Barnes & Company.

Parshall, K. H. (2003). Historical contours of the American mathematical research community. In G. M. A. Stanic and J. Kilpatrick (Eds.). *A history of school mathematics* (Vol. 1, pp. 113–158). Reston, VA: National Council of Teachers of Mathematics.

Peirce, B. (1837). *An elementary treatise on algebra; to which are added elementary equations and logarithms*. Boston, MA: James Munroe and Company.

Peirce, B. (1841). *An elementary treatise on curves, functions, and forces (Volume First): Analytic geometry and the differential calculus*. Boston, MA: James Munroe and Company.

Pelletreau, W. S. (1907). *Historic homes and institutions and genealogical and family history of New York*. New York, NY: Lewis Publishing.

Perkins, G. R. (1845). *The elements of algebra*. Utica, NY: B. S. Merrill.

Phillips, C. J. (2015). An officer and a scholar: Nineteenth-century West Point and the invention of the blackboard. *History of Education Quarterly, 55*(1), 82–108. https://doi.org/10.1111/hoeq.12093

Pike, N. (1788). *The new and complete system of arithmetic, composed for the use of the citizens of the United States*. Newbury-Port, MA: John Mycall.

Pike, N. (1793). *The new and complete system of arithmetic composed for the use of the citizens of the United States (abridged for the use of schools)*. Newbury-Port, MA: John Mycall, Isaiah Thomas.

Pike, S. (1822). *The teacher's assistant or a system of practical arithmetic; wherein the several rules of that useful science, are illustrated by a variety of examples, a large proportion of which are in Federal money. The whole is designed to abridge the labour of teachers, and to facilitate the instruction of youth*. Philadelphia, PA: Benjamin Warner.

Ray, J. (1848). *Ray's algebra: Part first*. Cincinnati, OH: Van Antwerp Bragg & Co.

Reisner, E. H. (1930). *The evolution of the common school*. New York, NY: Macmillan.

Roach, J. (1971). *Public examinations in England, 1850–1900*. Cambridge, England: Cambridge University Press. https://doi.org/10.1017/CBO9780511896309

Rotherham, W. (1852). *The algebraical equation and problem papers, proposed in the examinations of St. John's College, Cambridge from the year 1794 to the present time*. Cambridge, England: Henry Wallis.

Seybolt, R. F. (1921). *The evening schools of colonial New York City*. Albany, NY: The University of the State of New York.

Sherwin, T. (1842). *An elementary treatise on algebra*. Boston, MA: Benjamin B. Mussey.

Sherwin, T. (1845). *The common school algebra*. Boston, MA: Phillips and Sampson.

S. H. M. (1856). Written examinations. *Rhode Island Schoolmaster, 2*, 156.

Simons, L. G. (1923). A Dutch textbook of 1730. *The Mathematics Teacher, 16*(6), 340–347. https://doi.org/10.5951/MT.16.6.0340

Simons, L. G. (1924). *Introduction of algebra into American schools in the 18th century*. Washington, DC: Department of the Interior Bureau of Education.

Simons, L. G. (1931). The influence of French mathematicians at the end of the eighteenth century upon the teaching of mathematics in America. *Isis, 15*, 104–123. https://doi.org/10.1086/346540

Simons, L. G. (1936). *Bibliography of early American textbooks on algebra*. New York, NY: Scripta Mathematica, Yeshiva College.

Smith, D. E., & Ginsburg, J. (1934). *A history of mathematics in America before 1900*. Chicago, IL: The Mathematical Association of America. https://doi.org/10.1090/car/005

State of Massachusetts. (1848). *Report No. 12 to the Massachusetts School Board by Horace Mann*. Boston, MA: Author.

State of Massachusetts. (1855). *Eighteenth annual report of the Secretary of the Massachusetts Board of Education*. Boston, MA: Author.

Sterry, C., & Sterry, J. (1790). *The American youth: Being a new and complete course of introductory mathematics, designed for the use of private students*. Providence, RI: Authors.

Stocker, H. E. (1922). *A history of the Moravian Church in New York City*. New York, NY: Author.

The Kouroo Contexture. (2021). Image of Ebenezer Bailey retrieved on July 8, 2021, from http://www.kouroo.info/kouroo/thumbnails/B/EbenezerBailey

Thomas, M., & Kempis Kloyda, A. (1937). Linear and quadratic equations 1550–1660. *Osiris, 3*, 165–192.

Thomson, J. B. (1843). *Elements of algebra, being an abridgment of Day's algebra, adapted to the capacities of the young, and the method of instruction, in schools and academies*. New Haven, CT: Durrie & Peck.

Tillinghast, N. (1844). *Elements of plane geometry for the use of schools*. Boston, MA: Lewis & Sampson.

Venema, P. (1714). *Een korte en klare onderwysinge in de beginselen van de algebra ofte stel-konst*. Groningen, The Netherlands: Author.

Venema, P. (1725). Unpublished "precursor." Held in the Ellerton-Clements textbook collection, Bloomington, IL.

Venema, P. (1730). *Arithmetica of Cyffer-Konst, volgens de Munten Maten en Gewigten te Nieu-York, gebruykelyk als mede een kort Ontwerp van de Algebra*. New York, NY: Jacob Goelet.

Ward, J. (1719). *The young mathematician's guide: Being a plain and easie introduction to the mathematicks*. London, England: Thomas Horne.

Watson, F., & Kandel, I. L. (1911). Examinations. In P. Monroe (Ed.), *A cyclopaedia of education* (Vol. 2, pp. 532–538). New York, NY: The Macmillan Company.

Wayland, F. (1842). *Thoughts on the present collegiate system*. Boston, MA: Gould, Kendall & Lincoln.

Webber, S. (1801). *Mathematics compiled from the best authors and intended to be the textbook of the course of private lectures on these sciences in the University at Cambridge*. Boston, MA: Thomas and Andrews. https://doi.org/10.5962/bhl.title.17251

Webber, S. (1808). *Mathematics compiled from the best authors and intended to be the textbook of the course of private lectures on these sciences in the University at Cambridge* (2nd ed.). Cambridge, MA: William Hilliard.

White, E. (1886). *The elements of pedagogy*. New York, NY: American Book Company.

Wikipedia contributors. (2021, April 8). Jeremiah Day. In *Wikipedia, The Free Encyclopedia*. Retrieved July 14, 2021, from https://en.wikipedia.org/w/index.php?title=Jeremiah_Day &oldid=1016597452

Wikipedia contributors. (2021, May 9). Florian Cajori. In *Wikipedia, The Free Encyclopedia*. Retrieved July 11, 2021 from https://en.wikipedia.org/w/index.php?title=Florian_Cajori &oldid=1022259870

Zitarelli, D. E. (2019). *A history of mathematics in the United States and* Canada *(Vol. 1, 1492–1900)*. Providence, RI: MAA Press. https://doi.org/10.1090/spec/094

Chapter 6
Pre-College Geometry, Mensuration, Trigonometry, Surveying, and Navigation 1607–1865

Abstract This chapter analyzes pre-college education developments in geometry, mensuration, trigonometry, surveying, and navigation between 1607 and 1865, in the 13 colonies and then in the United States of America. Although throughout that period relatively few students prepared cyphering books which focused on anything other than *abbaco* arithmetic. Some school students did study one or more of algebra, geometry, trigonometry, astronomy, navigation, and surveying, but most of those who did had not previously studied topics like angles, decimals, fractions, logarithms, or elementary mechanics, and therefore it was extremely difficult for them to make good progress. Evidence will be presented showing that some students nevertheless managed to succeed. In particular, data from a cyphering book prepared by Thomas Willson in Pennsylvania in 1789 will be examined in detail, and the analysis will suggest what implemented curricula in post-*abbaco* forms of mathematics were like at that time. It has often been argued that so far as mathematics education was concerned much was achieved in the schools of that time, because there was an over-emphasis on mere memorization. In this chapter it is argued, however, that that contention rests on the untested assertion that students who prepared cyphering books did not understand and could not apply what they entered in their cyphering books. An important aim for the cyphering tradition was that students who prepared manuscripts would consult them if and when they felt the need to do so later in their lives.

Keywords Charles Davies • Cyphering books • Cyphering tradition • Decimal currency • Enoch Lewis • Euclidean geometry • Geometry education • Inequality of educational opportunity • Legendre • Metric system • Navigation education • Salem (Massachusetts) • Surveying education • Thomas Jefferson • Trigonometry • U.S. Military Academy (West Point)

How Much Geometry, Mensuration, Trigonometry, Surveying, and Navigation, Was Studied in Pre-College Education Institutions in North America, 1607–1865?

The Phillips Library within the Peabody Exeter Museum in Salem, Massachusetts, holds more than 200 cyphering books, some of which feature magnificent penmanship and calligraphy (Gaydos & Kampas, 2010). Many of the cyphering books were prepared in the eighteenth and early nineteenth centuries, when Salem was one of the largest and richest urban centers in North America

N. F. Ellerton, M. A. (Ken) Clements, *Toward Mathematics for All*, History of Mathematics Education,
https://doi.org/10.1007/978-3-030-85724-0_6

(Morison, 1921; Peabody Essex Museum, n.d.). Its wealth had been generated by the part it played in international trade with the Spice Islands, India and China. Ships left Salem and voyaged to the farthest ports in the East Indies. In mercantile centers, like Salem, many apprentices studied arithmetic, mensuration and navigation in evening classes because skilled reckoners and trained navigators were needed not only on the ships but also on the wharves and in the surrounding customs houses (Hertel, 2016).

Although navigation cyphering books were not usually prepared in the common schools in Salem during the eighteenth century there was a strong navigation course at nearby Harvard College. Those wishing to study navigation but did not qualify to enter Harvard could attend private evening classes in navigation in their home town. *The Salem Register* of March 29, 1802, for example, included an advertisement posted by a certain George Douglas offering "young gentlemen" instruction in "English, English grammar, writing, arithmetic, bookkeeping, mathematics, with their application to navigation" (p. 1). The standard navigation curriculum covered elementary practical geometry, trigonometry, logarithms, and various types of sailing (plain, Mercator's, parallel, traverse, great circle, etc.), and students usually prepared a log for an imagined or actual journey (e.g., from Boston to Nova Scotia). The demand was sufficiently great that Elias Hasket Derby—who, at one time, was reputedly North America's richest person—established a school of navigation in Salem from which many young men gained certificates which would help them obtain employment as midshipmen (Middlekauf, 1963; Phillips, 1947). According to Samuel Morison (1921), in the 1790s all seaport towns in Massachusetts had private navigation schools. Even in as small a village as Wellfleet, there were, in winter, a number of private schools at which young men often took courses in navigation.

In the eighteenth- and early-nineteenth centuries many cyphering books were prepared by midshipmen during voyages to distant locations (Durkin, 1942; Ellerton & Clements, 2012; Rawley, 1981; Taylor, 1966). The Phillips Library, in Salem, and the Houghton Library at Harvard, hold numerous cyphering books which were prepared by midshipmen as they travelled to and from Africa, Europe and Southern and Eastern Asia (Gaydos & Kampas, 2010; Rawley, 1981). Some of these were standard *abbaco*-arithmetic cyphering books, but others were navigation cyphering books (see, for example, the cyphering book in the Houghton Library prepared by William F. Allen in 1827, during a voyage from Salem towards India on the *Barque Pompey*).

After 1702, in England, anyone who wanted to be a naval schoolmaster needed a "naval schoolmaster's certificate," and there is evidence that this qualification was noticed, even sought after, in North America. Thus, for example, in the early 1740s Nathan Prince (1698–1748), a Harvard College graduate, a mathematics tutor at Harvard between 1723 and 1742, and someone who, in 1738, was a candidate for Harvard's Hollis Professor of Mathematics and Natural Philosophy (Zitarelli, 2019), went to London specifically for the purpose of qualifying for the certificate (see, Taylor, 1966, pp. 139–140). In 1801 a book was published in London "for instructors of sea youth" including "schoolmasters of the Royal Navy" (Morrice, 1801). Almost all ship-owners employed "naval schoolmasters" who assisted the navigators and instructed midshipmen in reading, writing, arithmetic and, for those acquainted with trigonometry and logarithms, in navigation (Durkin, 1942; Taylor, 1966).

During visits to Salem in 2009 and 2010 we (Nerida Ellerton and Ken Clements) gained the strong impression that navigation cyphering books were important in the history of mathematics education in North America, and that view has subsequently been confirmed through research by Joshua Hertel (2016). We arrived at that conclusion because we were impressed by the number of beautiful navigation cyphering books held in the Phillips Library at Salem and in the Houghton Library at Harvard University. However, we have now examined over 1500 cyphering books located in many places including the E-C Collection of 536 manuscripts prepared in North America during the period 1607–1865. It is clear that although the preparation of navigation cyphering books was very important for many Salem boys, and for many practicing or prospective midshipmen, it was not a priority for most young people in most other parts of North America.

Many Salem boys looked forward to a time when they might assume the responsibility of guiding large merchant ships, or even war vessels, across the Atlantic Ocean or to exotic far-away places in the East Indies. When they applied to be appointed as midshipmen it was usually expected that they would have already prepared navigation cyphering books which they could use as reference books when they were on sea voyages. Those who had passed through navigation classes might be employed to teach arithmetic and navigation to midshipmen during the long trips. Their navigation cyphering books offered technical navigation knowledge and skills that they could pass on to neophytes.

But, as reports spread of the brutal savagery of pirates, and of international hostilities which often resulted when ships of one nation were forcefully captured, as "prizes," by those of another, the lure of the sea diminished among many young men (Ellerton & Clements, 2014). Of the 536 cyphering books in the E-C collection prepared between 1607 and 1865, only 11 are navigation cyphering books (see Table 5.1 in Chapter 5). We reflected on the likelihood that those cyphering books would have been precious for the students who prepared them—the fact that they still survive, having been passed from family member to family member for two centuries, or more, testifies to that. But other youngsters, in other places, in or around 1800 might have found it more relevant to have prepared surveying cyphering books. There are 31 manuscripts in the E-C collection with sections on surveying, and there are 33 which focus on trigonometry. There are many others, of course, which focus on *abbaco* arithmetic which provided methodologies for many commercially-important calculations—e.g., for simple and compound interest, barter, tare and tret, alligation, fellowship, single or double false position, or mensuration.

In Table 5.1 (in Chapter 5) we reported that over 90 percent of the cyphering books (CBs) in the E-C collection focused solely on *abbaco* arithmetic topics. There were, however, other cyphering books which included sections on algebra (5% of all cyphering books up to 1865 in the E-C collection), geometry (7%), trigonometry (6%), surveying (6%) and navigation (2%). Almost all of the cyphering books in the E-C collection were prepared between about 1667 and 1861, with most of them being prepared before 1850. None of the individual cyphering books made any reference to calculus. Usually, those who prepared cyphering books which included material on surveying or navigation also included sections on Euclidean geometry and trigonometry. One cyphering book included sections devoted to all of geometry, plane and spherical trigonometry, surveying, and navigation. Many of the manuscripts devoted to trigonometry also included sections on mensuration and logarithms which were used to find measures of heights, distances, angles, time, etc. None of those who prepared cyphering books would have prepared them while they were attending common schools.

Thomas Willson's (1789) Composite Cyphering Book

In order to show what an implemented curriculum could have been with respect to non-*abbaco* arithmetic topics we now include a section based on a composite cyphering book prepared in Philadelphia in 1789 and 1790 by Thomas Willson. We adopted this form of data presentation and analysis because we wanted to help readers become *qualitatively* aware of the kinds of applied mathematics which were being studied in some pre-college North American schools during the early years of the United States of America. The major form of evidence, now available for the first time, comes from cyphering books. We (Ellerton and Clements) are among only a few historians to have had immediate access to a large collection of North American cyphering books which included manuscripts prepared in many parts of North America—and, in addition, to cyphering books from other countries—while writing on the history of applied mathematics in those parts of mainland America now known as the United States of America. Thus, for example, from the point of view of historical scholarship, it is no longer sufficient to assert —as, for example, Sinclair (2008) has asserted—that geometrical education in the United States really began in 1844 (the year when Harvard first required prospective students to demonstrate knowledge of geometry).

Thomas Willson's 185-page cyphering book had dimensions 12.75″ by 7.75″, marbleized hard covers, and a leatherette spine cover. Each page was numbered, and there were many inked diagrams. The penmanship and calligraphy were of a reasonably high standard and many of the handwritten pages had a very authentic appearance—with calculations obviously having been done by Thomas himself. Thomas prepared sections on gauging (18 pages), geometry (38 pages), mensuration (42 pages), trigonometry (17 pages), surveying (37 pages), and navigation (23 pages). He closed his manuscript with a log of a journey from Cape Henlopen (in Delaware) toward Barbados (10 pages). Often his spelling left something to be desired—for example, "polygon" was spelt as "pollygon." Parts of the text could have been copied from one or more textbooks but checking via Google did not enable any specific textbooks to be identified.

Thomas Willson's (1789) Section on Gauging

Thomas began what would become his composite cyphering book with a section on gauging. Gauging was an important part of the lives of those who packed or unpacked materials or prepared them for export. Those who made barrels were called "coopers" and they needed to know how to perform gauging calculations. Nathaniel Bowditch's father (see Chapter 8) became a cooper after retiring from being a sea captain. Thomas Willson defined gauging as "the art of measuring any kind of vessel and thereby fixing its true content." A similar definition of gauging was also given in William Hawney's (1775) *Complete Measurer*, but it is not known if Thomas Willson had access to any edition of Hawney's book.

Coopers needed to be able to make containers according to pre-specified dimensions, and to calculate the capacities of containers that they handled. If they worked on wharves receiving containers from other places they needed to be able to check how much a container held against what had been claimed by the sender. Barrels would come in different sizes, with different heights, and different maximum and minimum circumferences, and often the units used by the merchants from which the barrels originated were different from the units used by those receiving the barrels. Thus, coopers needed to measure, to calculate, to convert units, and to write brief reports. All of this had to be done quickly, and any errors could be costly.

Figure 6.1 shows page 6 of Thomas Willson's manuscript—it was concerned with finding "the content of a vessel whose diameters at the top and bottom were "parrallel (sic.) but unequal" That was "Case 3rd". **IRCEE** (**I**ntroduction, **R**ule, **C**ase, **E**xample, **E**xercise) and **PCA** (**P**roblem, **C**alculation, **A**nswer) genres were evident throughout (Ellerton & Clements, 2012), and it would appear to have been the case that Thomas did the calculations himself. Note the attractive penmanship and calligraphic headings.

Thomas Willson's (1789) Section on Geometry

Thomas began his section on geometry with a definition: "Geometry is the science of extension and is employed in the consideration of lines, surfaces, and solids as all extensions is (*sic.*) distinguished with length, breadth and thickness." Then followed a series of straight-edge/compass constructions, each of which was accompanied by a verbal description of the method used in order to complete the construction—but no discussion of the mathematics behind why the methods "worked" was given.

Figure 6.2 shows page 22 of the manuscript—it was concerned with the proposition, "To let fall a parpendicular (*sic.*) line upon a right line from a point assigned *C*". Let C' be the point from whence a parpendicular (*sic.*) is to be let fall upon the line of *B*." The section under "Practice" gave the method, and

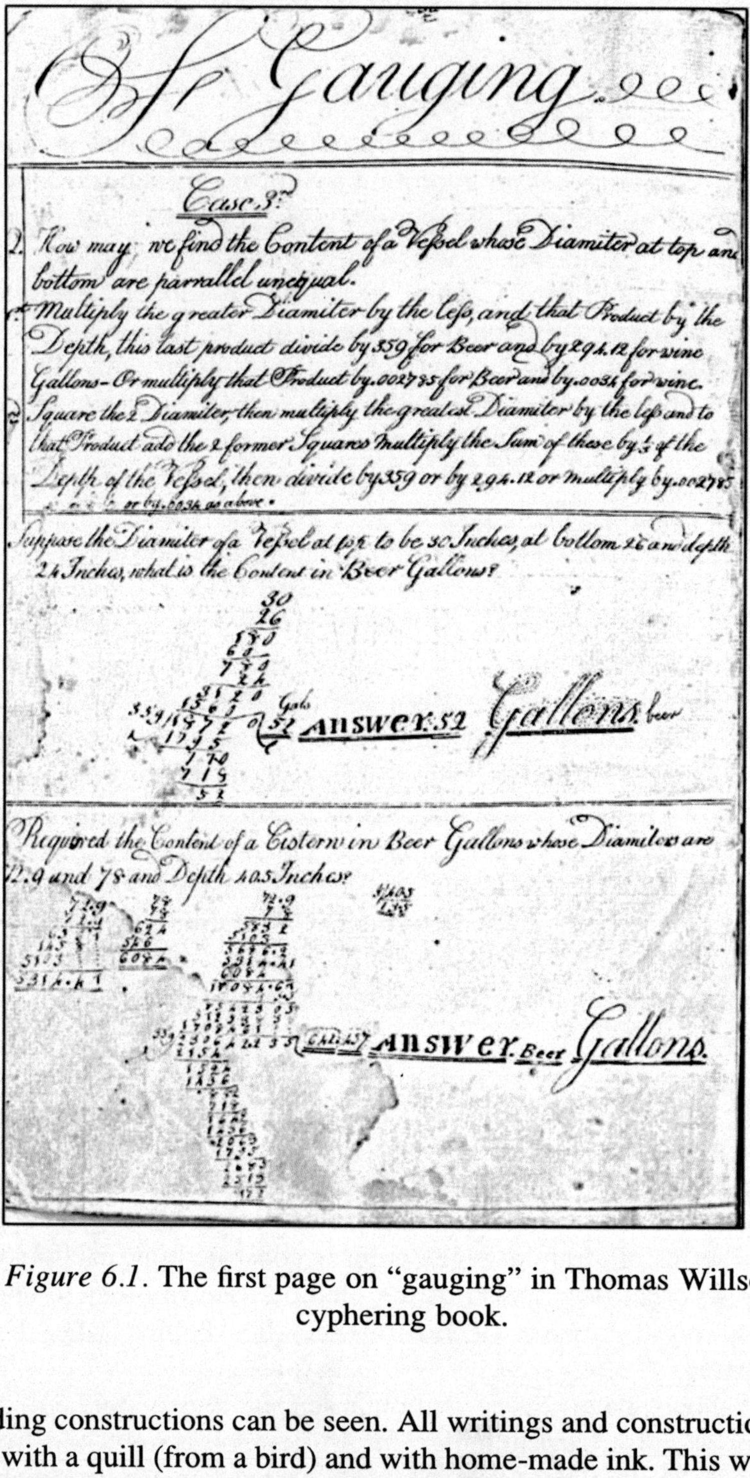
Of Gauging.

Case 3d

Q. How may we find the Content of a Vessel whose Diamiter at top and bottom are parralled unequal.

A. Multiply the greater Diamiter by the less, and that Product by the Depth, this last product divide by 359 for Beer and by 294.12 for wine Gallons – Or multiply that Product by .002785 for Beer and by .0034 for wine.

2. Square the 2 Diamiter, then multiply the greatest Diamiter by the less and to that Product add the 2 former Squares multiply the Sum of these by 1/3 of the Depth of the Vessel, then divide by 359 or by 294.12 or multiply by .002785 or by .0034 as above.

Suppose the Diamiter of a Vessel at top to be 30 Inches, at bottom 26 and depth 24 Inches, what is the Content in Beer Gallons?

Answer 52 Gallons beer

Required the Content of a Cistern in Beer Gallons whose Diamiters are 72.9 and 78 and Depth 40.5 Inches?

Answer Beer Gallons

Figure 6.1. The first page on "gauging" in Thomas Willson's cyphering book.

corresponding constructions can be seen. All writings and constructions would have been done with a quill (from a bird) and with home-made ink. This was one of many standard Euclidean constructions. Each page of the geometry section showed a new proposition and an associated construction.

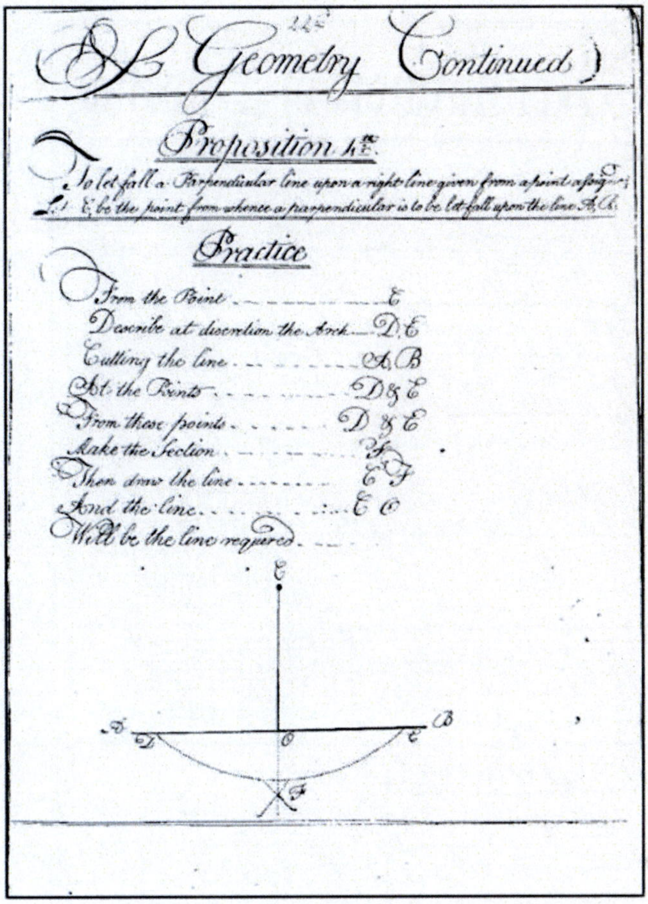

Figure 6.2. A Euclidean construction in Thomas Willson's (1789) cyphering book.

Thomas Willson's (1789) Section on Mensuration

Thomas began this section with the following vague "definition" of "superficial figures": "Superficial figures are all such as have only length and breadth, not having commensurable thickness." Although the meaning of that statement may not have been clear, Thomas proceeded immediately to define some simple geometrical figures—the first two being a square and a triangle (see Figure 6.3). Notice the evidence of **IRCEE** genre in Figure 6.3. Thomas defined a trapezium as "a figure of four unequal sides and oblique angles"—a definition which is not consistent with the concept of a trapezium as it is known in many parts of the world today. Later, in the section on mensuration, rules were framed in language which would be unfamiliar to most readers in the twenty-first century—like, for example, to find the area of a circle "multiply half the circumference by half the diameter." Figure 3.2 (in Chapter 3) showed a page from Thomas Willson's cyphering book which was dedicated to finding the "solid content" of a "cylindroid."

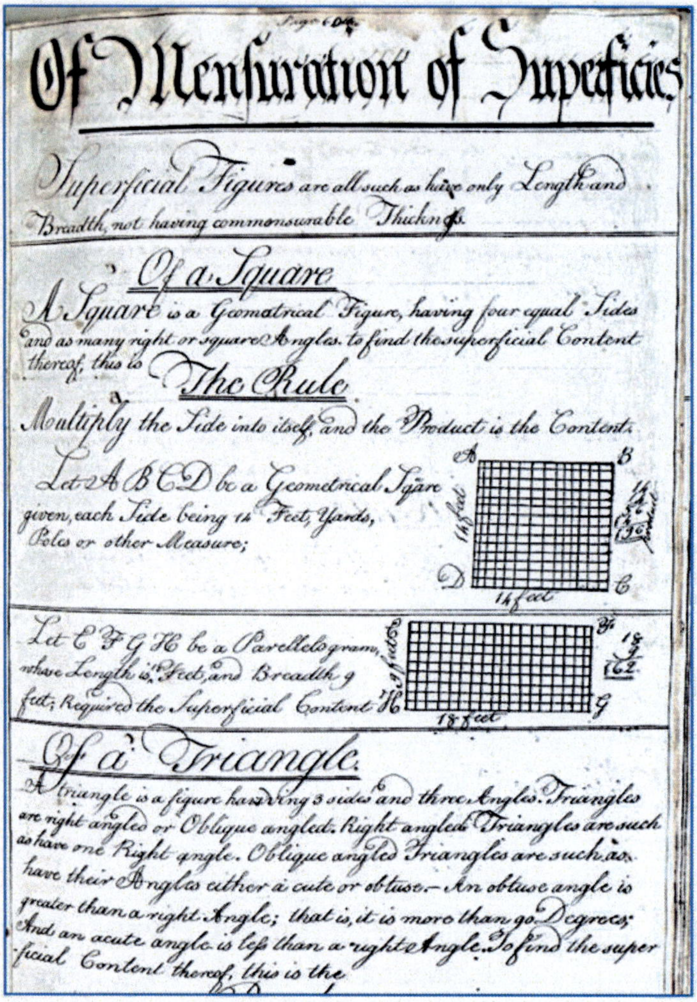

Of Mensuration of Superficies

Superficial Figures are all such as have only Length and Breadth, not having commensurable Thickness.

Of a Square

A Square is a Geometrical Figure, having four equal Sides and as many right or square Angles. to find the superficial Content thereof, this is

The Rule.

Multiply the Side into itself, and the Product is the Content.

Let A B C D be a Geometrical Square given, each Side being 14 Feet, Yards, Poles or other Measure;

Let E F G H be a Parellelogram, whose Length is 18 Feet, and Breadth 9 feet; Required the Superficial Content

Of a Triangle.

A triangle is a figure having 3 sides and three Angles. Triangles are right angled or Oblique angled. Right angled Triangles are such as have one Right angle. Oblique angled Triangles are such as have their Angles either acute or obtuse.– An obtuse angle is greater than a right Angle; that is, it is more than 90 Degrees; And an acute angle is less than a right Angle. To find the superficial Content thereof, this is the

Figure 6.3. An early page on "mensuration" in Thomas Willson's (1789) cyphering book.

Thomas Willson's (1789) Section on Trigonometry

The section on trigonometry introduced the idea that calculations could be simplified by the use of logarithms. Figure 6.4, which shows the first page in the section, was concerned with a situation for which two angles and the hypotenuse of a right triangle were given and it was required to find the length of "either of the legs." In Figure 6.4 the two given angles of the triangle were 90° and 54°30′ and the length of the hypotenuse was given as 121 leagues. The "direct rule of three" was used, with the logarithms expressed assuming that the length measure of the radius of the circle shown in Figure 6.5 was 10^{10}. Figure 6.5 was taken from Moore (1796).

Van Sickle (2011) has given the best summary of the history of the learning and teaching of trigonometry in North America during the eighteenth and nineteenth centuries. Basically, the main story is one of progressing from a traditional directed-line-segment approach to the trigonometric functions (see Figure 6.5) to a ratio approach based on ratios of sides of a right triangle. Van Sickle claimed that developments in North America were held back by the use of British textbooks in North America, and better mathematics and improved teaching of trigonometry slowly came as a result of more colleges and schools replacing textbooks which were written by British authors with those written by French and German authors.

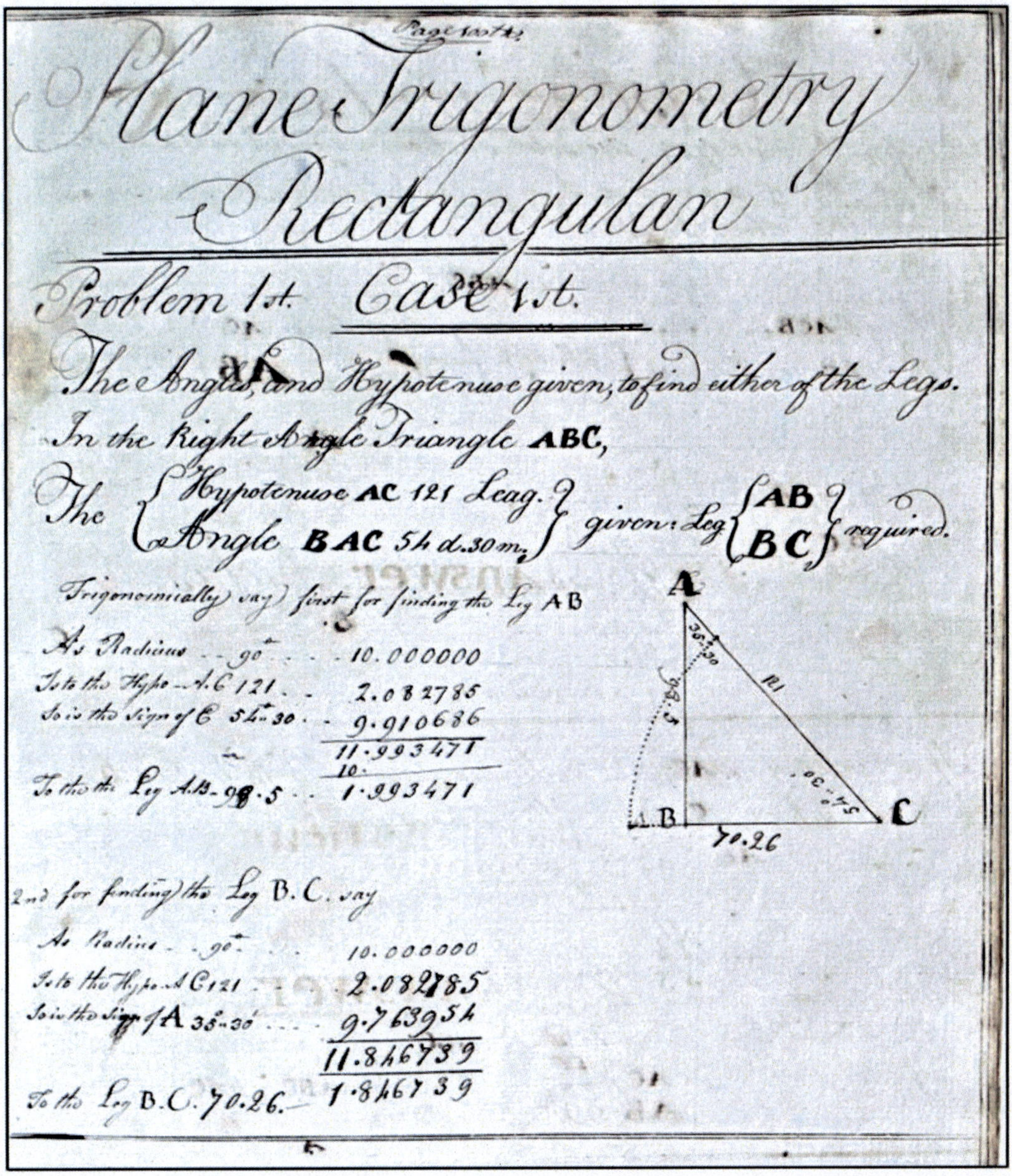

Plane Trigonometry
Rectangular

Problem 1st. Case 1st.

The Angles, and Hypotenuse given, to find either of the Legs.

In the Right Angle Triangle ABC,

The {Hypotenuse AC 121 Leag. / Angle BAC 54 d. 30 m.} given: Leg {AB / BC} required.

Trigonometrically say) first for finding the Leg AB

As Radius .. 90° ... 10.000000
Is to the Hypo -- A.C 121 .. 2.082785
So is the Sign of C 54-30 .. 9.910686
11.993471
10.
To the Leg A.B. 98.5 .. 1.993471

2nd for finding the Leg B.C. say

As Radius .. 90° .. 10.000000
Is to the Hypo A C 121 .. 2.082785
So is the Sign of A 35-30 ... 9.763954
11.846739
To the Leg B.C. 70.26 1.846739

Figure 6.4. Thomas Willson's (1789) introduction to plane trigonometry.

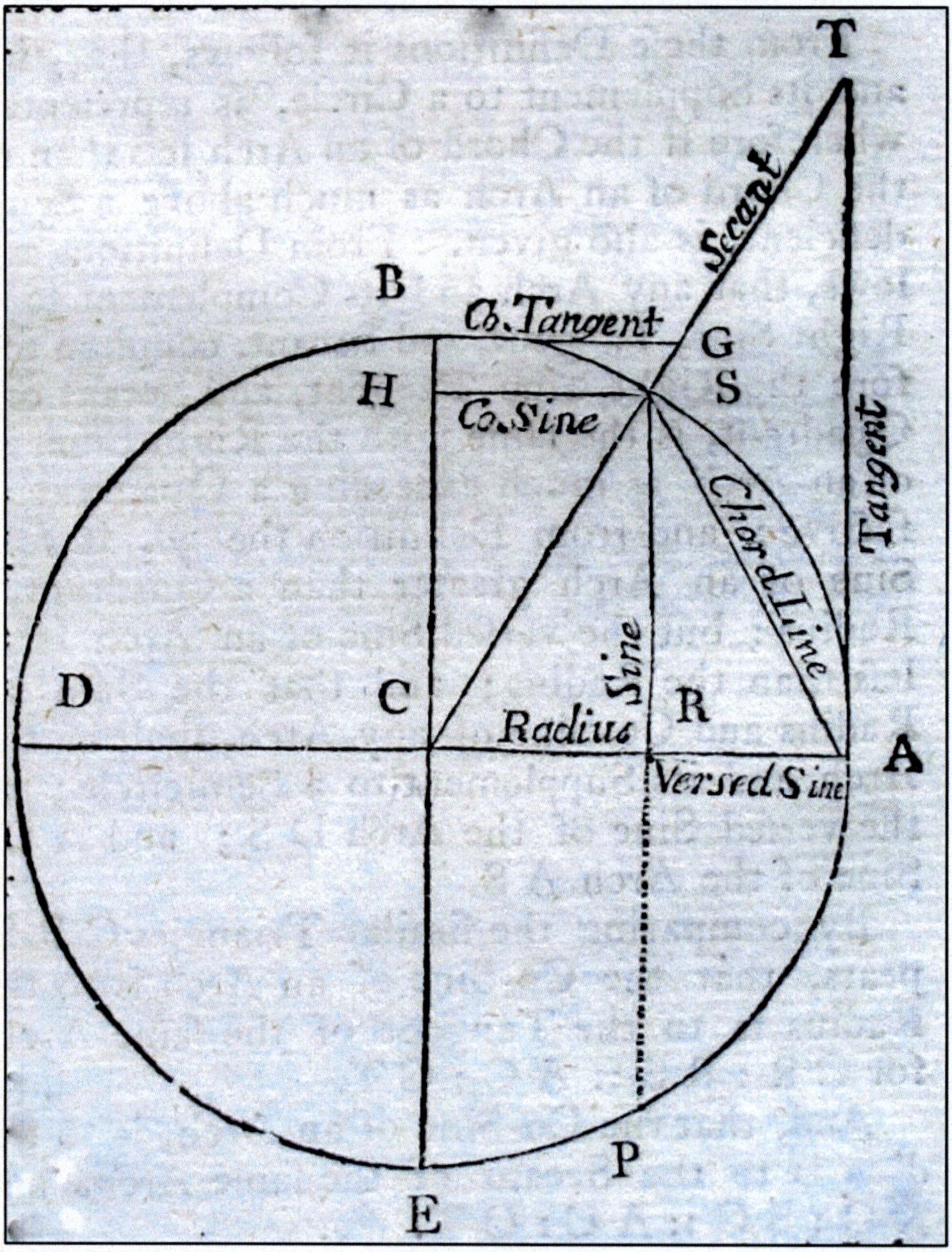

Figure 6.5. The directed line-segment approach to trigonometric functions (from Moore, 1796, p. 23). Until about 1850 it was often assumed that the radius length-measure for the circle was 10^{10}. Note that in that case, tan 45° would equal 10^{10}, not 1. If the radius length measure of the circle were 1 (i.e., we had a "unit circle"), then tan 45° would equal 1.

From both mathematical and educational perspectives, the combination of logarithms and the directed line-segment definitions for trigonometry often made it difficult of students to learn trigonometry well. On the fourth line of the calculations in Figure 6.4, for example, one finds "To AC =" and then 11.90092 appears. The fact was, the length measure of AC was *not* 11.90092—that was the logarithm (base 10) of the length measure. On the next line 2.13538 is given, which is, in fact, the logarithm, base 10, of 136.6, to five decimal places. In these calculations, the 11.90092 was obtained by adding 1.90309 and 9.99783; then, the 2.13538 was obtained by subtracting 9.76554 from 11.90092. Then, the 136.6 was probably found by reading a table of logarithms "backwards" (that is to say, by finding the "antilogarithm" of 2.13538). Thomas Willson did not formally state his answer to the original question, and there was no mention of the fact that the 2.07327 was the base 10 logarithm of the breadth (in leagues).

Most mathematics students of the 21st century will wonder how sin 54°30′ could be thought of as equal to 9.910686 because it has long been accepted that for any angle, $\theta°$ say, the value of $\sin\theta°$ must lie between -1 and 1 (or perhaps equal to -1 or 1). It was not until the mid-1820s in the United States of America that an argument was put forward for replacing the directed-line-segment approach to trigonometry with the now-familiar right-triangle ratio approach (Hassler, 1826).

If one looks up a modern table of sine values (or uses a calculator) one will find that sin 54°30′ is equal to 0.8141 (approximately). The point to be made here is that nowhere in Thomas Willson's cyphering book was there any explanation of why sin 54°30′ could be associated with 9.910686. In fact, the reason was that "directed-line-segment" definitions of trigonometrical functions were being used, and those were based on a circle which had a radius measure of 10^{10} (see Figure 6.5). 9.910686 was the value of the *logarithm* of sin 54°30′ assuming that the sine corresponded to the length of interval marked "sine" in the diagram in Figure 6.5, with $<TCA$ equal to 54°30′ and the radius length-measure being 10^{10}. Although that was complicated for students, at the time it was taken for granted by experienced teachers of mathematics. Another component of the reasoning was the so-called "direct rule of three" which, at that time was known as the "golden rule" (Ellerton et al., 2014, pp. 114–115; Jacoby, 1939).

It is reasonable to assume that very few students would have had any idea of the reasoning behind what was done. They would have just followed the rules, hoping to get right answers. Presumably, the introduction of the 10^{10} idea arose out of the need to avoid negative values for logarithms. Notice also that with the directed-line-segment approach, based on Figure 6.5, sin θ divided by cos θ would not equal tan θ (but, rather, tan θ divided by 10^{10}); notice, also, that tan 45° would be equal to 10^{10}.

The diagram shown in Figure 6.5 was shown on page 100 of Thomas Willson's cyphering book. Logarithms were used as calculational aids on many of the remaining 85 pages. Between 1607 and 1865, many applied mathematicians used double talk when referring to the logarithms of sines, cosines, tangents and secants. Little wonder, then, that almost certainly, Thomas Willson simply did what he was told to do. The E-C collection of cyphering books includes 33 manuscripts (prepared between 1607 and 1865) with sections on trigonometry and directed-line-segment definitions for trigonometric functions were used in almost all of them (see also, Bressoud, 2010; Hertel, 2016; Van Sickle, 2011).

As mentioned above, in urban centers like Salem, Boston, Providence, New York and Philadelphia there were special evening classes on navigation available to boys who aspired to becoming navigators (Seybolt, 1917, 1921, 1935), and in those classes calculations would often have been made using logarithms. Most of the boys who attended the classes were apprentices and would not have studied mathematics beyond mid-level *abbaco* arithmetic. For almost all of them, the likelihood that they understood the mathematics associated with logarithms was small. The students were taught to get right answers! Thomas Willson did not include any pages on spherical trigonometry in the section on "plane trigonometry" in his cyphering book, but later in his cyphering book he did—in the section on navigation.

Thomas Willson's (1789) Section on Surveying

Thomas began this section by describing the "art of surveying" as teaching "how to find the area, or superficial content or quantity of any field, close, wood, common, or any plot or parcel of ground, however situated, or in whatever form appearing, whether regular or irregular." After stating how the "content" of a triangular piece of land could be calculated, Thomas then moved to finding the content of land in the shape of a polygon by dividing the polygon into non-overlapping triangles (see Figure 6.6 for an example).

Thomas Willson's (1789) Section on Navigation

In the section on navigation Thomas once again relied heavily on the directed-line-segment, 10^{10} approach to trigonometry when performing calculations with the assistance of logarithms. Figure 6.7 shows a page from this section.

Thomas Willson's (1789) Log of a Journey

On the first page of this section, Thomas stated that this was "the journal kept by Joseph Clark, Chief Mate, begun on the 1st of the 3rd month, 1791." It is possible, but unlikely, that the log was created during an actual journey. Figure 6.8 reproduces a page from the log. Each page was supposed to summarize the progress of the ship on a particular day. Notice Thomas's use of the direct rule of three, and logarithms.

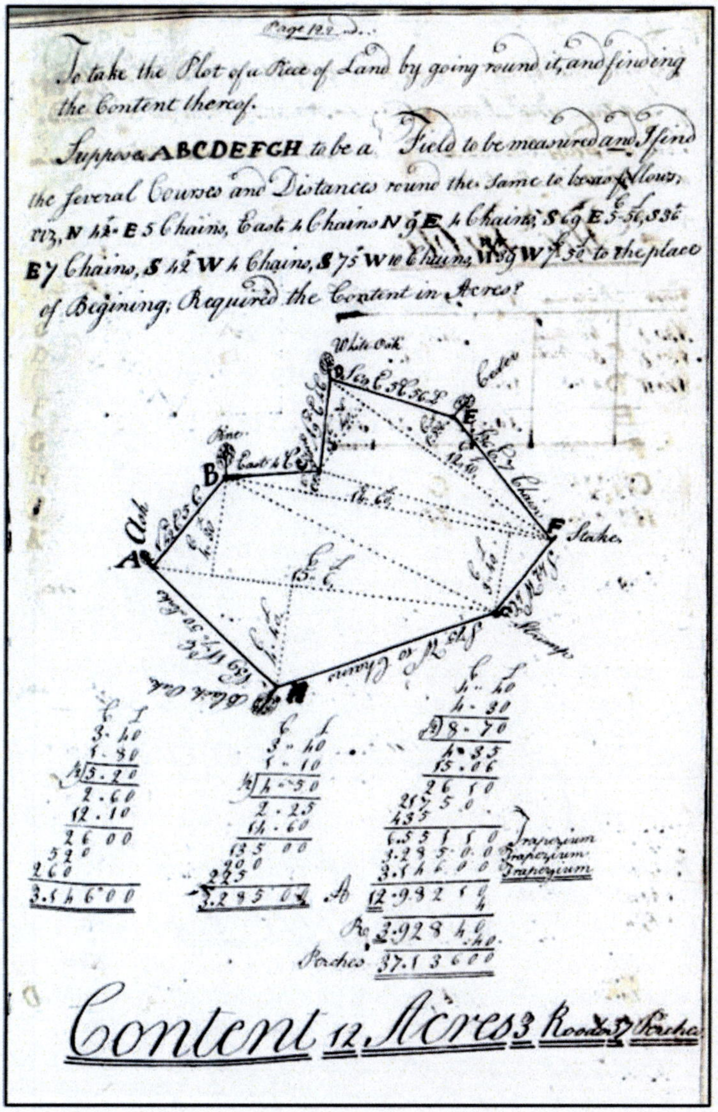

Figure 6.6. Thomas Willson's (1789) solution to a surveying task.

Shortage of Teachers for "Higher" Mathematics in Pre-College Schools

Of the 536 cyphering books prepared between 1607 and 1865 in the E-C collection which were analyzed for this book, only 55, or slightly more than 10 percent, had entries on one or more of algebra geometry, mensuration, trigonometry, astronomy. surveying or navigation. Of those 55 manuscripts, 21 were prepared before 1821 (the year when the first public school was created). We do not know how many of the 55 cyphering books were prepared by students at any of Harvard, William and Mary, Yale, Kings (later, Columbia), Brown, Princeton, the University of Pennsylvania, or Dartmouth, but there is evidence that after 1700 there were quite

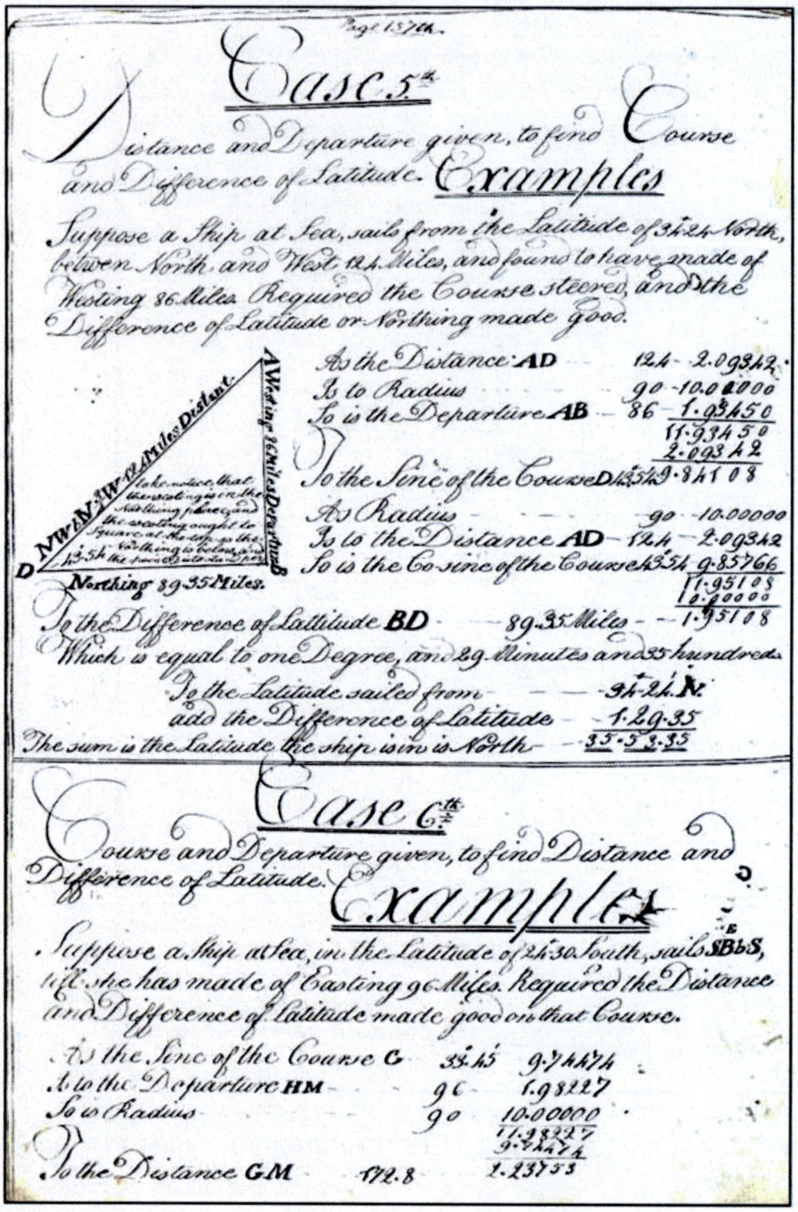

Figure 6.7. Thomas Willson's (1789) solution to a navigation task.

a few academies and grammar schools, and, after 1840, normal schools at which instruction in applied aspects of higher mathematics was available (Burke, 1982).

Thus, for example, in 1734, Alexander Malcolm, a Scot and author of a highly-regarded arithmetic textbook published in England in 1730, placed an advertisement in the *New York Gazette* stating that he was master at the "grammar school in the city of New York" and that his school was offering instruction in "all branches of the

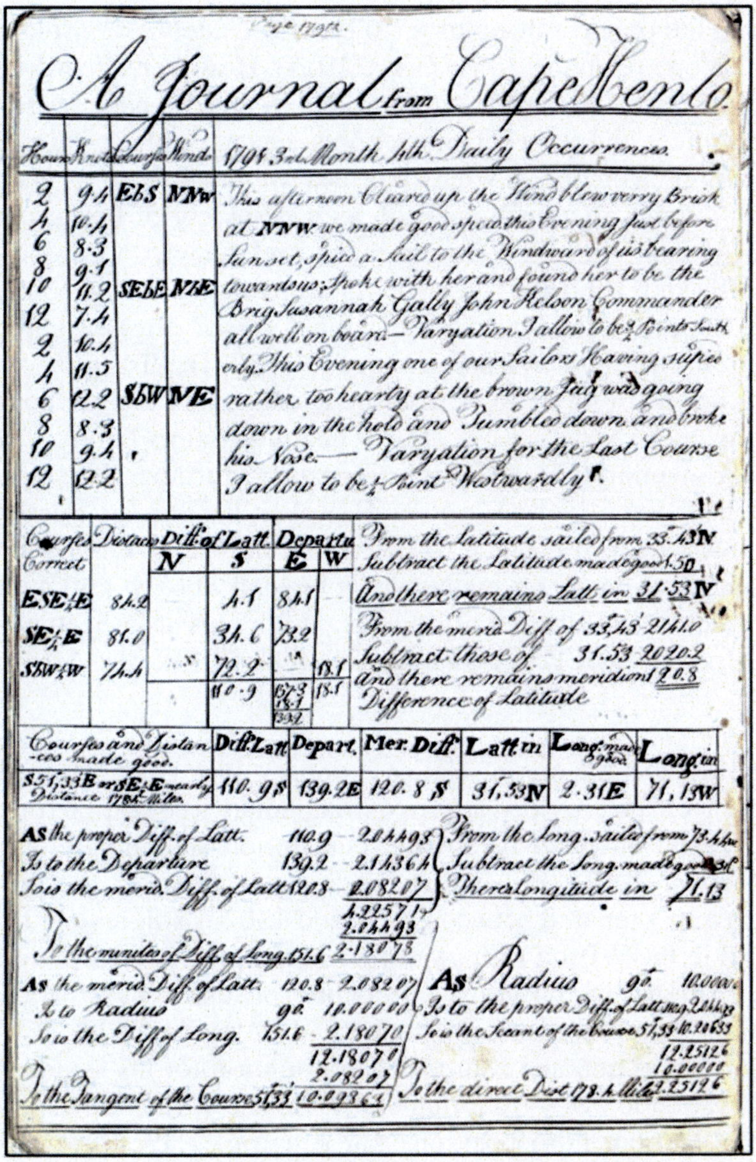
A Journal from Cape Henlo

Hours	Knots	Courses	Winds	1798 3d Month 4th Daily Occurrences
2	9.4	EbS	NNW	This afternoon Cleared up the Wind blew verry Brisk
4	10.4			at NNW we made good speed this Evening just before
6	8.3			Sunset, spied a Sail to the Windward of us bearing
8	9.1			towards us; spoke with her and found her to be the
10	11.2	SEbE	NbE	Brig Susannah Gally John Nelson Commander
12	7.4			all well on board — Varyation I allow to be ½ Points South
2	10.4			erly. This Evening one of our Sailors Having supped
4	11.5			rather too hearty at the brown Jug was going
6	12.2	SbW	NE	down in the hold and Tumbled down and broke
8	8.3			his Nose. — Varyation for the Last Course
10	9.4			I allow to be ½ Point Westwardly.
12	12.2			

Courses Correct	Distance	Diff of Latt N	Diff of Latt S	Departure E	Departure W
ESE½E	84.2		4.1	84.1	
SE½E	81.0		34.6	73.2	
SbW½W	74.4		72.2		18.1
			110.9	157.3	18.1
				139.2	

From the Latitude sailed from 33.43N
Subtract the Latitude made good 1.50
And there remains Latt. in 31.53N
From the merid Diff of 33.43 2141.0
Subtract those of 31.53 2020.2
And there remains meridional 120.8
Difference of Latitude

Courses and Distances made good	Diff Latt	Depart	Mer. Diff	Latt in	Long made good	Long in
S51,33E or SE½E nearly Distance 178½ Miles	110.9S	139.2E	120.8S	31.53N	2.31E	71.13W

As the proper Diff of Latt. 110.9 2.04493
Is to the Departure 139.2 2.14364
So is the merid Diff of Latt 120.8 2.08207
4.22571
2.04493
To the meridian Diff of Long 151.6 2.18078

From the Long sailed from 73.44W
Subtract the Long made good 2.31
There Longitude in 71.13

As the merid Diff of Latt 120.8 2.08207
Is to Radius 90 10.00000
So is the Diff of Long 151.6 2.18070
12.18070
2.08207
To the Tangent of the Course 51.33 10.09863

As Radius 90 10.00000
Is to the proper Diff of Latt 110.9 2.04493
So is the Secant of the Course 51.33 10.20633
12.25126
10.00000
To the direct Dist 178.4 Miles 2.25126

Figure 6.8. A page from a log of a journey, in Thomas Willson's (1789) cyphering book.

mathematicks, geometry, algebra, geography, navigation, merchants book-keeping after the most perfect manner" (quoted in Karpinski, 1940, p. 575). In 1822, the Reverend John Allen, who was Professor of Mathematics at the University of Maryland, included letters of recommendations from mathematics teachers at three different schools in Baltimore for his 500-page book on *Euclid's Elements and the Elements of Plain and Spherical Trigonometry*.

By the start of the nineteenth century, 40 private academies existed in the state of North Carolina alone (Connor, 1951; Coon, 1915). However, according to Myers and Nash (2006), nearly all of the students enrolled in those schools were boys from wealthy white families, and none of the schools accepted black children, free or slave. A few children from poor white families were allowed to attend. In the nineteenth century, some organizations such as the Society of Friends established schools at which African-American children were welcomed.

Between 1800 and 1860, 287 academies were chartered in North Carolina (Myers & Nash, 2006). A few endured, but most closed after a short time. The Presbyterian Church founded Davidson College in 1837; Baptists opened the Wake Forest Manual Labor Institute in 1834; and the Quakers and the Methodist Church combined to bring into existence the Union Institute (later Trinity College, and eventually Duke University). In 1851 the German Reformed Church established Catawba College in Newton, North Carolina. However, the primary mission for most of these institutions was to train men to be church ministers. Only a few of their teachers were capable of teaching mathematics beyond *abbaco* arithmetic. One of them was Robert Adrain (before and after his appointments at Princeton, Columbia, Rutgers and the University of Pennsylvania) (Zitarelli, 2019).

Robert Lazenby

Robert Lazenby was the son of a Revolutionary War soldier from Maryland. Between 1813 and 1827 he provided a form of mathematics education for children of plantation owners who attended his small subscription school in North Carolina (Ellerton & Clements, 2012). He died in 1828, aged only 42, but the 360-page cyphering book that he prepared between 1799 and 1802 is still among the Lazenby family papers held in the Wilson Library at the University of North Carolina (Doar, 2006). It is rather battered in appearance, probably because it was much handled by Lazenby and his subscription-school students. Lazenby's "book" would have provided the chief inspiration and sources of information for his senior students on a wide range of *abbaco* topics—from numeration to the various rules of three, and then fellowship, alligation, false position, and gauging. There were also sections dealing with elementary Euclidean geometry (definitions, and straight-edge and compass constructions), mensuration of superficies, and surveying (Ellerton & Clements, 2012, pp. 150–151).

We do not know how many "new" cyphering books were prepared by Lazenby's students, with Lazenby's book serving as the "parent." The idea of establishing lineage among early North American cyphering books is one which could usefully be taken up by future researchers.

Enoch Lewis

Enoch Lewis (1776–1856)—see Figure 6.9—dedicated much of his life to teaching mathematics to school students. He was a staunch Quaker, and an early advocate for the abolition of slavery. Born in Pennsylvania he displayed, from an early age, a great fondness and precocity for mathematics. He taught himself Latin in order that he might better be able to become acquainted with higher-level mathematical treatises.

Figure 6.9. Enoch Lewis (1776–1856) (From a photograph in Dewees & Dewees, 1899, p. 50).

While young, Lewis taught in pre-college grammar schools and he quickly earned a reputation as an outstanding teacher of mathematics. In the fall of 1799, he took charge of the mathematical department of the Westtown School—a new institution established by Quakers in Philadelphia (Dewees & Dewees, 1899). In 1808, a collection of articles on mathematics edited by Robert Adrian (which appeared in *The Analyst, or Practical Museum*) included the solution to a problem by Lewis (Zitarelli, 2019). Westtown became the pre-eminent pre-college mathematics school in Pennsylvania, educating many who would later be well-known teachers of mathematics.

In the spring of 1808, Enoch Lewis purchased part of a large farm, which had belonged to his father-in-law, in New Garden, Pennsylvania. In November of that year, he opened a boarding school for boys, limiting the number to 16, and admitting none under 15 years of age. Soon after opening the school, Lewis edited works of British mathematicians—John Bonnycastle, Thomas Simpson, and Robert Simson. Among the books he wrote on various branches of mathematics around that time were the following, which were extensively used in his school and in many other schools:

- 1824: *The arithmetical expositor; or a treatise on the theory and practice of arithmetic suited to the commerce of the United States.* Philadelphia, PA: Kimber and Sharpless;
- 1826: *Practical analyst, or a treatment on algebra, containing the most useful parts of that science, illustrated by a copious collection of*

examples, designed for the use of schools. Philadelphia, PA: Kimber and Sharpless;

- 1827: *Solutions of the most difficult questions in Lewis' algebra*. Philadelphia, PA: Kimber and Sharpless;
- 1844: *A treatise on plane and spherical trigonometry; including the construction of the auxiliary tables; a concise tract on the conic sections, and the principles of spherical projection*. Philadelphia, PA: H. Orr.

In his books, and in his mathematics classes, Lewis used calculus freely, preferring Leibniz's notation to Newton's fluxions (see, e.g., Lewis, 1848). As previously stated, none of the manuscripts in the Ellerton-Clements cyphering-book collection showed anyone using calculus.

Equity Considerations: Opportunity to Learn

Opportunity to Learn

Students preparing cyphering books which included sections on any of algebra, geometry, trigonometry, mensuration, surveying or navigation would normally have already studied mathematics at levels beyond elementary *abbaco* arithmetic. Given the need in the colonies for surveyors and navigators (Kiely, 1947), one might have expected more related manuscripts to be in the E-C cyphering book collection. However, the level of mathematics demanded of someone wishing to become an efficient surveyor or navigator was considerable, and although there were private tutors in urban centers like Boston, New York, Philadelphia, and Salem who claimed to be capable of teaching applied subjects (see, e.g., Karpinski, 1980; Kiely, 1947; Seybolt, 1921, 1935), one wonders how many of those tutors could have coped with what was needed to bring students to the stage where they could comprehend and apply the mathematics which was required for navigation or surveying. The following account, from a British text published in 1868, throws light on that issue:

> When a boy is too disobedient to be governed at home, too inattentive to learn at school, and too idle to work at "a place," he is then qualified for sea. He, perhaps, learnt while at sea that a knowledge of navigation would be useful, and he resolves to redeem 12 or 13 lost years of his life by the desperate efforts of a month. He betakes himself to the Mechanics Institution, and something like the following dialogue takes place in the mathematical department:
>
> *Teacher:* What do you wish to learn?
> *Sailor:* Double altitudes and lunars.
> *Teacher:* Do you understand trigonometry?
> *Sailor:* No!

Teacher:	Do you know anything of geometry?
Sailor:	No!
Teacher:	Do you understand decimals?
Sailor:	No!
Teacher:	What did you learn when you went to school?
Sailor:	I think I went as far as multiplication.

(*Chambers' Information for the People,* 1868. Vol. 1, pp. 733–734)

The point is that the mathematical expertise and practical experience that students needed in order to respond adequately to the curricular demands of any decent course in surveying or navigation were much more than would have been provided by a standard *abbaco* course in school arithmetic.

If "curricular importance" is measured by the number of enrolled students, then the data summarized in Table 5.1 (in Chapter 5) cast doubt on the accuracy of Edmond Kiely's (1947) claim that "surveying formed an important part of the American mathematical education program during its first two centuries" (p. 245). Although there can be little doubt that surveyors were much needed in early colonial society, in order to become a surveyor a student needed more prerequisite mathematical skills than those provided by the pragmatic and rushed forms of mathematics education that most common schools and evening classes provided during the period 1607–1865. Trigonometry, for example, was not taught in one-room schoolhouses because hardly any of the teachers had ever learned it and, in any case, almost all students who attended were not ready to learn it. In most of the academies and grammar schools much attention was given to Latin, and very little to mathematics or practical subjects. Apprentices who attended evening classes had rarely studied trigonometry before enrolling, and therefore were unlikely to be in a position to learn the theoretical content, or to acquire the skills, needed to become effective practicing surveyors.

Most of the manuscripts in the E-C cyphering-book collection which included material on one or more of algebra, geometry, trigonometry, astronomy, surveying or navigation show plenty of rules, cases, model problems, and exercises, and some of the students clearly solved most of the exercises by themselves. That said, during the cyphering era entries in cyphering books needed to be checked by teachers, and since there were no formal written examinations until the 1840s there must remain some doubt whether, for most of the period 1607–1865, the students who prepared the manuscripts actually understood most of what they wrote in them. In 1764, John Winthrop IV maintained that as Hollis Professor at Harvard he was expected to teach "geometry, algebra, conic sections and plane and spherical trigonometry" (Zitarelli, 2019, p. 39)—observe that calculus was not on his list, but "plane and spherical trigonometry" were, probably because Winthrop emphasized the mathematics of navigation in his teaching, and for that trigonometry was much needed.

But, even as late as 1865 very few students below college level were ready to learn any of algebra, geometry, trigonometry, mensuration, astronomy, surveying or navigation, meaningfully. That statement not only applied to students in European-background families, but also to any in African-American or Native American families. The nation was so far away from achieving "mathematics for all" that it did not even reflect on whether that was a desirable or achievable goal.

Gender Considerations

There are 536 cyphering books in the E-C collection which were prepared before 1865 and the earliest of those was probably prepared in the 1660s. About 20 percent of the 536 manuscripts were prepared by females. Of the manuscripts which dealt with any or all of algebra, geometry, trigonometry, mensuration, surveying and navigation, only one was definitely prepared by a female—that is a 13-page manuscript, prepared in 1847, which included Euclidean straight-edge and compass constructions, and solutions to elementary surveying tasks. The names of students who prepared seven other cyphering books were not given. What is clear is that during a 200-year period, which included over 35 years when public high schools existed, hardly any females studied any of algebra, geometry, mensuration, trigonometry, surveying, or navigation. Although North America was certainly not the only place where this occurred (Roach, 1971), the strength of gender-related curricular priorities needs to be noted and considered by anyone reflecting on progress toward "mathematics for all" in North America.

Patricia Cline Cohen (1982) summarized the situation regarding females and mathematics which existed in North America during the eighteenth and early nineteenth centuries in the following way:

> The crucial moment when women might have joined men in adapting their minds to quantitative reasoning came and passed, the opportunity missed. A gender boundary in numeracy was erected and defended. . . . The history of numeracy in the early nineteenth century illuminates the new American division of the sex roles; not only were men everywhere sent into the marketplace while women were isolated within a sentimental home, but quantification became masculinized, while its opposite, vague intuition, which resists pinning things down, came to be perceived as a desirable, natural, and exclusive attribute of woman. (p. 149)

Such a statement would seem to belittle emphases on art, cooking, sewing, and the preparation of samplers, which were such an important part of the education of females.

Is Cohen's (1982) statement, quoted above, a true summary of what transpired with respect to opportunities for females to study mathematics during the nineteenth century in the United States of America? To that question we offer the following five related comments:

1. The state of affairs between 1800 and 1820 with respect to gender-related aspects of quantitative reasoning did not suddenly come and go. It was the outcome of 200 or so years of colonial experience and early republican life in which females had not been invited to, or

expected to, study arithmetic. It is naïve to expect that the outcome of two centuries of social and educational expectations and practices could be turned around quickly, especially when in other nations a similar state of affairs had existed, and in most cases continued to exist.

2. During the period 1607–1865, or for much of that period, most females in North America and in many other parts of the world, were not permitted to enter college, and therefore had less need to prepare for higher mathematical studies.
3. Analyses of cyphering-book data reveal that a significantly higher percentage of manuscripts from the period 1800–1865 were prepared by females than was the case for manuscripts prepared between 1607 and 1799. That is to say, there did emerge a trend toward greater involvement of females with respect to the study of mathematics.
4. Between 1820 and 1834 Warren Colburn advocated strongly, and successfully, for the greater participation of young schoolgirls in school arithmetic.
5. There is evidence that in the 1820s there were other serious moves to advance the study of mathematics among U.S females. In 1828, for example, Catharine Beecher, a "distinguished member of a distinguished family" (Karpinski, 1980, p. 291), caused to have published her *Arithmetic Explained and Illustrated, for the Use of the Hartford Ladies Seminary*, and a later edition would appear in 1832. Beecher was the Principal of the Hartford Female Academy in Connecticut.

In a later statement, Cohen (2003) would offer a more balanced overview of advances in U.S. mathematics education during the first half of the nineteenth century.

With the above we did not mean to challenge Cohen's (1982, 2003) argument that moves, in the first half of the nineteenth century, toward greater active involvement of females in mathematics education were desirable. But, as she made clear in her 2003 chapter, that was only part of the story. We would argue, for example, that from an equity perspective, there was an equal, or even greater, need to bring the children of slaves (both African-American and white, and both male and female) as well as Native-American peoples into the mainstream of mathematics education as there was to have more European-background females, from well-to-do families, studying higher forms of mathematics.

Built-in Inequalities Which Threatened to Hold Back Needed Developments

Inequalities with respect to opportunity to learn mathematics extended well beyond gender considerations. The November 9th, 1867, issue of *Harper's Weekly* carried an article on "Education in the Southern States" (pages 706–707) which pointed out the following:

- Of the 8,000,000 Southern whites in 1860 only 300,000 "owned" slaves, and of those only 90,000 "owned" more than 10 slaves. Thus, there were

90,000 "great slaveholders," who, with their exceptional wealth, controlled the 7 million "poor" whites and the 4 million "blacks."

- One million whites owned the land and capital and monopolized the provisions for education. They ruled 11 million laborers with little or no property of their own, and without access to formal education.

Although the article drew attention to the fact that for well over a century many European-background "whites" had had little access to formal education, and almost all the children of African-American slaves had had no access, there was no mention of the plight of indigenous peoples (numbering about 0.5 million in 1850). Almost certainly, none of the students who prepared the cyphering books which included entries on applied forms of mathematics which were beyond *abbaco* arithmetic would have been a Native American. In addition, the *Harper's Weekly* article did not refer to the lack of education opportunities of the children of "whites" (in the North or in the South) who were not property owners.

What is clear is that between 1607 and 1865 only a small percentage of the North American population had any chance of studying higher levels of mathematics. Even in 1865 only a very small percentage of the population had studied, was studying, or would ever study, forms of mathematics beyond elementary forms of *abbaco* arithmetic. Of those who did have an opportunity to study "higher" mathematics, almost all were white, European-background males. That said, less than 5 percent of the European-background males would ever study any form of mathematics beyond middle-level *abbaco* arithmetic.

Although the inequalities relating to opportunity to study mathematics were fierce, they were largely hidden. They were more pertinent than most observers in 1865 would have recognized, or expected, because during the period 1776–1865 the United States had experienced a population explosion, fueled by European immigration, from about 2.5 million to about 28 million (Klein, 2012). A North American New World was rapidly created (Mayo, 1898). If one thought about it, the inequalities relating to mathematical opportunity had been constructed within the old New World, and those inequalities threatened to hold back bold structural progress which should have been possible in education.

Certainly, some *were* thinking about what might be achieved—although not specifically in the domain of mathematics. At a Woman's Rights Convention held in Akron, Ohio, in 1851, Sojourner Truth (see Figure 6.10) proclaimed, prophetically:

> I think that 'twixt the Negroes of the South and the women at the North, all talking about rights, the white men will be in a fix pretty soon. (Quoted, in Meacham, 2018, p. 23)

Sojourner Truth (1850) was born in New York as a slave but escaped to freedom with her infant child in 1826 (Whalin, 1997). She remained illiterate, never having an opportunity to study mathematics. She achieved so much in her lifetime—one can only speculate what opportunities may have been available to her if she had been able to become fluent in reading, writing and arithmetic.

Figure 6.10. Sojourner Truth (c. 1797–1883) Wikipedia contributors. (2021, July 4).

Our decision to reproduce so many pages from Thomas Willson's cyphering book in this chapter was made for two main reasons. The first was that the pages reveal just how much effort he put into the creation of "his book." The E-C cyphering book collection has many "prettier" manuscripts, but this was, indeed, *his* book, and his intention was to keep the book so that he could usefully refer to it for the rest of his life. It still survives, more than 230 years later. There is a genuine authenticity about what Thomas wrote, and to read his book is to come to recognize that most of the time, though certainly not all of it, he had some idea about the mathematics he was using to solve some often quite difficult, realistic problems. The second reason was to begin to convince the reader that at least one early North American student devoted much time to a range of different forms of mathematics. It is too simplistic to think that mathematics education was confined to *abbaco* arithmetic or, perhaps, to arithmetic and small dabs of algebra and geometry.

Reading entries in a single mathematics textbook written around 1789 (when Thomas Willson prepared his cyphering book) can never provide direct evidence of the range and qualities of *implemented* mathematics curricula at that time. But, we hope that our analyses of manuscripts in the E-C cyphering book collection have presented new insights into the history of mathematics curricula in schools and colleges in North America. Taken together, the cyphering books in the collection have provided unmistakable evidence with respect to what mathematics was learned, and how it was learned. As Lao Genevra Simons (1936) stated, more than 85 years ago:

> A great deal has already been said about the custom of keeping student notebooks during this period of difficulty in obtaining books from England and of printing books in the colonies. If all the notebooks now hoarded by descendants of graduates of the early American colleges or lying neglected

> and forgotten in attics and closets, if all these notebooks could be presented to the several college libraries or historical societies, the history of early American education would be greatly enriched. In these notebooks, there is found the content and scope of the curriculum of the day in evidence that is unmistakable. (Simons, 1936, p. 588)

We do not know much about Thomas Willson—almost all of what we do know comes from his 185-page manuscript which he prepared in Philadelphia in 1789 and 1790. That said, it is difficult to avoid the conclusion that Thomas really put his heart and soul into the mathematics he was doing and recording. He was *doing* mathematics, and therefore his manuscript attests *unmistakably* to an aspect of the history of mathematics in North America between 1607 and 1865. The mathematics that we have examined in his book related to gauging (see Figure 6.1), Euclidean geometry (Figure 6.2), mensuration (Figures 3.2 and 6.3), trigonometry (Figure 6.4), surveying (Figure 6.6), navigation (Figure 6.7), and the log of a journey (Figure 6.8). He was proud of "his book," and so were the members of his family and their descendants.

Geometry in North American Pre-College Educational Institutions, 1607–1865

It will now be useful to consider the extent of pre-college education with respect to geometry, mensuration, trigonometry, surveying and navigation during the 258-year period 1607–1865. Most of the analyses will be for the period 1700–1865 because although there was much activity in the New World during the seventeenth century with respect to surveying and navigation, there were very few attempts to get children to learn any of the associated forms of mathematics. Upon its establishment, in the late 1630s, Harvard College created a curriculum which focused on classical learning; mathematics being only a small part of the implemented curriculum.

We start with geometry, and then will move, successively, to trigonometry, mensuration, navigation and surveying. In this section the emphasis will be on geometry in pre-college institutions—part of the next chapter (Chapter 7) will be concerned with geometry in colleges during the period 1607–1865.

For over 2000 years, until the end of the eighteenth century, the reputation of *Euclid's Elements* as the ultimate model of logical soundness remained largely unquestioned. Although there had been many contributions to geometry by Chinese, European, Indian, and Arab thinkers it was *Euclid's Elements* which was thought to set the gold standard for logical purity—something which had been made possible by Arab mathematicians who translated the original Greek text of the *Elements* into Arabic (Stamper, 1909). In the late fifteenth century CE, translations into Latin were achieved, but it was not until the late sixteenth century that there was a translation into English (Cajori, 1907; Stamper, 1909).

During the eighteenth and nineteenth centuries, the legitimacy of Euclid's basic logical stance was challenged by European mathematicians such as Adrian Marie

Legendre (1752–1833), János Bolyai (1802–1860) and Nikolai Lobachevsky (1792–1856) (Cajori, 1907; Goldstein, 2000; Pycior, 1997; Sinclair, 2008; Smith, 1911). The clamor for change among many mathematicians grew louder as the nineteenth century progressed, with arithmetical/algebraic modifications to Euclid being put forward, and fundamentally different geometrical systems being proposed by those advocating non-Euclidean, projective, Kleinian and other forms of geometry whose logical assumptions and structures differed from Euclid's. That said, there was much resistance to this attack on Euclidean geometry, led by mathematicians like Isaac Todhunter, at the University of Cambridge in England (Cairns, 1934; Macfarlane, 1916).

During the first half of the nineteenth century the noise of battle with respect to *Euclid's Elements* echoed across the Atlantic and the works of Legendre, especially, became well known in the United States (Cajori, 1890). The time-honored axiomatic approach to geometry, as found in *Euclid's Elements*, was challenged, particularly around 1820 as a result of the French influence on the curriculum at the United States Military Academy at West Point (Sinclair, 2008; Rickey & Shell-Gellasch, 2019), and the influence of John Farrar at Harvard University. More will be said on that in the next chapter.

Thomas Carlyle, who would become a much-respected literary figure, translated Legendre's (1794) *Élements de Géométrie* into English for David Brewster, and the translation was published in Edinburgh, Scotland, in 1824. A revised and altered form of it was published for the use of USMA students in 1828 (Karpinski, 1980). The fourth edition, published in 1834, was "revised and abridged by Charles Davies" (Karpinski, 1980, p. 291), and in 1837 an edition "revised and adapted to the course of mathematical instruction in the United States, by Charles Davies" appeared. Slowly but surely Davies' (1838) text would become known as "Davies' Legendre," and 19 editions of it would be published by 1850 (Karpinski, 1980, pp. 292–293). During the 1840s it was much used in U.S. high schools.

In *The History of Geometry Curriculum in the United States,* Nathalie Sinclair (2008) stated, categorically, that until the middle of the nineteenth century in the United States, "geometry was generally taught only in the universities" (p. 14). Alva Stamper (1909), however, provided evidence that after 1818 geometry was taught in the higher classes at Phillips Exeter Academy. More damaging to Sinclair's position is well-documented evidence that in the eighteenth and early nineteenth centuries geometry was taught in many privately-conducted evening classes in New England and New York (Seybolt, 1917, 1921, 1935). Furthermore, the E-C cyphering book collection holds 25 cyphering books prepared before 1840 which include sections on geometry. In most of those 25 manuscripts the geometry sections were prepared immediately before sections on either surveying or navigation. In almost all cases the geometry sections included beautifully-drawn straight-edge/compass constructions. One such case is provided by Thomas Willson's cyphering book—a page of which was reproduced as Figure 6.2 in this chapter.

Alva Stamper (1909) and David Eugene Smith (1911) were correct, then, when they claimed that geometry was being taught in some pre-college schools well before 1844, the year when Harvard University mandated a knowledge of geometry for entrance. Stamper especially referred to geometry taught at Boston Latin School where *Catalogues* for the School indicated that it was taught in the fourth and fifth years of a five-year course. Seybolt (1921) provided evidence that John Walton (in 1723), Benjamin Leigh and Garrat Noel (in 1751), John Searson (in 1755), James and Samuel Giles (in 1759) offered instruction in geometry in evening classes in New York City. Advertisements in the *New York Mercury* on May 6, 13, 20, September 30, and October 7, 1865 indicated that a Mr. Thomas Carroll was teaching evening classes which included instruction in *Euclid's Elements*, algebra and "conick sections," mensuration of superficies and solids, surveying theory and all its different modes in practice, trigonometry, navigation, and instruments necessary for keeping a sea journal. Carroll also indicated that he taught the same subjects to all-girl classes in the morning (from 6 am to 9 am), and to all-boy classes from 9 am to 12 noon and in later classes in the day. It was also indicated in the advertisements that Mrs. Carroll would teach young ladies how to prepare samplers, quilting, and knotting for bed quilts.

Sinclair (2008) claimed that in the few secondary-school classrooms in which geometry was studied around 1820 the emphasis was on "memorizing the definitions, axioms, and propositions provided in the text" (p. 15). In most cases the textbook was a translation of *Euclid's Elements* which "contained no exercises." That is to say, according to Sinclair (2008), "no opportunities were provided for students to do applied or original work; for example, to apply the method of *reductio ad absurdum* to another proposition" (p. 15). Sinclair continued: "The goal was for students to learn and appreciate the work of Euclid" (p. 15). However, cyphering-book data do not support Sinclair's contention—Thomas Willson, and most of the other students who prepared sections on geometry in their cyphering books, actually *applied* Euclid's ideas, showing page after page of Euclidean constructions using straight edges and compasses. Surviving compass marks and arcs drawn have left us in no doubt that these students made the entries themselves and tried to make sense of what they were doing as they made them. The emphasis went beyond mere memorization.

It is not our intention to give the impression that geometry was a thriving part of the implemented curricula of post-elementary grammar schools and high schools before 1865. That was not the case. But the fact is, with the advent of high schools, from the 1820s onward, the opportunity was there to make geometry an integral part of school mathematics curricula in the United States. We would argue that that opportunity was missed because of the confusion arising from competing views so far as what should be important in high-school curricula. Before presenting that argument we would add that from our perspective this confusion had the unfortunate consequence of holding back the progress of geometrical education, and therefore geometry, in the United States of America.

The Push for Legendre's Geometry

Adrien-Marie Legendre was one of several great French mathematicians during the late eighteenth and early nineteenth centuries who not only carried out ground-breaking research but also prepared mathematics textbooks for schools and colleges. One of his most influential works was his *Élements de Géométrie* (Legendre, 1794) which, among other things, launched an attack on the logical soundness of *Euclid's Elements*. In *Élements de Géométrie*. Legendre re-ordered the propositions from Euclid's *Elements* and added diagrams that Euclid did not include. He attempted to give geometry a better logical basis than *Euclid's Elements* (Ackerberg-Hastings, 2000; Barbin & Menghini, 2014; Cajori, 1890, 1907).

The push to link the mathematics curricula of the United States' Military Academy (USMA) at West Point and Harvard University to French mathematics curricula reached its peak between 1820 and 1835 (Rickey & Shell-Gellasch, 2019). Immediately before his appointment as Superintendent of USMA in 1817, Colonel Sylvanus Thayer spent two years (from 1815 to 1817) studying at the Polytechnique in France. Thayer encouraged Claudius Crozet (Professor at USMA between 1816 and 1823), and Charles Davies (Professor between 1816 and 1837) to "upgrade" West Point mathematics by adopting recent French approaches. However, one of Crozet's early USMA students would recall that Crozet did not speak English well and that he was not very successful in his endeavors as a teacher of mathematics at West Point (Cajori, 1890).

Had it not been for Charles Davies, Professor of Mathematics at USMA (West Point), the push for French mathematics to exert greater influence on what was studied in mathematics classes in the United States might have faltered earlier than it did. Davies' genius with respect to geometry was to reach a compromise between Legendre's formalities and the state of geometrical education as it existed in the United States before the 1830s. In 1834 Davies wrote:

> The Editor [i.e., Davies himself], in offering to the public Dr. Brewster's translation of Legendre's *Geometry* under its present form, is fully impressed with the responsibility he assumes in making alterations in a work of such deserved celebrity.
>
> In the original work, as well as in the translations of Dr. Brewster and Professor Farrar, the propositions are not enunciated in general terms, but with reference to, and by the aid of, the particular diagrams used for the demonstrations. It is believed that this departure from the method of Euclid has been generally regretted. The propositions of Geometry are general truths, and as such, should be stated in general terms, and without reference to particular figures. The method of enunciating them by the aid of particular diagrams seems to have been adopted to avoid the difficulty which beginners experience in comprehending abstract propositions. But in avoiding this difficulty, and thus lessening, at first, the intellectual labor, the faculty of abstraction, which it is one of the primary objects of the

> study of Geometry to strengthen, remains, to a certain extent, unimproved. (Davies' 1834 Preface, reproduced in Davies, 1838, p. iii)

We believe that the points made by Davies were important from both mathematical and educational perspectives.

The study of geometry, whether it be based on any of Davies' (1838) *Legendre*, Playfair's (1814, 1822) *Euclid*, Simson's (1806) *Euclid*, or Lacroix and Bézout's (1826), *An Elementary Treatise on Plane and Spherical Trigonometry, and on the Application of Algebra to Geometry,* required a massive theoretical jump for pre-college students. They would have been less interested in the order of propositions, or the placement and details of diagrams, than in the meaning of words like "proposition," "axiom," "lemma," "corollary," "demonstration," "hypothesis," "scholia," "theorem," "conversely," "perpendicular," "oblique," "logarithm," and "rectilineal." Many more words could have been included in that list. The point being made is that words like those represented totally new concepts for most pre-college students, and in a geometry course based on a reputable geometry text the words and concepts came quickly—too quickly, in fact, for most pre-college students.

Even more puzzling would have been some of the theorems. Consider, for example, Proposition VIII as stated in Davies (1838):

> If from any point within a triangle, two straight lines be drawn to the extremities of either side, their sum will be less than the sum of the other two sides of the triangle. (p. 18)

This was one of the simpler theorems in Davies' book—it appeared on page 18 of a 360-page book. After the statement came a "demonstration" (i.e., a proof) which although beautiful in its logic would have been extremely difficult to understand for a high-school student studying geometry for the first time. On page after page there were theorems like Proposition IX:

> If two triangles have two sides of the one equal to two sides of the other, each to each, and the included angles unequal, the third side will be unequal, and the greater side will belong to the triangle which has the greater included angle.
>
> (Davies 1838, p. 18)

Although the wording of Proposition IX would have been easier to comprehend than the wording of Proposition VIII, its proof was considerably more difficult than that for Proposition VIII, with three "cases" needing to be considered, and a "scholarium" comprehended.

In this way Davies proceeded—proposition after proposition, new concept after new concept. On page 38 there is the statement of Proposition VIII for "Book II":

> Of four proportional quantities, if there be taken any equimultiples of the two antecedents, and any equimultiples of the two consequents, the four quantities will be proportional. (p. 38)

The wording was complex—perhaps unnecessarily so—and the proof of this proposition was given in algebraic terms, with the direct rule of three assumed. Figure 6.11 shows page 39 from Davies (1838). The wording for Proposition IX, X, and XI for Books II, together with the "demonstrations" for Propositions IX and X, can be seen.

BOOK II. 39

PROPOSITION IX. THEOREM.

Of four proportional quantities, if the two consequents be either augmented or diminished by quantities which have the same ratio as the antecedents, the resulting quantities and the antecedents will be proportional.

Let $M : N :: P : Q$, and let also
$M : P :: m : n$, then will
$M : P :: N \pm m : Q \pm n$.
For, since $M : N :: P : Q$, $M \times Q = N \times P$.
And since $M : P :: m : n$, $M \times n = P \times m$
Therefore, $M \times Q \pm M \times n = N \times P \pm P \times m$
or, $M \times (Q \pm n) = P \times (N \pm m)$:
hence $M : P :: N \pm m : Q \pm n$ (Prop. II.).

PROPOSITION X. THEOREM.

If any number of quantities are proportionals, any one antecedent will be to its consequent, as the sum of all the antecedents to the sum of the consequents.

Let $M : N :: P : Q :: R : S$, &c. then will
$M : N :: \overline{M+P+R} : \overline{N+Q+S}$
For, since $M : N :: P : Q$, we have $M \times Q = N \times P$
And since $M : N :: R : S$, we have $M \times S = N \times R$
Add $M \times N = M \times N$
and we have, $M.N + M.Q + M.S = M.N + N.P + N.R$
or $M \times (N+Q+S) = N \times (M+P+R)$
therefore, $M : N :: \overline{M+P+R} : \overline{N+Q+S}$.

PROPOSITION XI. THEOREM.

If two magnitudes be each increased or diminished by like parts of each, the resulting quantities will have the same ratio as the magnitudes themselves.

Figure 6.11. Geometrical propositions on page 39 of Davies (1838).

The important point to be made here is that cyphering-book evidence indicates that before 1865 this theoretical approach to geometry was not often adopted in pre-college schools. Indeed, of the 36 cyphering-book units which included sections on geometry in the E-C cyphering-book collection, only two showed geometrical proofs. Much more common were Euclidean constructions and practical applications taken from John Bonnycastle's (1818) *An Introduction to Mensuration and Practical Geometry*. Earlier in this chapter Figures 6.2, 6.3, and 6.4, showed pages which dealt with geometrical concepts and applications in Thomas Willson's (1789) cyphering book. We believe that those pages were representative of the admittedly small amount of geometry studied in pre-college classes before 1865 in North America.

Implications of issues concerning language in regard to geometrical education in the United States before 1865 would provide a promising research agenda. In 1865 only a tiny proportion of school-age children in the United States ever got to study geometry. Rarely did a girl, or a Native American, or an African-American, or a child in an economically poor white family study geometry. The move to adopt a geometry curriculum based on a text by Legendre, or Lacroix and Bézout, or Playfair, or Simson was certainly mathematically defensible, but from the point of view of equity what did such a move imply? Did mathematicians who concerned themselves with what was the best order of propositions, with the role of diagrams, with whether Euclid's fifth postulate was independent of other postulates, or whether it was wise to regard a circle as a limiting case of a regular polygon, etc., realize that very few students in North America had any idea of the notion of proof? Could not proof have been better introduced through algebra? There are many education-related questions which could, and should, be investigated, but have not been.

It has been unwise for historians to neglect the forms and extent of geometry education in schools during the period 1607–1865, for within that period norms were established for what forms of geometry needed to be studied, and how it should best be presented and learned. At the college level, after about 1800 excellent work was done in getting students to comprehend Euclidean geometry or geometry based on translations of Legendre's (1794) text. The process of recitation, whereby students were expected not only to memorize propositions but also to comprehend them to the point where they could answer searching questions about them and could apply what they had learned to solve new problems, became commonplace. Recitation, which was an integral part of the age-old cyphering tradition, was well used in mathematics departments in U.S. colleges, and in many grammar schools, and high schools throughout the nineteenth century (Ellerton & Clements, 2012; Minnick, 1921).

Trigonometry, Mensuration, Surveying and Navigation in North American

Pre-College Educational Institutions 1607–1865

Throughout the period 1607–1865 the word "mensuration" was often used instead of "measurement," especially in relation to lengths, areas, volumes and angles. Mensuration was typically the name given to a topic which was studied toward the end of the standard *abbaco* arithmetic sequence—it came after alligation, fellowship, false position, progressions, involution and evolution, and permutations and combinations. An elementary part of *abbaco* arithmetic was concerned with what was called "compound operations," which involved addition and subtraction of quantities, and the multiplication and division of quantities by a constant (e.g., the total weight of 5 objects, each weighing 6 pounds 8 ounces). Part of this elementary work on compound operations was known as "reduction" (e.g., "How many inches are there in a mile?"). A slightly more advanced part of *abbaco* arithmetic which might have been regarded as belonging to mensuration was "duodecimals"—for which algorithms employing base 12 multiplication, and multiplication with other bases, were routinely used by tradesmen seeking to calculate areas and volumes.

At the end of the last paragraph the words "areas" and "volumes" were used—of interest, here, is the fact that for much of the period being discussed most people did not know the meanings of those terms, and they certainly did not use them. The word "area" is a term which, today, is used in relation to the extent of surface within a well-defined closed region. It is a *general* term. In elementary *abbaco* arithmetic young learners learned to particularize the concept under consideration in measurement contexts. Thus, for example, the extent of a piece of cloth being purchased to make a garment was thought of quite differently from the extent of a table top, or of a piece of land. The units for measurement were different, and conceptually they were not thought of as being obviously related. It was unusual to find the word "area" in a cyphering book—although occasionally the term was used.

Figure 6.12 shows a page from a cyphering book, prepared by John Scott in Virginia, in 1797, on which the word "area" was used. The problem, stated at the top of the page, was "Required the area of a triangle, two sides of which are 49.2 and 40.8 perches and the contained angle 144 degrees?" After John had twice written the admonition "Time and tide wait for no man," he then showed a solution in which he used logarithms to arrive at his answer, 2 roods and 22 perches. The setting out suggested that John knew what he was doing.

Similarly, it was unusual to find any of the words "length," "volume" and "capacity" used in a cyphering book. Quite reasonably, the amount of fluid needed to fill a large barrel was not necessarily seen as conceptually related to the amount of wood that could be obtained by cutting up a large tree, or to the volume of a parallelepiped. Figure 6.13 shows the solutions given by Oliver Parry (1812), in his cyphering book, to two problems requiring the student to find the "solidities"—that is to say, the volumes—of a frustum of a cone and the frustum of a hexagonal pyramid. Oliver did not use logarithms to support his calculations.

Notice that in Figure 6.13, length units were not specified in the first problem but they were in the second; also, in the answers shown, units were not given. At the bottom of the page it was acknowledged that the rule to be applied to such problems could be found on page 135 of a book on mensuration by Bonnycastle

Distinctions between measurements applied to different real-life contexts were taken for granted by most people throughout the seventeenth and eighteenth century. So too was the fact that each nation had its own set of units, and sometimes units differed within a nation. With some indigenous counting systems, the same idea was applied to counting—for example, the word for "five" in an indigenous language might be different if one was referring to five dogs or to five houses or to five years.

For measuring weights there were three different systems—apothecaries, avoirdupois, and troy—depending on context. Toward the end of the eighteenth century a revolution in measurement, led by French mathematicians and supported by the French Revolutionary Government, attempted to standardize weights and other measures. The result was the formulation and introduction of the metric system. In fact, the base-10 fundamentals of that system had been anticipated by the introduction of decimal currency in the 1780s in the fledgling United States of America (Clements & Ellerton, 2015), and at that time Thomas Jefferson had already worked out—and wished to introduce in the United States of America—a measurement scheme which approximated, in its complexity, what the French would introduce 10 years later as the metric system (Clements & Ellerton, 2015).

A cyphering book which included material related to mensuration, surveying, or navigation would be likely to include a section on trigonometry. Although logarithms, to the base 10, were often used to assist calculations, only rarely was any explanation of the concept of a logarithm explained in a cyphering book. The directed-line-segment approach to trigonometry (Van Sickle, 2011) was preferred, by which the sine, cosine, and tangent of an angle were defined in terms of the length of directed-line segments on a circle (see Figure 6.5). This same approach was adopted in most U.S. textbooks which included sections on trigonometry (see, for example, Allen, 1822, p. 191; Davies, 1838, p. 214; Jacoby, 1939; Playfair, 1822, p. 312).

As with Figure 6.12, the setting out of solutions to mensuration problems often featured diagrams and calculations with logarithms based on the direct rule of three. With mensuration tasks, lengths, areas, volumes, capacities and angles were found using long-established formulae—which were only occasionally expressed algebraically.

Surveying in Pre-College Curricula

From the beginning of permanent European settlement (in 1607) in North America it was obvious that persons with knowledge of surveying would be much needed. Vast amounts of land had to be mapped and measured, townships planned and developed, lakes and mountains described quantitatively, boundaries of territories established, rivers tracked, and land disputes settled by authoritative means (Zitarelli, 2019). Surveying instruments brought from Europe as well as crude instruments constructed locally figured prominently in the development of pioneer settlements. During the colonial period they were used to identify and map

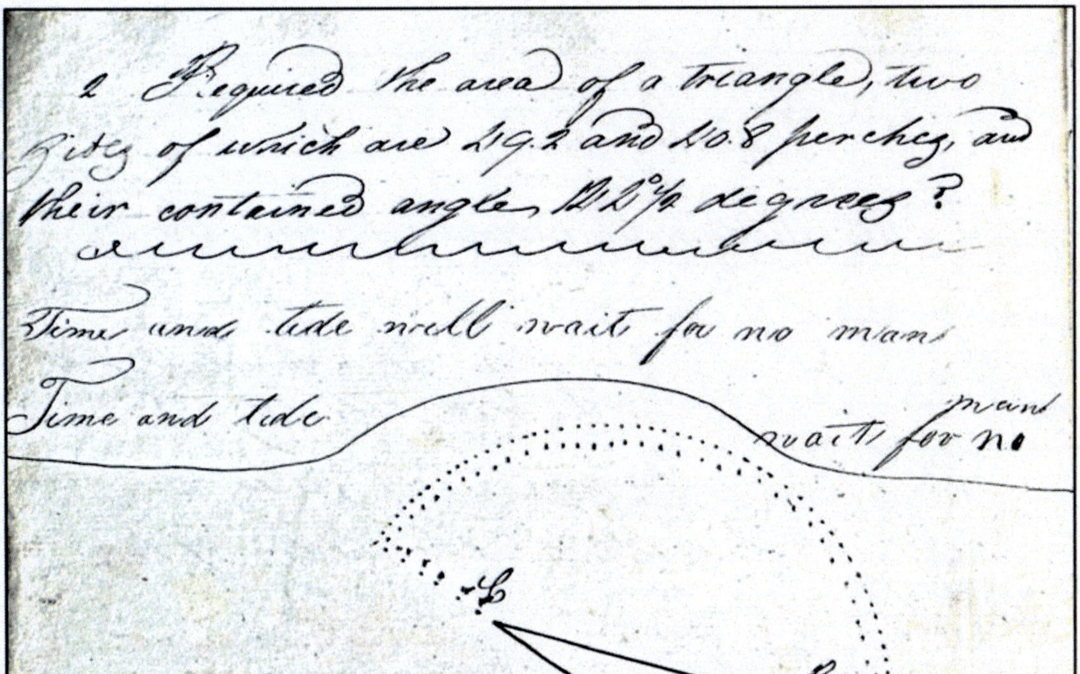
2 Required the area of a triangle, two sides of which are 49.2 and 40.8 perches, and their contained angle 124½ degrees?

Time and tide will wait for no man

Time and tide wait for no man

C

A

B

Given angle 124½° sine 9.76395

Containing sides {49.2 log 1.69196

{40.8 log 1.61066

13.06657

10.

3.06657

2)1165 log

582.5(3 A 2 R 22 P area

480

40)102 R

80

22 P

Figure 6.12. A mensuration problem for which the word "area" was used (from John Scott's (1797) cyphering book, which is held in the E-C cyphering-book collection).

the boundaries and extents of landed estates and farms, with the western population explosion after 1776 resulting in large amounts of public lands being divided into townships and sections. Then came the early builders of railroads who needed to determine accurately where they could, and could not, build. Persons who could

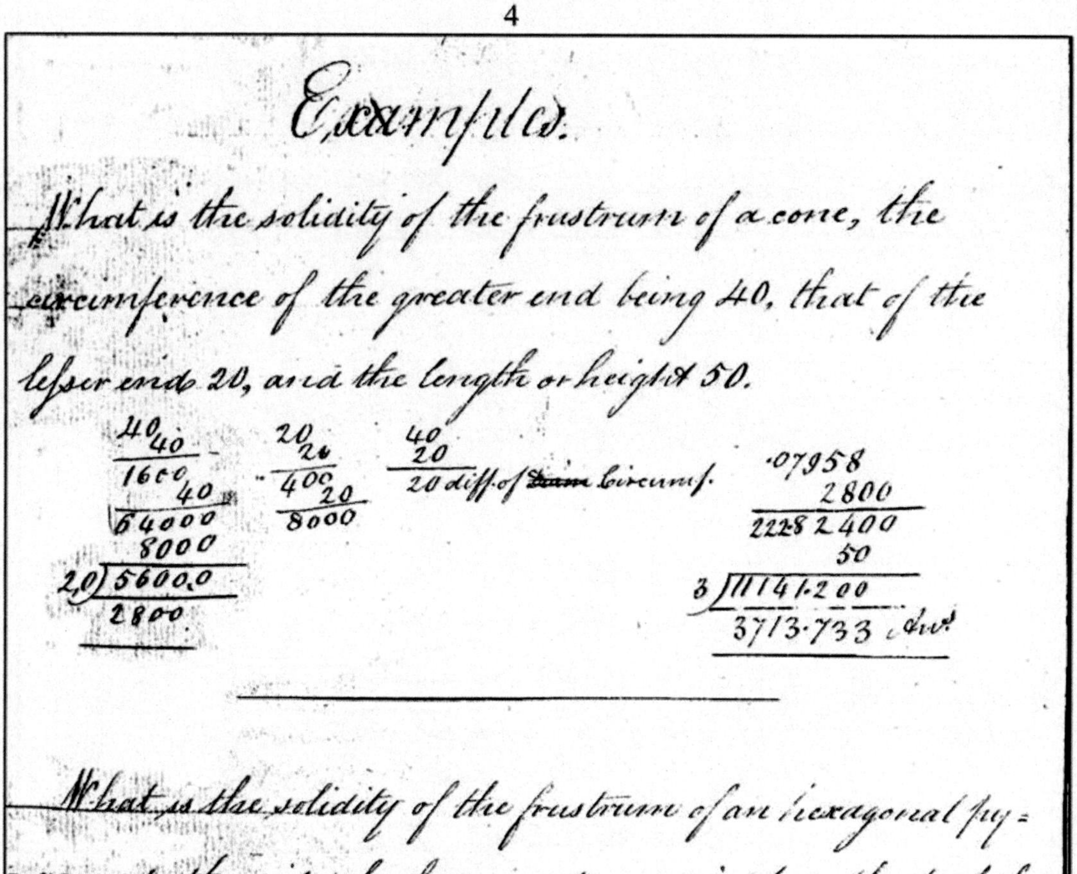

4

Examples.

What is the solidity of the frustrum of a cone, the circumference of the greater end being 40, that of the lesser end 20, and the length or height 50.

40 × 40 = 1600; 1600 × 40 = 64000
20 × 20 = 400; 400 × 20 = 8000
40 − 20 = 20 diff. of circumf.
64000 − 8000 = 56000; 56000 ÷ 20 = 2800

·07958 × 2800 = 222·82400
222·82400 × 50 = 11141·200
11141·200 ÷ 3 = 3713·733 Ans.

What is the solidity of the frustrum of an hexagonal pyramid, the side of whose greater end is 3 feet, that of the lesser end 2 feet, and the length 12 feet?

* 3 × 3 = 9; 2 × 2 = 4; 2 × 3 = 6; 9 + 4 + 6 = 19

2·598076 tabular number
× 19
23382684
2598076
49·363444
× 4
197·453776 Answer.

* for the rule for working this sum this method see a note under Problem 9. Page 135. Bonnycastles Mensuration —

Figure 6.13. Oliver Parry's (1812) solutions to two problems from Bonnycastle's *Mensuration*.

make accurate surveys in order to ascertain the limits of properties, and where cave-ins were likely to occur, were sought after. Those responsible for developing policies with respect to roads, bridges, rivers, and buildings, wanted to employ surveyors—especially if they were willing to work in remote regions (Uzes, 1980).

Early in his life the man who would become the first President of the United States of America became a practicing surveyor. As he grew toward maturity, young George Washington (1732–1799) looked beyond the meager prospects at his Ferry Farm plantation. After considering a career in the Royal Navy, he decided to become a surveyor, and began studying geometry and surveying, using a set of surveyor's instruments from the storehouse at Ferry Farm. In the 1740s he created a cyphering book—which is preserved to this day (Crackel, Rickey & Silverberg, 2017). Almost a century later, in 1833, future-President Abraham Lincoln responded to a need for a surveyor in New Salem, Illinois. But Lincoln had never studied surveying, so he sought help from Mentor Graham, a local pastor/teacher. Lincoln's first task was to learn about surveying. He borrowed two text books—Abel Flint's (1804) *A System of Geometry and Trigonometry with a Treatise on Surveying,* and Robert Gibson's (1814) *The Theory and Practice of Surveying*—and purchased some surveying equipment. With the help of Mentor Graham, he surveyed local roads, towns, and rivers. The main point to be made here is that although Lincoln had not previously studied mathematics beyond middle-level *abbaco* arithmetic (Ellerton et al., 2014), he was able to perform his surveying duties without much training.

Problems in elementary surveying were often easier to solve than problems in elementary navigation. Figure 6.14 reproduces a page from John Scott's cyphering book (prepared in Virginia in 1810) which shows the type of surveying problems which were typically posed in surveying cyphering books. The problem was:

> In a pentangular field beginning with the south side and measuring around towards the east, the first or south side is 2755 links, the second 3115, the third 2370, the fourth 2925, and the fifth 2220; also, the diagonal from the first angle to the third 2800 links and that from the third to the fifth 2010. Required, the area of the field.

John Scott's complete response to the task is shown. Observe that there was no explanation of the method given. It seems to have been assumed that the method was obvious, given the working shown. The diagram suggested that the areas of several triangles needed to be found and then added together. It seems that an area formula involving finding the semi-perimeter of a triangle, then subtracting the length measure of each side from the semi-perimeter before obtaining the square root of the product of the four numbers, was used. A corresponding algebraic formula—such as $A = \sqrt{s(s-a)(s-b)(s-c)}$—was not stated. The multiplications were performed using logarithms, with the square root being obtained by dividing the logarithm of the product by 2. This was repeated for each triangle.

The problem was realistic, and the method was one which surveyors actually used. In fact, the problem was conceptually simple—given the lengths of the sides, use the method to find the areas of the several triangles, and then add the several area measures. But the quality of explanation in the cyphering-book entry left much to be desired and if, at some later time, John referred to his cyphering book in order to be reminded how such problems were done it could have been difficult for him to work out what he had done. The working shown is an example of how the **PCA** ("Problem-Calculation-Answer") genre could be educationally inadequate. In fact, the answer was not even stated explicitly.

The most widely-used textbook on elementary surveying in North America around 1800 was that by Robert Gibson (1785) which—according to a statement at the front of the book—had been written according to methods commonly recommended by the surveyor general's office in Philadelphia (despite the fact that the editions of Gibson's text used in North America had been written and published in Great Britain). Gibson (1785) stated the rule used by John Scott in the example shown in Figure 6.14 in the following way: To find the area of a plane triangle given the lengths of the three sides:

> From half the sum of the three sides subtract each side severally; take the logarithms of the half sum and three remainders; and half their total will be the logarithm of the area; or take the square root of the continued product of the half sum and three remainders for the area. (p. 204)

Gibson (1785) offered a proof of the relationship used—although in that proof he did not state the formula in algebraic terms. However, John Scott did not show a proof on any page of his cyphering book.

Around 1800 the concept of proof was not something often mentioned in pre-college mathematics textbooks or cyphering books. The implemented curriculum included methods for obtaining answers to standard problems, but not proofs. Model examples were given showing how the methods should be applied. That was true for algebra, geometry, trigonometry, mensuration, surveying, navigation and for any other form of applied mathematics (e.g., gauging, dialling, fortification, gunnery) which might be under consideration. It is not clear from cyphering-book data whether teachers of mathematics knew how to prove the most important results, but it is likely that they did not.

Logarithms were regarded as quintessentially an aid to calculation. One never found in a surveying or navigation cyphering book any mathematical justification for taking half the value of a logarithm in order to get a square root. In directed-line-segment trigonometry, the logarithm of the sine of 40°26′ was equal to "9.811952" because the standard circular diagram had a radius of 10^{10}. Charles Davies (1838) did, in fact, explain why that was the case—he stated that "in log sine tables the values of the sines were "calculated for a radius of 10,000,000,000" (p. 234), and that "the logarithm of this radius is 10" (p. 234). The explanation was mathematically adequate but educationally hopeless for almost anyone reading it for the first time. It was not until the mid-1820s that trigonometry in the United States of America was

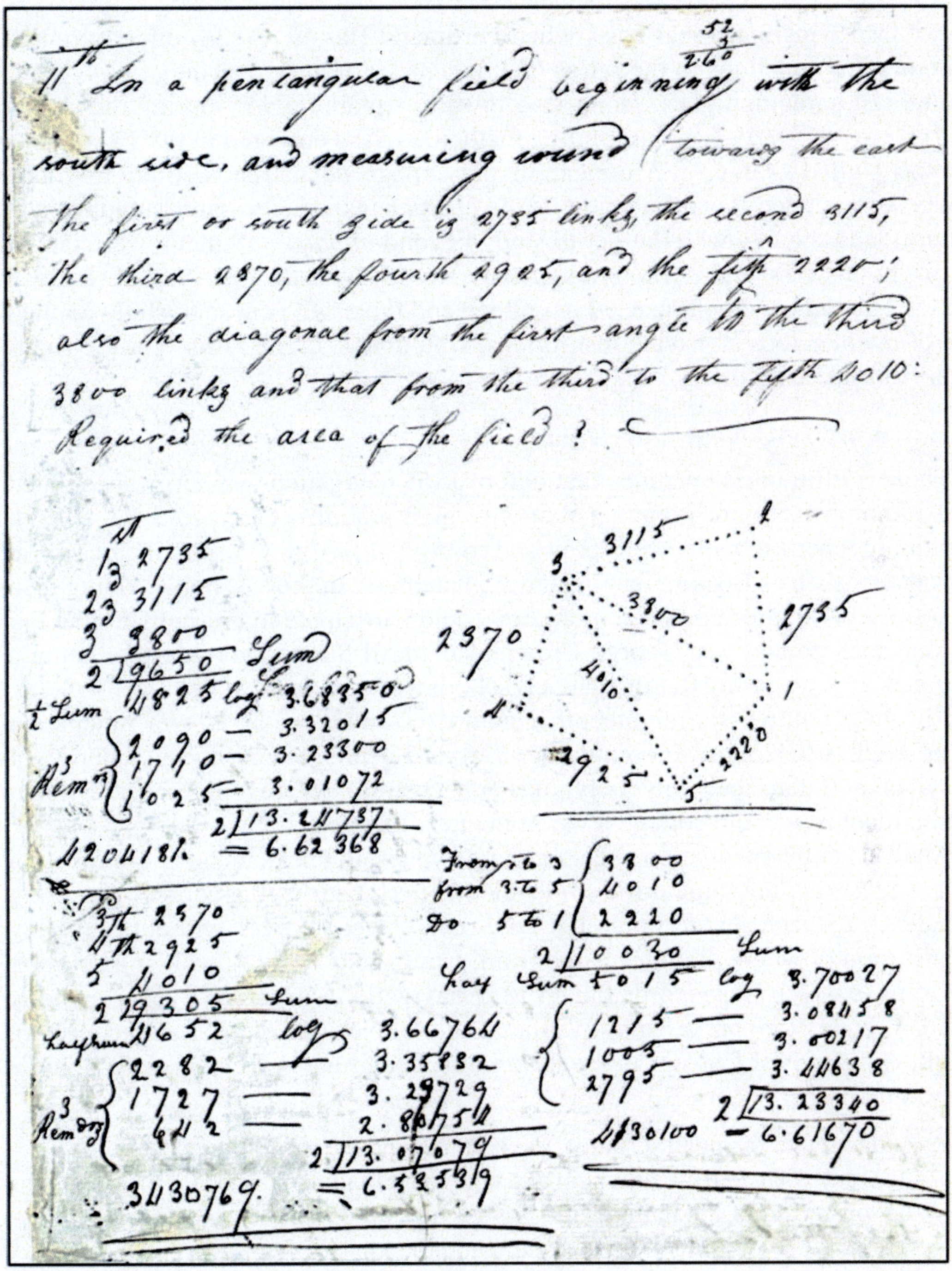

Figure 6.14. A solution to a surveying task in John Scott's (1810) cyphering book.

put on a firmer mathematical basis when Ferdinand Hassler (1826) directly linked trigonometrical functions to the ratios of side lengths of a right triangle.

The last paragraph goes a long way toward explaining why during the period from 1607 to 1865 very few outstanding mathematicians emerged in the 13 colonies or in the United States of America. In pre-college education institutions "top" students were guided toward a study of the classics (especially Latin language and literature), and the few who did get to study beyond-*abbaco* arithmetic were taught to get right answers, and not to wonder why any of the rules that they were given "worked." Students who proceeded to college and those who chose to study applied forms of mathematics at private institutions, would have carried this "get-the-right-answer" attitude with them.

Navigation in Pre-College Curricula

There is little-to-no evidence that before 1865 navigation was seriously studied in any common school or grammar school or high school. At the pre-college level, navigation cyphering books were prepared in private navigation schools located in urban centers such as Boston, Salem, and Philadelphia. In those schools, young men were trained to be midshipmen so that they could participate in trading with the Far East (one such person was George Crowninshield, of Salem, who, despite being a member of a very wealthy family, had left school when 11 years of age—see Wecter, 1937, p. 450). During the eighteenth century, however, college-level navigation courses were offered at, for example, Harvard College and the Academy of Pennsylvania (later, the University of Pennsylvania). More will be said about navigation courses at the college level in the next chapter.

Analysis of the navigation and surveying cyphering books in the E-C collection would suggest that the same criticism leveled above at surveying courses regarding the tendency to teach students merely to use rules to get answers, without providing any indication of where the rules came from, would also apply. Consider a solution to an exercise which was recorded in an 1837 navigation/astronomy cyphering book (in the E-C collection) prepared by William Hale (of Newbury-Port, Massachusetts). The following rule for finding "the true azimuth at any time" was given in William's cyphering book:

1. Add together the polar distance, the latitude, and the true altitude. Take the difference between the half-sum and note the remainder. Then add together the log-secant of the latitude, the log-secant of the altitude (rejecting each index), the log-cosine of the half-sum and the log-cosine of the remainder. Half the sum of these four logarithms will be the log-cosine of the true azimuth, which being doubled gives the true azimuth from the North in Northern latitudes and from the South in Southern latitudes.

(From William Hale's 1837 cyphering book)

One of the exercises which followed the statement of the rule was: "Given the sun's altitude corrected for dip refraction is 20°46′, his declination 17°10′ S and the lat. of the place 40°38′, required the true azimuth." William's solution, which was very neatly presented, was as follows:

Polar distance	107°10′	90.00 + 17.10 = 107.10
Latitude	40°38′	Secant 0.11982
Altitude	20°46′	Secant 0.02917
Sum	2/168°34′	
Half-sum	84°17′	Cosine 8.99830
Polar dist	107°10′	
Remainder	22°53′	Cosine 9.96440
		2 / 19.11169
½-sum log cosine	68°55′	9.55384
	2	
True azimuth	137°50′, from the North.	

Not knowing anything about technical aspects associated with such tasks, we admit to having had to consult the given rule to comprehend what William wrote. William did not offer a diagram to support his working. William Hale's navigation cyphering book throws light on the state of mathematics and mathematics education in the United States of America in 1837. The main aim, from a student's point of view, was to learn to get right answers to standard questions.

Joshua Hertel (2016) has provided an excellent overview of navigation cyphering books held in the Phillips Library (Salem, MA), the Houghton Library (Harvard University), the Wilson Library (in the University of North Carolina, Chapel Hill) and the Ellerton-Clements collection. His analysis revealed that many pages in navigation cyphering books were copied directly from pages in textbooks. Hertel's analysis showed that the most common textbooks consulted by those preparing navigation cyphering books were those authored by John Hamilton Moore (1796) and Nathaniel Bowditch (1802). Both authors offered the same rule as the one given above for finding the true azimuth, and neither showed an associated diagram.

Of one thing there can be no doubt: The quality of navigation education in the colonies and during the period from 1776–1865 paled in comparison with that available to 14- to 16-year-olds in England through the Royal Mathematical School (RMS), at Christ's Hospital in London (Ellerton & Clements, 2017), and through other specialist navigation schools in Great Britain. At the beginning of the eighteenth century the RMS intended curriculum for boys aged between 14 and 16 years was finalized after careful consultation with leading British mathematicians, including Isaac Newton. The following sequence of 10 topics was required to be taught, and the quality of student learning was evaluated through one-to-one interviews conducted by externally-appointed experts:

1. Arithmetic in integers, vulgar and decimal fractions, the extraction of roots, square roots and the use of logarithms.
2. The principles of geometry in the delineation and mensuration of planes and solids, with the application thereof.
3. Plane and spherical trigonometry, geometrically, arithmetically performed in the cases of rectangular and oblique-angular triangles.
4. The use of the globes, celestial and terrestrial, with the stereographical projection of the sphere upon the plane of any great circle.
5. Spherical triangles applied to the solutions of all useful problems in astronomy for finding the Sun's amplitude and azimuth, and the variations of the compass, as also the solutions of all propositions in geometry in all the four various situations of planes commonly called great circle sailing.
6. Plane sailing, viz. the construction and use of the plane sea chart, in all the cases thereof, and the working of traverses, both without ports and with ports, also the solution of all plane sailing questions, geometrically, arithmetically, and instrumentally, with absolute directions for keeping a journal at sea, and to correct the ship's dead reckoning, by observing the sun or any fixed star upon the meridian, with the application of plane triangles to oblique sailing and the doctrine of currents.
7. Mercator's sailing to be done in all respects in plane sailing in Article 6 with the true use of the log line and the minute glass.
8. To find the quantity of the degree under any great circle, the construction and use of instruments proper for observing the latitude at sea, as the cross staff, quadrants and other necessary instruments, as the sector and Gunter's rule.
9. The construction and use of right lines and circular maps, practice of drawing for laying down the appearances of lands, towns, and other objects worthy of notice.
10. The use of the calendar, with the common rules of finding the course of the Sun, Moon and tides, with so much of gunnery as is necessary for.

(Quoted in Ellerton & Clements, 2017, p. 95)

We believe that the RMS curriculum greatly influenced navigation studies at Harvard College in the eighteenth century, especially during John Winthrop IV's period as Hollis Professor. However, any influence at the *pre*-college level in North America was negligible.

What Was the Attained Mathematics Curriculum?

This chapter has 14 figures, most of them reproducing pages from cyphering books in the Ellerton-Clements collection. Examination of the mathematics on the pages shown in the figures suggests that although the students usually did the calculations themselves, the mathematics was often copied—from parent cyphering books or from textbooks. That raises the question, to what extent did the students "understand" what they wrote? The answer to that question is unknown because the word "understand" can have many meanings, and we do not have the benefit of interview data. Nevertheless, the cyphering tradition assumed that students would not make final entries into cyphering books until the material to be entered was approved by the teacher. Furthermore, good teachers were expected to make their students demonstrate, by answering well-constructed questions on what they had just done, that they comprehended what they would write, could explain it well, and could apply it by solving reasonably straightforward associated problems. But the quality of questions asked by teachers during what were called "recitation sessions" varied, and that was particularly the case with teachers who themselves did not have a full understanding of the mathematics under consideration.

Historians (e.g., Cohen, 1982) have typically claimed that most students preparing cyphering books merely memorized the material that they were studying. Such a claim hardly rings true when the material being studied was advanced *abbaco* arithmetic, or algebra, or geometry, or mensuration, or trigonometry, or surveying, or navigation. What sense does it make to say that the students who prepared the material shown in the figures in this chapter did nothing more than memorize what they had written on the pages? It is more likely that they had struggled to understand what they were being asked to learn, and had applied their emerging knowledge and skills when solving the problems shown. Part of the cyphering tradition was that the cyphering books would be used by students as reference texts later in their lives. One might have expected, therefore, that the students would do their best to do more than merely memorize the text. Sometimes, however, students would have had great difficulty arriving at and comprehending the mathematics associated with set tasks. In such circumstances, copying without any real level of understanding could well have become the order of the day.

Oliver Parry prepared a 160-page manuscript during the six-month period April through September 1812 when he was an 18-year-old student attending New Garden School in Chester County, Pennsylvania. At the start of the manuscript (on the inside front cover), Oliver wrote: "I began at geometry 4th month, 1812; in the 5th month; I went at Algebra; in the 7th month, 11th day; I finished geometry and pursued algebra; in the 7th month (22nd day); I finished algebra and went to surveying; In the 8th month (21st), I finished surveying and went to mensuration; in the 9th month (28th), I finished mensuration." Earlier in this chapter, Figure 6.13 showed a page from the mensuration section in Oliver's cyphering book. The standard of penmanship and calligraphy throughout the manuscript was high and the handwritten entries were mathematically accurate—although not always well explained.

It appears to have been the case that a small proportion of school students did get to *apply* mathematics which extended beyond *abbaco* topics. However, not many would have had the opportunity to remain at school throughout a whole year, including during non-winter months, like Oliver Parry did. Internal evidence from what was written in Oliver's cyphering book would strongly suggest that he worked on his mathematics steadily and conscientiously, and developed a strong understanding of the material he entered into his cyphering book (see Ziegler (2011) for more on the life and influence of the Parrys). Our analyses of cyphering-book data have revealed that he was the not the only student to do that.

References

Ackerberg-Hastings, A. (2000). *Mathematics is a gentleman's art: Analysis and synthesis in American college geometry teaching, 1790-1840*. PhD dissertation, Columbia University.

Allen, J. (1822). *Euclid's elements of geometry, the first six books are added, elements of plain and spherical trigonometry, a system of conick sections, elements of natural philosophy as far as it relates to astronomy, according to the Newtonian system, and elements of astronomy*. Baltimore, MD: Cushing and Jewett.

Barbin, E., & Menghini, M. (2014). History of teaching geometry. In A. Karp & G. Schubring (Eds.), *Handbook on the history of mathematics education* (pp. 473–492). New York, NY: Springer. doi:https://doi.org/10.1007/978-1-4614-9155-2_23

Bonnycastle, J. (1818). *An introduction to mensuration and practical geometry* (2nd ed.). Philadelphia, PA: Kimber and Sharpless.

Bowditch, N. (1802). *The new American practical navigator; being an epitome of navigation; containing all the tables necessary to be used with the Nautical almanac. . . . Also, the demonstration of the most useful rules on trigonometry; with many useful problems in mensuration, surveying and gauging; and a directory of sea-terms . . .* Newbury-Port, MA: Edmund Blunt.

Bressoud, D. (2010). Historical reflections on teaching trigonometry. *Mathematics Teacher, 104*(2), 106–112. doi:https://doi.org/10.5951/MT.104.2.0106

Burke, C. B. (1982). *American collegiate populations. A test of the traditional view*. New York, NY: NYU Press.

Cairns, W. D. (1934). "The Elements of Euclid, Thomas L. Heath." *The American Mathematical Monthly, 41*(6), 383. doi:https://doi.org/10.2307/2301562

Cajori, F. (1890). *The teaching and history of mathematics in the United States* (Circular of Information No. 3, 1890). Washington, DC: Bureau of Education.

Cajori, F. (1907). *A history of elementary mathematics with hints on methods of teaching*. New York, NY: The Macmillan Company.

Chambers' Information for the People. (1868). *Mechanics' institutions* (Vol. 1, pp. 713–744). New York, NY: United States Publishing Company.

Clements, M. A., & Ellerton, N. F. (2015). *Thomas Jefferson and his decimals 1775–1810: Neglected years in the history of U.S. school mathematics*. New York, NY: Springer. doi:https://doi.org/10.1007/978-3-319-02505-6

Cohen, P. C. (1982). *A calculating people: The spread of numeracy in Early America*. Chicago, IL: The University of Chicago Press.

Cohen, P. C. (2003). Numeracy in nineteenth-century America. In G. M. A. Stanic & J. Kilpatrick (Eds.), *A history of school mathematics* (pp. 43–76). Reston, VA: National Council of Teachers of Mathematics.

Connor, R. D. W. (1951). Genesis of higher education in North Carolina. *North Carolina Historical Review, 28,* 1–14.

Coon, C. L. (1915). *North Carolina schools and academies, 1790–1840: A documentary history*. Raleigh, NC: Edwards and Broughton Printing Co.

Crackel, T. J., Rickey, V. F., & Silverberg, J. S. (2017). Provenance lost? George Washington's books and papers lost, found and (on occasion) lost again. *Papers of the Bibliographical Society of America, 111*(2). doi:https://doi.org/10.1086/691826

Davies, C. (1838). *Elements of geometry and trigonometry from the works of A. M. Legendre* (revised ed.). New York, NY: A. S. Barnes & Co.

Dewees, W., & Dewees, S. B. (1899). *Centennial history of Westtown Board School 1799–1899*. Philadelphia, PA: Sherman & Co.

Doar, A. K. (2006). *Cipher books in the Southern Historical Collection*. University of North Carolina at Chapel Hill: Master of Science thesis, Wilson Library.

Durkin, J. J. (1942). Journal of the Revd. Adam Marshall, schoolmaster, *U.S.S. North Carolina*, 1824–1825. *Records of the American Catholic Historical Society of Philadelphia, 53*(4), 152–168.

Education. (1802, March 29). *The Salem Register*, p. 1.

Education in the Southern States. (1867, November 9). *Harper's Weekly, 11*(567), 706–707.

Ellerton, N. F., Aguirre, V., & Clements, M. A. (2014). He would be good: Abraham Lincoln's early mathematics, 1819–1826. In N. F. Ellerton & M. A. Clements (Eds.), *Abraham Lincoln's cyphering book, and ten other extraordinary cyphering books* (pp. 123–186). New York, NY: Springer. doi:https://doi.org/10.1007/978-3-319-46657-6

Ellerton, N. F., & Clements, M. A. (2012). *Rewriting the history of mathematics education in North America, 1607–1861*. New York, NY: Springer. doi:https://doi.org/10.1007/978-94-007-2639-0

Ellerton, N. F., & Clements, M. A. (2014). *Abraham Lincoln's cyphering book, and ten other extraordinary cyphering books*. New York, NY: Springer.

Ellerton, N. F., & Clements, M. A. (2017). *Samuel Pepys, Isaac Newton, James Hodgson and the beginnings of secondary school Mathematics: A history of the Royal Mathematical School at Christ's Hospital 1673–1868*. New York, NY: Springer. doi:https://doi.org/10.1007/978-3-319-46657-6

Flint, A. (1804). *A system of geometry and trigonometry together with a treatise on surveying*. Hartford, CT: Oliver and Cooke.

Gaydos, T., & Kampas, B. (2010). *American and Canadian ciphering books, 1727–1864*. Salem, MA: Phillips Library at the Peabody Essex Museum.

Gibson, R. A. (1785). *The theory and practice of surveying*. Philadelphia, PA. Joseph Crukshank.

Gibson, R. A. (1814). *A treatise on practical surveying*. New York, NY: Duyckinck.

Goldstein, J. (2000). A matter of great magnitude: The conflict over arithmetization in 16th-, 17th- and 18th-century English editions of *Euclid's Elements* Books I through VI (1561–1795). *Historia Mathematics, 27*, 36–53. doi:https://doi.org/10.1006/hmat.1999.2263

Hassler, F. R. (1826). *Elements of analytic trigonometry: Plane and spherical*. New York, NY: Author.

Hawney, W. (1775). *The complete measurer: Or, the whole art of measuring, in two parts. The first part teaching decimal arithmetick, with the extraction of the square and cube roots ... The second part teaching to measure all sorts of superficies and solids, by decimals; by cross multiplication, and by scale and* compasses (14th ed.). London, England: J. & F. Rivington.

Hertel, J. (2016) Investigating the implemented mathematics curriculum of New England navigation cyphering books. *For the Learning of Mathematics, 36*(3), 4–10.

Jacoby, H. (1939). *Navigation*. New York, NY: The Macmillan Company.

Karpinski, L. C. (1940). *Bibliography of mathematical works printed in America through 1850*. Ann Arbor, MI: University of Michigan Press.

Karpinski, L. C. (1980). *Bibliography of mathematical works printed in America through 1850*. New York, NY: Arno Press.

Kiely, E. R. (1947). *Surveying instruments: Their history and classroom use (19th yearbook)*. New York, NY: Teachers College Columbia University/National Council of Teachers of Mathematics.

Klein, H. S. (2012). *A population history of the United States*. Cambridge, United England: Cambridge University Press. doi:https://doi.org/10.1017/CBO9781139059954

Lacroix, S. F., & Bézout, É (1826). *An elementary treatise on plane and spherical trigonometry, and on the application of algebra to geometry from the mathematics of Lacroix and Bézout, for the use of students of the University at Cambridge, New England* (2nd ed.). Cambridge, MA: Hilliard and Metcalf.

Legendre, A-M (1794). *Élements de géométrie*. Paris, France: Fermin Didot.

Lewis, E. (1848). *A treatise on plane and spherical trigonometry including the construction of the auxiliary tables; a concise tract on the conic sections, and the principles of spherical projection* (2nd ed.). Philadelphia, PA: E. H. Butler & Co., and H. Orr.

Macfarlane, A. (1916). *Lectures on ten British mathematicians of the nineteenth century*. New York, NY: John Wiles & Sons.

Malcolm, A. (1730). *A new system of arithmetick, theoretical and practical wherein the science of numbers is demonstrated from its first principles through all the parts and branches thereof, either known to the ancients, or owing to the improvements of the moderns: The practice and application to the affairs of life and commerce being also fully explained: so as to make the whole a complete system of theory, for the purposes of men of science; and of practice, for men of business*. London, England J. Osborn and T. Longman.

Mayo, A. D. (1898). Horace Mann and the American common school. In *Report of the Commissioner of Education for 1896–97* (pp. 715–767). Washington, DC: Education Bureau.

Meacham, J. (2018). *The soul of America: The battle for our better angels*. New York, NY: Penguin Random House.

Middlekauf, R. (1963). *Ancients and axioms: Secondary education in eighteenth-century New England*. New Haven, CT: Yale University Press.

Minnick, J. H. (1921). The recitation in mathematics. *The Mathematics Teacher, 14*(3), 119–123. doi:https://doi.org/10.5951/MT.14.3.0119

Moore, J. H. (1796). *The new practical navigator, being an epitome of navigation, explaining the different methods of working the lunar observations, and all the requisite tables used with the nautical almanac, in determining the latitude and longitude, and keeping a complete reckoning at sea; illustrated by proper rules and examples; the whole exemplified in a journal kept from England to the island of Tenerife: also, the substance of the examination, every candidate for a commission in the Royal Navy, and Officer in the Honourable East India Company's service, must pass through previous to their being appointed; this with the sea terms, are particularly recommended to the attention of all young gentlemen designed for, or belonging to the sea* (12th ed.). Tower Hill, England: E. Law.

Morison, S. E. (1921). *Maritime history of Massachusetts 1783–1860*. Boston, MA: Houghton Mifflin Company.

Morrice, D. (1801). *The young midshipman's instructor (designed to be a companion to Hamilton Moore's Navigation): With useful hints to parents of sea youth, and to captains and schoolmasters of the Royal Navy*. London, England: Knight and Compton.

Myers, C. & Nash, J. D. (2006). Private education. In W. S. Powell (Ed.), *Encyclopedia of North Carolina*. Chapel Hill, NC: University of North Carolina Press.

Peabody Essex Museum (n.d.). *Salem: Maritime Salem in the age of sail*. Salem, MA: Author.

Phillips, J. D. (1947). *Salem and the Indies: The story of the great commercial era of the city*. Boston, MA: Houghton Mifflin Company.

Playfair, J. (1814). *Elements of geometry: Containing the first six books of Euclid, with a supplement on the quadrature of the circle and the geometry of solids; to which are added, elements of plane and spherical trigonometry*. New York, NY: Collins & Hannay.

Playfair, J. (1822). *Elements of geometry: Containing the first six books of Euclid, with a supplement on the quadrature of the circle and the geometry of solids; to which are added, elements of plane and spherical* trigonometry (6th ed.). Edinburgh, Scotland: Bell & Bradfute.

Pycior, H. (1997). *Symbols, impossible numbers and geometric entanglements: British algebra through the commentaries on Newton's Universal Arithmetic*. New York, NY: Cambridge University Press. doi:https://doi.org/10.1017/CBO9780511895470

Rawley, J. A. (1981). *The trans-Atlantic slave trade*. New York, NY: W. W. Norton.

Rickey, F., & Shell-Gellasch, A. (2019). Mathematics education at West Point: The first 100 years—Claudius Crozet. https://www.maa.org/book/export/html/116859 (viewed December 5, 2019).

Roach, J. (1971). *Public examinations in England, 1850–1900*. Cambridge, England: Cambridge University Press. doi:https://doi.org/10.1017/CBO9780511896309

Seybolt, R. F. (1917). *Apprenticeship and apprenticeship education in colonial New England and New York*. New York, NY: Teachers College Columbia University.

Seybolt, R. F. (1921). *The evening schools of colonial New York City*. Albany, NY: The University of the State of New York.

Seybolt, R. F. (1935). *The private schools of colonial Boston*. Cambridge, MA: Harvard University Press. doi:https://doi.org/10.4159/harvard.9780674599918

Simons, L. G. (1936). Short stories in colonial geometry. *Osiris, 1*, 584–605. doi: https://doi.org/10.1086/368442

Simson, R. (1806). *The elements of Euclid, viz, the first six books, together with the eleventh and twelfth, the errors by which Theon and others have long ago vitiated these books, and corrected, and some of Euclides' demonstrations are restored. Also the books of Euclid's data, like manner corrected*. Philadelphia, PA: Mathew Carey.

Sinclair, N. (2008). *The history of the geometry curriculum in the United States*. Charlotte, NC: Information Age Publishing.

Smith, D. E. (1911). *The teaching of geometry*. Boston, MA: Ginn and Company.

Stamper, A. W. (1909). *A history of the teaching of elementary geometry, with reference to present-day problems*. PhD dissertation, Columbia University, New York.

Taylor, E. G. R. (1966). *The mathematical practitioners of Hanoverian England 1714–1840*. Cambridge, England: Cambridge University Press.

Truth, S. (1850). *The narrative of Sojourner Truth* (edited by Olive Gilbert). Boston, MA: Author.

Uzes, F. D. (1980). *Illustrated price guide to antique surveying instruments and books*. Rancho Cordova, CA. Landmark Enterprises.

Van Sickle, J. (2011). *A history of trigonometry education in the United States: 1776–1900*. PhD dissertation, Columbia University.

Wecter, D. (1937). *The saga of American society: A record of social aspiration 1697–1937*. New York, NY: Charles Scribner's Sons.

Whalin, W. T. (1997). *Sojourner Truth: American abolitionist*. Uhrichsville, OH: Barbour Publishing.

Wikipedia contributors. (2021, July 4). Sojourner Truth. In *Wikipedia, The Free Encyclopedia*. Retrieved from https://en.wikipedia.org/w/index.php?title=Sojourner_Truth&oldid=1031850304

Ziegler, R. (2011). *The Parrys of Philadelphia and New Hope: A Quaker family's lasting impact on two historic towns*. Bloomington, IA: iUniverse.

Zitarelli, D. A. (2019). *A history of mathematics in the United States and Canada. Volume 1: 1492–1900*. Washington, DC: American Mathematical Society. doi: https://doi.org/10.1090/spec/094

Chapter 7
College Mathematics, 1607–1865

Abstract Throughout the period 1607–1865 most families had very few books other than a Bible in their homes, and most people did not know much mathematics beyond reading, writing, and counting with Hindu-Arabic numerals. Between 1636 and 1865 only a tiny proportion of the residents of that part of North America which is now mainland United States of America attended college and, of those who did, most had not previously studied mathematics beyond low-level *abbaco* arithmetic, elementary algebra, and the first few books of Euclid's Elements. It is not surprising, therefore, that the period did not produce more than three or four scholars who, by European standards, might be considered to have been "outstanding" mathematicians. The U.S. college curriculum had its origin in the classical curriculum traditions of the medieval universities of Europe and especially of Cambridge and Oxford Universities. However, many of those who attended North American colleges did study what we have called "applied mathematics"—embracing fields like astronomy, surveying, mensuration and navigation— while they were at college, and we argue that this aspect of the implemented curriculum had been successfully translated mainly from Great Britain.

Keywords Algebra • Applied mathematics • Benjamin Franklin • Benjamin Rush • Blackboards • Classical curriculum tradition • Classics • Conics • Euclidean geometry • Greek language and literature • Harvard College • Isaac Greenwood • Jeremiah Day • John Winthrop IV • Latin language and literature • Navigation • Proof • Pure mathematics • Royal Mathematical School at Christ's Hospital • Trigonometry • University of Cambridge • University of Oxford • Yale College • *Yale Report 1828*

> Mathematics in the eighteenth century . . . did not originate generally in the schools of this country . . . If we except such mechanical features as the elementary operations of arithmetic and algebra and consider the progress of real mathematics, neither the elementary schools of this country nor the colleges were much concerned in that period with the subject. (Smith & Ginsburg, 1934, p. 16)

> I should rejoice to see . . . Euclid honourably shelved or buried "deeper than did ever plummet sound" out of the schoolboys' reach. (Statement by J. J. Sylvester, 1870, p. 261).

N. F. Ellerton, M. A. (Ken) Clements, *Toward Mathematics for All*, History of Mathematics Education,
https://doi.org/10.1007/978-3-030-85724-0_7

The Classics Stranglehold 1607–1865

This book is primarily concerned with interpreting a history of progress toward "mathematics for all" in the Eastern colonies and in the United States of America during the period 1607–1865, and this chapter focuses on college mathematics during that period. We draw attention to elements of the cultural and educational milieu in which college mathematics was located, and in particular to what we call the "classics stranglehold" on higher education, which, we argue, was largely responsible for the attitudes of early colonial leaders toward education and mathematics.

This introductory section has four subsections

1. The first subsection summarizes the classics stranglehold on higher education which existed in Great Britain, in other parts of Western Europe, and in North America throughout the period. That stranglehold constrained the thinking of societal leaders with respect to both mathematics and mathematics education.
2. The second subsection discusses the strong negative reactions to the classics stranglehold by two colonial education leaders—specifically, Benjamin Franklin (1706–1790) and Benjamin Rush (1746–1813)—and the views that those two critics held with respect to mathematics.
3. The third subsection is concerned with the 1828 *Yale Report*, which defended the role of the classics in education.
4. The fourth subsection shows that there were determined attempts by some colonial educators to interpret mathematics so that it not only included arithmetic, algebra, geometry, trigonometry, and mensuration, but also navigation, surveying, physics, astronomy, optics and other aspects of what was regarded as "applied mathematics."

It will not be necessary here to tell the full story of how the defenders of the classics were able to resist the Philistine marauders. It is sufficient to say, perhaps, that even in 1865 those dedicated to the classics still controlled North American college education.

The European Background

Smith and Ginsburg's (1934) assessment of the behavior of early settlers in the New World with respect to mathematics was expressed in harsh terms. After asserting that "the seventeenth century in America produced no mathematician worth the namd" (p. 13), they went on to say:

> The century that saw the work of Galileo, Kepler, Gilbert, Napier, Fermat, Descartes, Pascal, Huygens, Newton, and Leibniz, in countries from which the settlers had come, saw among the intelligentsia no apparent appreciation of the discoveries of scholars of this class. Indeed, even the leaders seemed to sanction the very European religious persecution from which so

> many had fled, and actually to attempt to produce such erudite bigots as Cotton Mather who should have known the writings of these world leaders, but who wasted his abilities upon "daemons and witchcraft," "evil spirits," "evil angels," and other evidences of disordered minds that today pass the understanding of even the most reactionary. (pp. 13–14)

Smith and Ginsburg assumed that because Cotton Mather was "erudite" he should have understood the writings, and recognized the importance, of the writings of some of the greatest mathematicians in history. But, in the New World, Mather had neither the opportunity nor the time to keep up with such writings. In fact, there is direct evidence that he particularly admired Isaac Newton's works and encouraged Harvard students to keep up with what Newton was saying (Brasch, 1939). Indeed, in 1680 experimental observations made at Harvard were communicated to Newton across the Atlantic, who used them in his study on the gravitational influence of the sun and moon and comets and thereby assisted him at a time when he was writing his *Principia* (Brasch, 1939; Zitarelli, 2019).

The Inherited Bias Toward the Classics

In the process of gathering data when researching for *Samuel Pepys, Isaac Newton, and James Hodgson: A History of the Royal Mathematical School at Christ's Hospital 1673–1868* (Ellerton & Clements, 2017) we examined minutes of the main administrative body for Christ's Hospital—a school in central London—over a period of almost 200 years. That gave us insight into the "heart" of a school which was originally intended to provide educational opportunities for orphans and for children from very poor families. From the outset the school was especially proud of its "Grammar School," which was responsible for teaching the classics. Soon after the school was created, in 1552, scholarships began to be awarded, annually, to enable the top classics students from the school to attend either the University of Cambridge or the University of Oxford. That practice continued throughout the period 1607–1865 which frames this book.

In 1673, soon after the Bubonic Plague (in 1665), the Great Fire of London (in 1666), and a humiliating defeat of the British Navy by the Dutch Navy (in 1667), the British government funded the establishment of the Royal Mathematical School (RMS) within Christ's Hospital. Children aged about 14 years would be selected from within the school for placement in RMS and graduates would be required to take up apprenticeships in the Royal Navy or in the mercantile marine before their seventeenth birthdays. Under the circumstances, it was inevitable that the Grammar School at Christ's Hospital would compete with RMS for the school's "best" students.

Any Christ's Hospital graduate could attend meetings of the school's main administrative committee, the Committee of Almoners, and the result was that that Committee tended to make decisions in conformity with Grammar School traditions, especially in relation to the study of classics within the school. This bias was often hidden—for example, a student could enter the RMS program only after he had already studied the classics in the Grammar School for 4½ years, and then the selected students were expected to complete their RMS studies within 1½ years. The head of the Grammar School was automatically and officially the head of Christ's Hospital. The RMS curriculum was extensive and difficult, and RMS student achievement was externally examined by "experts." Be that as it may, by the 1770s RMS had developed what was, perhaps, the strongest secondary-school mathematics program in the world (Ellerton & Clements, 2017).

Within the Christ's Hospital community there was always a majority who believed that since the Grammar School had come first, the RMS was little more than an imposter, something forced upon the school by the British government. Despite the difficult circumstances, RMS became the world's first "secondary" school at which higher-level mathematics, including calculus, was taught and learned by children from impoverished families (Ellerton & Clements, 2017). The Grammar School's determination to maintain the importance of classical literature and language in the Christ's Hospital curriculum was only to be expected, of course, for in the sixteenth and seventeenth centuries the classical tradition was accepted as normal and required in any "decent" secondary school (Brown, 1721; Goldsmith, 1837; Lemprière, 1834; Martines. 1979) or college, not only in all European colleges and universities but also in all schools seriously preparing students for entry to colleges and universities. And, as Morison (1956) pointed out, many of "the university-trained founders of New England had been students at Oxford or Cambridge at a time when mathematical and other sciences, in those universities, were at their lowest ebb. University-bound English boys, in 1640, went through seven years of grammar school (and would then spend four years of college) without studying any more mathematics than the cyphering that they had learned in dame schools" (pp. 242–243). According to Morison (1956), "at Oxford and Cambridge, geometry was looked upon as a practical subject, and algebra was not studied" (p. 243).

Like the situation in Christ's Hospital, most of the administrators of the colonial colleges in the New World had academic backgrounds in which they had studied much Latin and Greek literature but hardly any mathematics (see, e.g., Fuess, 1917; Seybolt, 1969). It was not until 1711, more than a century after the initial settlement at Jamestown, that a senior academic appointment in mathematics was made, and that appointment, Professor Tanaquil Lefebre at the College of William of Mary, was short-lived (Zitarelli, 2019). The second professorial appointment in mathematics, Hugh Jones—also at William and Mary—occupied the period 1717–1722, but was also unsuccessful (Zitarelli, 2019). It was not until Isaac Greenwood's appointment at Harvard, late in 1727, that someone with strong up-to-date knowledge of mathematics was appointed to a senior position in mathematics.

From Table 7.1, it can be seen that the number of North American colleges grew steadily after 1745 (Crilly, 2008; Kraus, 1961). Although most of these "new" colleges had humble beginnings they hold an important place in the history of education in the United States.

Table 7.1
Leading U.S. Colleges Established Before 1770

Name of College	Colony/ Province	Year Estab- lished	Name of College	Colony/ Province	Year Estab- lished	Name of College	Colony/ Province	Year Estab- lished
New College, Harvard College, Harvard University	Massa- chusetts Bay	1636/ 1639	College of New Jersey/ Princeton University	New Jersey	1746	College of Rhode Island, Brown University	Rhode Island	1764
College of William & Mary	Virginia	1693 (opened 1699)	King's College, Columbia University	New York	1754	Queen's College, Rutgers	New Jersey	1766
Collegiate School, Yale College, Yale University	Connect- icut	1701	College of Philadelphia, University of Pennsylvania	Pennsyl- vania	1755	Dartmouth College	New Hampshire	1769

Source: Adapted from Dauben & Parshall (2014), Zitarelli (2019)

Some of the early colleges asked students to prepare *abbaco* arithmetic cyphering books, and most introduced their students to the first two books of *Euclid's Elements*. The students themselves did not usually own any textbooks (Zitarelli, 2019) and the curriculum focused on Greek, Latin, Euclidean geometry, history, ethics, and rhetoric. Scant attention was paid to "extras" such as algebra, trigonometry, surveying, or navigation. Colleges often took young students—some as young as 13 or 14 years—tuition fees were low, and scholarships were rarely available. Many of the teachers in the colleges were mere "passers-by" so far as higher education was concerned—they planned to become clergymen, lawyers, or physicians (Cremin, 1977; Geiger, 2016). Entries in Table 7.1, which show the major colleges which were established before 1776 (Dauben & Parshall, 2014) suggest that by the time that Independence was gained the Southern states had fallen well behind the Northern states so far as provision of tertiary colleges was concerned. Many Southern plantation owners sent their children to Europe, or to colleges in the North (Cremin, 1977).

In the colonial colleges, the study of Latin and Greek, and associated literatures, enabled teachers and students to explore Aristotle's three philosophies (natural, moral, and mental). At Harvard College the aim was for Latin and Greek to become living, spoken languages, for structure-of-the-mind arguments had persuaded administrators that this would help students to improve in logic, rhetoric, ethics, metaphysics, astronomy, physics, and mathematics. All of these subjects formed part of the basic course of study. Latin, Greek, Hebrew, logic, and rhetoric were studied in the first year, and in the following year, Greek and Hebrew held the pre-eminent place. In the junior sophister year, mental and moral philosophy were taught, which led to the study of economics, ethics, political science, and sociology. The senior sophister year provided a review in Latin, Greek, and logic, and mathematics was introduced.

At Yale College, during the first half of the eighteenth century, the sequence differed from that at Harvard, but the subject matter and the focus were similar (Dauben & Parshall, 2014; Rudolph, 1990). Quality of learning was assessed through one-on-one teacher-student recitations (Ackerberg-Hastings, 2014) or through disputations—a form of recitation employed as a means of stimulating discussion. Confronted with a thesis, or some claim to universal truth, one student would be the disputant and another the questioner, and the aim would be to explore the validity, or otherwise, of the thesis. These methods were also used to assess what students had learned over their four college years (Lucas, 1994; Rudolph, 1990).

Isaac Greenwood's and John Winthrop IV's Views on the Nature of Mathematics, 1727–1779

The story of Harvard's beginnings is well documented. In 1636, the general court of the Massachusetts Bay Colony appropriated funds for the establishment of a college in the village of Cambridge in Massachusetts (Lucas, 1994). Initially the name of the institution was "New College," but in 1639 it became "Harvard College." In the 1640s, under Henry Dunster, Harvard's second President, the curriculum revolved around the classics, especially Latin and Greek, and Biblical Studies, with Dunster himself teaching most of the subjects (Goodchild & Wechsler, 1989), including Mathematics (Zitarelli, 2019). Curriculum extras included Philosophy, Logic, Astronomy, Geometry, and Ethics (Gwynne-Thomas, 1981, Varcoe, 2002). The main aim was to establish an institution in which future ministers of religion, lawyers, and physicians would be trained (Rudolph, 1990), and not much attention was paid to mathematical studies.

With the number of European-background persons in the colonies increasing steadily to about 250000 by 1700 (Dexter, 1887), the Royal College of William and Mary was created in Williamsburg, Virginia, in 1693 (Tyler, 1897), and the Collegiate School in Killingworth, in Connecticut, in 1701. The Collegiate School would become Yale College, in New Haven, in 1716. Although during the period to 1720, Mathematics was not a major part of the curriculum at any of Harvard, William and Mary, and Yale, in each college time was given to *abbaco* arithmetic and to the first few books of *Euclid's Elements*. In most of the institutions, elementary algebra and trigonometry also featured in the mathematical studies. At Harvard, arithmetic and geometry were part of the course of study for the third year (which was, initially, the final year), with one day a week being allocated to those subjects for three-fourths of the year. When the course was extended to four years in 1655, mathematics remained part of the final year (Stamper, 1909; Zitarelli, 2019). For the first 100 years of Harvard's existence the study of geometry was largely confined to the first few books of *Euclid's Elements* (Stamper, 1909), or to the interpretation of those books in a textbook (e.g., in John Ward's (1719) *The Young Mathematician's Guide: Being a Plain and Easie Introduction to the Mathematicks*).

Toward the end of 1727, Harvard College appointed a former student, Isaac Greenwood, to the position of Hollis Professor of Mathematics and Natural and Experimental Philosophy (Simons, 1924). Greenwood held an M.A. from Harvard and had studied mathematics in England during the period 1723–1726. He was recommended to the position of Hollis Professor by Thomas Hollis himself. The document confirming the establishment of the Hollis chair stated:

1. That the Professor be a Master of Arts and well acquainted with the several parts of Mathematics and Experimental Philosophy.
2. That his province be to instruct the students in a system of Natural Philosophy and a course of Experimental Philosophy in which to be comprehended Pneumaticks, Hydrostaticks, Mechanicks, Staticks, Opticks, etc., the elements of Geometry together with the doctrine of Proportion, the principles of Algrebra (*sic.*), Conic Sections, Plain and Sperical (*sic.*) Trigonometry, with the general principles of Mensuration, Plain and Solids, in the principles of Astronomy and Geometry, viz, the doctrine of the Spheres and the use of the Globes, the motions of the Heavenly Bodies according to the different hypotheses of Ptolomy (*sic.*) Tycho, Brahe & Copernicus with the general principles of Dialling, the division of the world into its various kingdoms with the use of Maps, etc.

(Reproduced from Simons, 1924, p. 45)

Greenwood was Hollis Professor of Mathematics and Natural and Experimental Philosophy at Harvard between late 1727 and early 1738, when he was dismissed for intemperance (Zitarelli, 2019). During his time as Hollis Professor there was a strong Newtonian ring about the mathematics offerings, with Isaac Newton's *Principia Mathematica* and his research on optics and astronomy included within the ambit of mathematics. Arithmetic was not specifically taught, but algebra, geometry, trigonometry (both plain and spherical) were. Greenwood was given a generously light teaching load—one one-hour lecture each week, and the only persons entitled to attend were senior students whose parents had paid a special fee (Zitarelli, 2019).

The time Isaac Greenwood spent in England shortly before securing his appointment at Harvard College enabled him to look beyond the forms of mathematics that he himself had learned as a student at Harvard between 1717 and 1721. Books were in short supply in the colonies because they had to be imported from England (Gwynne-Thomas, 1981), and many of those directed at higher-level mathematics were written in Latin which, as Isaac Newton had liked to say, was the language of the law, church, medicine and science (Ellerton & Clements, 2017). Greek, on the other hand, was the language of the new humanism of Renaissance learning, something which, it was claimed, brought Homer and Hesiod, Greek lyrics and idylls into the experience of an educated man. Harvard was attempting to create a new "University of Cambridge," similar to the old one in England. The prime purpose of the curriculum implemented at the New Cambridge was to turn out gentlemen who were clergymen, scholars, landowners, public servants, and governors (Rudolph, 1990). Similar aims accompanied the creation of the College of William and Mary in Williamsburg, in Virginia, in 1693, and Yale College in Connecticut in the early 1700s (Cremin, 1977; Stoeckel, 1976).

Among the books which Greenwood brought back from England was a copy of Isaac Newton's *Principia Mathematica* and, according to Simons (1924) and Zitarelli (2019), during his time in the "mother country" he had obtained a strong knowledge of recent developments in algebra, including Newton's *Arithmetic Universalis,* which was first published in Latin in 1707 but was then translated into English by Joseph Raphson in 1720. That Greenwood had the intellect and teaching power to go well beyond a curriculum which addressed mainly the old forms of *abbaco* arithmetic, algebra, and Euclidean geometry was made obvious in the following advertisement which appeared in the *Boston Weekly News Letter* in both 1726 and 1739.

> Such as are desirous of learning any parts of the Mathematics whether theoretical, as the demonstrating of Euclid, Apollonius, &c., or practical, as arithmetic, geometry, trigonometry, navigation, surveying, gauging, algebra, fluxions, &c. Likewise any of the branches of natural philosophy, as mechanics, optics, astronomy, &c may be taught by Isaac Greenwood, A.M. &c. at the Duke of Marlborough's Arms in King Street over against the Golden Fleece, Boston, where attendance is given from 9 to 12 a.m. and 3 to 6 p.m.
>
> N.B. If any gentlemen or particular company of such are desirous of private instruction relating to the premises at their respective places, attendance will be given out of the aforesaid hours of teaching.
>
> (Quoted in Karpinski, 1980, p. 572)

The services which Greenwood offered were similar to those offered in London's coffee houses during the period 1700–1730 (Stewart, 1999). At the time when the advertisement was first placed (in 1726), Greenwood had just returned after spending three years in and around London, in England, and it is highly likely that while he was there he consulted with James Hodgson, who, as head of the Royal Mathematical School at Christ's Hospital, held the most important appointment in mathematics in England outside of Cambridge and Oxford Universities (Ellerton & Clements, 2017). Hodgson was a master teacher, a mover and shaker who, before taking up his appointment at Christ's Hospital, had had much experience giving talks and performing demonstrations in London's coffee houses (Ellerton & Clements, 2017; Iliffe, 1997). It is likely that Greenwood attempted to reproduce in Boston part of his coffee-house mathematics experiences in London.

The colorful but sad career of Greenwood as Hollis Professor at Harvard College has been well summarized by Zitarelli (2019) in more detail than is provided here. However, we now provide additional support for our contention that it was Greenwood at Harvard and Jeremiah Day at Yale who took on the responsibility of defining a "British approach" to college mathematics which would be suited to the educational needs of the North American New World. The first point is that Greenwood and Day relied almost completely on recent British developments in mathematics, and only rarely referred to writings of Continental European mathematicians. Simons (1924) maintained that Greenwood's early teaching of algebra at Harvard was directly based on John Wallis's (1685) *A Treatise of Algebra both Historical and Practical*, Greenwood and was also probably influenced by the writings on algebra by other British mathematicians—including Edmund Halley, Isaac Newton, William Oughtred, Joseph Raphson, James Hodgson, and William Whiston, The second point is that Greenwood took seriously a perceived need to address the applied mathematics component of the curriculum specified in his appointment agreement. He engaged his students in experiments, using equipment which had been donated to the University by Thomas Hollis. Between 1727 and 1732 he succeeded in having three "applied mathematics" papers, based on his analyses of data collected at Harvard, published in the prestigious *Philosophical Transactions*, the journal of the British Royal Society. He also authored a book, *Arithmetick Vulgar and Decimal; with the Application Thereof to a Variety of Cases in Trade and Commerce*, which appeared in 1729 and was the first mathematics textbook written by a North American scholar to be published in North America. It would be difficult to sustain any argument that Greenwood's appointment was a failure. He worked hard, and achieved much, in his all-too-brief academic career.

The following list summarizes the algebra topics in two cyphering books (now held in the Houghton Library at Harvard University) prepared by Harvard students in the 1730s—one by James Diman in 1730 and the other by Samuel Langdon in 1739. In fact, entries in the two cyphering books were almost (though not exactly) the same.

> Notation; algebraical characters: addition of integers; subtraction; multiplication of algebraical integers; division; algebraical fractions; addition and subtraction of fractions; multiplication of fractions; division of fractions; involution of whole quantities; fractional evolution; binomial quantities; involution of binomial quantities; promiscuous examples; involution of binomial fractional quantities; multinomial quantities; fractional compound quantities; evolution of multinomial compound quantities; surd quantities; addition and subtraction of surds; compound surds; multiplication of binomial surds; division in compound surds; reduction of equations; reduction by addition; reduction by subtraction; reduction by multiplication; reduction by division; reduction by involution; reduction by evolution; reduction by analogies to equations; the method of resolving

> algebraical questions; general rules concerning reduction of equations; simple equations; the solution of adfected quadratic equations; Mr. Oughtred's method of solving adfected quadratics; the solution of adfected quadratic equations by taking away the second term; the solution of adfected quadratic equations by the method completing the square; questions producing adfected quadratic equations; the resolution of cubic equations; cubic equations by substitution; cubic equations by trials and depression; the solution of irregular cubics; the method of converging series and approximation; Mr Raphson's theorems for simple powers; Mr. Raphson's theorems for adfected equations; Dr. Halley's theorems for solving equations of all sorts; concerning the method of resolving geometrical problems algebraically.

References to "Mr. Oughtred's method," "Mr. Raphson's theorems" and "Dr. Halley's theorems" in this summary suggest that, as stated previously, the greatest curricular influence on Harvard mathematics at this time was from England, and not from the Continent.

John Winthrop IV Builds on the Work of Isaac Greenwood at Harvard

Toward the end of 1737, Isaac Greenwood was removed from his position at Harvard and was replaced by 23-year-old John Winthrop the Fourth. Winthrop, a descendant of the first Massachusetts Bay governor, John Winthrop the first, had been a brilliant student under Greenwood and would remain Hollis Professor of Mathematics and Natural and Experimental Philosophy until his death in 1779. During his tenure he mastered Newton's fluxions, and Newton's *Principia,* and had 12 articles published in *Philosophical Transactions* (the journal of the Royal Society) between 1753 and 1775. Only one of those 12, however, was in the realm of "pure mathematics." According to Smith and Ginsburg (1934), he displayed an impressive knowledge of the works of the top European mathematicians of his day, especially those of British mathematicians. But he correctly interpreted his appointment as requiring him to instruct his students in subjects like pneumatics, hydrostatics, mechanics, statics, optics, mensuration, astronomy, navigation, and surveying—as well as in algebra, geometry, plane and spherical trigonometry, calculus, and conic sections. In his heart, he was an applied, not a pure, mathematician, and his great love, and major achievements, were in the field of astronomy. Such was his influence that professors in other colleges would follow his example (Dauben & Parshall, 2014; Zitarelli, 2019). Almost all his best publications—those which appeared in *Philosophical Transactions*—were in the realm of astronomy. He enjoyed a strong relationship with Benjamin Franklin, and this resulted in his paying careful attention to theoretical aspects of electricity and magnetism. He interpreted data that he obtained in terms of Franklin's and his own theories. We would definitely regard Winthrop as a "mathematician."

Evidence for Winthrop's commitment to applied aspects of mathematics can be found in the 170-page cyphering book prepared at Harvard College in 1769 and 1770 by his son, William Winthrop. This manuscript, which is held in the Houghton Library at Harvard University (manuscript number fMS Am 1565 Am 550), included sections on elementary geometry, mensuration, navigation, dialing, spherical trigonometry, spherical geometry, and gauging. What was entered in this manuscript, and in some other cyphering books in the Harvard collection, is reminiscent of entries in navigation cyphering books prepared in the Royal Mathematical School at Christ's Hospital, in London (Ellerton & Clements, 2017). It is possible that William Winthrop's cyphering book was modeled on a manuscript that Isaac Greenwood brought back to Harvard from England, and which was made available to John Winthrop IV when he was studying under Greenwood.

Benjamin Franklin's and Benjamin Rush's Attacks on What They Perceived to Be an Over-Emphasis on Classics

Most leading colonists accepted the value of the classics. They themselves had studied the classics in European universities and some had even sent their own sons back to England to study at Cambridge University or Oxford University (Martines, 1979). But, even if personally they harbored misgivings about its suitability in the New World, they could not afford to question too strongly the value of classical learning.

However, the world was changing. Names like Copernicus, Descartes, Galileo, Kepler, Pascal, Newton, and Leibniz had become known, and although their works may not have been well understood, it was recognized by some in the upper echelons of colonial society that in the new order the curricula of higher education institutions needed to be broadened beyond the classics. In 1647, for example, President Dunster asked Harvard officials for funds to purchase suitable books "especially in law, *physicke*, philosophy and mathematics" to be made available to scholars "whose various inclinations to all professions might thereby be encouraged and furthered" (quoted in Morison, 1956, p. 32). The Reverend Jonathan Mitchell, a senior fellow at Harvard College, asked for funds to establish chairs in history, languages, law, mathematics, and medicine to train up "choise and able schoolmasters," "able, eminent and approved physicians," and to provide education "to accomplish persons for the magistracy and other civil offices" (quoted in Morrison, 1956, p. 32). The requests were not heeded, but the fact that they were made was a sign of the times.

Benjamin Franklin. In Pennsylvania in the 1740s a young Benjamin Franklin was among a group of notable citizens who decided to take the matter into their own hands. In 1749 they established a school which would subsequently be chartered as the College and Academy of Philadelphia. Franklin was the first President of the Board. Initially there was an associated "charity school" which taught reading, writing, and arithmetic, with the Academy offering a more advanced curriculum.

William Smith was appointed Provost (Education) in 1756. Seven men graduated in May 1757, six with a bachelor degrees and, intriguingly, one with a Master of Arts.

In his "Proposals Relating to the Education of Youth in Pensilvania," Franklin (1749) made it clear that he believed that, in a well-devised curriculum, English, arithmetic, and practical subjects were just as important as classical studies (Beadie, 2010). One of Franklin's most notable curriculum proposals was that courses be taught solely in English and not in Latin. The decision to emphasize teaching in the English language instead of Latin or Greek set Franklin and co-reformers of like mind apart. They wanted the college curriculum to emphasize subjects they considered valuable (Beadie, 2010), and although Franklin had an enormous scholarly interest in language issues (Looby, 1984), he also greatly valued mathematics and science (Cohen, 1982).

When, in 1753, the Academy opened, it comprised just two schools—a Latin school which offered a classics-based curriculum, and an English school which offered "practical" courses including history, geography, navigation and surveying, taught in the English language. The first Provost of the Academy/University, the Reverend William Smith, slowly but surely directed its curriculum away from Franklin's English school toward the Latin school (Blinderman, 1976). Franklin's practical curriculum was designed so that students would benefit from arithmetic, accounting, geometry, astronomy, English grammar, writing, public speaking, and histories of mechanics, natural philosophy, and agriculture. Latin and Greek were included but only for those who desired them. According to Blinderman (1976), President Smith favored his classical masters—for example, the Latin master was paid 200 pounds annually to teach 20 students whereas the English master was paid only 100 pounds annually to teach 40 students; the Latin master was given 100 pounds to spend on books and maps, but the English master was given nothing to spend on books and maps. After 40 years, the English school was closed.

In 1791 the Academy became the University of Pennsylvania (Dauben & Parshall, 2014). The original College educated many persons who would become leaders within the United States—indeed, 21 members of the Continental Congress were graduates of the College, and 9 persons who signed the Declaration of Independence were either alumni or trustees.

Franklin (1793) believed that whereas those wishing to be clerics should be taught Latin and Greek, those aiming to be physicists should study Latin, Greek, and French, and those destined to be merchants should study French, German, and Spanish. Although no-one should be compelled to learn any language other than English, anyone wishing to study a particular language should be able to do so. This was an early foreshadowing of the elective principle in college administration.

Theophilus Grew, a friend of Franklin's, was appointed Professor of Mathematics, and his priorities can be judged by the title of his 1753 textbook—*The Description and Use of the Globes, Celestial and Terrestrial ... To Which is Added Rules for Working all the Cases in Plain and Spherical Triangles*. According to Karpinski (1980), this was "the first trigonometric treatise of the Americas" (p. 63). Before taking up his appointment, Grew had taught mathematics in

Philadelphia and Maryland, and had become well known for teaching not only basic arithmetic but also applied forms of mathematics such as surveying, navigation, astronomy, accounting, and the use of globes. He had also published a description of an approaching eclipse of the sun and served as one of the commissioners who established the boundary between Pennsylvania and Maryland.

Benjamin Rush. Benjamin Rush (1746–1813) was a University of Pennsylvania physician and educator, someone who signed the Declaration on Independence, and a friend of Benjamin Franklin, Thomas Jefferson, and John Adams. He believed that insistence on teaching the ancient languages held back the development of the country, and wrote: "To spend four or five years in learning two dead languages is to turn our backs on a gold mine in order to amuse ourselves catching butterflies" (quoted in Pioariu, 2011, pp. 169–170). He asked the poignant question:

> Who are guilty of the greatest absurdity—the Chinese who press the feet into deformity by small shoes, or the Europeans and Americans who press the brain into obliquity by Greek and Latin? Do not men use Latin and Greek as the cuttlefish emit their ink, on purpose to conceal themselves from an intercourse with the common people?
>
> (Benjamin Rush, 1789, quoted in Richard, 1994, p. 200)

He described the study of Greek and Latin languages by the English nation as "one of the greatest obstructions that has been thrown in the way of the propagation of useful knowledge" (Rush, 1806, pp. 30–31). Clearly, British scholars like Isaac Barrow and Isaac Newton would not have agreed with him for they wrote their mathematics-related textbooks in Latin, arguing that that was the language of science. But Rush was interested in propagating the concept of "mathematics for all."

Maintaining the Status Quo with College Mathematics, 1776–1810

If ever there was a time when the dominance of the classical tradition in North American colleges might have been expected to be seriously challenged, the Federalist period (1776–1810) was it. Benjamin Franklin and Benjamin Rush had prepared the nation's thinking for such a challenge, and George Washington and Thomas Jefferson seemed to be prepared to listen to reason. Noah Webster took up the challenge of establishing norms for American spelling, writing, and speaking, and what was needed was someone who not only recognized the stifling power of the classical tradition on American education, but was also prepared to do something in order to change the situation. Instead, the leadership for change in college mathematics was taken on by a schoolteacher, Nicolas Pike. But he would move away from education quickly, to become a judge, and any momentum for change which had been established quickly dissipated. Pike's mind-set was captured by his use of the word "arithmetic" in the title of his book—yet he included sections on algebra, trigonometry, mensuration, geometry, and gauging (Albree, 2002).

Having just emerged victorious from the Revolutionary War with its former colonial master, England, the fledgling nation now looked forward to facing, and conquering, many educational challenges (Ogg, 1927). Sufficient, perhaps, to recall the rapid heightening of national consciousness, and the feeling that from that moment onward everything in the curricula of the nation's schools should reflect the nation's achievement of independence. George Washington, in writing to Nicolas Pike (George Washington to Nicolas Pike, June 20, 1788) could not have been clearer on the matter:

> I hope and trust that the Work [i.e., Pike's text on arithmetic] will prove not less profitable than reputable to yourself. It seems to have been conceded, on all hands, that such a System was much wanted. Its merits being established by the approbation of competent judges, I flatter myself that the idea of its being an American production, and the first of the kind which has appeared, will induce every patriotic and liberal character to give it all the countenance and patronage in his power. In all events, you may rest assured that, as no person takes more interest in the encouragement of American genius, so no one will have more highly the unfeigned pleasure to subscribe himself.

Although education historians have not provided a ready summary of the shifts of attitude and perspective which distinguish the period, it will be readily acknowledged that in those years the young nation's leaders, particularly Franklin, Jefferson, Rush, and Washington, were keenly aware of a need to reshape their education institutions, to change what went on in those institutions in the name of "education." In particular, they were prepared to fashion and introduce structural alterations to society which would have important implications for school curricula of a kind which would have been entertained by only a vanguard of reformers in the colonial era.

Thus, for example, in 1786 Congress officially introduced decimal currency, with the United States becoming the first nation in the world to decimalize its currency fully (Clements & Ellerton, 2015; Pike, 1788; Robinson, 1870; Schlesinger, 1983). One might have expected that those responsible in the states for developing school arithmetic curricula would have thought carefully about how best to assist the young nation to make decimal currency "normal" in the minds and day-by-day activities of the people. It might have been expected, too, that they would also have scrutinized the old system of measuring lengths, areas, volumes, capacities, and time (Clements & Ellerton, 2015; Cohen, 1982, 2003; Halwas, 1990). But Congress decided to decimalize currency only. From the outset of the post-colonial era it was recognized that what was needed was a system of education which was clearly superior to that which had been available in the old colonial days. There was now a reluctance to cling to colonial vestiges. The change to decimal currency indicated that. But Congress did not act on the other measures, even though, in the early 1790s, France was introducing its metric system.

Clearly, in the 1780s one of the big challenges was for North American teachers, scholars and writers to produce textbooks which could replace those which had previously been used in American schools and academies. Those earlier books had been written by English authors who had never been to America. Following the Revolutionary War, school texts written by American authors began to appear. Perhaps the most prolific of the publishers of those texts was Isaiah Thomas (Tebbel, 1972), who would publish Noah Webster's (1787) famous *The American Speller* and later editions of Pike's *Arithmetic*.

In the period before the Revolutionary War, authors of almost all of the arithmetics used in American schools lived in England, and employed that country's sterling currency (involving "pounds, shillings, pence and farthings") in questions involving money, as well as the so-called imperial measurement system for lengths, areas, volumes, capacities, weights, time, etc. In fact, that would continue well into the nineteenth century because the imperial measurement system had been translated into the American colonies and had been adopted fully. One obvious area in which an important change might have been achieved quickly was the arithmetic textbooks used in private and grammar schools, and in colleges. Such a change was needed to support Congress's momentous 1786 decision to introduce decimal currency.

But, in the 1780s, there existed great diversity and confusion with respect to the currency of the American colonies. "At the time of the adoption of decimal currency by Congress, in 1786, the colonial currency or bills of credit issued by the colonies, had depreciated in value, and this depreciation, being unequal in the different colonies, gave rise to the different values of the State, currencies" (Robinson, 1870, p. 190). These local currencies continued to be used for almost 100 years after Congress formally adopted decimal currency in 1786. Even in the 1870s there was Georgia currency, Canada currency, New England currency, Pennsylvania currency, and New York currency (Robinson, 1870).

In 1788, at the time of the publication of Nicolas Pike's *A New and Complete System of Arithmetic Composed for Use of Citizens of the United States,* Isaac Greenwood's (1729) book seemed to have been completely forgotten, and a textbook by McDonald (1785) was barely known. It is not surprising, therefore, that some historians have regarded Pike's (1788) *Arithmetic* as the first arithmetic textbook written in English by an American author ever to be used in America's schools. Certainly, Pike's text had been a long time in the making. In the early 1780s, Pike, a schoolteacher in Newbury-Port—a seaport some 30 miles northeast of Boston—composed his manuscript and in 1785 he began submitting it to certain "men of prominence" for endorsements. Supported by his publisher, John Mycall, he traveled as far as Baltimore to hand-deliver "letters negotiating these recommendations" (Albree, 2002).

Nicolas Pike's (1788) *Arithmetic*

Prior to the Revolutionary War, mathematics texts originally published in England (e.g., those authored by Bonnycastle, Cocker, Dilworth, Fisher, Hawney, Hill, Hodder, Moore, Walkingame, and Ward) were regularly used throughout the American colonies. Thus, for example, in his autobiography, Benjamin Franklin mentioned that he read Cocker's (1720) *Arithmetick* after he moved from his home to Pennsylvania. Franklin (1793) wrote:

> And now it was that, being on some occasion made asham'd of my ignorance in figures, which I had twice failed in learning when at school, I took Cocker's book of Arithmetick, and went through the whole by myself with great ease. I also read Seller's and Shermy's books on Navigation and became acquainted with the little geometry they contain; but I never proceeded far in that science. (p. 21)

Immediately after the Revolutionary War there was a surge of activity in American publishing for schools and colleges, and Pike's was the first major school arithmetic to appear. Nicolas Pike (1743–1819) was a native of New Hampshire, who had graduated from Harvard College in 1766 and after that had been a schoolteacher for about 20 years.

"Old Pike," as Pike's *Arithmetic* would come to be known, would go through six editions between 1788 and 1843 (Karpinski, 1980). Around 1790 it sold for about $2.50—at that time a substantial price for a textbook, and one which placed it out of the reach of most college students (Monroe, 1917). Originally, Pike hoped that it would be suitable for schools and colleges, but it soon became obvious that it was too difficult for most school students. It was in general use in the United States until around 1840, but mainly in the colleges. The original 1788 publication was a large volume of 512 pages, almost encyclopedic in its arithmetical range. Besides arithmetic proper, it introduced the student to algebra, geometry, trigonometry, geometry, gauging, mensuration, and conic sections. Applications were made to problems in mechanics, gravity, pendulum, mechanical powers, and to problems in astronomy requiring calculations of the moon's age, the times of its phases, the time of high water, and the date of Easter (Albree, 2002). In 1793, an abridged version directed at schools was published. This abridgement omitted the subjects of logarithms, trigonometry, algebra and conic sections.

The first edition of Pike's Arithmetic included a copy of the Act of Congress of 1786 which created the U.S. Federal Money System with denominations of mills (1/1000th of a dollar), cents, dimes, dollars, and eagles (ten dollars). None of the problems in the book referred to the new American money, but instead were based on English sterling currency. Units to be memorized included measures for cloth, wine, and beer (the last two had different units). Beer measures consisted of pint, quart, gallon, firkin, kilderkin, barrel, hogshead, puncheon, and butt. Avoirdupois, troy and apothecary weight units were mentioned, and their usage distinguished.

The popularity of Pike's book in colleges suggests that the colleges were still requiring their students to focus on arithmetic rather than on other forms of mathematics. Pike's approach was traditional, aimed at assisting college students to copy material into cyphering books. Florian Cajori (1907), a respected historian of mathematics and mathematics education, defended Pike's approach. Referring to criticisms of Pike's textbooks by George Martin (1897), Cajori wrote:

> A recent writer makes Pike responsible for all the abuses in arithmetical teaching that prevailed in early American schools. [Martin's (1897) book was indicated in a footnote.] To us, this condemnation of Pike seems wholly unjust. It is unmerited, even if we admit that Pike was in no sense a reformer among arithmetical authors. Most of the evils in question have a far remoter origin than the time of Pike. Our author is fully up to the standard of English authors to that date. He can no more be blamed by us for giving the aliquot parts of pounds and shillings, for stating rules for "tare and trett," for discussing the "reduction of coins," than the future historian can blame works of the present time for treating of such atrocious relations as that 3 ft. = 1 yd., 5 yds. = 1 rd., 30 sq. yds. = 1 sq. rd., etc. So long as this free and independent people chooses to be tied down to such relics of barbarism, the arithmetician cannot do otherwise than supply the means of acquiring the precious knowledge. (p. 218)

Cajori's (1907) defense of Pike, then, went something like this: Nicolas Pike, and no doubt his publisher, John Mycall, recognized that in 1788 the most likely users of his text would be students at higher educational institutions like Harvard and Yale, and students in pre-college academies. Of course, the author and the publisher maintained that the book was "suitable for schools," and so it was (if, by schools, we include only those "academies which concentrated on preparing students for higher study"). But in 1793, a publisher's preface to the abridgement stated that the original [1788] *Arithmetic* was "now used as a classical book in all the New England universities," and excelled "everything of the kind on this content" (Pike, 1793, p. ii).

According to Cajori (1907), any faults in Pike's (1788) book had originated with deficiencies in English arithmetics. Cajori maintained that it was not Pike's responsibility to attempt to do for mathematics what Webster had done for American English. The issue was whether Pike should have accepted the existing education settings of his day, and proceeded cautiously, taking into account contextual constraints; or should he, as a person acting at a pivotal period of history, have seized the moment and attempted to achieve fundamental, even radical, change, not only in the content but also in the methods of teaching and learning mathematics. One could argue that it was his responsibility to be brave, to set a new tone, to break away from colonialist fetters which had strangled teaching and learning in the schools before the Revolutionary War.

Cajori (1907) believed that it was not an arithmetic author's task seek to change the way people used currencies and units of measurement within society. Rather, the task was to make sure that students learned to cope, arithmetically, with the ways currencies were being used on a daily basis. With respect to pedagogy, according to

Cajori (1907), Pike's (1788) emphasis on rules was in line with the "best thinking of the day" on teaching and learning.

We agree with Martin's views rather than Cajori's. Pike knew that his text was important, and so did all the notable personalities who allowed their names to be used to provide supporting statements in the recommendations section at the front of Pike's (1788) book. Pike wanted his book to be the first English-language arithmetic text written by a U.S. citizen and widely used in the schools and colleges of the new nation. Cajori seemed to argue that it was unfair to have expected Pike to move towards methods other than those Pike himself would have employed as a teacher.

Was it unreasonable to have expected Pike to see beyond the horizons surrounding his world and context at the time? That question raises intriguing issues of historiography. What principles can historians look to if they want to generate faithful yet historical accounts of events and offer penetrating, insightful interpretations of those events? Under what circumstances is it fair to criticize a writer for "silence" about ideas and practices of which he was either totally unaware, or only dimly aware? Those kinds of questions are fiercely contested within the world of academic history today (see e.g., Macintyre & Clark, 2004; Windschuttle, 1996).

Puzzling Events Surrounding Mathematics at Yale College in the 1820s

Jeremiah Day's Textbooks and His Mathematics Program

In Chapter 5 of this book we argued that Jeremiah Day (see Figure 5.3), who was Professor of Mathematics and Natural Philosophy at Yale College between 1803 and 1817, and President of the College between 1817 and 1846, exerted a strong influence on U.S. college and pre-college mathematics. Here we elaborate upon his influence on college mathematics, at Yale and elsewhere, and also discuss two puzzling events which occurred during the period of his presidency.

Day's (1814) *An Introduction to Algebra Being the First Part of a Course of Mathematics* was the first textbook entirely dedicated to algebra written by a North American author. It was a substantial book (of approximately 300 pages) and took advantage of British approaches to the subject as shown in textbooks by scholars like Isaac Barrow, Isaac Newton, and William Whiston. He also took account of works by Euler and Bézout, but the lineage was definitely British. *An Introduction to Algebra* was extremely successful, being used in many colleges—Karpinski (1980) pointed to 67 editions of the book having been published by 1850, and it is likely that were many more after that. According to Karpinski (1980), "no other American mathematical work to 1850 has so long a series of consecutively numbered editions" (p. 202).

But Day's *Algebra* was not his only influential mathematics textbook. Among others were:

- (1815). *A Treatise of Plane Trigonometry . . . Being the Second part of a Course of Mathematics, Adopted to the Method of Instruction in the American Colleges*. New Haven, CT: There would be six editions by 1850.
- (1817). *The Mathematical Principles of Navigation and Surveying, with the Mensuration of Heights and Distances*. There would be six editions by 1850.
- (1836). *The Teacher's Assistant in the "Course of Mathematics."* A second edition would be published in 1845.

Day used these texts to define a special "Yale" course of mathematics which, he claimed, was "adopted to the method of instruction in American colleges" (Karpinski, 1980, p. 380).

We believe that during the period 1814–1865 Day's "Yale" course had a greater influence on the teaching and learning of mathematics in U.S. colleges than any other program. The mathematics in the books was straightforward, with rules and cases being provided, model examples shown, and exercises provided. The textbooks provided sets of notes which could be copied into cyphering books and provided the basis for what Day (1814) claimed was the "method of instruction in the American colleges" (p. iii).

The Extent of French Influence on Mathematics Curricula in North American Colleges

It is noteworthy that within his "course of mathematics," Day included books he had written on navigation, surveying, and mensuration as well as books dedicated to algebra, trigonometry, and geometry. It is arguable that he inherited his "applied" *and* "pure" mathematics emphasis from Oxford and Cambridge. By contrast, at Harvard, John Farrar, Hollis Professor of Mathematics and Natural and Experimental Philosophy, instituted a Harvard course of mathematics by having books written by Continental European mathematicians—a textbook on algebra by Euler, an elementary treatise on arithmetic by Lacroix, an *Elements of Geometry* by Legendre, an elementary treatise on plane and spherical trigonometry by Lacroix and Bézout, and an *Elements of Algebra* by Bourdon—translated into English. Much has been written about the use of these at Harvard.

Farrar (1818) earnestly believed that the more difficult parts of mathematics had been more fully and more clearly explained by the French authors, than by British authors, of mathematics textbooks. Like Thomas Jefferson, Farrar believed that Lacroix's books held a very distinguished place. He believed them to be the most complete of all available mathematics textbooks for upper-secondary school and college students. Jefferson, like Farrar, noted that Lacroix's texts had received the sanction of the French government and had been adopted in the principal schools of France (Thomas Jefferson, to John Farrar, November 10, 1818). From our vantage point, however, the educational worth of Lacroix's (1818) *Elementary Treatise of*

Arithmetic was nothing special. The book followed a traditional *abbaco* arithmetic sequence: numeration, the four operations on whole numbers, fractions, decimals, money and measures for weight, liquids, cloth, length, time, reduction, compound numbers, simple and compound proportion, fellowship, alligation, and the new French weights and measures.

Cajori (1890) described authors like Lacroix as "giants" (p. 98) and argued that in North America "improvements in mathematical text-books and reforms in mathematical instruction were due to French influences" (p. 99). Although Cajori hedged his bets on the influence of the French by adding that "many of the works which were adopted in North America were beginning to be 'behind the times' when introduced in America" (p. 99), he devoted 195 pages of his book *The Teaching and History of Mathematics in the United States* to a section titled "Influx of French Mathematics."

Despite Thomas Jefferson's support of Continental and, in particular, French approaches to mathematics and mathematics education, we do not accept the contention that, overall, the French exerted a positive influence on college mathematics in the United States between 1776 and 1865. The books written by Continental mathematicians and translated into English for Farrar's "Cambridge course" were not popular and decisions were made soon after Farrar's retirement from Harvard (in 1836) to discontinue publication of almost all of them. At West Point, Charles Davies reported that the algebra text by Bourdon was too complicated for use in pre-college schools and academies. He admitted that it would need to be modified so that it would be less discursive before it would be found useful for North American college students.

If, indeed, the introduction of the blackboard (sometimes referred to as "chalkboard" (Krause, 2000; Wylie, 2012)) into American education can be attributed to French influences then something good did come from France, but even that claim may not be true. James Pillans, headmaster and geography teacher at the Old High School in Edinburgh, Scotland, is credited with inventing the first modern blackboard when, in 1801, he hung a large piece of slate on the classroom wall (Pillans, 1856). In America, the first recorded use of a wall-mounted blackboard occurred in 1801 at West Point, in the mathematics classroom of instructor George Baron. That was 15 years before Sylvanus Thayer and Claudius Crozet made use of connected slates—surely conceptually similar to blackboards—at West Point (Adams, Russell et al., 1965; Buzbee, 2014).

Releasing the Education Potential of Recitation Through Blackboards

Before 1865, lectures in mathematics were rarely the main basis for mathematical instruction in schools and in colleges in North America. Students copied notes into personal cyphering books, from "parent" cyphering books or from textbooks. During "recitation sessions" they gave answers to questions posed to them by professors and tutors on what they intended to enter, or had already entered, into their cyphering books. This composite private-study/recitation approach was appropriate at a time when there were relatively few well-qualified mathematics teachers in the colleges. It became easier to implement when small blackboards (often with dimensions about 3 feet by 3 feet) became available from about 1820 onward (Roberts, 2014).

In the 1820s the United States Military Academy (USMA) at West Point instituted a blackboard-assisted recitation system which would become widely used in American college mathematics (Ackerberg-Hastings, 2014; F. D. B, 1911; Phillips, 2015; Rickey & Shell-Gellasch, 2010). The blackboard-assisted approach was brought to West Point from France and Claudius Crozet adopted it totally in his Descriptive Geometry classes. Initially, Crozet's English was not strong, and students struggled to comprehend what he was saying. That issue was compounded by the great difficulty and novelty of his Descriptive Geometry course—which had been "created" in Paris by Gaspard Monge for highly-selective groups of students. Some idea of the complexity of the content for Monge's course can be gained from Figure 7.1, which is from a plate in Monge's (1811) *Géométrie Descriptive*.

In 1821, "Part 1" of Crozet's (1821) *A Treatise on Projective Geometry, for the Use of the Cadets of the United States Military Academy* was published in New York by A. T. Goodrich and Co, but Part II was never published, and no further editions of Part I appeared. Although Sylvanus Thayer, the Director of West Point, had arranged for tutors to be available to help with the time-consuming recitation sessions, Crozet's teaching was not well regarded by West Point students and he resigned his position there in 1823. Thereupon, Charles Davies, forever the entrepreneur, seized the opportunity to write his *Elements of Descriptive Geometry, with Their Application to Spherical Trigonometry, Spherical Projections, and Warped Surfaces* (Davies, 1826). That textbook was first published in Philadelphia in 1826, and would go through many editions, being published until the 1870s (Karpinski, 1980, p. 270).

The above Monge/Thayer/Crozet/Davies story is of a type that has often been repeated in histories of mathematics and mathematics education. A well-respected mathematician would introduce an idea which embodied new forms of mathematics; this which would be picked up enthusiastically by those responsible for defining mathematics curricula in colleges and schools; when implemented, however, the idea would be found to involve mathematics which was too difficult for most "ordinary" students; and it would be either abandoned, or the complexity of its application in education settings would be reduced substantially. In the Monge/Thayer/Crozet/Davies story, however, an extra dimension was added—Thayer and Crozet had observed the success of the "blackboard" in French education settings around 1815, and brought that idea to West Point where it was applied to a "recitation" tradition which was already widely used at Harvard, Yale and other colleges (Kidwell, Ackerberg-Hastings, & Roberts, 2008).

Crozet was not the first Continental-trained mathematician to experience difficulties teaching students at West Point. The same happened with Ferdinand Hassler, originally from Switzerland, who was acting professor at West Point between 1807 and 1810. Apparently, Hassler's teaching was disliked by USMA students who were not exceptionally capable in mathematics, and the same thing was true when he was Professor of Natural Philosophy and Mathematics at Union College in Schenectady, New York, between 1810 and 1812 (Cajori, 1980).

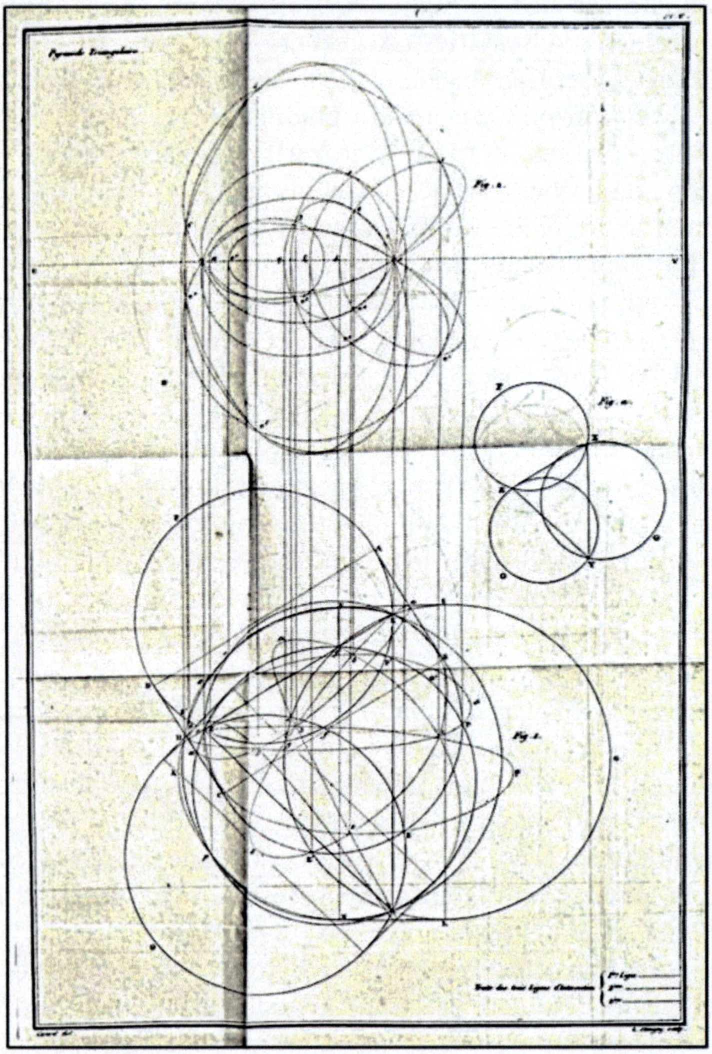

Figure 7.1 A diagram from Gaspard Monge's (1811) *Géométrie Descriptive*.

The blackboard extension to the normal recitation approach met with some success at West Point, possibly because Thayer was willing to provide additional resources, in the form of funds to employ tutors who helped with the resource-intensive, typically one-on-one, recitation sessions. One highly positive outcome was that West Point students were forced to come to grips with formal mathematical proofs. In their recitations they were expected to demonstrate that they fully understood the logic involved. This was a hugely important step for U.S. mathematics education, particularly because Charles Davies applied the same recitation approach to other subjects that he taught, making use of numerous textbooks that he authored.

It did not take long before other colleges felt the need to purchase, or construct, blackboards and to use them in recitations for mathematics, especially for geometry. During the first half of the nineteenth century, blackboards were increasingly used in U.S. college mathematics classes. Certainly, with geometry, "proof and proving" became much more a part of implemented mathematics curricula than ever before (Anderson, 1962; F. D. B, 1911; Molloy, 1975; Phillips, 2015). Although historians have recognized the importance of what happened, it should also be noted that teething troubles accompanied the revolution which took place. Serious protests by mathematics students at Yale College in 1825 and 1830, about the use of blackboards, resulted in many students being expelled from the College. We now draw attention to details associated with the Yale protests, as they were summarized by Clarence Deming (1904).

The Yale "Conic-sections" rebellions. Apparently, at Yale College around 1820 the sophomore year, and in particular the studies of mathematics during that year, were not liked by students because of the level of difficulty experienced in recitation sessions. Matters came to a head in 1825 when, it was claimed, a sophomore tutor agreed that if his class would accept the idea that during a recitation a student should not have access to what was written in the textbook or cyphering book, but would make explanations purely from figures provided in appendices on conic sections in the textbook, then corollaries to theorems would not be subject to examination in the final "major recitations" from which students' grades were allocated. However, according to students, later the tutor later reversed his decision (or was not allowed to carry it out) and insisted on including corollaries. This resulted in 38 of the 87 sophomore students refusing to participate in the final recitations, and the students were promptly suspended. After much heated contestation, the faculty won the day, and the suspended students returned after signing letters of submission. Each had to sign the following statement:

> We, the undersigned, having been led into a course of opposition to the government of Yale College, do acknowledge our fault in this resistance, and promise on being restored to our standing in the class, to yield a faithful obedience to the laws.
>
> (Quoted in Deming, 1904, p. 668)

But that was not the end of the story. Five years later, in 1830, an even more serious conic-sections dispute occurred (Green, 2015). Deming (1904) described this dispute "as the most grave in Yale history" (p. 668).

Once again, the dispute arose in connection with the system of recitation which was used with the conic-sections class. This time the dispute was over whether students, in recitation sessions, could be asked to demonstrate proofs "from the book" (without having access to the diagrams in the appendices). The students wanted to have access to the figures during recitations, so that they would not have to remember the figures. There were about eight pages altogether, and associated diagrams would also need to be remembered. The students asserted that "the subject of conic sections had been crowded into too brief a period" and that had meant that lessons were "unattainable during the time prescribed by the laws to any ordinary intellect" (quoted in Deming, 1904, p. 668). Forty-four of the ninety-five sophomores continued the rebellion, and were dismissed from the College, permanently, and, according to Deming (1904), "not one of them returned" (p. 669). Apparently Rutgers admitted several of the rebels, and so did a couple of minor colleges. Harvard refused to admit any. Almost 50 years later, in 1879, Yale agreed that the 18 rebels who were still alive should be granted the M.A. degree. Although there were no more conic-sections rebellions at Yale, Deming (1904) would write: "The terrors of sophomore mathematics continued to reach a climax in the hated biennial examinations of a later decade" (p. 669).

Largely hidden from the above account is how the existence of blackboards at Yale helped to fuel the flames of dispute. In recitations, which were conducted between tutor and student on a one-to-one basis, the sophomore was required to prove theorems relating to conic sections, and each had to reproduce, from memory, the diagram(s) relating to nominated theorems. Before the advent of the blackboard the student could hold a textbook open at pages in the appendices showing relevant diagram(s) and explain the proof(s) to the tutor "from the book."

Muttappallymyalil, Mendis, John, Shanthakumari, Sreedharan and Shaikh (2016) have claimed that by the mid-19th century in the United States "every class room had a blackboard to teach students" (p. 588). Although that may not be true, we would argue that the decision by Yale College in the first half of the nineteenth century to expect students to be able to reproduce and explicate proofs represented an extremely positive advance in the implemented curricular forms of college mathematics. Many college mathematics students and their teachers were forced not only to think about meanings and relationships, but also to illustrate them, and be able to articulate them. From our perspective, that was a more important advance than other often-mentioned content changes (e.g., teaching geometry according to Legendre rather than Euclid, or teaching algebra according to Bézout or Bourdon, rather than Bonnycastle). The method of recitation had been inherited from the cyphering tradition, and now its education potential was unlocked following West Point's and Yale's emphases on proof and proving. That development was facilitated by the introduction of blackboards (Association of Members of the Boston Public Schools, 1844).

What is puzzling for us, though, is that the Yale conic-sections disputes (Jackson, 2002) occurred during the presidency of Jeremiah Day, who, it is claimed, was highly respected by students. If, indeed, he continued to be respected by students during and after what were obviously extremely bitter disputes then that achievement was, in itself, amazing.

The *1828 Yale Report*, and Yale College's Confirmation of the Pre-eminence of the Classics

There was another major event at Yale during the 1820s which affected the position of mathematics within college curricula—and that event was the preparation of what is known as the *1828 Yale Report*. Any peculiarities within the domestic situations of those in the American New World did not seem to change the attitudes of those among the settlers who wanted their children to receive a college education. Those who could afford a classics education for their children—at a grammar school and then at a college—sought to establish conditions by which that would be possible. Paradoxically, although, by its content and by its methods of implementation, the classical curriculum was not directly "Christian," many leaders in colonial societies saw it as offering the best medium for training pastors as well as lawyers, and physicians. That resulted in the colonial leaders creating colleges in which the classics stranglehold was maintained. Although that was true, it was often the case, especially in the South, that some "leaders," in well-to-do families, sent their boys to prestigious British boarding schools and universities.

During part of the seventeenth century Harvard College students had been required to converse in Latin all day long, in classrooms, in corridors, and in college grounds. Another indicator of the valuing of the classics within the upper-echelons of North-American society was the fact that early in the eighteenth century, at both Harvard College and Yale College, graduating students were expected to prepare "mathematical theses" in Latin (Simons, 1924). The topics themselves were elementary (e.g., a 1721 thesis at Harvard was on the following: "Arithmetic proceeds from given to required quantities; algebra, however, from quantities sought to those given"; and a 1718 thesis at Yale dealt with "All rectilinear triangles contain two right angles"). The most difficult part of preparing the theses was, almost certainly, making sure that the Latin in which they had to be written and presented was of an acceptable standard.

In North America the educational preparation for schoolboys destined to study at higher education institutions was reasonably well defined by the middle of the eighteenth century. After the hornbook, a boy would first study Latin and English grammar and perhaps arithmetic, either in school or through private tuition, and in college there would be Greek, Latin, Hebrew, rhetoric, logic, astronomy, arithmetic, geometry according to Euclid, music according to Pythagoras, and finally moral philosophy, natural philosophy, and theology (Crilly, 2008; Plimpton, 1916).

The initial course of study at Harvard in the 1640s was for three years, and included philosophy, logic and physics for the first year, ethics and politics for the second year, with arithmetic, geometry, and philosophical disputations for each class. Wednesday was "Greek day" for all classes, and Thursdays were devoted to the theory of Hebrew, Chaldee and Syriac grammar in the morning, and practice in associated biblical texts in the afternoon. Friday was given over to rhetoric, in which students prepared English compositions and participated in declamation. On Saturdays, students were taught "divinity" and engaged in discussion on doctrine, and also history. The most remarkable aspect of the curriculum was the fact that on all days, Latin was not only the language of instruction but also the language which had to be used, even in the college yard. A reasonable competence in Latin was the main requirement for admission to Harvard College (Rudolph, 1990). The expectation was that "when a scholar was able to understand Tully (Cicero), or some other classical Latin author, and make and speak true Latin in verse and prose *suo ut aunt Marte*, and decline perfectly the paradigms of nouns and verbs in the Greek tongue, then and not before then, should he be considered for admission into the college" (Hurd, 1890, p. 165). Some local clergymen prepared prospective college students by arranging regular instruction and conversations with them in Latin (Leacock, 1970; Morison, 1956; Quincy, 1860).

It is well documented that most college administrators in colonial North America (and later in the United States of America) during the period 1607–1865 had an unwavering belief that the study of Latin and Greek texts generated intellectual power and provided the best preparation for those wishing to be clergymen, politicians, physicians, and lawyers (Campbell, 1968; Ellerton & Clements, 2012, 2017). One result of this valuing of the classics above all else in higher education was that one-half to two-thirds of curriculum time in the early North American colleges was dedicated to the classics.

As was the case at Christ's Hospital, in London, courses of study in North American colleges allowed much time for the study of the classics, but not much time for mathematics. Quite simply, the period 1607–1865 was a time when mathematics was seen, by most interested observers, as merely an adjunct to classical studies within any well-regarded system of higher education. Although a brief consideration of Euclidean geometry was included in the courses of study of most colleges, and algebra was studied by Harvard and by William and Mary students from the 1720s (Crilly, 2008; Simons, 1924), they were "extras," designed to fill the space left after the "most important" work, the classics, and related literary studies, had been given most of the available curriculum time. "Mathematics" itself was assumed to include physics, astronomy, navigation and surveying (as well as other things, depending on the interest and expertise of those who were available to teach). According to Zitarelli (2019), in North America it was not until 1784 that a person at the level of professor taught pure mathematics only (see, p. 29). The belief that classics should be the pre-eminent force in the curriculum was only occasionally challenged during the period (Meyer, 1968).

A major challenge to the assumed pre-eminence of the classics in pre-college and college education came between 1827 and 1828 and culminated in what is known as the *Yale Report* (Herbst, 2004). Responding to claims that because the

Latin and Greek languages were "dead languages" it made little sense to have classical literatures as the focus of college learning (Cremin, 1977), the *Report* maintained that in order to generate well-rounded graduates, a college curriculum had to be based around Latin and Ancient Greek literatures. The situation was all the more interesting because the President of Yale at the time was Jeremiah Day, a much-respected mathematician and author of numerous mathematical textbooks.

According to the *Yale Report*, "the two great points to be gained in intellectual culture, are the *discipline* and the *furniture* of the mind—that is to say, expanding its powers, and storing it with knowledge." The *Report* was based on what came to be known as the theory of "faculty psychology"—which asserted that the mind was made up of different faculties, and each faculty needed to be exercised before a proper and balanced development could occur. Those branches of study which would achieve the best development had to be carefully woven into both intended and implemented curricula, so that due attention would be given to directing the students' trains of thought in order that they would correctly analyze what was needed, avoid unnecessary extras, balance evidence to reach well-formed judgments, stimulate the imagination within proper bounds, and direct memory toward what was important. This would take time, and was not likely to occur outside a well-organized, suitably-rich course of study. The *Report* believed that the idea of preparing undergraduates for specific professional work was of secondary importance and argued for the laying of a classical foundation which was common to all.

Although the ideas were convincing for those who supported the emphasis on traditional classical education, no real evidence was ever presented showing that the claimed benefits of a traditional classics education were more real than imaginary (Richard, 1994; Rush, 1806; Rudolph, 1990). Thus, *Part II* of the *Yale Report* argued that, knowledge of Ancient Greek and Latin literature was the foundation for a "rounded liberal education," and those who prepared the *Report* regarded that as axiomatically correct. The authors of the *Report* believed that the study of the classics was useful because it laid "the foundations of a correct taste, and furnished the student with those elementary ideas which are found in the literature of modern times," and formed "the most effectual discipline of the mental faculties" (Yale College, 1828, p. 36).

The authors of the *1828 Yale Report* conceded that not all students were satisfied with Yale's classical curriculum, and suggested the possibility of a plan "to confer degrees on those only who have finished the present established course"—but to allow "other students, who do not aim at the honors of the college, to attend on the instruction of the classes as far as they shall choose" (p. 42). In the end, however, the *Report* recommended that more stringent College-entrance requirements be gradually introduced, with higher levels of Greek and Latin being required (Broome, 1903; Yale College, 1828). One result of the *Yale Report* was that the validity of the classics' stranglehold over U.S. college curricula was confirmed. That stranglehold would remain for the next 70 years, at least.

The *1828 Report* attested to a recognition, among some influential people, that the age of enlightenment was under attack. Useful subjects, including pure and applied forms of mathematics, were being called for, and the *Report* represented an uncompromising, powerful reaction from those who wanted to maintain the status quo. We are amazed that at Yale College, President Jeremiah Day was able to retain the respect of he various competing groups in his community at that time. For not only had he taught pure and applied forms of mathematics, but he was the author of the nation's most influential set of college mathematics textbooks. On the other hand, he was also an ordained minister of religion who, as an undergraduate, had experienced a full Yale education.

Benjamin Peirce and the "Functions" Breakthrough in U.S. College Mathematics

Benjamin Peirce (1809–1880) (see Figure 7.2) was a mathematical prodigy who graduated from Harvard College in 1829. After teaching in a school for two years he was appointed a mathematics tutor at Harvard in 1831 and was then appointed University Professor of Mathematics and Natural Philosophy in 1833. In 1842 his title was changed to Perkins Professor of Astronomy and Mathematics (Hill, 1880). He remained at Harvard as a professor until his death in 1880 (Archibald, 1925; Matz, 1895; Peterson, 1955).

Figure 7.2 Portrait of Benjamin Peirce (in 1857).
Retrieved from Peirce (2019).

Between 1831 and 1846 Peirce wrote a series of text-books on geometry, trigonometry, and "curves, functions, and forces" (Karpinski, 1980, p. 646; Peirce, 1846). They were aimed at undergraduate college students, but the approach Peirce took with respect to "standard" topics was highly algebraic and strikingly different from that of other authors of college mathematics textbooks. The two books which proved to be the most controversial were:

(1841). *Curves, Functions, and Forces, Book I: Application of Algebra to Geometry*;
(1846). *Curves, Functions, and Forces, Volume Second: Calculus of Imaginary Quantities, Residual Calculus, and Integral Calculus.*

These books were published in Boston by James Munroe and Company, and their content was so different from that of other undergraduate mathematics textbooks of the time that it is hardly surprising that they were not best sellers. Each had approximately 300 pages, and their high-quality binding, hard covers, and appearance were striking. The copies of the books in the Ellerton-Clements collection have foldout illustrations inserted toward the end of each volume. Thomas Hill (1880), a former student of Peirce's (and someone who would later become President of Harvard), made some quite sweeping statements in his evaluation of the books:

> They were so full of novelties that they never became widely popular ... but they have had a permanent influence upon mathematical teaching in this country; most of their novelties have now become commonplace in all textbooks. The introduction of infinitesimals or of limits into elementary books; the recognition of direction as a fundamental idea; the use of Hassler's definition of a sine as an arithmetical quotient free from any entangling alliance with the size of the triangle; the similar deliverance of the expression of derivative functions and differential co-efficients from the superfluous introduction of infinitesimals; the fearless and avowed introduction of new axioms, when confinement to Euclid's made a demonstration long and tedious—in one or two of these points European writers moved simultaneously with Peirce, but in all he was an independent inventor, and nearly all are now generally adopted. (p. 91)

It is worth drawing special attention to Hill's (1880) claim that Peirce's textbooks were ahead of European textbook writers on many things—and that Peirce was "an independent inventor"!

Peirce required his undergraduate students to become thoroughly familiar with *Curves, Functions, and Forces*, and that caused a storm of protest. Careful inspection of the texts revealed that the standard of mathematics was high, with uncompromisingly difficult notations being introduced and maintained throughout (Cajori, 1928). Cajori reported that freshmen who were required to study Peirce's texts, were particularly unhappy, and complained of overwork. In 1839, a committee reported that the mathematical studies of the Freshman class were so extensive that they encroached upon the time and attention that could be given to other subjects.

Repeated and loud complaints were made at Harvard that the mathematical teaching was poor. The majority of students said they disliked mathematics, and they dropped it as soon as possible. In 1848, only five Harvard students passed *Curves and Functions*, and a committee was set up to find out why mathematics at Harvard had become "so very decidedly unpopular." A majority of the committee reported that Peirce's textbooks were simply too abstract and too difficult, but a minority of mathematicians closed ranks and defended Peirce, describing his texts as seeking to reform mathematics in the United States. In 1848, another committee was impaneled and charged with investigating the mathematics program at the University. A majority on the committee reported that "the text-books were abstract and difficult," and that "there are other mathematical works of no small merit, which embraced the same subjects as the text-books now used, which were much less difficult of comprehension."

However, a two-person minority on the committee came to a different conclusion. These two (Thomas Hill and J. Gill) claimed:

> These text-books, by their beauty and compactness of symbols, by their terseness and simplicity of style, by their vigor and originality of thought, and by their happy selection of lines of investigation, offer to the student a beautiful model of mathematical reasoning, and lead him by the most direct route to the higher regions of the calculus. For those students who intend to go farther than everyday applications . . . this series of books is, in the judgment of the minority, by far the best series now in use.
>
> (Quoted in Cajori, 1890, p. 141)

Beauty is in the eye of the beholder. What was beautiful to Hill and Gill was obviously not beautiful, or even appropriate, to the other committee members or to most students.

Before making further comments on the negative views expressed by many with respect to Peirce's *Curves, Functions, and Forces,* it will be useful to comment on what was to be found on the same themes in other textbooks of the time. In the Ellerton-Clements textbook collection we hold many textbooks used for college mathematics classes in the eighteenth and in the first half of the nineteenth century (e.g., Ward, 1758), and *none* of them show the graph of a function on a Cartesian plane on *any* page. Indeed, rarely is the word "function" used. Expressions like "linear function" or "quadratic function" or "sine function" are not to be found. But in Volume 1 of *Curves, Functions, and Forces* Peirce defined a function and then immediately introduced concepts such as algebraic functions (including linear, integral, and irrational functions), exponential and logarithmic functions, trigonometric or circular functions, compound functions, and continuous functions (see Peirce, 1841, pp. 163–171). Then followed definitions of infinitesimals, differentials, indeterminate forms, leading, quickly to Taylor's Theorem and MacLaurin's Theorem (p. 184). Along the way, theorems were proved, corollaries stated, and graphs drawn. The speed at which definitions, concepts, theorems, etc., were introduced was, from an educational perspective, breathtaking. Volume 2, which dealt with integration, and the "calculus of imaginary quantities," among other things, could be described as difficult and terse. Both books were absolutely different from anything that had ever previously appeared in mathematics textbooks for college students in the United States of America. A typical page from Volume 2 is shown in Figure 7.3.

In the 1840s and 1850s very few of the students in U.S. high schools or academies ever drew a Cartesian graph—we can be sure of that because no Cartesian graph, of any kind, can be found on any page in any of the 536 North-American cyphering books prepared between 1667 and 1865 in the Ellerton-Clements cyphering book collection. There can be no doubt, therefore, that initially, at least, college students would have had great difficulty comprehending Peirce's textbooks. Furthermore, Cartesian graphs of linear, quadratic, trigonometric, logarithmic and exponential functions were *not* to be found in textbooks used in U.S. secondary schools at the time. Chateauneuf's (1930) analysis showed that functions and graphs appeared in textbooks slowly, and only after about 1880.

Having written all of the above, we now present another side of the story. From our vantage point, the two volumes of Peirce's (1841, 1846) *Curves, Functions, and Forces* were the first books in a new era which ultimately transformed secondary-school and college algebra curricula in many parts of the world—certainly, that was the case in English-speaking nations. Unfortunately, Peirce was not a good teacher himself, and his books were written so tersely that they did not do justice to the powerful mathematics, and the transformative messages about mathematics, that he attempted to convey. Within 30 years his approach would transform college mathematics in the United States, but it was only after 1900 that the approach to mathematics that they embodied began to touch secondary-school mathematics in the United States of America (Chateauneuf, 1930). Ultimately, the approach would change both college and school mathematics for the good.

In fact, in 1837 Peirce had published an earlier book, the 276-page *An Elementary Treatise on Algebra; to Which are Added Elementary Equations and Logarithms,* in which he followed a similar line of argument to what he would use in in *Curves, Functions, and Forces*. Definitions of a function and the derivative of a function were given, and derivatives of sums, products and powers of functions considered. From a historical perspective, the following excerpt from Peirce's Preface to that book is significant:

> The excellent treaties on Algebra which have been prepared by Professor Smyth and Professor Davies, containing as they do the best improvements of Bourdon and other French writers, would seem to leave nothing to be desired in this department of mathematics. The form, however, adopted in the English works of instruction, of dividing the subject as much as possible into separate propositions, is probably the best adopted to the character of the English pupil. This form has, therefore been adopted in the present treatise, while the investigation of each proposition has been conducted according to the French system of analysis. (p. iii)

§ 44.] IRRATIONAL FUNCTIONS. 95

Integration of binom al irrational functions.

3. Reduce the integral of $(1+x^5)^{-\frac{12}{5}}$ to depend upon one, in which the exponent of the binomial is positive and less than unity.

Ans. $\frac{1}{7}\, x\,(1+x^5)^{-\frac{7}{5}}\,(3+x^5-x^{10})+\frac{4}{7}\int.\,(1+x^5)^{\frac{3}{5}}.$

4. Reduce the integral of $x^{\frac{2}{3}}\,(a+b\,x^{\frac{4}{5}})^p$ to depend upon one of the same form, but in which the exponents are integral, except that of the binomial.

Solution. In (383) we have, for this case,

$$l=15, \quad x=y^{15},$$

so that (385) gives

$$\int.\, x^{\frac{2}{3}}\,(a+b\,x^{\frac{4}{5}})^p = 15\int.\, y^{24}\,(a+b\,y^{12})^p.$$

5. Reduce the integral of $x^{\frac{4}{5}}\,(a+b\,x^{\frac{2}{3}})^p$ to depend upon one of the same form, but in which the exponents are integral, except that of the biomial.

Ans. $15\int.\, y^{26}\,(a+b\,y^{10})^p.$

6. Reduce the integral of $x^{\frac{3}{4}}\,(a+b\,x^{-2})^{\frac{2}{3}}$ to depend upon one in which the exponent of x in the binomial is positive.

Ans. $\int.\,(b+a\,x^2)^{\frac{2}{3}}.$

44. *Problem. To find the value of the definite integral*

$$\int_0^c x^m\,(a+b\,x^n)^p \qquad (387)$$

in which

$$c=\sqrt[n]{-\frac{a}{b}} \qquad (388)$$

and m, n, and p are positive.

Figure 7.3 Page 95 from Peirce (1841).

Peirce seemed to be saying that English mathematicians had best identified the structures of the main arguments, and that the discursive style of French mathematicians provided the best way of explaining salient features of, and relationships between, the components of the structures.

Although James Sylvester (1870), the much-vaunted English mathematician who has been much praised for his success in launching a mathematics research community at Johns Hopkins University in the 1870s, might have recommended the French approach to geometry, while severely criticizing those who preferred Euclid to Legendre ("I should rejoice to see . . . Euclid honourably shelved or buried 'deeper than did ever plummet sound out of the schoolboys' reach'," p. 261), the fact was that some highly-regarded mathematicians disagreed with him. For example, writing in 1862, Isaac Todhunter (1955), a Mathematics Professor at the University of Cambridge, had this to say on the matter:

> It cannot be denied that defects and difficulties occur in the *Elements* of Euclid, and that these become more obvious as we examine the work more closely; but probably during such examination the conviction will grow that these defects and difficulties are due in great measure to the nature of the subject itself, and to the place which it occupies in a course of education; and it may be readily believed that an equally minute criticism of any other work on Geometry would reveal more and graver blemishes. (p. xii)

Time would point to Todhunter's stance on the matter as having been prophetic (Crilly, 2008). Despite almost constant criticisms, over the next 150 years, of *Euclid's Elements*, it would retain its status as the epitome of pure logic (Hirsch & Van Haften, 2019). Nigel Wilson (2006) would even write that it could be argued that *Euclid's Elements* has been "the most successful textbook ever written" (p. 278).

Maintaining a Wider View of Mathematics

Within the collections of cyphering books held at Harvard University, Yale University, the University of Pennsylvania, the Phillips Library at Salem (Massachusetts), and in the Ellerton-Clements collection, now held in the Library of Congress, there are some manuscripts which deal with navigation and surveying (Hertel, 2016). One of the manuscripts in the Ellerton-Clements collection was prepared in 1760 by Galparis Yeates, who was a student at the "College of Philadelphia." According to a note on the first page of this 61-page cyphering book, it contained material "collected from the most approved masters on each subject." We could not find on the Internet any records relating to Galparis's life, but there were plenty on Jasper Yeates Senior (whom we presume to have been Galparis's father) and on Jasper Yeates Junior (whom we presume to have been Galparis's brother). Jasper Junior also attended the College of Philadelphia around 1760. The father was a successful merchant and an associate justice of the Supreme Court of Pennsylvania.

The pages in Galparis Yeates' manuscript dealt with trigonometry, navigation ("plain sailing"), surveying, and mensuration ("heights and distances"). The penmanship and calligraphy were exquisite throughout, and 45 of the 61 pages included hand-drawn diagrams, some of them quite intricate and beautifully colored. It is difficult to imagine how Galparis could have prepared his manuscript with a quill, but almost certainly he did. To examine Galparis's manuscript is to learn that this was something which was very important to him. It was *his* book. The mathematics was quite sophisticated in places, and all calculations were carefully presented.

In the trigonometry section of Galparis's manuscript the directed-line-segment definition was assumed, with the circle having a radius length measure of 10^{10}. The direct rule of three was employed throughout, and proportional aspects of similar figures were assumed and applied. Logarithms were used to assist calculations. It seemed that, for Galparis, mathematics had become much more than memorizing rules and making calculations. It was solving problems which related to real-life situations. For example, one of the problems was introduced in this way:

The following table for square measure will be of general standing. Use [it] for finding the content of fields in any assigned way of mensuration:

Links	Yards				
20 2/3	1	Perches			
625	30¼	1	Chains		
10000	484	16	1	Roods	
25000	1210	40	2½	1	Acres
100000	4840	160	10	4	1

Then followed exercises which revealed that the table was to be related to an island close to Philadelphia. Galparis seemed to be learning technical aspects of real-life surveying. The Houghton Library at Harvard University holds a 30-page manuscript prepared in 1770 by Samuel Hanson, also at the College of Philadelphia, which was devoted to notes on conic sections. The penmanship and calligraphy were not as impressive as Galparis's, but Samuel would later serve as surgeon in George Washington's life guards.

The Ellerton-Clements cyphering book collection includes some manuscripts devoted to surveying, navigation, and mensuration. We do not know how many of those were prepared in colleges, and how many were prepared in pre-college institutions or in evening classes. But in the minds of most of the students who prepared them, mathematics was not confined to making calculations and solving contrived problems. In the eighteenth and nineteenth centuries there were college students across North America who learned that mathematics was something which could be applied, something more than merely doing algebraic manipulations, or memorizing Euclidean proofs from "Playfair," or from "Davies" or from "Hutton," or from "Bonnycastle."

When, in 1711, the Reverend Tanaquil Lefevre was appointed to the College of William and Mary as a professor, his official title was "Professor of Mathematics and Philosophy" (Phalen, 1946). At Harvard College, Isaac Greenwood took up his appointment as Hollis Professor of "Mathematics and Natural and Experimental Philosophy" at the beginning of 1728, and his successor, John Winthrop IV, had the same title. At Yale, Thomas Clap, a Harvard graduate, brought a Newtonian emphasis to Yale's mathematics program between 1740 and 1766 when he was Yale's President (Dauben & Parshall, 2014; Zitarelli, 2019), and this tradition was continued for most of the period 1770–1817 when Nehemiah Strong, Josiah Meigs, and Jeremiah Day each held the position of "Professor of Mathematics and Natural Philosophy" (Dauben & Parshall, 2014). In 1758, William Small, a Scot who would have a strong influence on the mathematical development of Thomas Jefferson, was appointed as "Professor of Natural Philosophy and Mathematics" at William and Mary; Robert Patterson was "Professor of Natural Philosophy and Mathematics" at the University of Pennsylvania between 1810 and 1813, and that was also the title of Charles Bonnycastle's position at the University of Virginia between 1825 and 1840 (Dauben & Parshall, 2014; Smith & Ginsburg, 1934).

The point being made in the last paragraph is that in colonial and the Federalist era, mathematics and experimental philosophy (including physics, chemistry, astronomy, navigation, surveying) were intimately linked in the minds of most college administrators and those holding senior college appointments in mathematics. Indeed, that way of thinking about the nature of mathematics was standard within the United States of America throughout much of the period 1607–1865. William Churchill Houston and Walter Minno, for example, were appointed "Professor of Mathematics and Natural Philosophy" at Princeton in 1771 and 1788, respectively (Smith & Ginsburg, 1934), and between 1836 and 1844 Elias Loomis was "Professor of Mathematics and Experimental Philosophy" at Western Reserve College, in Ohio, and then between 1844 and 1860 "Professor, Natural Philosophy and Mathematics" at the University of the City of New York. According to Frederick Rudolph (1990), "by 1776 six of eight colonial colleges supported professorships of mathematics and natural philosophy" (p. 29).

Nearly all of the persons who were professors of mathematics regarded it as part of their work to be academically and experimentally concerned with scientific matters—such as supervising student participation in college observatories, or leading teams to observe astronomical phenomena. This link had been translated from Great Britain where the experimental works and associated research of Isaac Newton, Robert Hooke, Christopher Wren, Edmond Halley, James Hodgson, William Whiston, Charles Babbage, and George Boole, received serious academic attention. And those were not the only British scholars who straddled the fields of pure mathematics and architecture, mechanics, astronomy, optics, electricity, navigation, and logic (Ellerton & Clements, 2017). Of course, that broader view of mathematics was not confined to Great Britain—after all, the Frenchman Pierre-Simon Laplace (1749–1827), "the French Newton," was recognized as the world's greatest astronomer of his time—but it was especially to be found in Great Britain, and that continued to be the case into the twentieth century.

Perhaps the greatest embodiments of this wider pure- *and* applied-view of mathematics in North America during the period 1607–1865 were to be seen in the lives of Nathaniel Bowditch and his protégé, Benjamin Peirce. More will be said about Bowditch in the next chapter. Regarding Peirce, the following list reveals some of the areas of applied mathematics with which he became vitally concerned:

- He proof-read—when he was only 19 and 20 years of age—much of Bowditch's detailed review of Laplace's *Traité de Mécanique Céleste*;
- In 1842 he wrote articles on the motion of a top, a theory of storms, and adaption of the epicycles of Hipparchus to meteorological theory and practice;
- In 1843 he gave lectures which resulted in a decision to establish the Observatory at Cambridge, Massachusetts;

- He wrote scientific papers on the discovery of the planet Neptune, on Uranus, and on the rings of Saturn;
- He involved himself heavily in the work of the U.S. Coast Guard, and in fact was Superintendent of the Coast Guard between 1867 and 1874;
- He became a universally-recognized authority on analytical mechanics.

Many other areas of applied mathematics occupied Peirce's attention, and he liked to engage his students—undergraduate and graduate—as assistants in data-gathering and data-analysis aspects of his applied projects. And yet, while he was doing all this he was also heavily involved in his pure mathematical research (e.g., on quaternions) (Hill, 1880).

Although the two display quotations given later in this present paragraph were part of a speech made well outside the period 1607–1865, they point to a broad conception of mathematics, one which had been gradually transferred into colonial North America from Europe, and *especially* from Great Britain. In August 1912, the University of Cambridge hosted the Fifth International Congress of Mathematicians and in his opening address Sir George Darwin—President of the Cambridge Philosophical Society, past-President of the Royal Astronomical Society, and a son of Charles Darwin—put forward the view that Great Britain led the world of *applied* mathematics. He told an audience of more than 800 international mathematicians:

> The science of mathematics is now so wide and is already so much specialized that it may be doubted whether there exists to-day any man fully competent to understand mathematical research in all its many diverse branches. I, at least, feel how profoundly ill-equipped I am to represent our Society as regards all that vast field of knowledge which we classify as pure mathematics. I must tell you frankly that when I gaze on some of the papers written by men in this room I feel myself much in the same position as if they were written in Sanskrit. But if there is any place in the world in which so one-sided a President of the body which has the honour to bid you welcome is not wholly out of place it is perhaps Cambridge. It is true that there have been in the past at Cambridge great pure mathematicians such as Cayley and Sylvester, but we surely may claim without undue boasting that our University has played a conspicuous part in the advance of applied mathematics. Newton was a glory to all mankind, yet we Cambridge men are proud that fate ordained that he should have been Lucasian Professor here. But as regards the part played by Cambridge I refer rather to the men of the last hundred years, such as Airy, Adams, Maxwell, Stokes, Kelvin, and other lesser lights, who have marked out the lines of research in applied mathematics as are studied in this University. (Darwin, 1913, pp. 33–34)

Darwin (1913) added:

> Both the pure and the applied mathematicians are in search of truth, but the former seeks truth in itself and the latter seeks truth about the universe in which we live. (p. 35)

It is worth adding that, like Isaac Newton, Stephen Hawking (1942–2018) was Lucasian Professor of Mathematics at the University of Cambridge. It has been too easy for historians to claim that for much the eighteenth, nineteenth and twentieth centuries, Great Britain was a backwater so far as mathematics was concerned. Usually such statements are based on evaluations of the quality of publications of British mathematicians compared with those made on publications by mathematicians in other nations. Answers to the question "What is mathematics?" would support the viewpoints of those persons or groups asking and answering the question relating to which nation produced the best mathematics and the best mathematicians.

In the 1740s John Winthrop IV presided over the first laboratory of experimental physics in America, and in fact his pure- *and* applied-interpretation of mathematics was maintained at Harvard to 1865 (and beyond that year). In 1734, Yale College imported a telescope, microscope, and barometer from Europe and began exposing students to Lockean, Newtonian, and Copernican theory (Brasch, 1939). In 1745, Yale made mathematics an entrance requirement, thereby ending the exclusive reign of Latin and Greek (Rudolph, 1990). Many North American college students of the period 1607–1865 learned, through processes of osmosis, that mathematics could be more than arithmetic, algebra, geometry, trigonometry and calculus—in fact, one might say that, in the words of Galileo Galilei, mathematics might be thought of as "the language with which God has written the universe" (quoted in Lial, Miller & Hornsby, 1992, p. 2).

Education in the New Nation, 1776–1865

Immediately after the Revolutionary War there emerged a belief among legislators that colleges should be responsible for preparing the people for new ways of thinking about the democratic distribution of power. Franklin, Jefferson, and Rush were home-grown products of the enlightenment and the Revolution, and they led a rising tide of opinion that college curricula should give much greater attention to mathematics and science (Rudolph, 1990).

Thomas Jefferson, as governor of Virginia, had attempted to reorganize the structures and curricula of his State's education facilities so that they would emphasize the practical sides of life (Clements & Ellerton, 2015; Rudolph, 1990). In 1792, New York's Columbia College boasted professors in economics, natural history, and French, and in 1795, the University of North Carolina (which in 1789 had been established as the nation's first public university), moved to establish professorships in agriculture, mechanics, chemistry, and languages, including English. The time was ripe for a wider view of mathematics to be given a more prominent place in the curricula of schools (Clements & Ellerton, 2015; Rudolph, 1990).

During the period 1776–1865 a series of fragmented, vocational and specialized curricula were established across the new nation, and the clash between those giving priority to classical studies, on the one hand, and to practical studies, on the other hand, became increasingly evident. As the number of universities expanded rapidly, and college enrollments grew, the intellectual community was called on to embrace the coming-together of different religions and thought patterns (Cohen, 1998). Nineteen new liberal arts colleges—a new component of higher education in America—were established between 1782 and 1802, with each introducing its own course of study (Gwynne-Thomas, 1981). The Regents of the State of New York chartered Union College in 1795, and in 1845 it became the first liberal arts college in the United States to include engineering in its curriculum (Cohen, 1998). It became common for two types of teachers—temporary tutors and regular professors—to be appointed as instructors within the new colleges (Lucas, 1994), with the title "professor" being employed much more liberally than it ever had been in Great Britain. As would have been expected, graduates of Harvard, Yale, and Princeton were appointed to many of the leading administrative and teaching positions in the new liberal arts colleges—including Amherst, Bates, Bowdoin, Colby, Colgate, Dennison, Hamilton, Middlebury, Oberlin, St. Lawrence, Wooster, and Williams (Barnard, 1875–1876; Cremin, 1977).

Methods of Teaching Mathematics in U.S. Colleges, 1636–1865

Amy Ackerberg-Hastings (2014) has claimed that although the method of learning independently from a textbook "began to develop when textbooks became accessible to many students in the eighteenth century" and that although it persists in the twenty-first century "it was apparently never as popular as formal forms of instruction provided in classrooms" (p. 528). It is interesting to compare that claim with data given in responses by 168 U.S. colleges and universities in the 1890s to Cajori's (1890) questions on the forms of mathematics instruction which then prevailed across the United States of America. One particular question that Cajori asked was "Is the mathematical teaching by text-book or lecture?" and responses are summarized in Table 7.2.

Admittedly, the summary analysis shown in Table 7.2 comes from data collected in the second half of the 1880s, whereas this book is concerned with the 258-year period 1607–1865. Textbooks were more available in the United States from the 1880s onward than they were in the period 1607–1865, so between 1865 and the 1880s things might have changed. We believe, however, formal lectures in mathematics were much less common during the earlier period.

Table 7.2

Responses to the Question "Is the Mathematical Teaching by Textbook or by Lecture?" by 168 North American Universities or Colleges in the 1880s (from Cajori, 1890, pp. 301–302)

Type of Instruction Indicated	Number $n = 168$
"By textbook only"	46
"Mainly by textbook"	65
"Both textbook & lectures"	55
"By lectures mainly"	1
"By lectures only"	1

We contend that before 1865 most college students taking mathematics prepared cyphering books which were based on older parent cyphering books or on textbooks available in the colleges, and the cyphering tradition included a strong recitation component. The following statement by Lao Genevra Simons (1924) is pertinent:

> It must be very desirable . . . to relieve the professors from these duties. This may be done . . . by employing a number of young graduates who would not only act as assistant professors but also under the instruction of the President would perform the more active executive duties. This class of persons would as teachers be eminently useful even now but will be found indispensable whenever the number of students shall amount to several hundreds. A professor can deliver lectures to many more than he can thoroughly teach. I will illustrate the idea I would convey by supposing a case. A class of 80 students is to be taught Mathematics or Natural Philosophy devoting three hours of each day to the study of the subject at their rooms and three other hours with the professor. One hour is to be taken up in the lecture but this alone is not sufficient. Each student should demonstrate a proposition or explain an investigation at the blackboard and also be interrogated to see that he thoroughly understands the principles. This will require, as experience proves, not less than 15 minutes on an average for each student. Now it is evident that only eight students can be examined in the remaining few hours so that each can be examined only about once a fortnight which in effect is merely equivalent to no examination at all. What is to be done? Let the class be divided into at least four parts or sections and let each section attend 3 hours daily with an assistant professor to be examined upon the subject of the lecture or lessons given on the preceding day. The Professor besides lecturing may either have the recitations of one section himself or what would be the better practice, he might without taking the immediate charge, be present at the recitations, visiting each section in turn and only occasionally putting questions and giving explanations. You know that this is the system of instruction which has been practiced at West Point during the last ten years with what success I leave it for others to say. (p. 2)

It seems to us that the most common scheme adopted in the colleges to assist student learning of mathematics was for the student to read a set piece about a nominated topic (either in a textbook or in a parent cyphering book), then to make notes on what was read, and then to attend a recitation session at which a tutor attempted to assess, and improve, the student's understanding of the topic. Then, the student would make an entry on the topic in a personal cyphering book. Sometimes, *but certainly not always*, a lecture on the topic by a professor would be given.

The gender factor in college mathematics. For most of the period 1607–1865 relatively few women attended college, and of those who did, only a small proportion studied mathematics beyond arithmetic and the first few books of *Euclid's Elements*. In 1833 the Oberlin Collegiate Institute opened in Ohio, and in 1837 it admitted four women, thereby becoming the first coeducational college in the United States. Soon women comprised between one-third and one-half of the students at the Institute. Strongly opposed to slavery, Oberlin admitted African-American students in the 1830s (Fletcher, 1943). The move toward "mathematics for all" was accelerating. Five years after the Civil War, in 1870, 9,100 women attended college, comprising 21% of all U.S. college students. The gender balance was changing and would improve, gradually, over the next 100 years (Else-Quest, Hyde, & Linn, 2010).

Concluding Comments

In this chapter we have not provided an overview of courses offered in colleges between 1607 and 1865—readers wishing for more details on such matters should consult books by authors like Smith and Ginsburg (1934) and Zitarelli (2019). We have been concerned to draw attention to overarching issues affecting college mathematics curricula, especially the priority given by college administrators to classical literature and to the Latin and Greek languages.

Historians have tended to think of college mathematics between 1607 and 1865 mainly in terms of an *intended* curriculum comprising algebra, geometry, trigonometry and, toward the end of the period, calculus. Equally important, however, were changes in the *implemented* curriculum—that is to say, what and how college teachers decided to teach, and how students went about learning the intended curriculum. Throughout most of the period a cyphering tradition prevailed whereby students copied notes from textbooks (which, before 1776, were mainly imported from England) or from parent cyphering books. This was complemented by a recitation component. Recitation provided a method for assessing the attained curriculum—that is to say, investigating what students had actually learned. The effectiveness of recitations improved dramatically during the period 1800–1865 as a result of the introduction of blackboards—although, the greater demands on the students were not always appreciated by the students, as the remarkable conic-sections episodes in the 1820s at Yale College illustrated. One of the major advances arising as a result of improved recitation techniques was associated with student learning of what "proof" meant, and how one could go about proving.

One of the most important aspects of college mathematics in North America during the period, has gone largely unnoticed by historians. That aspect was the emphasis on what might be called "applied" forms of mathematics—particularly mensuration, surveying, navigation, astronomy and mechanics. This was especially in line with British mathematics, and the stimulus for it probably came from Isaac Newton's work, which was introduced to Harvard College students following the appointment of the first Hollis professor, Isaac Greenwood, in the 1720s. Although there is no evidence that Greenwood spoke directly with Newton himself during his three years in London in the early 1720s, there can be no doubt that he introduced Newtonian perspectives into college mathematics at Harvard during the 10 years or so he was professor there soon after his return from England.

Throughout the period, the proportion of U.S. college students who were female and who studied mathematics gradually increased, but even in 1865 that proportion was still small—much less than 50 percent. That would change steadily over the next 150 years, so that by 2015 almost 40 percent of all students enrolled in undergraduate mathematics classes in the United States would be female (Cowley & Williams, 1991; Else-Quest, Hyde & Linn, 2010; Hu, 2016).

David Zitarelli's (2019) summary of U.S. undergraduate mathematics programs in 1850 is worth quoting here, because it should be interesting for the reader to compare what has been described in this chapter with Zitarelli's overview:

> The undergraduate program in 1850 differed from the year 2018 in three vital ways: the curriculum, the manner of instruction, and the academic emphasis. First, all 100 students [at Yale] took the same classes through the first semester of their junior year. For mathematics this meant that all students, regardless of major interest, studied algebra, Euclidean geometry, trigonometry, conic sections and spherical geometry. It is impressive that every student had to be proficient in these areas (spherical geometry has not been part of the curriculum for almost a century). Yet, aspiring mathematics scholars interested in pursuing science would be handicapped by a program that did not go beyond the rudiments of analytic geometry. It was still possible for science-oriented students to elect to take a calculus course in the second semester of their junior year, but most courses in calculus had to be taken independently because formal courses in the subject were rarely offered.
>
> A second major difference for today's college students was the nineteenth-century emphasis on rote learning. Students stood in class and recited material they had memorized; in mathematics, they went to the chalkboard to solve problems using techniques that had been drilled in previous class meetings. Few questions were posed, and independence was hardly nurtured, although critical thinking was developed in debating societies. There was no such thing as collaborative learning.

> Thirdly, the emphasis was on the classical languages, Latin and Greek, where long passages had to be memorized and regurgitated. Can today's students imagine a college experience without athletics? No sports teams yet existed. Instead, debating societies flourished. (Zitarelli, 1919, pp. 221–222)

Clearly, Zitarelli struggled to see positive effects on learning deriving from the recitation system. He seemed to think that that system came in addition to a systematic lecture program. By contrast, we do not think that there was a systematic lecture program in mathematics—at Yale, or at any other U.S. college—in 1850. We would be much more generous in relation to the benefits derived from blackboard-supported recitations, especially in relation to the development of the concept of proof, and to understanding connections between the lines memorized in a challenging piece of mathematics. The emphasis on memorization and being able to justify going from one step to another logically did not imply poor teaching in a poor program. But Zitarelli was right when he maintained that in 1850 the quality of implemented mathematics curricula in colleges was still compromised by the need for students to spend much time on the classical languages and literatures. The emphasis on memorization in Latin and Greek was much harder to justify, educationally, than the emphasis on memorizing content and on understanding in mathematics.

To conclude this chapter, it will be appropriate to reproduce the cover page (see, Figure 7.4) of a book on "Mathematical Tradition in the North of England" (Wallis, Wallis, Ransom, & Fauvel, 1991), produced for the annual conference of the Mathematical Association (in England) held at Newcastle-upon-Tyne in 1991. One of the authors of that book was the late John Fauvel, a noted historian of mathematics and mathematics education. Fauvel, who was very interested in the contributions of Thomas Jefferson to mathematics (see Fauvel, 1999), clearly recognized that in Northern England mathematics was traditionally seen as much more than arithmetic, algebra, trigonometry, geometry and calculus. His applied conception of mathematics was connected to a "pure and applied" vision of the subject and formed a support structure at the base of the tree. It was that tradition which was passed on to North America during the period 1607–1865.

The diagram, shown in Figure 7.4, first appeared in John Draper's *The Young Student's Pocket Companion,* published in Great Britain in 1772 by the Newcastle Literary and Philosophical Society. The diagram itself, and the year, place of publication, and the publisher's name, are all relevant to the main argument summarized in this chapter.

The diagram depicts the thinking which prevailed at the time. The trunk of the tree represents Mathematics and Experi[mental] Philosophy and emerging from this trunk one sees major branches—arithmetic, geometry, trigonometry, and mechanics.

From the branch of arithmetic, the main attachments are logarithms and mensuration, on one side and, interestingly, bookkeeping, algebra, and fluxions on the other side; then, from the geometry branch the attachments are surveying, conic sections, and architecture; from the trigonometry branch emerged navigation, use of globes, geography, dialing, astronomy and altimetry. Then, emerging as if from new growth, one can find hydrostatics, hydraulics, magnetism on one side, and pneumatics, electricity, and optics, on the other side.

This was the prevailing view of mathematics in Great Britain during much of the seventeenth and eighteenth centuries (Johnston, 1996; Stedall, 2012), and it was this view of mathematics that Isaac Greenwood brought back to Harvard from London in the mid-1720s. Isaac Newton researched and published papers on many of the named areas—like for example, mechanics, fluxions, optics, and algebra. In the 1750s, Theophilus Grew, Benjamin Franklin's Professor of Mathematics at what would become the University of Pennsylvania, published a book on the use of globes.

The idea that mathematics included both pure and applied forms was held to be philosophically appropriate—the labels on the trunk of the tree are Mathematics and Experimental Philosophy. The publisher was the Newcastle Literary and Philosophical Society, and the place of publication was Newcastle-Upon-Tyne, a major coal-mining city in which the livelihoods of ordinary people relied on the findings of applied mathematics. And, somehow, no elements of the tree embraced the classical curricular traditions and emphases on Latin, Greek and Hebrew languages and literatures.

The schools and colleges in the North American New World flirted with the composite pure/applied view of mathematics. Isaac Greenwood and John Winthrop IV at Harvard were great supporters and so too were Theophilus Grew in Pennsylvania, and Walter Minto (1788), in his inaugural oration at Princeton. But John Farrar, at Harvard, was inclined to favor the more abstract approaches of the French. Cyphering book data show that the implemented mathematics curricula in most schools rarely stretched further than "mathematics as arithmetic" with perhaps a little of algebra, geometry, geometry, mensuration, surveying and navigation added for the occasional student. However, if the classical viewpoint embodied in the Yale Report of 1828 were to continue to be the dominant curricular philosophy in colleges and academies, very little time would be left for anything other than the "pure" side of mathematics.

For Bowditch and Peirce there were new worlds to conquer—and they succeeded in leading the way by showing that by combining the pure and applied sides of mathematics a major weapon was being unleashed for unlocking the mysteries of the universe.

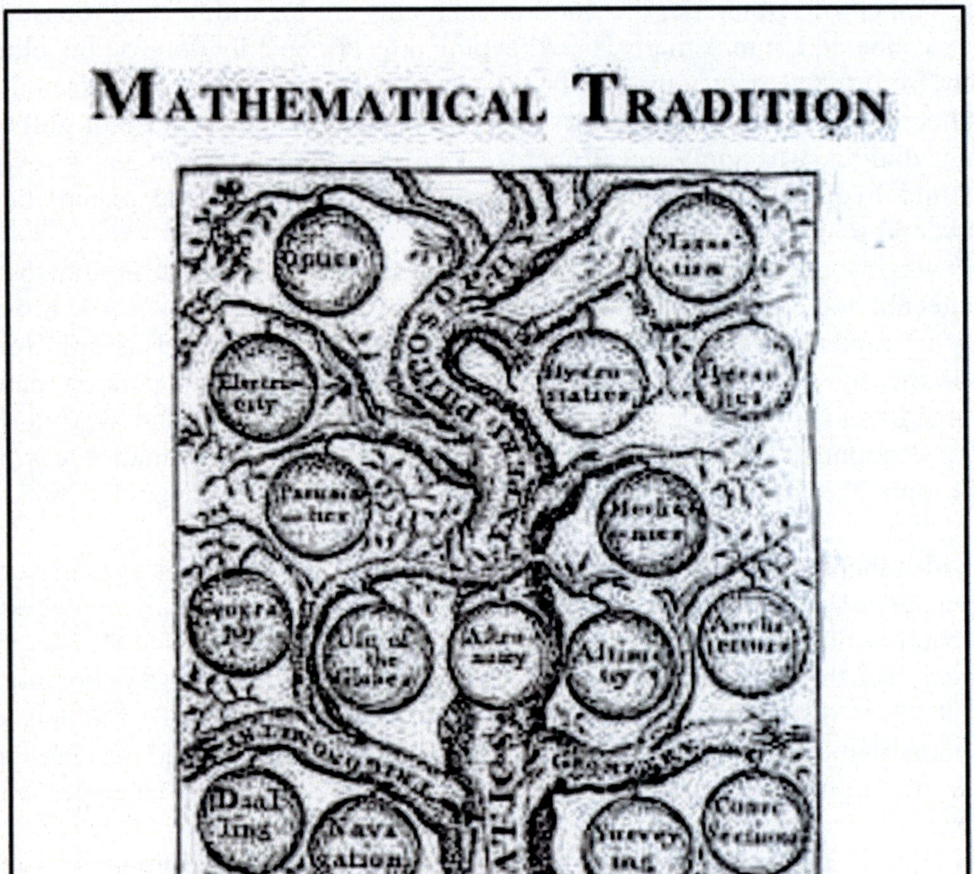

Figure 7.4 A mathematics tradition developed in, and passed on from, Great Britain (front cover of Wallis, Wallis, & Fauvel, 1991).

References

Ackerberg-Hastings, A. (2014). Mathematics teaching practices. In A. Karp & G. Schubring (Eds.), *Handbook on the history of mathematics education* (pp. 525–540). New York, NY: Springer. https://doi.org/10.1007/978-1-4614-9155-2_26

Adams, C., Russell, T. et al. (Ed.). (1965). *The West Point Thayer papers 1808–1872*. West Point, NY: Association of Graduates.

Albree, J. (2002). Nicolas Pike's Arithmetic (1788) as the American *Liber Abbaci*. In D. J. Curtin, D. E. Kullman, & D. E. Otero (Eds.), *Proceedings of the Ninth Midwest History of Mathematics Conference* (pp. 53–71). Miami, FL: Miami University.

Anderson, C. (1962). *Technology in American education: 1650–1900* (Report No. OE-34018). Washington, DC: Office of Education, U.S. Department of Health, Education, and Welfare.

Archibald, R. C. (Ed.). (1925). *Benjamin Peirce*. Oberlin, OH: The Mathematical Association of America. https://doi.org/10.1080/00029890.1925.11986401

Association of Masters of the Boston Public Schools. (1844). *Remarks on the Seventh Annual Report of the Hon. Horace Mann, Secretary of the Massachusetts Board of Education*. Boston, MA: Charles C. Little & James Brown.

Barnard, F. A. P. (1875–1876). Progress of the exact sciences. *Harper's New Monthly Magazine, 52*, 82–100.

Beadie, N. (2010). Education, social capital and state formation in comparative historical perspectives: Preliminary investigations. *Paedagogica Historica, 46*(1–2), 15–32. https://doi.org/10.1080/00309230903528439

Blinderman, A. (1976). *Three early champions of education: Benjamin Franklin, Benjamin Rush, and Noah Webster*. Bloomington, IA: Phi Delta Kappa Educational Foundation.

Brasch, F. E. (1939, October). *The Newtonian epoch in the American colonies (1680–1783)*. Paper presented to the American Antiquarian Society.

Broome, E. C. (1903). *A historical and critical discussion of college admission requirements*. New York, NY: Columbia University.

Brown, C. (1721). *The geography of the ancients so far described as it is contain'd in the Greek and Latin classicks*. London, England: Author.

Buzbee, L. (2014). *A personal history of the classroom*. Minneapolis, MN: Gray Wolf Press.

Cajori, F. (1890). *The teaching and history of mathematics in the United States* (Circular of Information No. 3, 1890). Washington, DC: Bureau of Education.

Cajori, F. (1907). *A history of elementary mathematics with hints on methods of teaching*. New York, NY: Macmillan.

Cajori, F. (1928). *A history of mathematical notations*. La Salle, IL: Open Court Publishing Company.

Cajori, F. (1980). *The chequered career of Ferdinand Rudolph Hassler*. New York, NY: Arno Press.

Campbell, F. (1968). Latin and the elite tradition in education. *The British Journal of Sociology, 19*(3), 308–325. https://doi.org/10.2307/588835

Chateauneuf, A. O. (1930). *Changes in the content of elementary algebra since the beginning of the high school movement as revealed by the textbooks of the period*. PhD dissertation, The University of Pennsylvania.

Clements, M. A., & Ellerton, N. F. (2015). *Thomas Jefferson and his decimals 1775–1810: Neglected years in the history of U.S. school mathematics*. New York, NY: Springer. https://doi.org/10.1007/978-3-319-02505-6

Cocker, E. (1720). *Decimal arithmetic, wherein is shewed the nature and use of decimal fractions in the usual rules of arithmetic, ...* (5th ed.). London, England: J. Darby for M. Wellington.

Cohen, A. M. (1998). *The shaping of American higher education*. San Francisco, CA: Jossey-Bass.

Cohen, P. C. (1982). *A calculating people: The spread of numeracy in early America*. Chicago, IL: The University of Chicago Press.

Cohen, P. C. (2003). Numeracy in nineteenth-century America. In G. M. A. Stanic & J. Kilpatrick (Eds.), *A history of school mathematics* (Vol. 1, pp. 43–76). Reston, VA: National Council of Teachers of Mathematics.

Cowley, W. H., & Williams, D. (1991). *International and historical roots of American higher education*. New York, NY: Garland Publishing, Inc.

Cremin, L.A. (1977). *Traditions of American education*. New York, NY: Basic Books.

Crilly, T. (2008). Arthur Cayley and mathematics education. *HPM Newsletter, 68*,1–4.

Crozet, C. (1821). *A treatise on projective geometry, for the use of the cadets of the United States Military Academy. Part 1*. New York, NY: A. T. Goodrich and Co.

Darwin, G. H. (1913). Opening address. In E. W. Hobson & A. E. H. Love (Eds.), *Proceedings of the Fifth International Congress of Mathematicians* (pp. 33–36). Cambridge, England: Cambridge University Press.

Dauben, J. W., & K. H. Parshall (2014). Mathematics education in North America to 1800. In A. Karp & G. Schubring (Eds), *Handbook on the history of mathematics education* (pp. 175–185). New York, NY: Springer.

Davies, C. (1826). *Elements of descriptive geometry, with their application to spherical trigonometry, spherical projections, and warped surfaces*. Philadelphia, PA: H. C. Carey and I. Lea.

Day, J. (1814). *An introduction to algebra, being the first part of a course of mathematics, adapted to the method of instruction in the American colleges*. New Haven, CT: Howe & Deforest.

Day, J. (1815). *A treatise of plane trigonometry . . . being the second part of a course of mathematics, adopted to the method of instruction in the American colleges*. New Haven, CT: Yale College.

Day, J. (1817). *The mathematical principles of navigation and surveying, with the mensuration of heights and distances*. New Haven, CT: Yale College.

Day, J. (1836). *The teacher's assistant in the "course of mathematics."* New Haven, CT: Yale College.

Deming, C. (1904). Yale wars of the conic sections. *The Independent, 56,* 667–669.

Dexter, F. B. (1887). Estimates of population in American colonies. *Proceedings of the American Antiquarian Society* (pp. 22–50). Boston, MA. American Antiquarian Society.

Draper, J. (1772). *The young student's pocket companion*. Newcastle-Upon-Tyne, England: Newcastle Literary and Philosophical Society.

Ellerton, N. F., & Clements, M. A. (2012). *Rewriting the history of mathematics education in North America, 1607–1861*. New York, NY: Springer. https://doi.org/10.1007/978-94-007-2639-0

Ellerton, N. F., & Clements, M. A. (2014). *Abraham Lincoln's cyphering book, and ten other extraordinary cyphering books*. New York, NY: Springer. https://doi.org/10.1007/978-3-319-02502-5

Ellerton, N. F., & Clements, M. A. (2017). *Samuel Pepys, Isaac Newton, James Hodgson and the beginnings of secondary school mathematics: A history of the Royal Mathematical School at Christ's Hospital 1673–1868*. New York, NY: Springer. https://doi.org/10.1007/978-3-319-46657-6

Else-Quest, N., Hyde, J. S., & Linn, M. (2010). Cross-national patterns of gender differences in mathematics: A meta-analysis. *Psychological Bulletin, 136*(1), 103. https://doi.org/10.1037/a0018053

Farrar, J. (1818). Introduction to Farrar's translation of Lacroix, S. F., *An elementary treatise on arithmetic and an introduction to the elements of algebra*. Cambridge, MA: Hilliard & Metcalf.

Fauvel, J. (1999, April 15). *Thomas Jefferson and mathematics*. Lecture given at the University of Virginia.

F. D. B. (1911). Blackboards. In P. Monroe (Ed.), *A cyclopedia of education* (Vol. 1, pp. 390–394). New York, NY: The Macmillan Company.

Fletcher, R. S. (1943). *A history of Oberlin College from its foundation through the Civil War*. Oberlin, OH: Oberlin College.

Franklin, B. (1749). *Proposals relating to the education of youth in Pensilvania*. Philadelphia, PA: Author.

Franklin, B. (1793). *The private life of the late Benjamin Franklin, LL.D*. London, England: J. Parsons.

Fuess, C. M. (1917). *An old New England school: A history of Phillips Academy Andover*. Boston, MA: Houghton Mifflin Company.

Geiger, R. L. (2016). *The history of American higher education*. Princeton, NJ: Princeton University Press.

George Washington to Nicolas Pike, 20 June 1786. *Founders Online*, National Archives, accessed December 29, 2019 https://founders.archives.gov/documents/Washington/04-04-02-0119.

Goldsmith, O. (1837). *The Grecian history, from the earliest state to the death of Alexander the Great*. Hartford, CT: Judd, Loomis & Co.

Goodchild, L. F., & Wechsler, H. S. (1989). *The history of higher education* (2nd ed.). Boston, MA: Pearson Custom Publishing.

Green, A. (2015, September 15). The Yale chalkboard rebellion of 1830. From https://www.mentalfloss.com/article/68749/yale-chalkboard-rebellion-1830, viewed on April 9, 2020.

Greenwood, I. (1729). *Arithmetick, vulgar and decimal, with the application thereof to a variety of cases in trade and commerce*. Boston, MA: Kneeland & Green.

Grew, T (1853). *The description and use of the globes, celestial and terrestrial, with variety of examples for the learner's exercise: Intended for the use of such persons who would attain to the knowledge of those instruments: but chiefly designed for the instruction of young gentlemen at the Academy of Philadelphia. To which is added rules for working all the cases in plain and spherical triangles without a scheme*. Germantown, PA: Christopher Sower.

Gwynne-Thomas, E. H. (1981). *A concise history of education to 1900 A.D.* Washington, DC: University Press of America, Inc.

Halwas, R. (1990). *American mathematics textbooks 1760–1850*. London, England: Author.

Herbst, J. (2004). The Yale Report of 1828, *International Journal of the Classical Tradition, 11*(2), 213–231. https://doi.org/10.1007/BF02720033

Hertel, J. (2016) Investigating the implemented mathematics curriculum of New England navigation cyphering books. *For the Learning of Mathematics, 36*(3), 4–10.

Hill, T. (1880). Benjamin Peirce. *The Harvard Register, 6*(1), 91–92.

Hirsch, D., & Van Haften, D. (2019). *The tyranny of public discourse*. El Dorado Hills, CA: Savas Beatie.

Hobson, E. W., & Love, A. E. H. (Eds.). (1913). *Proceedings of the Fifth International Congress of Mathematicians*. Cambridge, England: Cambridge University Press.

Hu, J. C. (2016, November 4). Why are there so few women mathematicians? *The Atlantic*.

Hurd, D. H. (1890). *History of Middlesex County, Massachusetts, with biographical sketches of many of its pioneers and prominent men*. Philadelphia, PA: J. W. Lewis & Co.

Iliffe, R. (1997). Mathematical characters: Flamsteed and Christ's Hospital Royal Mathematical School. In F. Willmoth (Ed.). *Flamsteed's stars: New perspectives on the life and work of the first Astronomer Royal (1646–1719)* (pp. 115–144). Woodbridge, England: The Boydell Press.

Jackson, A. (2002). Teaching math in America: An exhibit at the Smithsonian. *Notices of the American Mathematical Society, 49*(9), 1082–1083.

Johnston, S. (1996). The identity of mathematical practitioners in 16th-century England. In I Hantsche (Ed.), *Der "mathematicus": Zur entwicklung und beturdung einer neuen berufsgruppe in der zeit Gerhard Mercators* (pp. 93 – 120). Bochum, Germany: Brockmeyer.

Karpinski, L. C. (1980). *Bibliography of mathematical works printed in America through 1850* (2nd ed.). New York, NY: Arno Press.

Kidwell, P. A., Ackerberg-Hastings, A., & Roberts, D. L. (2008). *Tools of American mathematics teaching, 1800–2000*. Baltimore, MD: Johns Hopkins University Press.

Kraus, J. W. (1961). The development of the curriculum in the early American colleges. *History of Education Quarterly, 1*(2), 64–76. https://doi.org/10.2307/367641

Krause, D. A. (2000). "Among the greatest benefactors of mankind": What the success of the chalkboard tells us about the future of computers in the classroom. *Computers and the Future of the Humanities, 33*(2), 6–16. https://doi.org/10.2307/1315198

Lacroix, S. F. (1818). *An elementary treatise of arithmetic, and an introduction to the elements of algebra, comprehending the mathematics required for admission to the University of Cambridge, New England.* Cambridge, MA: Hilliard and Metcalf.

Leacock, S (1970). *Feast of Stephen.* Toronto, Canada: McClelland & Stewart, Inc.

Lemprière, J. (1834). *Lemprière's classical dictionary for schools and academies.* Boston, MA: Carter, Hender & Co.

Lial, M. L., Miller, C. D., & Hornsby, E. J. (1992). *Beginning algebra.* Boston, MA: Addison Wesley.

Looby, C. (1984). Phonetics and politics: Franklin's alphabet as a political design. *Eighteenth-Century Studies, 18*(10), 1–34. https://doi.org/10.2307/2738304

Lucas, C. J. (1994). *American higher education: A history.* New York, NY: St. Martin's Griffin.

Macintyre, S., & Clark, A. (2004). *The history wars.* Melbourne, Australia: Melbourne University Press.

Martin, G. H. (1897). *The evolution of the Massachusetts public school system: A historical sketch.* New York, NY: D. Appleton and Company.

Martines, L. (1979). *Power and imagination: City states in renaissance Italy.* New York, NY: Alfred A. Knopf.

Matz, F. P. (1895). B. O. Peirce: Biography *American Mathematical Monthly, 2*, 173–179. https://doi.org/10.1080/00029890.1895.11998647

McDonald, A. (1785). *The youth's assistant* . . . Norwich, CT. John Trumbull.

Meyer R. (1968). Opponents of classical learning in America during the revolutionary period. *Proceedings of the American Philosophical Society, 112*, 221–234.

Minto, W. (1788). *An inaugural oration, on the progress and importance of the mathematical sciences*. Manuscript, held in the Clements Library, The University of Michigan.

Molloy, P. M. (1975). *Technical education and the young Republic: West Point as America's École Polytechnique, 1802–1833*. PhD dissertation, Brown University.

Monge, G. (1811). *Géométrie descriptive* (New edition). Paris, France. J. Klostermann.

Monroe, W. S. (1917). *Development of arithmetic as a school subject*. Washington, DC: Government Printing Office.

Morison, S. E. (1956). *The intellectual life of colonial New England*. Ithaca, NY: Great Seal Books.

Muttappallymyalil, J, Mendis S, John L. J., Shanthakumari N., Sreedharan J, & Shaikh R. B. (2016). Evolution of technology in teaching: Blackboard and beyond in medical education. *Nepal Journal of Epidemiology, 6*(3), 588–592. https://doi.org/10.3126/nje.v6i3.15870

Ogg, F. A. (1927). *Builders of the Republic*. New Haven, CT: Yale University Press.

Peirce, B. (1837). *An elementary treatise on algebra; to which are added elementary equations and logarithms*. Boston, MA: James Munroe and Company.

Peirce, B. (1841). *An elementary treatise on curves, functions, and forces (Volume First): Analytic geometry and the differerential calculus*. Boston, MA: James Munroe and Company.

Peirce, B. (1846). *An elementary treatise on curves, functions, and forces (Volume Second): Calculus and imaginary quantities, residual calculus, and integral calculus*. Boston, MA: James Munroe and Company.

Peirce, B. (2019, May 13). *Wikiquote*, Retrieved July 5, 2021 from https://en.wikiquote.org/w/index.php?title=Benjamin_Peirce&oldid =2596188.

Phalen, H. R. (1946). The first professorship of mathematics in the colonies. *The American Mathematical Monthly, 53*(10), 579–582. https://doi.org/10.1080/00029890.1946.11991755

Peterson, S. R. (1955). Benjamin Peirce: Mathematician and philosopher. *Journal of the History of Ideas, 16*, 89–112. https://doi.org/10.2307/2707529

Phillips, C. J. (2015). An officer and a scholar: Nineteenth-century West Point and the invention of the blackboard. *History of Education Quarterly, 55*(1), 82–108. https://doi.org/10.1111/hoeq.12093

Pike, N. (1788). *The new and complete system of arithmetic, composed for the use of the citizens of the United States*. Newbury-Port, MA: John Mycall.

Pike, N. (1793). *Abridgement of the new and complete system of arithmetick composed for the use, and adapted to the commerce of the citizens of the United States*. Newbury-Port, MA: John Mycall, Isaiah Thomas.

Pillans, J. (1856). *First steps in the physical and classical geography of the ancient world*. London, England: Longman, Brown, Green & Longmans.

Pioariu, R. (2011). Cross-cultural issues in teaching English to Romanian students. In T. Popescu, R. Pioariu, & C. Herteg (Eds.), *Cross-disciplinary approaches to the English language: Theory and practice* (pp. 150–160). Newcastle-on-Tyne, England: Cambridge Scholars Publishing.

Plimpton, G. A. (1916). *The hornbook and its use in America*. Worcester, MA: American Antiquarian Society. https://doi.org/10.5479/sil.258315.39088004241238

Preveraud, T. (2015). American mathematics journals and the transmission of French textbooks to the United States. In K. Bjarnadottir, F. Furinghetti, J. Prytz, & G. Schubring (Eds.), "*Dig where you stand 3*" (pp. 309–325). Uppsala, Sweden: Uppsala, Universitet

Quincy, J. (1860). *The history of Harvard University*. Boston, MA: Crosby, Nicholls, Lee & Co.

Richard, C. J. (1994). *The founders and the classics: Greece, Rome, and the American enlightenment*. Cambridge, MA: Harvard University Press.

Rickey, V. F., & Shell-Gellasch, A. (2010, July). Mathematics education at West Point: The first hundred years—Teaching at the Academy. *Convergence* (Publication of the Mathematical Association of America).

Roberts, D. L. (2014). History of tools and technologies in mathematics education. In A. Karp & G. Schubring (Eds), *Handbook on the history of mathematics education* (pp. 565–578). New York, NY: Springer. https://doi.org/10.1007/978-1-4614-9155-2_28

Robinson, H. (1870). *The progressive higher arithmetic for schools, academies, and mercantile colleges*. New York, NY: Ivison, Blakeman, Taylor & Co.

Rudolph, F. (1990). *The American college and university: A history*. Athens, GA: The University of Georgia Press.

Rush, B. (1806), *Essays, literary, moral and philosophical* (2nd ed.). Philadelphia, PA: Thomas and William Bradford.

Schlesinger, A. M. (Ed.). (1983). *The almanac of American history*. New York, NY: G. P. Putnam.

Seybolt, R. F. (1969).*The public schools of colonial Boston 1635–1775*. New York, NY: Arno Press.

Simons, L. G. (1924). *Introduction of algebra into American schools in the 18th century*. Washington, DC: Department of the Interior Bureau of Education.

Smith, D. E., & Ginsburg, J. (1934). *A history of mathematics in America before 1900*. Chicago, IL: The Mathematical Association of America.

Stamper, A. W. (1909). *A history of the teaching of elementary geometry, with reference to present-day problems*. PhD dissertation, Columbia University. https://doi.org/10.1017/S0007087499003556

Stewart, L. (1999). Other centres of calculation, or where the Royal Society didn't count: Commerce, coffee-houses and natural philosophy in early modern London. *British Journal of the History of Science, 32*, 132–153.
Stedall, J. (2012). *The history of mathematics: A very short introduction*. New York, NY: Oxford University Press.
Stoeckel, A. (1976). Presidents, professors, and politics: The colonial colleges and the American revolution. Conspectus of History, 1(3), 45–56.
Sylvester, J. J. (1870, January 6). A plea for the mathematician II. *Nature, 1*, 261–263.
Tebbel, J. (1972). *A history of book publishing in the United States*. New York, NY: R. R. Bowke.
"Thomas Jefferson to John Farrar, 10 November 1818," *Founders Online*, National Archives, https://founders.archives.gov/documents/Jefferson/03-13-02-0343
Todhunter, I. (Ed.). (1955). *Euclid's elements*. London, England: J. M, Dent and Sons.
Tyler, L. G. (1897). *Education in colonial Virginia. Part II, private schools and tutors. William and Mary Quarterly, 6, 1–6*. https://doi.org/10.2307/1914792
Varcoe, K. E. (2002). A historical review of curriculum in American higher education: 1636–1900. PhD course paper, Nova Southeastern University.
University of Michigan (1967). *Education in early America*. Ann Arbor, MI: The University of Michigan Library.
Wallis, P., Wallis, R., Ransom, P., & Fauvel, J. (1991). *Mathematical tradition in the north of England*. Newcastle-Upon-Tyne, England: University of Newcastle.
Ward, J. (1719). *The young mathematician's guide: Being a plain and easie introduction to the mathematicks*. London, England: Thomas Horne.
Ward, J. (1758). *The young mathematician's guide: Being a plain and easy introduction to the mathematicks*. London, England: C. Hitch and L. Hawes.
Webster, N. (1787). *The American speller*. Boston, MA: Isaiah Thomas.
Wilson, N. G. (Ed.). (2006). *Encyclopedia of ancient Greece*. New York, NY: Routledge
Windschuttle, K. (1996). *The killing of history: How literary critics and social theorists are murdering our past*. San Francisco, CA: Encounter Books.
Wylie, C. D. (2012). Teaching manuals and the blackboard: Accessing historical classroom practices. *History of Education, 41*(2), 257–272. https://doi.org/10.1080/0046760X.2011.584573
Yale College. (1828). *Reports on the course of instruction in Yale College by a committee of the Corporation, and the academical faculty*. New Haven, CT: Author.
Zitarelli, D. A. (2019). *A history of mathematics in the United States and Canada. Volume 1: 1492–1900*. Washington, DC: American Mathematical Society. https://doi.org/10.1090/spec/094

Chapter 8
Different Perspectives on Mathematics in North America 1607–1865

Abstract It would be unreasonable to expect the inhabitants of North America to have produced great works of mathematics—judging by European standards—during the period 1607–1865. At that time a New World began to be constructed in North America by the European "invaders"—houses, schools, and towns were built, administrative structures were created, and lands were cleared for farming. But very few books other than bibles and, perhaps, almanacs were to be found in homes or schools, and most of the relatively few settlers who knew enough mathematics to teach it had other things to do. It is not surprising, therefore, that the 258-year period did not produce more than three or four mathematicians who, by the European standards of the time, might be regarded as "outstanding." Between 1775 and 1820 U.S. college curricula drew their inspiration from the classical curricular traditions of the medieval universities of Europe and especially of Cambridge and Oxford Universities. However, many students who attended the North American colleges did enroll in "applied mathematics" subjects—embracing fields like astronomy, surveying, mensuration, and navigation. Interest in those forms of mathematics had been successfully translated mainly from Great Britain.

Keywords Abraham Lincoln • Benjamin Banneker • Benjamin Franklin • Benjamin Peirce • Benjamin Rush • Classics (Greek and Latin) • College of William and Mary • Conics • David Rittenhouse • Decimal currency • Declaration of Independence • Dollar • Euclidean geometry • Harvard College • Jeremiah Day • Magic squares • Nathaniel Bowditch • Navigation • Proof • Robert Adrain • Thomas Jefferson • University of Pennsylvania • Yale College

Applying Mathematics in North America 1607–1865

Originally it was planned that this chapter would deal with the extent of, and developments with respect to, mathematics research in North America during the period 1607–1865. However, after considering the question "what do we mean by the term "mathematics research?", and after examining what seemed to be relevant historical data, we decided that it would be more appropriate to provide an account of some of the most creative developments in ways mathematics was thought about, and used in everyday life in North America during the period.

Karen Hunger Parshall (2003) stated that it was Morris Kline (1972) who influenced her to accept the judgment that "the United States produced essentially no noteworthy mathematics prior to 1900 but began to figure more prominently in

N. F. Ellerton, M. A. (Ken) Clements, *Toward Mathematics for All*, History of Mathematics Education,
https://doi.org/10.1007/978-3-030-85724-0_8

the history of the discipline sometime after that date" (p. 114). Parshall (2003) reasoned this way:

> From the historiographical point of view that Kline adopted in his study, mathematical results merited inclusion in the historical narrative provided they formed a weight-bearing link in the great chain of mathematical ideas that stretches across time from the present to the past. (p. 114)

Recognizing that that is the traditional perspective of mathematicians we decided to change our focus for this chapter, so that we could include everyday applications and other developments in mathematics by colonial and early U.S. scholars—not necessarily by persons regarded as "mathematicians"—which might *not* fit the idea of "research" among people who regard themselves as "mathematicians" in the twenty-first century.

As Parshall (2003) pointed out, "nineteenth-century European mathematicians rarely used or favorably commented on the work of their American contemporaries" (p. 115). That statement was consistent with her claim that no eighteenth-century American contributed significantly to the technical development of mathematics, and that the only early nineteenth-century U.S. scholar who did was Nathaniel Bowditch (1773–1838), the self-taught New England translator of, and commentator on, the first four (of five) volumes of the notoriously complex *Traité de Mécanique Céleste* by Pierre-Simon Laplace (1749–1827).

Our first two examples of persons who, we shall argue, were responsible for noteworthy mathematical developments will be two "founding fathers," Benjamin Franklin (1706–1790) and Thomas Jefferson (1743–1826). Until recently, neither Franklin nor Jefferson featured in the literature as having been responsible for significant mathematical developments (Weems, 1820; Zitarelli, 2019), but our reasons for arguing that they deserve to be remembered for their mathematical innovations should become clear after we discuss their cases. Following that discussion, we will return to a question associated with the meaning of the term "noteworthy mathematics."

Benjamin Franklin's Contributions to Mathematics

In his old age, Benjamin Franklin (1706–1790) was convinced that when he was a boy he had struggled to cope with mathematics. In his *Autobiography,* Franklin (1917) made it clear that later in his life he had come to believe that he had much more mathematical talent than had been revealed during his time at school. Reminiscing on his school days, he recalled that he had "twice failed in learning [arithmetic]." That demands comment. Pencil-and-paper arithmetic tests were not used in North America early in the eighteenth century (Ellerton & Clements, 2012), so it is not clear what Franklin meant when he wrote that he "failed in learning" arithmetic. However, it *is* clear that late in his life, when he wrote his *Autobiography,* he had changed his mind and had come to believe that when he was young he *could* do arithmetic well. He acknowledged that his negative school experiences with the subject had affected his psyche so much that during his early years he had believed that he was not proficient.

Benjamin Franklin's magic squares. During the second quarter of his life Benjamin Franklin found himself getting more and more interested in what today are known as "magic squares" (Behforooz, 2012; Blindeman, 1976). Perhaps the simplest magic square appears in the form of a 3 by 3 grid, with each element of the grid having a different numeral from among 1, 2, 3, . . ., 8, and 9 placed in it so that the sums of the numbers in each row of the grid are equal to the sums of the numbers in each column of the grid and to the sums of the numbers in each of the two main diagonals of the grid. Figure 8.1 shows a 3 by 3 magic square and Figure 8.2 a 5 by 5 magic square.

Can you construct a 7 by 7 magic square with the numbers 1, 2, 3, . . . , 49? An 11 by 11 magic square with the numbers 1, 2, . . ., 121? A 4 by 4 magic square with the numbers 1, 2, . . ., 16? Suppose you constructed an n by n magic square with the numerals 1, 2, 3, . . ., n^2 (where n represents any natural number greater than or equal to 3). What would be the sum of the numbers in each row, each column, and in each of the two main diagonals? How many different 3 by 3 squares with the numerals 1, 2, . . ., 9, other than the one shown in Figure 8.1, can you construct? Suppose you multiplied each number in Figure 8.1 by 3 and then subtracted 1, would you get another magic square? Generalize.

8	1	6
3	5	7
4	9	2

Figure 8.1. A 3 by 3 magic square.

17	24	1	8	15
23	5	7	14	16
4	6	13	20	22
10	12	19	21	3
11	18	25	2	9

Figure 8.2. A 5 by 5 magic square.

We apologize to readers already familiar with much of the rapidly expanding literature on magic squares, because for them the above questions will be trivial. For others, the questions should provide entrée into an elementary but nevertheless magical mathematical world.

There is evidence that Benjamin Franklin was introduced to magic squares by an older friend, James Logan (1674–1751), who alerted Franklin to a book on the subject by a French scholar (Franklin, 1749, 1793). From the perspective of this chapter, that is interesting because it links early colonial North American mathematics to European mathematics of the seventeenth century.

The British mathematician Godfrey H. Hardy wrote in his *Mathematician's Apology* (2004) that "the best mathematics is serious as well as beautiful" (p. 89). Hardy went on to say that "the 'seriousness' of a mathematical theorem lies not in its practical consequences ... but in the significance of the mathematical ideas which it connects" (p. 90). In the context of this chapter that raises the question whether magic squares should be described as serious mathematics. Many mathematicians would say "no," but we would answer "emphatically yes," and our belief has received support from Paul C. Pasles (2007), in his detailed study of magic squares, and from Christopher Henrich (1991) and Hossein Behforooz (2012). Their investigations with magic squares have suggested that serious mathematics surrounds them (Zitarelli, 2019). Pasles (2007), in telling the story of Franklin's almost fanatical fascination with magic squares (and with other similar structures which he called "magic circles") retold the apocryphal story that in the late 1730s, when Franklin was a clerk in the Pennsylvania Assembly, he often became bored with the proceedings and would amuse himself by constructing magic squares and magic circles (Garcia, Meyer, Sanders, & Seitz, 2009; Wunsch, 2007).

We now look at a remarkable 8 by 8 ,"almost-magic," square which Franklin created. For readers not familiar with Franklin's efforts with respect to magic squares we are including here a brief interlude on the square shown in Figure 8.3 (which is not quite a magic square because the sums of the numbers in the diagonals differ from the sums of the numbers in the rows and columns). We would also comment that Franklin constructed a 16 by 16 square which he described in a letter to a friend as "the most magically magical of any magic square ever made by any magician" (quoted in Zitarelli, 2019, p. 78).

52	61	4	13	20	29	36	45
14	3	62	51	46	35	30	19
53	60	5	12	21	28	37	44
11	6	59	54	43	38	27	22
55	58	7	10	23	26	39	42
9	8	57	56	41	40	25	24
50	63	2	15	18	31	34	47
16	1	64	49	48	33	32	17

Figure 8.3. A remarkable 8 by 8 "almost-magic" square constructed by young Benjamin Franklin.

The numbers in each row and column of the square shown in Figure 8.3, but *not* the numbers in either of the two main diagonals, sum to 260. Franklin also noted that half of the numbers in each row or column sum to 130 (i.e., half of 260). What is even more remarkable, the numbers in each of what Franklin called "bent rows" (see Figure 8.4) sum to 260. Each "bent row" has 8 numbers with four possible orientations—see Figure 8.4.

There is also a "wrap around" effect. If we keep translating the "bent row" in the top-left orientation (in Figure 8.4) one unit to the left, wrapping the ends around, we obtain the patterns shown in Figure 8.5. In addition, Franklin noted that the "shortened bent rows" plus the "corners" also sum to 260 (Ahmed, 2004). An example of this pattern is shown in Figure 8.6.

As with the previous patterns, this template can be rotated in any of the four directions and translated into any of the eight positions (with wrap-around), and the sum of the highlighted numbers is always 260. Finally, Franklin noted that the two sets of eight numbers depicted in Figure 8.7 also sum to 260. These patterns can also be translated (with wrap-around), both horizontally and vertically.

Where did Franklin's ideas on magic squares come from? Many commentators have tried to answer that question (see, e.g., Ahmed, 2004) but, according to Zitarelli (2019), the source of Franklin's inspiration with respect to magic squares "is not known" (p. 77). Some have arrived at the squares by using mathematical techniques which were not yet developed in Franklin's time. Others have worked backwards from completed squares to look for patterns, There has been much recent mathematical research on magic squares (Ahmed, 2004; Pasles, 2007), but we still do not know details of the kind of thinking corresponding to the algorithm(s) and spatial transformations Franklin developed and used 280 years ago.

```
ó - - - - - - -     - - - - - - - ó
- ó - - - - - -     - - - - - - ó -
- - ó - - - - -     - - - - - ó - -
- - - ó - - - -     - - - - ó - - -
- - - ó - - - -     - - - - ó - - -
- - ó - - - - -     - - - - - ó - -
- ó - - - - - -     - - - - - - ó -
ó - - - - - - -     - - - - - - - ó

- - - - - - - -     ó - - - - - - ó
- - - - - - - -     - ó - - - - ó -
- - - - - - - -     - - ó - - ó - -
- - - - - - - -     - - - ó ó - - -
- - - ó ó - - -     - - - - - - - -
- - ó - - ó - -     - - - - - - - -
- ó - - - - ó -     - - - - - - - -
ó - - - - - - ó     - - - - - - - -
```

Figure 8.4. Franklin's "bent rows" for his 8 by 8 almost-magic square.

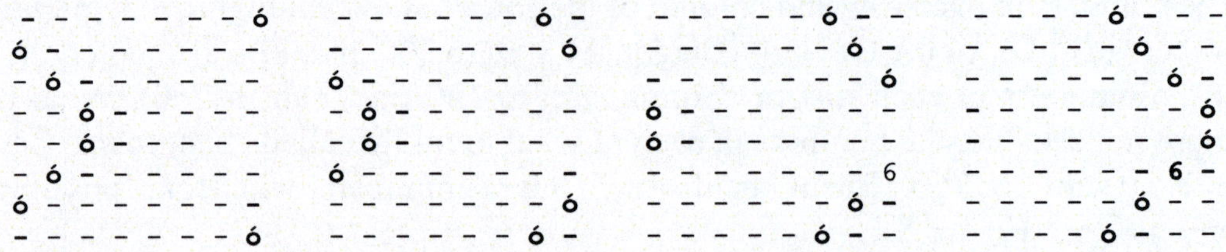

Figure 8.5. Translating the bent rows.

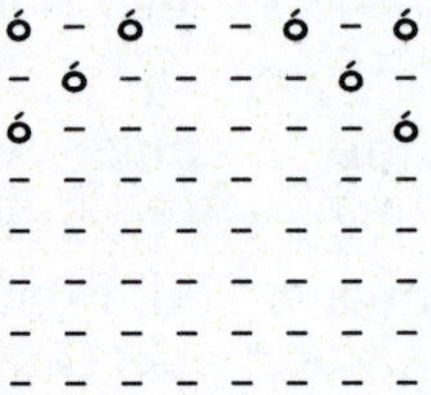

Figure 8.6. Shortened bent rows plus corners.

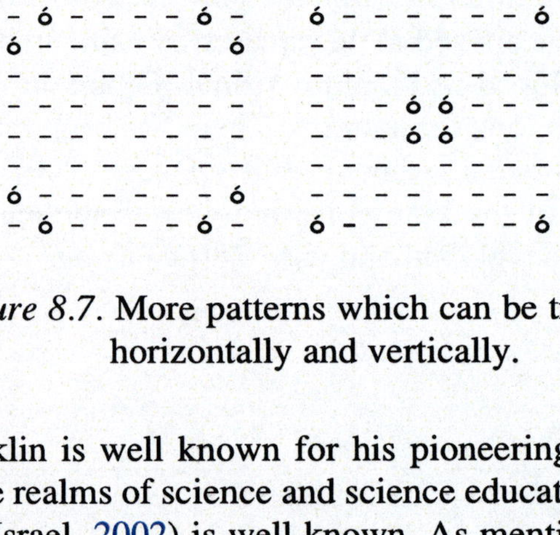

Figure 8.7. More patterns which can be translated horizontally and vertically.

Benjamin Franklin is well known for his pioneering work on electricity and more generally in the realms of science and science education (Anderson, 1997) and radical philosophy (Israel, 2002) is well known. As mentioned in Chapter 7, in the 1740s in Pennsylvania, Franklin was among a group of notable citizens who decided to attempt to modernize college curricula. In 1749 the group established a school which would subsequently be chartered as the College and Academy of Philadelphia, with Franklin as the first President of the Board. Initially there was an associated "charity school" which taught reading, writing, and arithmetic (Dauben & Parshall, 2014; Turner, 1953). When the Academy opened, in 1753, it comprised just two schools—a Latin school which offered a classics-based curriculum, and an English school which offered "practical" courses including history, geography, navigation and surveying, taught in the English language. Franklin's practical curriculum, with the English School, was designed so that students would benefit from arithmetic, accounting, geometry, astronomy, English grammar, writing, public speaking, and histories of mechanics, natural philosophy, and agriculture (see Grew, 1853). Latin and Greek could be studied but would not be compulsory.

Franklin's creative work on magic squares and magic circles revealed a "mathematical mind," but it is only in recent times (Pasles, 2007) that it has been given the mathematical credit some believe it deserves. But as stated earlier, many mathematicians have tended to reject the idea that anyone playing around with magic squares and circles is *really* doing mathematics.

Thomas Jefferson's Usage of Mathematics in His Attempts to Improve the Everyday Lives of U.S. Citizens

Jefferson and the Declaration of Independence

In 1776, a youthful Thomas Jefferson, together with Benjamin Franklin, John Adams, Roger Sherman, and Robert R. Livingston, were handed the responsibility of drafting a Declaration of Independence, a document which would justify why the 13 British colonies in North America needed to become independent of their colonial master, Great Britain. There is strong evidence (see Lucas, 1989) that the first draft of the document was mainly conceived and written by Jefferson.

Jefferson, a graduate of William and Mary College, in Williamsburg, Virginia, came from a wealthy Virginia family and had had a privileged undergraduate education at the College of William and Mary, where he developed a particularly close relationship with a Scot, William Small, who, between 1758 and 1764, was an articulate, widely-read Professor of Natural Philosophy and Mathematics. Small introduced Jefferson to members of Virginian society, and Small, Jefferson, George Wythe (a leading colonial jurist) and Francis Fauquier (the Governor of Virginia) regularly dined together (Boyd, 1950a; Ganter, 1947; Wiencek, 2012; Wilson, 1992). The discussions they had would have an important influence on young Jefferson's intellectual development.

Small influenced Jefferson to develop a strong love for mathematics, which Jefferson retained throughout his life—indeed, in 1812 Jefferson wrote that when he was young, mathematics had been the "passion of his life" (Thomas Jefferson to William Duane, October 12, 1812). In a letter to John Adams in the same year, the third President of the United States informed the second President that he had "given up newspapers for Tacitus and Thucydides, for Newton and Euclid" and added that that had made him "the happier." Jefferson had long been an admirer of Euclid's logical structures (Zitarelli, 2019), and it is highly likely that he, with Benjamin Franklin and John Adams, were responsible for the final form of the introduction to the Declaration which pointed to a determination to adopt Euclidean logical structure in the Declaration (Lucas, 1989; Zitarelli, 2019). Consider, for example, the following passage:

> When in course of human events, it becomes necessary for one people to dissolve the political bonds which have connected them with another, and

> to assume among the powers of the earth, the separate and equal station to which the Laws of Nature and Nature's God entitle them, a decent respect to the opinions of mankind requires that they should declare the causes which impel them to the separation.

This was a neat statement of "what had to be proved." Immediately following came:

> We hold these truths to be self-evident, that all men are created equal, that they are endowed by their Creator with certain inalienable Rights, that among these are Life, Liberty, and the pursuit of Happiness—that to secure these rights, Governments are instituted among Men, deriving their just powers from the consent of the governed—That whenever any form of Government becomes destructive of these ends, it is the Right of the People to alter or to abolish it, and institute new Government, laying its foundation on such principles and organizing its powers in such form, as to them shall seem most likely to affect their safety and happiness.

These words in the Declaration have a very Euclidean ring about them (Hirsch & Van Haften, 2019b). There is a statement of "a self-evident truth," or axiom—"All men are created equal and have inalienable Rights." This statement, and other statements in the above passages, include undefined terms—such as "rights," "equal," "liberty," "happiness," "inalienable," "men," and axioms such as "governments derive their powers from the consent of the governed." From these emerged a "proof" of the proposition that the people had the right to abolish a government if the people did not approve of what was happening; finally, came the conclusion that the existing government should be replaced by a government which aimed to institute a system of government which would be likely to achieve safety and happiness for all people. Jefferson, and the others who prepared the Declaration concluded that King George III had not established an appropriate government for his North American subjects: he had refused to give his assent to laws which were wholesome and necessary for the public good and had forbidden his governors to pass laws of immediate and pressing importance. And, by submitting facts pointing to the violation of rights, Jefferson completed his proof of the case for a Declaration of Independence, the *logical* force of which would make it a document based on the ultimate form of law—universal law.

Recent analysis by David Hirsch and Dan Van Haften has suggested that when preparing the Declaration, Jefferson *consciously* adopted Euclidean logic by which valid proofs were established. According to Hirsch and Van Haften (2019a), the following "six elements light the path to reasoned persuasion":

1. *Enunciation*

   (a) Given
   (b) Sought

2. *Exposition:* This takes separately what is given, and prepares it in advance for use in the investigation—it answers the question: "What additional facts are needed to be known about what needs to be investigated?"
3. *Specification:* This takes separately the thing that is sought and makes clear precisely what it is. It answers the question: "What must be demonstrated to resolve what is sought?"
4. *Construction:* This adds what is lacking in the "given" for finding what is sought.
5. *Proof:* This draws the proposed inference by reasoning scientifically from the propositions that have been admitted. It answers the question: "How does the admitted proof confirm the proposed inference?"
6. *Conclusion:* This reverts to the enunciation, confirming that what was to be proved has in fact been proved. It answers the question: "What was demonstrated?" (p. 1)

These six elements of a proposition collectively define a *demonstration*. The image is a regular tetrahedron, with the conclusion at the top and with six edges representing pathways from enunciation to conclusion. Hirsch and Van Haften (2019a, 2019b, 2019c) attributed this model to Proclus (412–485 CE), a Greek philosopher who offered a commentary on the first book of *Euclid's Elements* (see Morrow, 1970).

The claim is that when Jefferson was writing the Declaration of Independence, he *consciously* adopted Euclid's approach to proof—as that could be ascertained from the concept of proof celebrated in *Euclid's Elements*. He wanted the *structure* of Euclid's proofs to be embodied in the Declaration.

Abraham Lincoln applies Euclidean structure. When preparing key speeches Van Haften and Hirsh (2018) have also claimed that the "Enunciation → Exposition → Specification → Construction → Proof → Conclusion" sequence for logically structuring an argument was *consciously* used by Abraham Lincoln, the sixteenth President of the United States of America. It is well known that during the 1840s Lincoln, as a traveling attorney on the Eighth Judicial Circuit in Illinois, made a concentrated study of *Euclid's Elements*, memorizing texts of proofs while on horseback or in a carriage between towns, or in hotel rooms in the evenings (see, e.g., Ketcham, 1901). Hirsch and Van Haften (2019c) have argued that, like Jefferson, Lincoln was determined to apply the ways Euclid structured geometrical proofs in his public addresses when he was trying to convince audiences that his ways of thinking were more reasonable than those of his opponents. Hirsh and Van Haften have carefully analyzed the texts of famous Lincoln addresses (such as the Gettysburg address) and have attempted to show that he followed the "Enunciation → Exposition → Specification → Construction → Proof → Conclusion" sequence.

Hirsch and Van Haften were not the first authors to draw attention to Lincoln's fascination with Euclidean geometry and his attempts to make use of Euclidean logical structure when planning important public addresses (see, e.g., Levin & Levin, 2010). Readers are invited to reflect on that claim as they read the text of the sixteenth President's most famous public address, the Gettysburg Address—delivered at Gettysburg, in Pennsylvania, on November 19th, 1863:

> Four score and seven years ago our fathers brought forth on this continent, a new nation, conceived in Liberty and dedicated to the proposition that all men are created equal.
>
> Now we are engaged in a great civil war, testing whether that nation, or any nation so conceived and so dedicated, can long endure. We are met on a great battlefield of that war. We have come to dedicate a portion of that field, as a final resting place for those who here gave their lives that that nation might live. It is altogether fitting and proper that we should do this.
>
> But, in a larger sense, we cannot dedicate—we cannot consecrate—we cannot hallow—this ground. The brave men, living and dead, who struggled here, have consecrated it, far above our poor power to add or detract. The world will little note, nor long remember what we say here, but it can never forget what they did here. It is for us the living, rather, to be dedicated here to the unfinished work which they who fought here have thus far so nobly advanced. It is rather for us to be here dedicated to the great task remaining before us—that from these honored dead we take increased devotion to that cause for which they gave the last full measure of devotion—that we here highly resolve that these dead shall not have died in vain—that this nation, under God, shall have a new birth of freedom—and that government of the people, by the people, for the people, shall not perish from the earth.

The speech comprised only 10 sentences and 272 words, but it struck a chord that would resonate across time. One should ask—Why was this short speech so powerful (Roberts, 2019)?

Jefferson and the introduction of a system of decimal currency. In the United States during the period 1783–1785, immediately after the Revolutionary War, there was a strong national feeling that, from the outset, the nation's political structures, and its schools and colleges, should take full advantage of the opportunity and challenge to create a unique and model democracy (see Peden, 1955). In particular, a major question which needed to be resolved quickly, was: "What should be the system of currency, and the associated coinage, for the new nation?" The two principal figures in the debate on this issue became Robert Morris—who, since 1781 had held the post of Superintendent of Finance for the Continental Congress (Bordley, 1789; Clements & Ellerton, 2015; Frost, 1846; Hepburn, 1915; Linklater,

2003; Morris, 1782; Rappleye, 2010; Seaman, 1902)—and the former Governor of Virginia, Thomas Jefferson (1784b).

Morris, a Philadelphia merchant, land speculator, and slave owner who had a large reputation for economic wisdom, had shown leadership with respect to financial matters during the Revolutionary War (Frost, 1846; see McCusker, 1992). But Jefferson, as the framer of the Declaration of Independence, was also widely respected. Both Morris and Jefferson favored decimal systems of currency. At that time the only two decimalized currencies in the world (McCusker, 1992)—and they were only partly decimalized—were in Russia and Japan. In 1704, in Russia, Peter the Great had created a *rouble* (or *ruble*) equal to 100 *kopeks* (Brekke, 1977; Fenzi, 1905); and in Japan—where silver money was basically money by weight—1000 *mommes* were equal to 1 *kan* (Nishikawa, 1987).

The fundamental unit within Morris's proposed system would have had much less value than the Spanish dollar, which Jefferson put forward as the basis for the fundamental unit in his system. Jefferson proposed that decimalization should be formulated around a dollar roughly equal in value to the Spanish dollar because that had been much used in the United States during the period 1775–1784 (Goodwin, 1953).

Jefferson's (1784c) two-page document, "Some Thoughts on a Coinage," is printed in Volume 7 of Julian P. Boyd's (1953), *The Papers of Thomas Jefferson* (Vol. 7 March 1784 to February 1785), published by Princeton University Press. Like Boyd (1950a, b, 1961), we believe that *before* he went to France in 1784, Jefferson had already worked out a comprehensive decimalized system of weights and measures, and that during his five years in France (1784–1789) he probably taught the French thinkers on weights and measures more about the possibilities of a coordinated system of weights and measures than they taught him. Jefferson's original handwritten pages were undated, but Boyd (1953) indicated that they were prepared around March 1784—that is to say, *before* Jefferson left for Paris, in July 1784, to serve as Minister Plenipotentiary to France. According to Boyd, "Some Thoughts on a Coinage" was written entirely in Jefferson's hand, and was "erroneously placed with the rough draft and notes of Jefferson's report to the House of Representatives of a plan for establishing uniformity in currency, weights and measures" dated July 4, 1790 (p. 175). The document is in the Library of Congress, Washington, DC, Thomas Jefferson Papers, 233, 41972.

"Some Thoughts on a Coinage" needs to be distinguished from the better-known and much longer document "Notes on Coinage," which was written between March and May 1784 (and is printed in Boyd, 1953, pages 175–185). With respect to "Some Thoughts on a Coinage," Boyd (1953) wrote:

> It is now known that Jefferson considered at this time a comprehensive plan for the decimalization of weights and measures as well as money. [Jefferson (1984c)], never before published, shows that Jefferson's "Some

> Thoughts on a Coinage" was in reality an outline of "Notes on Coinage." It was probably drawn up as early as March, or even February 1784. At that time Jefferson must have intended to advocate the dollar as the money unit as well as a decimalized coinage, and once these points were established, to make a transition to a decimal reckoning in weights, measures, and perhaps time. But he must have concluded that the country was not ready for such a thorough-going innovation and that the latter parts of his program could be more readily accomplished after the coinage had been settled. In this sense, then, Jefferson's "Notes on Coinage" must be regarded merely as the preliminary expression of a plan to which he returned six years later, immediately on assuming office as Secretary of State. (pp. 155–156)

Clements and Ellerton (2015) reprinted Jefferson's (1784b) "Some Thoughts on a Coinage" as Appendix *A* in their book *Thomas Jefferson and his Decimals 1775–1810: Neglected Years in the History of U.S. School Mathematics*, and we now reprint the same document below. Jefferson's misspellings and idiosyncratic use of upper-case first letters have been retained. For the original document, *see* Jefferson, 1784a.

III. Some Thoughts on a Coinage [ca. March 1784]

Some Thoughts on a Coinage and the Money Unit for the U.S.

1. The size of the Unit.
2. It's (*sic.*) division.
3. It's (*sic.*) accommodation to known coins.

The value of a fine silver in the Unit.
The proportion between the value of gold and silver.

The alloy of both 1. oz in the pound. This is Brit. standard of gold, and Fr[ench]
Ecu of silver.

The Financier's [i.e., Robert Morris's] plan.
A table of the value of every coin in Units.

Transition from money to weights.

10 Units to the American pound. 3650 grs = 152 dwt. 2 grs. = 7 oz. 12 dwt. 2 grs

Transition from weights to measures.
Rain water weighing a pound, i.e. 10 Units, to be put in a cubic vessel and one side of that taken for the standard or Unit of measure.

Note. By introducing pure water and pure silver, we check errors of calculation proceeding from heterogeneous mixtures with either.

Transition for measures to time.

I find new dollars of 1774, 80, 81 (qu. Mexico Pillar) weigh 18 dwt. 9 grs. = 441 grs. If of this there be but 365 grs. Pure silver, the alloy would be 2.1 oz. in the lb. instead of 19 dwt. The common Spanish alloy, which is 1 dwt. worse than the Eng. Standard. Whereas if it is of 19 dwt in the lb. troy, it will contain 406 grs. Pure silver. The Seville peice (*sic.*) of eight weighing 17½ dwt. By Sir I. Newt's assay contained 387. grs. Pure silver. The Mexico peice (*sic.*) of 8 [weighing] 17 dwt. 10 grams. (alloy 18 dwt. as the former) 385½. The Pillar peice (*sic.*) of 8 [weighing] 17–9 (alloy 18 dwt.) 385¾. The old ecu of France or peice (*sic.*) of 6, gold Tournois is exactly of the weight and fineness of the Seville peice (*sic.*) of 8. The new ecu is by law 1. oz. alloy, but in fact only 19½ dwt. 19 dwt. 14½ grs. pure metal is 432¼ grs.

Dollars	Weight	In water	Loss
*1773	17—8½	15—15	1—17½
*1774	17—8	15—14	1—18
*1775	17—8½	15—15½	1—17
1776			
1777			
*1778	17—9½	15—16	1—17½
*1779	17—9½	15—16	1—17½
*1780	17—10	15—17½	1—16½
*1781	17—8½	15—15½	1—17
1782			

*These average 417. grs. weight in air 41.3 grs. loss in water. i.e. 1/10 or nearly 1/1000 or ten times the weight of water.
Cassini makes a degree in a great mile contain

Miles	D
69	864 = 365,184 feet

Then a geographical mile will be 6086.08 feet.

a Statute mile is 5280 f.

A pendulum vibrating seconds is by Sr. I. Newton 39.2 inches = 3.2666 &c. feet

Then a geographical mile of 6086.1 f. = 1863 second pendulums.

Divide the geometrical mile into 10. furlongs

each furlong 10 chains

each chain 10 paces

Then the American mile		= 6086.4 f.	English = 5280 f.
furlong	= 608.64 f.		= 660
chain	= 60.864		= 66
pace	= 6.0864	fathom	= 6.
Russian mile			.750 of a geographical mile
English mile			.8675
Italian mile			1.
Scotch and Irish do.		1.5	
old league of France		1.5	
small league of do.			2.

great league of do.	3.
Polish mile	3.
German mile	4.
Swedish mile	5.
Danish mile	5.
Hungarian mile	6.

A rod vibrating seconds is nearly 58½ inches.

Comments on "Some Thoughts on a Coinage." "Some Thoughts on a Coinage," which should be read in conjunction with Julian Boyd's (1953) editorial notes (pp. 150–160 and pp. 185–188), is a historically remarkable document which reveals Jefferson's struggles to achieve a decimalized scientific and mathematical link between coinage and weights and measures. Reading the document makes it obvious that *before he went to Paris,* Jefferson was extremely well read with respect to the key issues, and that he thought a totally integrated system of weights and measures and money was what the newly-created United States of America should strive to achieve (Clements & Ellerton, 2015). It is also obvious that in "Some Thoughts on a Coinage" he was not merely reiterating what someone else had advised him. He had taken on the challenge of devising a coordinated system which fitted the demands of science and mathematics, yet at the same time would be convenient in day-by-day practices. Within a decade, (Ellerton & Clements, 2019) the same task would be handed to a select group of France's top mathematicians and scientists by the French Revolutionary Government, and the "metric system" would be the outcome. But at the beginning of 1784 Jefferson devised and responded to the same challenge all by himself.

From the perspective of the present book, Jefferson was someone who had enthusiastically studied mathematics and science at a university and was brave enough to try to employ his mathematical knowledge and talents to create a fully decimalized monetary system for his nation. Furthermore, he wanted to link this new system with an entirely new decimalized system of weights and measures.

With time, Jefferson's introduction of the decimalized dollar would change the world's thinking about forms of currency. In the early 1790s, post-Revolutionary France would change the world's thinking about the measurement of weights and measures by its introduction of the metric system. However, like Christopher Wren and Bishop John Wilkins in the 1660s in Great Britain (Wilkins, 1668), Jefferson has not been given the credit he deserves for his anticipation of that system—which, ironically, has never been fully accepted by his own country.

Jefferson did not introduce his proposals on coinage and weights and measures from the vantage point of a practicing scientist or mathematician, but rather from that of a public servant. He earnestly believed that if his proposals were to be implemented they would lead to a marked improvement in the everyday arithmetic performances of U.S. citizens (see Boyd, 1953; Clements & Ellerton. 2015, pp. 68–70; Ellerton & Clements, 2019; Fauvel, 1999; Lee, 1797; Honeywell, 1931; Jefferson, 1784a).

A Textbook Which Conquered "the Horn"

The title of this book, *Toward Mathematics for All: Reinterpreting History of Mathematics in North America 1607–1865*, was always meant to suggest that we had in mind national progress toward "mathematics for all." An important theme would be how mathematics in North America gradually reached beyond a "mathematics-for-the-minority" mentality, away from the position that the study of theories and applications of mathematics was relevant only to mainly high-ability, white European-background males from well-to-do families. We wanted to go beyond the history of "groundbreaking research" in mathematics in North America, during the period 1607–1865.

Some progress toward "mathematics for all" did take place during the period 1607–1865. About 20 percent of the 549 cyphering books in the Ellerton-Clements collection of North American cyphering books were prepared by females (Ellerton & Clements, 2021). However, not much progress was made during the period so far as making formal mathematics, beyond elementary *abbaco* arithmetic, available to Native Americans, to African-American slaves, or to white, European-background indentured servants (Ellerton & Clements, 2021).

The task of making mathematics available to everyone who lived within the eastern colonies and states was not an easy one, but the problem was exacerbated for those who lived in, or ventured to, the gold-rich western states of North America during the period 1840–1860.

That fact is well illustrated by handwritten inscriptions on two front endpapers of an old arithmetic textbook authored by Charles H. Mattoon, titled *Common Arithmetic Upon the Analytic Method of Instruction*. This 386-page book, originally published in 1850, is now part of the Ellerton-Clements textbook collection and was originally published in Medary, Ohio. On its title page it is stated that the book was "designed for the use of schools."

There are two handwritten inscriptions on the front endpapers (see Figure 8.8) indicating that there is something special about this book. The first was:

> This book was brought by clipper ship around the Horn to Oregon Territory about 1852. It belonged to Marrietta Walker who was one of the first schoolteachers in the Oregon Territory. Marietta Walker was related to Rachel and Robert Walker who settled in Beaverton [Oregon] in 1850—They established the "Old Meadow Farm."

The other inscription stated "Love to Katie, Jan 23, 1973. To remember the good times we all had as children at the Old Meadow Farm. Dona"

The first inscription offers a reminder of how difficult it was for pioneer parents to make sure that their children obtained a satisfactory education. In the West there were very few qualified teachers prepared to do anything other than dig for gold, and very few textbooks. This textbook, by Mattoon, had to survive a lengthy treacherous journey before it would serve the purpose for which it was written. In his preface, Mattoon wrote that "the work was designed for *use*, and not merely for *show*. It was

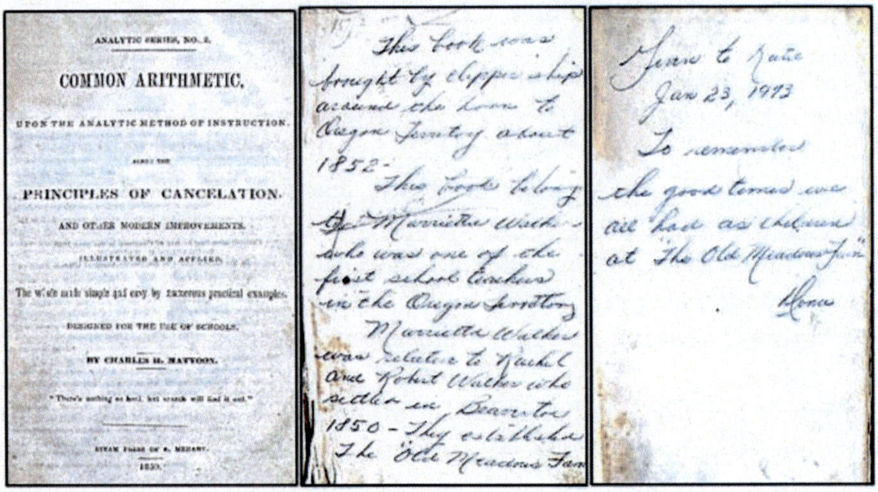

Figure 8.8. An arithmetic textbook (Mattoon, 1850), with inscriptions, which survived the arduous trip around the Horn, to Oregon Territory.

prepared by the author while he was engaged as a teacher, and was the result of arduous labor, and hard study, aided by practical observation and experience in the school room" (p. 9). Hopefully, pupils at the Old Meadow farm who used the textbook appreciated the effort required to bring to them the knowledge which it made available.

William Cook Pease Seeks "to Make Himself a Better Man"

William Cooke Pease (1819–1865) was a U.S. Coast Guard pioneer (Kern, 1982). He was born on Martha's Vineyard Island in 1819 and, in the 1850s, would become the youngest-ever captain in the Revenue Cutters service of the Army (Educational Unit, 1966). He took two revenue cutters around Cape Horn during the Gold Rush days. He maintained a diary for many years, and that greatly assisted Florence Kern (1982) to write a biography of him (see Figure 8.9 and Figure 8.10). The Ellerton-Clements cyphering book collection includes a cyphering book which Pease prepared in 1844 and 1845, when he was aged 25 and 26 years of age (in Figure 8.11 a page from the cyphering book is shown). Careful checking has revealed that entries in his cyphering book were based on Nathaniel Bowditch's *The New American Practical Navigator*, and were made during a period in 1844 when Pease was serving on the cutter *Van Buren,* which was having a new mast erected on it when it was harbored in Charleston, South Carolina (Kern, 1982).

From the perspective of this chapter, in this book, Pease's cyphering book attests to an ambitious young man who was already married and was wishing to make himself "a better man" (quoted from Pease's diary in Kern, 1982, p. 11). Pease *voluntarily* prepared his cyphering book—he was not attending a formal education institution or preparing for an examination. He *wanted to know more mathematics*, because he believed that that would help him to become a better person by improving his performance in the career that he had chosen for himself. Many of the cyphering books in the Ellerton-Clements were prepared by young adults like Pease, and like Gertrude Deyo (see later in this chapter), who *chose* to study mathematics *by themselves*. We believe that stories like theirs should be recognized, and reported, as part of the history of U.S. mathematics during the period 1607–1865.

Figure 8.9. Captain William Cooke Pease, c. 1860 (Reproduced from Kern, (1982, p. ii).

Figure 8.10. The title page of Florence Kern's (1982) biography of William Cooke Pease.

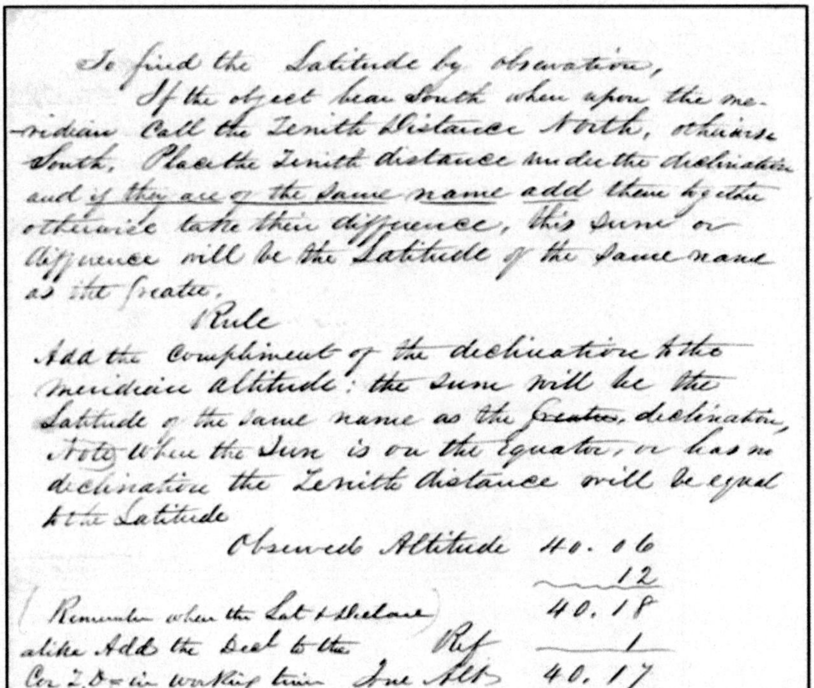
To find the Latitude by observation,
If the object bear South when upon the me-
-ridian Call the Zenith Distance North, otherwise
South, Place the Zenith distance under the declination
and if they are of the same name add them together
otherwise take their difference, this Sum or
difference will be the Latitude of the same name
as the greater,

Rule

Add the Compliment of the declination to the
meridian altitude: the sum will be the
Latitude of the same name as the greater declination,
Note When the Sun is on the Equator, or has no
declination the Zenith distance will be equal
to the Latitude

Observed Altitude	40.06
	12
	40.18
Ref	1
True Alt	40.17
Ze dist	90.00
True Z D	49 43 N
Suns Decl	23.28 N
Latitude	73.11 N

(Remember when the Lat & Declin
alike Add the Decl to the
Cor Z.D & in working time
they are Subtracted)

In getting this Lat the Eye is 26 feet above
the horizon,

Alt	40.20
Semidiam	16
	40.36
Dip 5. Ref 1	6
	40.30
	90.00
	49.30 N
	9.56 N
Lat	59.26

Remember Dip & Ref always Subtractive,

Figure 8.11. An early page of a 60-page, lightly pre-lined cyphering manuscript (dimensions 13" by 8") which was prepared by William C. Pease of Edgartown, Martha's Vineyard, Massachusetts. On another page it is stated: "W. C, Pease, U. S. Rev Schooner, Van Buren, Charleston, S. C., 23rd March 1844. Bought in Charleston, S. C.".

In the 18th and early-19th centuries, cyphering books were often prepared by midshipmen during voyages to distant locations—like, for example, when ships sailed between Western European nations and the East Indies, or between North America and the East Indies (Durkin, 1942; Ellerton & Clements, 2012; Rawley, 1981; Taylor, 1966). The Phillips Library, in Salem, Massachusetts, for example, holds about 100 cyphering books which were prepared by midshipmen employed on merchant ventures to Africa, Europe and the East Indies (Gaydos & Kampas, 2010; Rawley, 1981). Some of these were arithmetic cyphering books, and others were navigation cyphering books (Hertel, 2016). It would be wrong to think of these as having been prepared under duress. Some of the manuscripts feature penmanship and calligraphy of the highest order (see, e.g., Ellerton and Clements, 2014, Chapters 9 and 10). In Great Britain, William Beattie, King William IV's physician and private secretary, personally spent much time preparing a beautiful navigation cyphering book (based on John Hamilton Moore's (1796) *The New Practical Navigator*) as a gift for his friend, William IV (King of England), who loved the sea so much that he was known as the "Sailor-King" (Ellerton & Clements, 2014, p. 300). That cyphering book, and the copy of Moore's textbook used by Beattie, are now part of the Ellerton-Clements collections.

A Young Huguenot Woman's Incomplete Cyphering Book

Our vision of "mathematics for all" includes mathematics for young adults, male or female, who were not studying in formal educational institutions. In 1844, a seriously ill, 26-year-old, pregnant, woman, Gertrude Bogardus Deyo, who was living in, or near, the Huguenot settlement of New Paltz, in the state of New York, prepared an authentic 150-page cyphering book. Figure 8.12 shows a photograph of

Figure 8.12. Portrait of Gertrude Bogardus Deyo, shortly before her death in 1844. Her portrait hangs in the Deyo house in Huguenot Street, New Paltz.

an oil portrait of Gertrude (the artist is not known) which now hangs on the wall of the main room of the Deyo House in Huguenot Street, New Paltz. Gertrude died, possibly of tuberculosis, before she had completed her cyphering book, and the question arises: why did a young woman like Gertrude put so much effort into preparing her cyphering book at such a challenging time in her life? Figure 8.13 shows a page from Gertrude's cyphering book. Interested readers may like to explore online commentary on Gertrude's life and legacy.

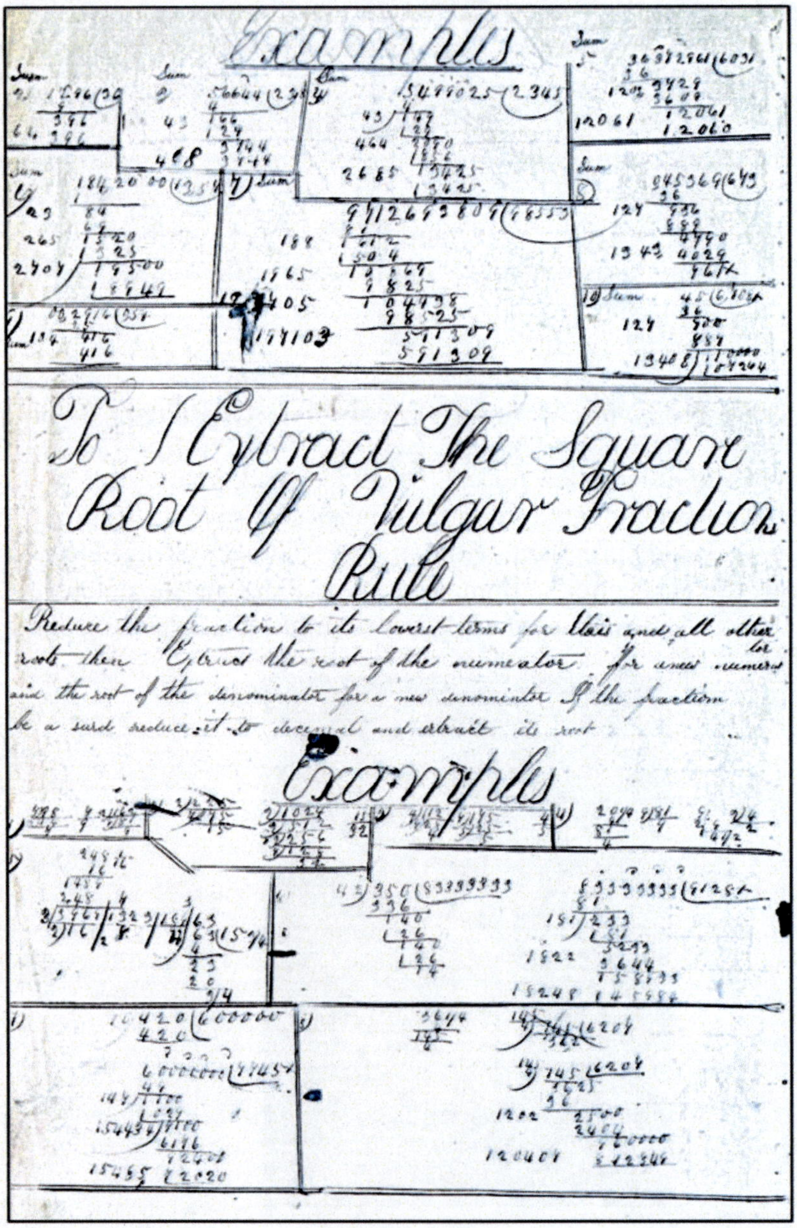

Figure 8.13. A page in Gertrude Bogardus Deyo's (1844) cyphering book.

The Cases of David Rittenhouse (1732–1796), Benjamin Banneker (1731–1806) and Robert Adrain (1775–1843)

The cases of William Pease and Gertrude Deyo (outlined above) concerned two young adults who prepared cyphering books outside of formal education institutions. This was a common phenomenon, and we now discuss three more cases, also involving adults, who became seriously engaged with mathematics outside of schools or colleges. The three cases—involving David Rittenhouse, Benjamin Banneker, and Robert Adrain have drawn much more comment in histories of U.S. mathematics than the cases of William Cooke Pease and Gertrude Deyo. Rittenhouse, Banneker, and Adrain were largely self-educated persons from poor families who, as a result of determination, love of mathematics, and patronage from influential persons were able to create mathematics of sufficiently high quality that their work is mentioned in almost all histories of U.S. mathematics (see e.g., Zitarelli, 2019).

David Rittenhouse

Zitarelli (2019) believed that Rittenhouse was the U.S.'s "most accomplished colonial mathematician" (p. 74). He was one of only two Americans elected to the Royal Society between 1776 and 1800 (Zitarelli, 2019, p. 74), and had several papers published in the prestigious *Transactions of the American Philosophical Society*. Between 1791 and 1796 he was President of the American Philosophical Society. When telling of Rittenhouse's early years on a farm in Pennsylvania, Zitarelli mentioned that he may have "attended school at a Presbyterian church located near the farm" (p. 67), but emphasized that although he "was fascinated with mathematics from his early years he had little opportunity for schooling, and was largely self-taught from books on elementary arithmetic and geometry and a box of tools inherited from an uncle David Williams" (p. 67). The interested reader is referred to Zitarelli's (2019, pp. 67–73) excellent summary of Rittenhouse's impressive contributions to astronomy and surveying as well as to pure mathematics, especially integral calculus.

One wonders how Rittenhouse managed to achieve what he did in mathematics. He certainly worked hard to become a very skillful builder of mathematically-inspired artifacts, and that probably helped him to "see" connections which might be made in mathematics. Later in his life he became associated with influential figures in U.S. mathematics, especially with Robert Patterson, a mathematics professor at the University of Pennsylvania (see Patterson, 1819). He loved mathematics and worked hard to advance it. He deliberately set out to identify unsolved problems in mathematics, and then worked hard to solve them. He created his own opportunities.

Benjamin Banneker

Banneker (1731–1806) was another self-educated person of the colonial era who contributed to the development of mathematics in North America. He was a largely self-educated clockmaker, astronomer, compiler of almanacs and author. It is usually claimed that he was the grandson of Molly Welsh, an English woman who was found guilty of theft and transported to North America as an indentured servant (Wadsworth, 2005). She married an African-American slave, and Benjamin's mother was one of their children (Wadsworth, 2005). Benjamin's youth was spent working on a small farm near Baltimore, Maryland, which his parents had purchased. He initially showed flair as a clockmaker, but after his grandmother, Molly, taught him to read and write he became interested in science and mathematics (Wadsworth, 2005).

In the 1780s, he studied mathematics from books which he had borrowed. In his private world he conducted astronomical studies, the results of which became the basis for an almanac which he prepared for possible publication. Although when young, he had attended school, the time he could spend on formal study was restricted after his father died and he was the only remaining male in his family. With the help of several prominent citizens—Thomas Jefferson, Andrew Ellicott, and David Rittenhouse, to name just three—he was able to gain employment as a member of a team charged with the responsibility of surveying the recently created District Capital (Washington). But his most creative work was in the field of astronomy.

From a social vantage point, there can be little doubt that Banneker's successes in the field of science, surveying, astronomy, and mathematics provided a tentative answer to Thomas Jefferson's question whether African-Americans could be as mentally endowed as white European-background persons (Wadsworth, 2005; Zitarelli, 2019). The answer was "Yes," and the challenge to the fledgling United States of America was to create an education system which provided equitable forms of mathematics education for *all* its citizens.

Robert Adrain

Another of the North American mathematicians in the colonial and Federalist periods to be largely self-educated was Robert Adrain who, according to Zitarelli (2019), was "the first creative mathematician in the United States" (p. 122). Adrain (1775–1843), originally from Belfast in Ireland, was the son of a teacher which meant that he received the benefit of an early school education. However, both his parents died when he was 15, and in 1798, at the age of 23, he chose to emigrate to New York. He must have had recognizable leadership abilities for he was appointed headmaster of three schools—Princeton Academy in New Jersey, then York Academy in Philadelphia, and then a school in Reading, Pennsylvania.

Adrain gained a reputation for being able to solve challenging mathematics problems, and after he had authored two articles which appeared in the *Mathematical Correspondent* it was clear that he had entered the world of mathematics. One of his early articles was on Diophantine algebra and it has been claimed that he was the first person to introduce Diophantine analysis to the United States (Cajori, 1890, p. 94). In 1807 he edited a volume of the *Mathematical Correspondent* and, in 1808, following a decision to cease publication of that journal, he created and began editing his own journal the *Analyst or Mathematical Museum*. His contributions to the new journal revealed that he had an impressive knowledge of the works of many leading European mathematicians. He included curious problems in the only four issues of the *Mathematical Museum* which appeared, and wrote higher-level articles for the journal (such as "Research Concerning the Probabilities of the Errors which Happen in Making Observations"—in that article he introduced to America the least squares method, which had originated in the work of Carl Friedrich Gauss, Adrien-Marie Legendre, and Pierre-Simon Laplace.

Adrain's leadership in the mathematics world, and his obviously strong knowledge of contemporary European research in mathematics, raises the question: How did he do it, considering the mathematics he knew was largely self-taught and he had not studied mathematics at a university? In his prefaces and articles in his *Mathematical Museum* he showed that he had devoured much of the writings on mathematics by Europe's leading mathematicians of his time. It was obvious that mathematics was his passion. According to Zitarelli (2019):

> The fact that an unknown person working in a virtual American mathematical wasteland can also be mentioned [in the same breath as Gauss, Legendre, and Laplace] . . . is also impressive, though admittedly Adrain's accomplishments "do not put him in the first rank of nineteenth-century mathematicians." A recent account of Adrain described the necessity of a *community* by asserting, "It takes more than a village to raise a scientist. It takes a village full of scientists." (p. 124)

But during the Federal period there *was* one U.S. citizen who did come to deserve to be listed among the first rank of nineteenth-century mathematicians—that citizen was Nathaniel Bowditch, whose life and achievements we now summarize.

The Remarkable Mathematical Contributions of Nathaniel Bowditch

Nathaniel Bowditch's Life—A Summary

Like Franklin, Rittenhouse, and Banneker, Nathaniel Bowditch (1773–1838) was largely self-educated. He was born in Salem, Massachusetts, the son of Habakkuk, a former sea captain who had become an alcoholic and, falling on hard times, became a cooper (i.e., a maker and repairer of casks and barrels). Nathaniel was the fourth of seven children. When he was just 10 years old his mother died and, soon after that, he left school. Initially he worked with his father as a cooper, but in 1785 he was indentured, for nine years, as a bookkeeping apprentice, and his main task was to look after a maritime-related shop. During this time, he managed to gain access to mathematics books written by English and French authors and in the evenings he taught himself algebra, then calculus, then Latin, then French.

In 1795, having completed his apprenticeship, he went to sea as a ship's clerk. He would make five trips to India, initially as a clerk, then as a supercargo, and finally as a captain, and part-owner. In 1798 he married Elizabeth Boardman, but she died soon after, while Nathaniel was at sea. Nathaniel showed precocity with mathematics and languages and, having taught himself to read and write Latin and French, he began to read Isaac Newton's *Principia,* John Hamilton Moore's, *The New Practical Navigator,* and Pierre-Simon Laplace's *Mécanique Céleste*. His brilliant mathematical mind was recognized and, remarkably, in 1806 he was offered the prestigious Hollis Chair in Mathematics and Natural Philosophy at Harvard College—quite an honor for someone who had left school when 10 (Roberts, 2019).

Bowditch turned down the Harvard offer in order to continue his work as President of the Essex Fire and Marine Insurance Company (in Salem). In 1823 he moved from Salem to Boston to become actuary of the Massachusetts Hospital Life Insurance Company. It could be argued that he was North America's first outstanding actuary, and he continued to work as an actuary until his death—declining to accept offers from the University of Virginia and the United States Military Academy (at West Point) to be professor of mathematics. Arguably, it was he who defined the role of a professional actuarial scientist in North America. He presided over a number of very successful companies and was inducted into numerous scientific and learned societies across the United States and Europe. His personal library (donated to the Boston Public library after his death) was larger than Thomas Jefferson's—which had been used to establish the Library of Congress (Thornton, 2016).

Although the argument has hardened into received tradition that the American mathematical research community effectively began with the appointment of the Englishman, J. J. Sylvester to Johns Hopkins University in 1876 (see, e.g., Parshall, 2003; Sylvester, 1870), we think the honor rightfully belongs to Nathaniel Bowditch. Each of Morris Kline (1972), Karen Hunger Parshall (2003) and David Zitarelli (2019) recognized Nathaniel Bowditch as an outstanding mathematician, but fell short of acknowledging Bowditch's efforts in laying a foundation for a strong U.S. mathematical research community. However, Bowditch's contemporaries recognized his greatness, and in 1827 they elected him President of the American Academy of Arts and Sciences, a position he held for the rest of his life.

During his long voyages to India in the late 1790s and early 1800s Bowditch identified and corrected many errors in John Hamilton Moore's *The New Practical Navigator*, to the point where Edmund Blunt the well-known Massachusetts publisher of Moore's text, decided to publish Bowditch's revised version of it (which

Blunt titled *The New American Practical Navigator*), and to name Bowditch as the sole author—see Karpinski, 1980, p. 142). Subsequently, "The Bowditch" would become the most widely used navigation guide in the world and was issued to all new U.S. Navy trainees until the 1960s. In 1800, Nathaniel married his cousin, Mary Ingersoll, and they had six sons and two daughters (Thornton, 2016).

Tamara Thornton's (2016) biography of Bowditch (see Figure 8.14) captures the feel of Nathaniel's early life within the major shipping port of Salem, during and after the Revolutionary War. By the time Bowditch was 30 he was regarded, in Salem and in Boston, as the foremost academic scholar in the United States and also as one of nation's most successful businessmen. He was inducted into numerous scientific and learned societies throughout the United States and Europe. The extent of the recognition of the quality of Bowditch's mathematical prowess might be best judged by Florian Cajori's (1890) summaries of the astonished reactions of leading European mathematicians and scientists to Bowditch's extensive review of Pierre-Simon Laplace's notoriously complex five-volume *Mécanique Céleste*. According to Cajori, (1890, p. 105), in 1836 Sylvestre François Lacroix wrote, in a letter to Bowditch, "I am more and more astonished at a task so laborious and extensive." In 1832, Adrien-Marie Legendre informed Bowditch that his work on *Mécanique Céleste* was "not merely a translation with a commentary" but "a new edition" (quoted in Cajori, 1890, p. 105). And, from England, Charles Babbage recognized that Bowditch had begun the task of getting the United States to be seen as capable of contributing seriously to international science and mathematics. Babbage wrote: "It is a proud circumstance for America, that she has preceded her parent country in such an undertaking" (quoted in Cajori, 1890, p. 105).

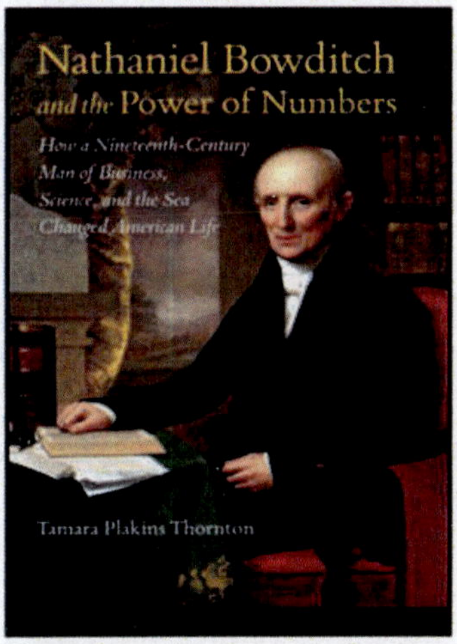

Figure 8.14. Dust jacket of Tamara Thornton's (2016) biography of Nathaniel Bowditch.

Cajori (1890) still held back a little in his assessment of the quality of Bowditch's research. After asking the question: "How should Bowditch be ranked as a mathematician?" he acknowledged that in his time Bowditch stood at the head of the scientific men of his country, and "contributed more to his country's reputation than any other contemporary scientist" (p. 90). He then added "although Bowditch was the first in line of American mathematicians, his work was not original enough to be at the same level as Newton or Gauss or Laplace." David Zitarelli (2019) did not regard Bowditch as a "legitimate mathematician" (pp. 3–4). Although one can understand how such assessments might be justified, there can be no doubt that in the early 1800s no-one else in North America could have achieved what Bowditch achieved mathematically. And it should be remembered that whereas Newton had attended the University of Cambridge in England, Bowditch had left school at the age of 10, and after working as a cooper had been an apprentice maintaining a maritime shop in Salem. Note, though, that Zitarelli (2019) did acknowledge that "Nathaniel Bowditch contributed to almost every development that occurred in American mathematics over the first third of the nineteenth century" (p. 111). But Bowditch's contributions in two very practical areas of life—as a supercargo, and as the author of a major guidebook on navigation—are worthy of special comment (Roberts, 2019).

Nathaniel Bowditch's Work as a Supercargo

Many purchases made by residents of the 13 North American colonies during the seventeenth and eighteenth centuries did not involve the use of paper money or coins, because *bartering* for goods was commonplace (Hadden, 1994). In order to overcome the dilemma experienced by someone who wanted to obtain a commodity (for example, a bushel of oats) but did not have a form of currency the seller wanted to exchange for that commodity, the practice of using popular products (especially tobacco) as a kind of currency was adopted. Successful bartering often demanded the carrying out of tricky arithmetical calculations mentally, and that probably explains why the topic "barter" was part of the *abbaco* curriculum in schools—although it was sufficiently difficult that relatively few students got to study it formally.

Ship owners who traded in far-off locations (like France, or India, or China, or the Spice Islands) employed "supercargoes" whose specific on-board task was to make sure that trading for goods was done profitably (Bartlett, 1933). The quality of the work of a supercargo was inevitably measured by how much profit was made on a voyage. Nathaniel Bowditch was, simultaneously, one of the greatest-ever supercargoes and one of the world's greatest navigators.

The work of a supercargo demanded a high degree of what is now called "number sense." Imagine what was required—a trading ship berths in a port on an island somewhere in the Indian Ocean where the local language is not English, or even a major European language. The local currency could be unlike any other, with ratios between the local coins (if such objects existed) and the American dollar fluctuating greatly. The units for measuring weights, and relative values of different objects, could be totally unique to the island. The supercargo's task of trading profitably was linguistically, socially, and above all, arithmetically challenging, but the success of the captain and his crew would be measured almost entirely in terms of profits made. Furthermore, the supercargo had to anticipate the most tradeable goods readily available on the island, and to work out what goods he would be able to "get rid of" (i.e., to trade) by bartering with local traders on the island. All of that would need to be done very quickly, because "time represented money."

Nathaniel Bowditch (1797), prepared a handwritten journal (Log 1301, Ship *Astrea*, now held in the Phillips Library, Peabody Essex Museum, Salem, MA) summarizing his trades, and the algorithms and principles he used when, as a supercargo, he bartered during a voyage from Salem, Massachusetts to Manila between March 1796 and May 1797. *Astrea* was owned by the notoriously demanding Elias Hasket Derby, a Salem resident who was one of the world's wealthiest men. Bowditch's handwritten record of the trades he made, and the details he recorded in relation to local currencies, indigenous counting systems, and local measures of weights, was, almost literally, "worth its weight in gold" for rival ship captains and traders. Remarkably, during its voyage the *Astrea* traveled from Salem, to Lisbon (Portugal), Madeira, Saint Paul Island, *Terres Australes et Antarctiques Françaises*, Java (Indonesia), Banca (on the Bangka Island, Indonesia); *Pato Sapata* (probably, based on coordinates, an island in the South China Sea), Manila (in The Philippines), the Isle of France (i.e. Mauritius), the Cape of Good Hope (South Africa), Saint Helena, Ascension Island, the Bermuda Islands, and finally back to Salem.

"The Bowditch"

However impressive Nathaniel Bowditch's academic achievements may have been there can be little doubt that he is best known for his book *The New American Practical Navigator*, which was first published by Edward Blunt in Newburyport, Massachusetts, in 1802. *Wikipedia's* May 2020 matter-of-fact description of the book was as follows:

> *The American Practical Navigator,* originally written by Nathaniel Bowditch, is an encyclopedia of navigation. It serves as a valuable handbook on oceanography and meteorology and contains useful tables and a maritime glossary. In 1867 the copyright and lithography were bought by the United States Hydrographic Office of the United States Navy. As of 2019 it is still published by the U.S. Government and is available free online from the National Geospatial-Intelligence Agency, the modern successor agency to the 19th Century Hydrographic Office.
>
> (https://en.wikipedia.org/wiki/Main_Page)

According to *Wikipedia,* the U.S. government purchased the copyright for the book in 1867 and, since then, 52 new editions have been published, with an all-digital version being released in 2017. For well over a century, *The American Practical*

Navigator has been known, simply, as "The Bowditch." Ironically, the original version of Bowditch's *The American Practical Navigator* was largely plagiarized from John Hamilton Moore's (1799) *The New Practical Navigator and Daily Assistant*—although perhaps the word "plagiarized" overstates the realities of the case, which we now consider.

John Hamilton Moore (1738–1807) was a well-known teacher of navigation. He was born in Edinburgh and, after being educated in Ireland, joined the British Royal Navy. In 1770, he established a nautical academy at Brentford, Middlesex (England) and in 1772 he published his popular epitome of navigation under the title *The New Practical Navigator and Daily Assistant*. Later, he taught navigation to "gentlemen designed for, or belonging to the sea," and with the help of his family, carried on a business, as a chart-seller and purveyor of nautical instruments. In the preface to the twelfth (1796) edition of his *Practical Navigator,* Moore claimed that he had been responsible for many progressive improvements in navigation which he had documented as his "knowledge became extended through investigation," and as his "judgement had matured with experience." On the title page and in the preface to this 1796 edition, he styled himself as "Teacher of Navigation, Hydrography and a Chart-seller to His Royal Highness the Duke of Clarence." In the 1820s the Duke (who, in 1830, became King William IV), owned a navigation cyphering book based on Moore's *Practical Navigator* (Ellerton & Clements, 2014)

It is not an exaggeration to say that *The Practical Navigator* quickly became a navigational classic with a twentieth edition of it being brought out just 26 years after the publication of the first edition. But its influence would reach well beyond Great Britain.

According to Eva Taylor (1966), a British naval historian, Moore had the flattering (if unrewarded) experience of having his book printed almost verbatim in the United States. The author named on the title page of new editions was not Moore but Nathaniel Bowditch. The first American edition of Moore's book had been published in 1799 with Moore being named as the author. The circumstances surrounding the strange event by which the authorship of new editions (from 1802 onward) was attributed to Nathaniel Bowditch are worth considering.

Without detracting from the genius of Nathaniel Bowditch—who was undoubtedly an exceptional authority on all matters navigational (Thornton, 2016)—while at the same time attempting to do justice to the memory of John Hamilton Moore, it is appropriate to consider the exaggerated claims that have sometimes been made on behalf of Bowditch, especially in relation to the unfairness of the situation as it related to Moore who had to endure claims that Bowditch corrected more than 8000 errors "found in the *Practical Navigator*" (Cotter, n.d., p. 323).

In 1800, Moore's *The New Practical Navigator*, which had first been published in 1772, was the most popular English-language guidebook on navigation, both in North America and Great Britain (Hertel, 2016). A new edition of *The New Practical Navigator* had been published in 1799, with Moore named as the sole author.

Another edition, with Moore once again named as the sole author, appeared in 1800. Then, Edward March Blunt, the Newbury-Port publisher of the American version of the work, asked Bowditch to correct and revise the book while on his fifth voyage to India. When preparing the revised American edition of the book, Nathaniel Bowditch recomputed the numbers in all of Moore's tables, and rearranged and expanded the work. The task proved to be so extensive that Bowditch agreed to claim authorship of the revised book, and to put in it "nothing that he couldn't teach his crew" (Navigation Division, Defense Mapping Agency, Hydrographic/Topographic Center, 1995, p. viii). On that fifth trip to India, it has been said that every man of the crew of 12, including the ship's cook, became competent to take and calculate lunar observations and to plot the correct positions of the ship.

By 1802, Blunt was ready to publish a third American edition of *The New Practical Navigator*. Because Nathaniel Bowditch and others had corrected so many errors in Moore's text, he named Bowditch, not Moore, as the author. Bowditch had proof-read and edited the American editions of Moore's *Practical Navigator* which had been published in 1799 and 1800. It is true that Bowditch's (1802) version was a significant improvement on Moore's original, for there had been many errors in tabulated quantities in Moore's book. Moore's table of maritime positions needed considerable attention, for at that time the use of the chronometer was fast becoming commonplace, and the longitudes of harbors, headlands and other navigational marks were now being found with accuracy hitherto impossible (Cotter, n.d.).

A large number of errors discovered by Bowditch in Moore's *Practical Navigator* could be traced to two large traverse tables, for points and degrees. According to Charles Cotter (n.d.) a random check on a single page of the degree traverse table for 14^0 (i.e., 14 degrees) in Moore's *Practical Navigator* revealed 27 discrepancies. Moore had reproduced Nevil Maskelyne's table of proportional logarithms which, although previously regarded as authoritative, also contained several errors which, like those in the traverse tables, were mostly in the decimal quantities. The tables of "log–trig" functions also contained many errors, as did the table of amplitudes, although none of the errors in the amplitude table exceeded half-a-degree. The table of "latitudes and longitudes," given in Bowditch's book was more extensive, and more accurate, than Moore's corresponding table, but Bowditch was cautious in remarking: "Notwithstanding the care taken in correcting the table, it must, from the nature of it, be in a degree erroneous, owing to the uncertainty of the observations on which it is founded" (quoted in Cotter, n.d., p. 325).

Current editions of Bowditch's *The American Practical Navigator* trace their pedigrees to the 1802 edition. However, it was undoubtedly the case that most of the paragraphs in the first American version could be linked directly to paragraphs in Moore's *The New Practical Navigator*, and so a charge of plagiarism might have been sustained. Edmund Blunt continued to publish the book until 1833; upon his retirement, his sons, Edmund and George, became owners of the copyright for the publication. Upon Edmund Blunt Senior's death in 1862 and his son Edmund

Junior's death in 1866, George Blunt (Junior) sold the copyright to the U.S. government for $25,000, and the government has published "The Bowditch" ever since.

Although it would be easy to excuse Moore for the numerous small errors in his original text, it should be recognized that any small error could be magnified when a ship's navigator calculated his ship's location. Such errors could result in shipwreck or serious loss of time. Given that Bowditch identified and corrected the errors at a time when calculations were performed by hand, it is easy to understand why Bowditch agreed with Blunt (Senior) that he should be regarded as the author of the revised versions. A case for plagiarism against Bowditch was never prepared or taken to court and, given the copyright laws of the time, and that Moore lived in England and Bowditch in North America, would probably have had little chance of succeeding in an American court of law (Solberg, 1905; Tebbel, 1972).

Bowditch's navigational abilities became the stuff of legend. At one time when his ship was in Manila Bay, a captain of a ship other than Bowditch's stated, pointing at Bowditch's ship, "there is more knowledge of navigation aboard that ship than there ever was in all the vessels that ever floated in Manila Bay" (quoted on p. 38 of State Street Trust Company, 1917).

Benjamin Peirce

Benjamin Peirce (1809–1880) was probably the greatest home-grown mathematician in the United States in the nineteenth century (see Chapter 7 for a brief summary of his life work). He was a school friend of Nathaniel Bowditch's son, and some idea of his prodigious mathematical talent can be gauged from the fact that Bowditch Senior asked him, when he was just 19 years of age, to proof-read his reviews and commentaries on Laplace's *Mécanique Céleste*. Peirce would be appointed a mathematics tutor at Harvard in 1831 and then University Professor of Mathematics and Natural Philosophy in 1833 (Hill, 1880; Matz, 1895; Peterson, 1955). Almost certainly he had the advantage of being privately tutored, when he was young, by Nathaniel Bowditch, and it would be unreasonable to think that most of the mathematics that he learned was something he had been taught at school or at Harvard. The text of *Mécanique Céleste* was far more difficult than anything he would have been asked, at that time, to study at Harvard.

Concluding Comments

As we wrote this chapter, we could not help but notice how almost all of the most impressive works in mathematics in North America during the period 1607–1865 were created by persons who lacked opportunities to be taught much mathematics at school or in colleges. Benjamin Franklin, David Rittenhouse, Nathaniel Bowditch, and Benjamin Banneker were all in that category and, to a lesser extent, so too were Robert Adrain and Benjamin Peirce. Indeed, we have often reflected on how those persons (excluding Peirce) were able to achieve so much, given the absence of opportunities each had to interact with others with extensive mathematical knowledge and keen mathematical minds. Think of Nathaniel Bowditch, for example—he left school when 10 years old and taught himself Latin and French so that he could read books which contained very high-level mathematics but were written in Latin or French.

Any decent history of mathematics in North America for the period 1607–1865 should include reflections on (a) how and why the phenomenon described in the last paragraph came to be, and (b) why the formal mathematics courses in the colleges did not produce more outstanding mathematicians than they did (Kraus, 1961). Were the intended and implemented mathematics curricula so seriously deficient in the sense that they did not inspire young persons with mathematical talent to investigate, creatively, problems in mainstream mathematics? Were most of the teachers too lacking in knowledge to inspire outstanding young minds? And how can one explain the remarkable mathematical accomplishments of Franklin, Rittenhouse, and Bowditch?

The case of Benjamin Franklin and his magic squares raises another question. Why has it taken centuries for Franklin's incredible inventiveness with respect to magic squares finally to be recognized as making a genuine contribution to mathematics (Pasles, 2007; Sesiano, 2019; Zitarelli, 2019)? Since we began to write this chapter we introduced magic squares to our 9-year-old grand-daughter, who was staying with us for three weeks. She was absolutely intrigued. We gave her, as a parting gift, a copy of Pasles' (2007) book. Soon after returning home, we received an email message from her telling us how much she loved reading Pasles' book. For *her*, magic squares are fascinating and represent important mathematics!

Anyone writing a history of mathematics in the United States of America might benefit from adopting a "mathematics-for-all" way of thinking. We need to look beyond the mathematics taught in schools and colleges, and beyond the "weight-bearing research" criterion for inclusion demanded by mathematicians who think that histories of mathematics should be mainly about the achievements of outstanding adult mathematicians who are now recognized as having contributed to the development of modern directions for research in the subject. Think of how Thomas Jefferson consciously applied Euclidean logical structure when he was developing the Declaration of Independence; or how he changed the world's thinking about forms of currency when he developed a mathematical basis for the introduction of the decimalized dollar (Clements & Ellerton, 2015), and then tried to link that with a decimalized system for weights and measures (Boyd, 1961; Garrett & Guth, 2003). Think, too, about the validity of the recent claim by David Hirsch and Dan Van Haften (2015, 2019a, b, c) that Abraham Lincoln *consciously* applied Euclidean structure and logic when preparing his greatest speeches—including his Gettysburg address (Hirsch & Van Haften, 2019c) What is mathematics, and can "ordinary" people be expected to make significant developments not only in pure mathematics but also with respect to everyday living?

So far as progress toward "mathematics for all" was concerned, in 1865 there was still a very long way to go. Between 1865 and 1880, Indigenous Americans and African-Americans were rarely if ever to be found studying high-level mathematics in colleges, and although 20 percent of all extant cyphering books from the period 1607–1865 were prepared by females, there was still only a small percentage of the persons enrolled in college mathematics courses—much less than 20 percent—who were female (Ellerton & Clements, 2021). That said, it can be argued (see, e.g., Clements, Keitel, Bishop, Kilpatrick & Leung, 2013) that in the Reconstruction period which followed immediately after the Civil War a new day would dawn so far as recognition that "mathematics *should be* for all," not only in the United States of America but everywhere. But in order to achieve that goal fresh ways of thinking would be needed. And the features of the desired new day, and decisions on which steps would be taken in order to generate it, need to be researched—by mathematicians, by mathematics educators, by sociologists, and by general historians. In 1776 and in 1865 there was, "a challenge to change." However, the journey is still ongoing—a similar challenge remains today (Ellerton & Clements, 1989).

References

Ahmed, M. M. (2004). How many squares are there, Mr Franklin? Constructing and enumerating Franklin Squares. *American Mathematical Monthly, 111*, 394–410.

Anderson, D. (1997). *The radical enlightenments of Benjamin Franklin*. Baltimore, MD: Johns Hopkins University Press.

Bartlett, J. R. (1933). *Letter of instructions to the captain and the supercargo of the brig "Agenoria," engaged in a trading voyage to Africa*. Philadelphia, PA: Howard Greene and Arnold Talbot.

Behforooz, H. (2012). Weighted magic squares. *Journal of Recreational Mathematics, 36*(4), 283–286.

Blinderman, A. (1976). *Three early champions of education: Benjamin Franklin, Benjamin Rush, and Noah Webster*. Bloomington, IA: Phi Delta Kappa Educational Foundation.

Bordley, J. B. (1789). *On monies, coins, weights and measures*. Philadelphia, PA: Daniel Humphreys.

Bowditch, N. (1797). *Journal of a voyage from Salem to Manila in the ship Astrea, E. Prince, Master, in the years 1796 and 1797*. Handwritten manuscript held in the Bowditch Collection, Boston Public Library. https://doi.org/10.5479/sil.274043.39088000385815

Bowditch, N. (1802). *The new American practical navigator* . Newbury-Port, MA: E. M. Blunt.

Boyd, J. P. (Ed.) (1950a). *The papers of Thomas Jefferson, Volume 1, 1760–1776*. Princeton, NJ: Princeton University Press. https://doi.org/10.1515/9780691184661-006

Boyd, J. P. (Ed.). (1950b). *The papers of Thomas Jefferson, Volume 2, January 1777 to June 1779*. Princeton, NJ: Princeton University Press.

Boyd, J. P. (Ed.). (1953). *The papers of Thomas Jefferson, Volume 7, March 1784 to February 1785*. Princeton, NJ: Princeton University Press.

Boyd, J. P. (1961). Report on weights and measures: Editorial note. In J. P. Boyd (Ed.), *The papers of Thomas Jefferson 16, November 1789 to July 1790)* (pp. 602–617). Princeton, NJ: Princeton University Press. https://doi.org/10.1515/9780691185224-014

Brekke, B. F. (1977). *The copper coinage of Imperial Russia, 1700–1917*. Malmoe, Sweden: Forlagshuser.

Cajori, F. (1890). *The teaching and history of mathematics in the United States*. Washington, DC: Government Printing Office.

Clements, M. A., & Ellerton, N. F. (2015). *Thomas Jefferson and his decimals 1775–1810: Neglected years in the history of U.S. school mathematics*. New York, NY: Springer. https://doi.org/10.1007/978-3-319-02505-6

Clements, M. A., Keitel, C., Bishop, A. J., Kilpatrick, J., & Leung, F. (2013). From the few to the many: Historical perspectives on who should learn mathematics. In M. A. Clements, A. J. Bishop, C. Keitel, J. Kilpatrick, & F. Leung (Eds.), *Third international handbook of mathematics education* (pp. 7–40). New York, NY: Springer. https://doi.org/10.1007/978-1-4614-4684-2_1

Cotter, C. (n.d.). John Hamilton Moore and Nathaniel Bowditch. *Forum, 30,* 323–326. https://doi.org/10.1017/S0373463300044003

Dauben, J. W., & K. H. Parshall (2014). Mathematics education in North America to 1800. In A. Karp & G. Schubring (Eds), *Handbook on the history of mathematics education* (pp. 175–185). New York, NY: Springer.

Defense Mapping Agency, Hydrographic/Topographic Center. (1995). *The American practical navigator: An epitome of navigation*. Bethesda, MD: Author.

Durkin J. J. (1942). Journal of the Revd. Adam Marshall, schoolmaster, *U.S.S. North Carolina,* 1824–1825. *Records of the American Catholic Historical Society of Philadelphia, 53*(4), 152–168.

Educational Unit (U.S. Merchant Marine Cadet Corps). (1966). *Americans who have contributed to the history and traditions of the United States merchant marine*. Washington, DC: U.S. Cadet Corps.

Ellerton, N. F., & Clements, M. A. (Eds.). (1989). *School mathematics: The challenge to change*. Geelong, Australia: Deakin University.

Ellerton, N. F., & Clements, M. A. (2012). *Rewriting the history of mathematics education in North America, 1607–1861*. New York, NY: Springer. https://doi.org/10.1007/978-94-007-2639-0

Ellerton, N. F., & Clements, M. A. (2014). *Abraham Lincoln's cyphering book, and ten other extraordinary cyphering books*. New York, NY: Springer. https://doi.org/10.1007/978-3-319-02502-5

Ellerton, N. F., & Clements, M. A. (2019, September 22). *Major influences on U.S. school mathematics in the nineteenth century*. Paper presented to a meeting of the HPM/AMS Sectional Meeting in Madison, Wisconsin.

Ellerton, N. F., & Clements, M. A. (2021). *Cyphering books prepared in the North American colonies (but not Canada), or in the United States of America*. Perth, Australia: Meridian Press.

Fauvel, J. (1999, April 15). *Thomas Jefferson and mathematics*. Lecture given at the University of Virginia.

Fenzi, G. (1905). *The rubles of Peter the Great*. Moscow, Russia: Open Library.

Franklin, B. (1749). *Proposals relating to the education of youth in Pensilvania*. Philadelphia, PA: Author.

Franklin, B. (1793). *The private life of the late Benjamin Franklin, LL.D*. London, England: J. Parsons.

Franklin, B. (1917). *The autobiography of Benjamin Franklin*. Boston, MA: Houghton Mifflin.

Frost, J. (1846). *Lives of American merchants*. New York, NY: Saxon & Miles.

Ganter, H. B. (1947). William Small, Jefferson's beloved teacher. *William and Mary Quarterly, 4*, 505–511. https://doi.org/10.2307/1919640

Garcia, R., Meyer, S., Sanders, S., & Seitz, A. (2009). Construction and enumeration of Franklin circles. *Involve, 2*(3), 357–370. https://doi.org/10.2140/involve.2009.2.357

Garrett, J., & Guth, R. (2003). *100 greatest U.S. Coins*. Atlanta, GA: H.E. Harris & Co.

Gaydos, T., & Kampas, B. (2010). *American and Canadian ciphering books, n.d., 1727–1864*. Salem, MA: Phillips Library at the Peabody Essex Museum.

Goodwin, J. (2003). *Greenback: The almighty dollar and the invention of America*. New York, NY: Henry Holt and Company.

Grew, T (1853). *The description and use of the globes, celestial and terrestrial, with variety of examples for the learner's exercise: intended for the use of such persons who would attain to the knowledge of those instruments: but chiefly designed for the instruction of young gentlemen at the Academy of Philadelphia. To which is added rules for working all the cases in plain and spherical triangles without a scheme*. Germantown, PA: Christopher Sower.

Hadden, R. W. (1994). *On the shoulders of merchants: Exchange and mathematical conception of nature*. New York, NY: SUNY Press.

Hardy, G. H. (2004). *Mathematician's apology*. Cambridge, England: Cambridge University Press.

Henrich, C. J. (1991). Magic squares and linear algebra. *American Mathematical Monthly, 98*(6). 481–488.b https://doi.org/10.1080/00029890.1991.11995746

Hepburn, A. B. (1915). *A history of currency in the United States with a brief description of the currency systems of all commercial nations*. New York, NY: The Macmillan Company.

Hertel, J. (2016) Investigating the implemented mathematics curriculum of New England navigation cyphering books. *For the Learning of Mathematics, 36*(3), 4–10.

Hill, T. (1880). Benjamin Peirce. *The Harvard Register, 6*(1), 91–92.

Hirsch, D., & Van Haften, D. (2015). Abraham Lincoln and the structure of reason. El Dorado Hills, CA: Savas Beatie.

Hirsch, D., & Van Haften, D. (2019a). *The tyranny of public discourse*. El Dorado Hills, CA: Savas Beatie.

Hirsch, D., & Van Haften, D. (2019b). *The ultimate guide to the Declaration of Independence*. El Dorado Hills, CA: Savas Beatie.

Hirsch, D., & Van Haften, D. (2019c). *The ultimate guide to the Gettysburg address*. El Dorado Hills, CA: Savas Beatie.

Honeywell, R. J. (1931). *The educational work of Thomas Jefferson*. Cambridge, MA: Harvard University Press. https://doi.org/10.4159/harvard.9780674337299

Israel, J. (2002). *Radical enlightenment: Philosophy and the making of modernity, 1650 –1750*. London, England: Oxford University Press.

Jefferson, T. (1784a). *Notes*. In W. Peden (Ed.), *Notes on the State of Virginia*. Chapel Hill, NC: University of North Carolina Press for the Institute of Early American History and Culture, Williamsburg, Virginia.

Jefferson, T. (1784b). *Notes on the establishment of a money unit and of a coinage for the United States* (handwritten manuscript). Washington, DC: Library of Congress.

Jefferson, T. (1784c) *Some thoughts on a coinage* (handwritten manuscript). [ca. March 1784]. Founders Online, National Archives (Jefferson/01-07-02-0151-0004, ver. 2014-05-09). Sou, In J. P. Boyd (Ed.), *The papers of Thomas Jefferson, Vol. 7, 2 March 1784–25 February 1785*, Princeton, NJ: Princeton University Press, 1953, pp. 173–175. Also, in T. Jefferson (1785). *Notes on the establishment of a money unit and of a coinage for the United States*. Paris, France: Author. The notes are reproduced in P. F. Ford (Ed.). (1904). *The works of Thomas Jefferson* (Vol. 4, pp. 297–313). New York, NY: G. P Putnam's Sons.

Jefferson, Thomas to Edward Carrington, May 27, 1788. In H. A. Washington (Ed.), *The writings of Thomas Jefferson*, New York, NY: H. W. Derby, 1861).

Karpinski, L. C. (1980). *Bibliography of mathematical works printed in America through 1850* (2nd ed.). New York, NY: Arno Press.

Kern, F. (1982). *Captain William Cooke Pease: U.S. Coast Guard pioneer*. Bethesda, MD: Alised Enterprises.

Ketcham, H. (1901). *The life of Abraham Lincoln*. New York, NY: Perkins Book Company.

Kline, M. (1972). *Mathematical thought from ancient to modern times*. New York, NY: Oxford University Press.

Kraus, J. W. (1961). The development of the curriculum in the early American colleges. *History of Education Quarterly, 1*(2), 64–76. https://doi.org/10.2307/367641

Lee, C. (1797). *The American accountant; being a plain, practical and systematic compendium of Federal arithmetic* ... Lansingburgh, NY: William W. Wands.

Levin, J. E., & Levin, M. R. (2010) *Abraham Lincoln's Gettsysburg address*. New York, NY: Simon & Schuster.

Linklater, A. (2003). *Measuring America: How the United States was shaped by the greatest land sale in history*. New York, NY: Plume.

Lucas, S. E. (1989). Justifying America: The Declaration of Independence as a rhetorical document. In T. W. Benson (Ed.), *American rhetoric: Context and criticism* (pp. 67–130). Carbondale, IL: Southern Illinois University Press.

Mattoon, C. H. (1850). *Common arithmetic upon the analytic method of instruction*. Medary, OH: Steam Press of S. Medary.

Matz, F. P. (1895). B. O. Peirce: Biography. *American Mathematical Monthly, 2*, 173–179. https://doi.org/10.1080/00029890.1895.11998647

McCusker, J. J. (1992). *Money and exchange in Europe and America 1600–1775*. Chapel Hill, NC: University of North Carolina Press.

Moore, J. H. (1796). *The new practical navigator, being an epitome of navigation, explaining the different methods of working the lunar observations and all the requisite tables used with the nautical almanac, in determining the latitude and longitude, and keeping a complete reckoning at sea; illustrated by proper rules and examples; the whole exemplified in a journal kept from England to the island of Tenerife; also the substance of the examination, every candidate for a commission in the Royal Navy, and Officer in the Honourable East India Company's service, must pass through previous to their being appointed; this, with the sea terms, are particularly recommended to the attention of all young gentlemen designed for, or belonging to the sea* (12th ed.). Tower Hill, England: F. Law.

Moore, J. H. (1799). *The new practical navigator*. Newbury-Port, MA. Edmund March Blunt.

Morris, R. (1782, January 15). Robert Morris to the President of Congress, January 15, 1782. In J. P. Boyd (Ed.), *The papers of Thomas Jefferson 16, March 1784 to February 1785* (pp. 160–169). Princeton, NJ: Princeton University Press.

Morrow, G. R. (1970). *Proclus's commentary on the first book of Euclid's Elements*. Princeton, NJ: Princeton University Press.

Navigation Division, Defense Mapping Agency (1995). *American practical navigator, Bowditch*. Hydrographic/Topographic Center. ASIN: B001B4884O.

Nishikawa, S. (1987). The economy of Chōshū on the eve of industrialization. *The Economic Studies Quarterly, 38*(4), 209–222.

Ogg, F. A. (1927). *Builders of the Republic*. New Haven, NJ: Yale University Press.

Parshall, K. H. (2003). Historical contours of the American mathematical research community. In G. M. A. Stanic & J. Kilpatrick (Eds.), *A history of school mathematics* (Vol. 1, pp. 113–158). Reston, VA: National Council of Teachers of Mathematics.

Pasles, P. C. (2007). *Benjamin Franklin's numbers: An unsung mathematical odyssey*. Princeton, NJ: Princeton University Press.

Patterson, R. (1819). *A treatise of practical arithmetic intended for the use of schools* (2 parts). Pittsburgh, PA: R. Patterson and Lambdin.

Peden, W. (Ed.). (1955). *Notes on the State of Virginia by Thomas Jefferson*. Chapel Hill, NC: University of North Carolina Press.

Peterson, S. R. (1955). Benjamin Peirce: Mathematician and philosopher. *Journal of the History of Ideas*, *16*, 89–112. https://doi.org/10.2307/2707529

Rappleye, C. (2010). *Robert Morris: Financier of the American Revolution*. New York, NY: Simon & Schuster.

Rawley, J. A. (1981). *The trans-Atlantic slave trade*. New York, NY: W. W. Norton.

Roberts, D. L. (2019). *Republic of numbers: Unexpected stories of mathematical Americans through history*. Baltimore, MD: Johns Hopkins University Press.

Seaman, W. H. (1902, March). How Uncle Sam got a decimal coinage. *School Science*, 232–236. https://doi.org/10.1111/j.1949-8594.1902.tb00443.x

Sesiano, J. (2019). *Magic squares: Their history and construction from ancient times to AD 1600. Studies in the history of mathematics and physical sciences*. Gewrestrasse, Switzerland: Springer Nature. https://doi.org/10.1007/978-3-030-17993-9

Smith, D. E., & Ginsburg, J. (1934). *A history of mathematics in America before 1900*. Chicago, IL: The Mathematical Association of America. https://doi.org/10.1090/car/005

Solberg, T (1905). Copyright in Congress 1789–1904. Washington, DC: Government Printer.

State Street Trust Company. (1917). *Some events of Boston and its neighbors*. Boston, MA: Author.

Sylvester, J. J. (1870, January 6). A plea for the mathematician. *Nature, 1,* 261–263.

Taylor, E. G. R. (1966). *The mathematical practitioners of Hanoverian England 1714–1840*. Cambridge, England: Cambridge University Press.

Tebbel, J. (1972). *A history of book publishing in the United States*. New York, NY: R. R. Bowke.

Thornton, T. P. (2016). *Nathaniel Bowditch and the power of numbers: How a nineteenth-century man of business, science and the sea changed American life*. Chapel Hill, NC: University of North Carolina Press. https://doi.org/10.1038/001261a0

Turner, W. L. (1953). The charity school, the Academy and the College Fourth and Arch Streets. *Transactions of the American Philosophical Society, 43*(1), 179–186. https://doi.org/10.2307/1005670

Wadsworth, G. (2005). *Benjamin Banneker: Pioneering scientist*. Minneapolis, MN: Lerner.

Weems, M. L. (1820). *The life of Benjamin Franklin*. Baltimore, MD: Author.

Wiencek, H. (2012). *Master of the mountain: Thomas Jefferson and his slaves*. Minneapolis, MN: Lerner Publishing.

Wilkins, J. (1668) *Essay towards a real character and a philosophical language*. Held in the Balliol Library, University of Oxford.

Wilson, D. L. (1992). Thomas Jefferson and the character issue. *The Atlantic Monthly, 270*(3), 37–74.

Wunsch, J. (2007). Magic squares and circles. *Nature, 450*(1162). https://doi.org/10.1038/4501162a

Zitarelli, D. E. (2019). *A history of mathematics in the United States and Canada (Vol. 1, 1492–1900)*. Providence, RI: MAA Press and American Mathematical Society. https://doi.org/10.1090/spec/094

Chapter 9
Toward Mathematics for All: Answers to Research Questions, Limitations, and Possibilities for Further Research

Abstract This final chapter begins by answering the six research questions which were posed towards the end of the first chapter. Those questions were:

1. What were the intended, implemented and attained mathematics curricula for young children aged less than 10 years (in North America) (a) during the seventeenth century? And (b) during the period 1700–1865? And, to what extent did the answers to those questions vary across North America, and in different groups of children (e.g., boys versus girls, European-background children versus Native American children, and European-background children versus African-American children)?
2. What were the intended, implemented and attained mathematics curricula for North American children aged between 10 and 15 years during (a) the seventeenth century, and (b) the period 1700–1865? And, to what extent did the answers to those questions vary across different parts of North America, and across different groups?
3. What were the intended, implemented and attained mathematics curricula for North American pre-college students aged between about 15 and 18 years during (a) the seventeenth century, and (b) the period 1700–1865? And, to what extent did the answers to those questions vary across different parts of North America, and across different groups?
4. What were the intended, implemented and attained mathematics curricula for North American college students during (a) the seventeenth century, and (b) the period 1700–1865? And, to what extent did the answers to those questions vary across different parts of North America, and across different groups?
5. What perspectives on the status of mathematics in college curricula were held in the North American colonies during the period 1607–1865?
6. What are the implications of the answers to the first five questions (above) for those investigating the history of mathematics in North America? What future research is needed, and to what extent will it be feasible to conduct that research?

N. F. Ellerton, M. A. (Ken) Clements, *Toward Mathematics for All*, History of Mathematics Education,
https://doi.org/10.1007/978-3-030-85724-0_9

While carrying out the research for this book we came to recognize that authors of general histories of mathematics have tended to view the history of mathematics in terms of whether an event or person(s) associated with an event contributed to a "weight-bearing link" (Parshall, 2003, p. 114) with the present state of knowledge for key areas of mathematics. In other words, they have looked back from the present situation with respect to high-level mathematics in an attempt to identify persons who, and events which, progressed mathematics toward what is now regarded as important. In this book, however, we have considered the history of mathematics in North America (excluding Alaska and Canada) between 1607 and 1865 from a more inclusive, bottom-up, mathematics-for-all perspective (Clements et al., 2013). In this final chapter the above questions are answered from analyses provided in the preceding eight chapters. The chapter closes with a discussion of limitations of the research, and how a consideration of those limitations draws attention to various questions which need to be the subject of further research.

Keywords History of mathematics • History of mathematics education • Mathematics for all • Mathematics in the 13 colonies • Pestalozzi • Warren Colburn

Answering the Research Questions

Answer to Research Question 1: What Were the Intended, Implemented and Attained Mathematics Curricula for Young Children (Aged Less than 10 Years) in North America During (a) the Period 1607–1820? and (b) the Period 1820–1865?

1607–1820 Although we cannot say precisely what mathematics young European-background children living in North America learned during the seventeenth century, we do know that issues associated with the teaching and learning of mathematics to children aged less than 10 years were rarely a focus of attention for anyone other than the small number of teachers and learners involved. Children could learn to count with Hindu-Arabic numerals in some dame schools and in most schools supported by local taxation, or in privately-operated subscription-schools, but the main curricular foci in those schools were reading, writing, and religion. Some children were taught to read and write Hindu-Arabic numerals, but in most schools there was either no paper at all, or only a sparse supply, and written mathematics was only occasionally a main focus. When arithmetic involving Hindu-Arabic numerals was studied, not much attention was given to the concepts of addition, subtraction, multiplication, and division because most teachers of children aged 10 years or less knew very little mathematics. That changed, especially after the publication of Warren Colburn's (1821) *An Arithmetic on the Plan of Pestalozzi with Some Improvements*, which not only included more appropriate content for young children, but also provided guidance on how it might be taught effectively.

The best evidence we have for the lack of attention given to mathematics in the schools of the seventeenth century is to be found in William Heard Kilpatrick's (1912) doctoral dissertation on the "Dutch schools of New Netherland and Colonial New York." Kilpatrick reported that in early New York (called New Amsterdam 1664) some Dutch-background children who were attending "dame schools" they learned to read and a few some learned to write. Sometimes, learning to "cypher" was also a possibility but cyphering carried an additional fee to the standard fee to be paid to the teacher. Given the conditions in New Amsterdam (and, later in seventeenth-century New York) it is unlikely that many families would have been able to afford to pay the additional fee needed for their children to be taught "to cypher." The lack of attention to arithmetic for young children continued to be the case throughout the seventeenth century and into the eighteenth century (Ellerton & Clements, 2012).

The Colburn-inspired revolution with respect to school mathematics for young children, 1820–1865. In the 1820s, serious questions began to be asked about the need for, and purposes of, school arithmetic, and there seemed to be a distinct possibility that teachers and schools would move beyond the 600-year-old tradition that the only mathematics that young children needed to learn was to count, and to read and write the Hindu-Arabic numerals. Some historians have suggested that the force driving this movement was Warren Colburn (1793–1833), who graduated from Harvard College in 1820. In 1821, Colburn caused to have published a little book aimed at assisting teachers to help young children (from around six years of age) to learn the first principles of numeration, counting, and the four operations. Through the title of his book (*An Arithmetic on the Plan of Pestalozzi, with some Improvements*), Colburn (1821) acknowledged the influence on his thinking of Johann Heinrich Pestalozzi, the great Swiss educator (Biber, 1831).

Although Colburn only taught in schools for a few years after 1821, *An Arithmetic on the Plan of Pestalozzi* was destined to become a classic—one of the most influential texts in the history of U.S. education (Cajori, 1890; Doar, 2006). The traditional argument is that from the 1820s, American teachers of the young, inspired by the writings of Colburn and Pestalozzi, increasingly began to invite students to construct their own mathematical knowledge. This not only resulted, so the tradition goes, in students at about 6 years of age (rather than at 10 or 11) beginning to learn more arithmetic than had been the case for children of the same age in earlier times. Furthermore, young girls began to study arithmetic in much greater numbers than previously. A new cadre of Colburn-inspired teachers quickly energized the teaching and learning of arithmetic. Right across the nation, teachers began to ask young learners to figure out, mentally, answers to carefully sequenced sets of questions which referred to real objects (like fingers, stones, amounts of money, etc.), and then to articulate those answers in well-formed sentences. In this way, teachers were asked to play a far more active role in facilitating arithmetic learning than ever before.

Although we believe that Colburn's influence has been exaggerated (Ellerton & Clements, 2012), there can be no doubt that during the period 1820–1865 Colburn inspired many other writers to adopt a Pestalozzian approach to school arithmetic—e.g., Fowle, 1849; Ray, 1834; Ruter, 1828; Smith, 1827. By 1850 it had become clear that although the "inductive" approach that Colburn and Pestalozzi recommended to teachers of children aged between 6 and 10 was not always successful (Ellerton & Clements, 2012), it had challenged many educators to think about, *how* much mathematics should be taught but also *what mathematics* should be taught to young children, and also *how* it should be taught. However, the Pestalozzian approach did not find its way into most of the relatively few schools for indigenous children which had been established. In most of those schools the old hornbook/battledore-approach to teaching young children to learn to read, write, and speak the alphabet and the Hindu-Arabic numerals prevailed (Ellerton & Clements, 2012; Littlefield, 1904), although in a few schools the new Lancaster "monitorial" approach was adopted (see, e.g., Fowle, 1849). The situation was less promising for the children of African-American slaves, the great majority of whom were not permitted to attend schools at all until after the Civil War (Butchart, 2010; Guasco, 2014). "Mathematics for all" remained a long way off.

Answer to Research Question 2: What Were the Intended, Implemented and Attained Mathematics Curricula for North American Children Aged Between 10 and 15 Years During (a) the Seventeenth Century, and (b) the Period 1700–1865?

During the seventeenth century William Heard Kilpatrick (1912) reported that in New Amsterdam many Dutch-background children attended "dame schools" where some of them learned to read and a few learned to write. Given the pioneering conditions in New Amsterdam (and, later, in New York), it is unlikely that many families were willing, or could afford, to pay "extra" to have their children learn "to cypher." A similar situation prevailed in most other parts of North America during the seventeenth century. It is almost certain that most young North American children were not given the opportunity to prepare cyphering books.

There is only one extant manuscript in the E-C cyphering book collection which was prepared during the seventeenth century. Of that book there are only seven pages which remain. They are reproduced in Ellerton et al. (2014—see pages 13–21). The arithmetic in that manuscript was at a very low level. Although the name of the person who prepared it, and the name of the place where it was prepared, are unknown the seven pages are watermarked and, based on the watermark, the year when it was prepared could be as early as 1666. The last owner indicated that the manuscript was purchased from a very long-established family in Maine, New England. Two interesting features of the manuscript are: (a) the person who prepared it seemed to want to be able to claim that he or she had cyphered to the rule of three—we say that because, all of a sudden, that rule is mentioned on the last remaining page; and (b) one of the topics mentioned was "halving," which was *not* a normal part of the *abbaco* curriculum sequence for arithmetic, but *was* part of the rival Sacrobosco curriculum sequence for arithmetic (Ellerton & Clements, 2014).

Early North American settlements and states were not ready for a problem-solving mathematics regime like that followed by Thomas Dixson in England in the 1630s (see Ellerton & Clements, 2014, chapter 7). Many of the settlers had had very little formal training in anything to do with mathematics and there were relatively few persons with sufficiently strong mathematical backgrounds to be in a position to teach mathematics beyond "hornbook arithmetic" effectively. Of the few who might have known something about mathematics, the demands of daily life in a new and difficult world could have precluded their having the time to take seriously the challenge of teaching formal mathematics to children, including to their own children (Cohen, 2003). That state of affairs would have had an ongoing effect: because children did not learn much mathematics beyond mere counting, the next generation, the children's children, would not have had easy access to persons who could teach them mathematics beyond counting. The only available mathematics textbooks were those which had been brought over from the "mother country" and, in the New World, these were both difficult to get and expensive. Little wonder, then, that very few of the children growing up in New England or New York or Virginia learned more than the absolute rudiments of elementary mathematics—certainly not beyond being able to count and to read and write the Hindu-Arabic numerals which appeared on some hornbooks. Some North American indigenous children did get instructed in hornbook arithmetic by missionaries, but most did not.

That would explain why the quality of mathematics present in the only extant cyphering book from the period was so low. Putting the matter bluntly, it is likely that neither the student nor his or her teacher (if a teacher was involved) would have known much about the four operations on numbers, especially multiplication and division. But even in those early times, cyphering "up to the rule of three" was thought to be something desirable to do, and so that single extant cyphering book from the seventeenth century closes with the student being able to say to himself (or herself) that he (or she) had "cyphered to the rule of three."

Undoubtedly, the approach to mathematics in the seven pages emphasized form rather than substance. With division, for example, the emphasis was on knowing the meanings of terminology like "dividend," "divisor," "quotient" and "remainder," but not on learning how to find the actual values of the quotients and remainders for a given division task.

The best available information regarding *implemented* curriculum is in the form of a hornbook held by Ellerton and Clements which could have been used by indigenous children in the seventeenth century. The hornbook shows both Hindu-Arabic numerals (1, 2, 3, 4, 5, 6, 7, 8, 9 and 0) and Roman numerals (I, II, III, IV, V, VI, VII, VIII, IX, and X), but we have no firm data on how children used it. Probably children attending dame schools would have learned to count the objects in small collections of less than 10, and to read and write the Hindu-Arabic numerals, but we cannot even be sure of that. There are no records, or key indicators like the extent of arithmetical knowledge of teachers in privately-operated dame or subscription schools.

The situation with respect to European-background children whose parents were indentured servants also needs to be carefully investigated (Chessman, 1965). For much of the seventeenth century such children probably outnumbered indigenous Native American children and also children of African American slaves (Dobyns, 1983; Galenson, 1984). But only a tiny proportion of the children in all three groups received much formal education with respect to *abbaco* arithmetic—although in a few regions, missionaries did offer an elementary hornbook-style education to some children.

During the period 1700–1776. Data summarized from cyphering books in the Ellerton-Clements cyphering book collection during this period (Ellerton & Clements, 2021), together with data gleaned from manuscripts held in the Phillips Library (Salem, MA), the Houghton Library (at Harvard University), the Clements Library (at the University of Michigan), the New York Public Library, the David Eugene Smith Collection (held in Columbia University, New York), the Beinecke Rare Book Library at Yale University, the Wilson Library at the University of North Carolina at Chapel Hill, the Swem Library at the College of William and Mary, the Rockefeller Library in Williamsburg, and the Huguenot Historical Society Library in New Paltz, New York, provide compelling evidence that the cyphering tradition (see Chapter 3) defined and controlled the selecion of content and the dominant instructional method in mathematics throughout the period 1700–1776. Although most cyphering books dealt solely with *abbaco* arithmetic (Ellerton & Clements, 2012), occasionally the content related to any of algebra, gauging, geometry, navigation, surveying, or trigonometry. Usually, students were not permitted to begin cyphering until they had reached the age of 10.

During the period 1776–1865. The concept of the one-room "local" school was translated from Europe into colonial America and even as late as the second half of the nineteenth century most of the rural schools dotted across the United States of America were still of the one-room variety (Mydland, 2011). As the nineteenth century progressed, larger schools were built, with more space and more students than had been the case with the earlier log-cabin, one-room, often dirt-floored, schools. During the nineteenth century some students and their teachers were able to sit at desks and have access to blackboards (Ackerberg-Hastings, 2014; Roberts, 2014), and teachers increasingly tended to adopt more group-based approaches to learning than had been the case during earlier periods. From about 1830, in some schools, children of roughly the same age were asked to work together in small groups. Recitations, and student learning became less individualized and a methodology evolved by which small groups of children worked together on common themes and took recitations together. From about 1850 in the cities, students were increasingly placed in grades and whole-class teaching became the norm—but that was not feasible in most one-room rural schools (Monaghan, 2007).

So far as intended curriculum was concerned, the content of the mathematics to be studied during the period 1700–1865 was defined by the *abbaco* sequence of arithmetic, which for the purposes of this book has been divided into three levels—notice, however, although, that the term "level" is used here for convenience, and was *not* used during the period 1607–1865.

Level 1: For 10- to 14-year-olds (for details, see Chapter 3 of this book)

- Numeration (Hindu-Arabic system)
- Four operations, using the Hindu-Arabic system
- Weights and measures (often referred to as "compound operations")
- Federal currency (dollars, cents) and "legal" currency (sterling—pounds, shillings, pence, farthings)
- Reduction (e.g., "How many inches in a mile?")
- Practice

Level 2: For 14- to 17-year-olds (see Chapter 4 of this book)

- Exchange
- Barter
- Rule of three (direct)
- Rule of three (inverse)
- Double rule of three (sometimes called the "rule of five")
- Equation of payments
- Tare and tret(t)
- Interest—simple and compound
- Loss and gain
- Brokage, commission, annuities
- Discount
- Common fractions
- Decimal fractions
- Involution and evolution

Level 3: For advanced students (see Chapter 6 of this book)

- Fellowship (the arithmetic of partnerships)
- Alligation (the arithmetic of mixing)
- Single and double position
- Duodecimals (often called "cross multiplication")
- Arithmetical and geometrical progressions
- Permutations and combinations
- Mensuration

The cyphering tradition incorporated the above sequence of topics which had been brought from the reckoning schools of Europe to the New World (Grendler, 1989). Cyphering books typically featured an **IRCEE** (Introduction-Rules-Cases-Examples-Exercises) development for a topic and, when solutions to word problems were shown, a **PCA** (Problem-Calculation-Answer) genre was adopted by students. Often, cyphering books featured fine penmanship and calligraphy. The strength of **IRCEE** and **PCA** genre expectations is evident in many school mathematics textbooks and in many extant cyphering books, and those genres still seem to influence the thinking of present-day mathematics teachers, students, and textbook writers (see Voigt, 1995).

The order of topics presented in an arithmetic cyphering book was usually fairly consistent with—although only occasionally perfectly consistent with—the above list. Many students began to study simple interest without having previously studied common or decimal fractions, or even percentages. One such student was Abraham Lincoln, who attended one-room schools in Pigeon Creek, Indiana, in the 1820s (Ellerton et al., 2014).

Although only a small proportion of school students got to study any of algebra, geometry or trigonometry during the period 1607–1820, the numbers doing so steadily increased during the period 1820–1865 when colleges began to require prospective students to be able to demonstrate a knowledge of elementary algebra and elementary geometry (see Chapters 5 through 7 of this book).

Initially, the emphasis in algebra was on standard manipulations. It was not until the late 1830s, when Benjamin Peirce began to emphasize a functions approach to algebra at Harvard, that there was a move toward thinking about algebra as a formal study of relationships. Not one page in any of the cyphering books in the E-C cyphering book collection shows a Cartesian graph, or any attempt to deal with differential or integral calculus. Most students who prepared navigation or surveying cyphering books utilized logarithms and a directed-line-segment approach to trigonometry, with a circle-with-radius-measure-10^{10} assumption being evident (see Chapter 6 of this book, and also Hertel, 2016; Van Sickle, 2011). Throughout most of the period 1607–1820 the only kind of geometry formally studied in the schools was of the elementary Euclidean variety, mainly constructions using straight edges and passes. But, with the beginning of public high schools more schools began to introduce a “Legendre textbook approach” (see Chapter 6).

The concept of “proof” was not well handled in school mathematics classes. With arithmetic, the word “proof” became synonymous with “check” (so, for example, a multiplication might be shown to “prove” the result of an elementary division task, or a “casting-out-nines” check might be used to “prove” a result of a calculation). At the time of writing we have examined about 1500 cyphering books prepared in North American schools during the period 1607–1865 and have not found what might be regarded as a genuine proof in more than a handful of them.

The implemented curriculum. In 1800, very few U.S. school students owned a mathematics textbook. Most of the textbooks used before the 1780s were authored by British writers. Between 1780 and 1865, books by U.S. authors (like Nicolas Pike, Daniel Adams, Nathan Daboll, Michael Walsh, Stephen Pike, Warren Colburn, Roswell Smith, Joseph Ray, Charles Davies, Benjamin Greenleaf, and John Stoddard) were increasingly used. Relatively few textbooks written by French authors, or translations into English of such books, found their way into U.-S. schools—although they could be found in U.S. colleges, especially Harvard College and the United States Military Academy (USMA) at West Point. If a textbook was used, the intended curriculum as indicated by the "contents" page at the beginning of the textbook often did not correspond to the order of topics studied by students. Thus, for example, in schools the introduction of common fractions was often delayed until well after the rules of three had been studied.

Teachers in one-room schools did not teach arithmetic, or indeed any subject, from the fronts of their rooms. Students would prepare for a recitation individually throughout the school day while teachers engaged with individual students, in one-on-one recitations. Entries in cyphering books were often based on "parent" cyphering books owned by the teachers (or by members of the students' families). If 8 boys, say, aged at least 10 years, were attending a one-room school, then at any particular time those 8 boys would be likely to be studying up to 8 different aspects of mathematics. An important assumption within the cyphering tradition was that the teaching and learning of mathematics should be individualized, with arrangements between teachers and individual students constantly being negotiated recitation sessions.

The attained curriculum. Before the 1840s there were no written examinations to enable teachers to assess *attained* curricula (that is to say, what the children actually learned). Recitation sessions between the teacher and individual students were intended to serve that purpose. It was not until about 1830 that blackboards began to appear in some schools—for most of the period 1700–1865, blackboards were not available. Often slates were used by students when working out "rough" solutions to problems. That was done, and slates were checked by teachers before students would be permitted to enter "correct, approved solutions" into their cyphering books. Penmanship and calligraphic headings were expected to be of a high standard (Ellerton & Clements, 2012, 2021). When a student came to a college for possible enrollment, any assessment of his mathematical knowledge was likely to be based on a brief interview with a college official, and it was expected that the student's cyphering book(s) would be available for inspection.

It is not an exaggeration to say that before 1840, the cyphering tradition, complemented by recitation, *controlled* school mathematics in North America (Ellerton & Clements, 2012).

Answer to Research Question 3: What Were the Intended, Implemented and Attained Mathematics Curricula for North American Pre-College Students Aged Between About 15 and 18 Years During (a) the Seventeenth Century, and (b) the Period 1700–1865? And, to What Extent Did the Answers to Those Questions Vary Across North America, and Across Different Groups?

During the seventeenth century. When Harvard College opened its doors to students in the late 1630s, but mathematics was not a very important component of its curriculum and that remained the case throughout the remainder of the seventeenth century. According to Zitarelli (2019), Henry Dunster, Harvard's President, "read mathematics and astronomy to third-year students" (p. 18), but "10 hours per week were devoted to philosophy, 7 to Greek, 6 to Rhetoric (i.e., disputations in Latin), 4 to Oriental Languages, and just 2 to Mathematics" (p. 18). During the period 1636–1700 there were only a few students studying mathematics in post-elementary ("pre-college") schools, mainly because there were hardly any schools which offered college-preparation instruction in mathematics.

It was not until 1693 that the second college (within the "13 colonies") was established—namely, the College of William and Mary, in Williamsburg, Virginia—and Yale College (initially known as "The Collegiate School") was the third college, created in Connecticut in 1701.

The mathematics studied in seventeenth-century schools preparing students for college rarely went beyond *abbaco* arithmetic and elementary aspects of Euclidean geometry. Occasionally, the geometry was applied to surveying or navigation tasks (Zitarelli, 2019). At Harvard, instruction in mathematics was sometimes in the Latin language, and a cyphering/recitation approach was the order of the day. This made it difficult for schools to prepare their students well for college mathematics.

Native-American children and children of indentured servants hardly ever studied "European-background" *Level 1 or Level 2 abbaco* forms of mathematics, although it is known that at least four Native American students attended Harvard College in its early days, and Caleb Cheeshahteaumuck and Joel Hiacoomes, two members of the Wampanoag tribe from Martha's Vineyard, graduated from Harvard in 1665 (Lopenzina, 2012; Silverman, 2006). Before entering college, Cheeshahteaumuck and Hiacoomes had attended a preparatory school in Roxbury, Massachusetts, where, presumably, they studied some mathematics.

During the period 1700–1865. In the E-C cyphering book collection there are 63 manuscripts which were prepared in North America (excluding Canada) before 1800. These seventeenth- and eighteenth-century manuscripts were not easy to find and were expensive to purchase. The E-C collection of seventeenth- and eighteenth-century North American cyphering books is the second largest in existence, with only the Phillips Library (in Salem, Massachusetts) having more. Taken together, the cyphering books in the E-C collection and in the Phillips Library provide overwhelming evidence not only with respect to what mathematics was learned at the time, but also to how it was learned. As Lao Genevra Simons (1936) stated, many years ago:

> A great deal has already been said about the custom of keeping student notebooks during this period of difficulty in obtaining books from England and of printing books in the colonies. If all the notebooks now hoarded by descendants of graduates of the early American colleges or lying neglected and forgotten in attics and closets, if all these notebooks could be presented to the several college libraries or historical societies, the history of early American education would be greatly enriched. In these notebooks, there is found the content and scope of the curriculum of the day in evidence that is unmistakable. (p. 588)

From our analyses of cyphering books we can say that, without doubt, the implemented mathematics curriculum for those 15- to 19-year-olds attending schools which paid attention to mathematics was consistent with the sequence of *abbaco* topics listed earlier as belonging to Level 2, and Level 3. The biggest variation occurred with respect to fractions, which were usually covered early in most textbooks, but much later (if at all) in cyphering books—that is to say, the implemented curriculum differed from the intended curriculum. Fractions were already becoming a "weeping sore in school mathematics" (Ellerton & Clements, 1994).

Entries in Table 5.1, in Chapter 5, reveal that between 1607 and 1829, only 6 of 339 cyphering books (CBs) in the E-C collection prepared during that period included entries on algebra (i.e., 2%), 28 included entries on geometry (i.e., 8%), 21 included entries on trigonometry (i.e., 6%), 20 included entries on surveying (i.e., 6%), and 8 included entries on navigation (i.e., 2%). Of the CBs prepared between 1830 and 1865, 19 (i.e., 10%) included entries on algebra; 14 (i.e., 7%) included entries on geometry; 12 (i.e., 6%) included entries on trigonometry; 11 (i.e., 6%) included entries on surveying; and 3 (i.e., 2%) included entries on navigation. Those data attest to the *content*-side of the implemented curriculum.

Up to about 1840 the pedagogy-side of the implemented curriculum was consistent with the cyphering tradition's individualized, one-on-one recitation-supported, cyphering approach to learning and teaching. However, the influence of normal schools from about 1840 onward led to a rapid abandonment of that form of pedagogy in high schools and academies, with whole-class teaching supported by textbooks and an assessment system based on public "group recitations"—often at blackboards—becoming more widely used. Also, written tests and examinations quickly became a commonly accepted way of assessing the attained curriculum.

Neither differential nor integral calculus appeared in the intended or implemented mathematics curricula in the early schools. Any algebra which was taught usually emphasized elementary manipulation of symbols, and geometry and trigonometry usually appeared only in sections on surveying, navigation, or mensuration (which were *not* part of the implemented curricula in most schools). Methods of measuring the attained curriculum in the high schools changed from the 1850s onward as a result of the introduction, largely though the influence of Horace Mann, of externally-set written examinations.

Cyphering-book data indicate that throughout the period 1607–1865, Native-American children, African-American children, and children of European-background indentured servants rarely studied Level 2 or Level 3 *abbaco* topics, or any of algebra, geometry, trigonometry, surveying or navigation (Ellerton & Clements, 2021).

Answer to Research Question 4: What Were the Intended, Implemented and Attained Mathematics Curricula for North American College Students During (a) the period 1607–1776? and (b) the Period 1776–1865? And, to What Extent Did the Answers to Those Questions Vary Across North America, and Across Different Groups?

During the period 1607–1776. Mathematics at Harvard College in the seventeenth century was pitched at a low level, and it was not until Isaac Greenwood's appointment as Hollis Professor of Mathematics and Natural Philosophy, in 1728, that someone with strong up-to-date knowledge of the more advanced "European" forms of mathematics was appointed to a senior position in mathematics in a college in North America (excluding Canada).

Although the number of North American colleges grew steadily after 1745, immediately before the Declaration of Independence in 1776 there were only nine colleges established in what would become the United States of America—Harvard College (established in 1636), College of William and Mary (est. 1693), Yale College (est. 1701), College of New Jersey (Princeton, est. 1746), King's College (Columbia, est. 1754), College of Philadelphia (est. 1755), College of Rhode Island (Brown, est. 1764), Queen's College (Rutgers, est., 1766) and Dartmouth (est. 1769). Some asked students to prepare *abbaco* arithmetic cyphering books, and most introduced their students to the first two books of *Euclid's Elements*. For the first 90 years of Harvard's existence the study of geometry was largely confined to the first few books of *Euclid's Elements* (Stamper, 1909), or to the interpretations of those books in textbooks (e.g., in John Ward's (1719) *The Young Mathematician's Guide: Being a Plain and Easie Introduction to the Mathematicks*).

College students did not usually own any mathematics textbooks, and those made available in college libraries were usually written by European (especially British) authors. That was largely because sending textbooks across from England made a great deal of economic sense for Great Britain, because that practice provided useful balancing currency exchange for "high-demand" imports such as tobacco (Australia and Van Diemen's Land, 1868). North American college students were drilled in Greek, Latin, geometry, history, ethics, and rhetoric, but little attention was given to "extras" such as high-level *abbaco* arithmetic, algebra, trigonometry, surveying, navigation, or probability. Colleges often accepted young students—some as young as 13 or 14 years—and tuition fees were low, with scholarships only rarely being available. Many of the teachers were mere "passers-by" so far as mathematics was concerned—they planned to become clergymen, lawyers, or physicians (Cremin, 1970, 1977; Geiger, 2014). In the seventeenth and eighteenth centuries the Southern colonies fell well behind the Northern colonies so far as provision of colleges was concerned, with many Southern plantation owners sending their older children for higher studies to Europe, or to colleges in New England, Pennsylvania, or New York (Cremin, 1977; Mayo, 1898).

During the period 1776–1865. In 1788 the intended mathematics curriculum in mathematics at Harvard College comprised arithmetic, algebra, Euclidean geometry, trigonometry, conic sections, and spherical geometry (Zitarelli, 2019. p. 139). Between 1815 and 1830 John Farrar, the Hollis Professor of Mathematics and Natural Philosophy at Harvard, devoted considerable time and effort into revising the College's mathematics curriculum—the best-known aspect of his revision being his introduction of translations into English of textbooks written by famous Continental-European mathematicians. In particular, he thought that if Harvard students constantly referred to textbooks written by French mathematicians (e.g., by, Étienne Bézout, Louis Bourdon, Sylvestre François Lacroix, or Adrien-Marie Legendre) then the overall quality of mathematics at Harvard would immediately be improved. However, his hopes were not realized—the European-background texts were not well received by many Harvard students (or by students in the small number of high schools and academies where the books were sometimes used), possibly because the texts were too abstract. Farrar's intended curriculum was associated with the books, and comprised higher forms of arithmetic, Euclidean and solid geometry, algebra, trigonometry, and calculus.

College mathematics curricula between 1776 and 1830 did not inspire students, although the presence of calculus in a few colleges by 1830 at least represented progress. Aside from curricular innovations, especially those in algebra introduced by Benjamin Peirce at Harvard in the 1830s and 1840s (see Chapter 5 and Chapter 7 in this book), and Robert Adrain's introduction of the least squares method in statistics to his students, it is our judgement that throughout the period 1700–1865 college-level mathematics in North American colleges was usually of an elementary nature and was behind-the-times when compared with mathematics studied in European colleges. New developments in analysis or in probability, for example, were conspicuous by their absence.

A realistic perspective on the state of college mathematics in the North American colleges in the early 1800s can be seen from the fact that in 1801, Samuel Webber, then Hollis Professor at Harvard, caused to have published his two-volume, *Mathematics Combined from the Best Authors and Intended to be the Textbook of the Course of Private Lectures on These Sciences in the University of Cambridge*. Volume 1 had 426 pages and volume 2, 610 pages (Webber, 1801). The volumes provided mostly plagiarized forms of chapters in an English textbook by Charles Hutton (Karpinski, 1980, p. 140), and their lack of originality was obvious—

although they certainly offered a wide-ranging set of standard topics (arithmetic, logarithms, algebra, geometry, plane trigonometry, mensuration of surfaces, gauging, heights and distances surveying, navigation, conic sections, dialing, spherical geometry and spherical trigonometry) (Zitarelli, 2019, p. 131). However, the treatment proved to be too difficult for most of the Harvard students. According to Zitarelli (2019), the volumes were Webber's "only contribution to mathematics" (p. 131). Be that as it may, they were widely adopted in U.S. colleges between 1802 and 1810, before they were superseded by Jeremiah Day's texts on geometry, algebra, plane trigonometry, navigation, surveying, and mensuration (Karpinski, 1980, p. 630). Day's books were written by Day, not plagiarized like the texts in Webber's volumes seen as.

At the risk of being seen to be judgmentally harsh, we believe that by the mid-1820s, Webber, Day, and Farrar had succeeded in establishing college mathematics in the United States as a subject for a minority. Webber and Farrar referred students to mathematics texts used in European colleges and universities and, arguably, the main outcome experienced by students who relied upon those texts was to view mathematics as something which they could not do very well.

There was a positive side, however. At Yale, Harvard, and USMA (West Point) during the 1820s the attained curriculum was increasingly measured through one-on-one recitations, with blackboards being widely used—especially at Yale College and at West Point. For the first time in the history of college mathematics in the United States there emerged a strong emphasis on formal proof, especially in relation to geometry and conic sections. Not all students liked this development, especially at Yale (see Chapter 7), but we believe that from a mathematical perspective it was an important step in the right direction.

Answer to Research Question 5: What Perspectives on the Status of Mathematics in College Curricula Were Held in North America During the Period 1607–1865?

In October 1916, George A. Plimpton began a presentation to the American Antiquarian Society by showing an illustration of an early sixteenth century figure, called the "tower of knowledge," taken from Gregor Reisch's (2002) *Margarita Philosophica* (Plimpton, 1916). At the bottom of the tower a teacher was shown handing a hornbook to a young boy who wished to enter the tower, and once having entered he would learn the alphabet and the Hindu-Arabic numeration system before climbing stairs leading to rooms representing 12 key areas of knowledge—Latin grammar, dialectics, rhetoric, arithmetic, music, geometry, astronomy, physics, natural history, physiology, psychology, and ethics. According to Plimpton, this list incorporated the prevailing ideal of what an educated person should know. Notice that the list not only included arithmetic and geometry, but also astronomy and physics—all of which were part of the curricula offered in most of the early North American colleges. After its establishment in the late 1630s, Harvard College devised a curriculum which included arithmetic and Euclidean geometry, and other subjects which dealt with aspects of physics and astronomy. So, from the beginning, there was a nod in the direction of applied forms of mathematics.

It was assumed from the start that beginning students at Harvard were acquainted with the Hindu-Arabic numeration system, based on the numerals 1, 2, 3, 4, 5, 6, 7, 8, 9, and 0. It was also assumed that they could use established algorithms to add, subtract, multiply and divide, and could apply these to everyday commercial activities involving money, and tasks associated with the measurement of weights, time, and distances. Not much more was expected however, of beginning students, so far as mathematical knowledge was concerned. In fact, around 1640 most European-background people in and around Boston were not functionally familiar with Hindu-Arabic numerals or the operations which could be used with those numerals by someone attempting to solve real-life problems.

The "tower-of-knowledge" curricular position was bolstered by adherence to a theoretical position based on the ideas of Anicius Manlius Severinus Boethius, a fifth-century CE Italian philosopher who during his lifetime had been an enthusiast for number theory (Dürr, 1951; Høyrup, 2014). Boethius's ideas on curricular design became the basis for the so-called "quadrivium," a medieval curricular theory which combined the "mathematical arts" of arithmetic, geometry, astronomy, and music with classical traditions (involving Latin, Greek, and Hebrew languages and literatures) to create an education philosophy which controlled the thinking of decision-makers associated with European universities for centuries (Gilman et al., 1905; Høyrup, 2014; Schrader, 1967). Mathematics was deemed to have only a minor role in higher education—a much less important role than the classical languages and literature (Burton, 1996; Elliott & Rossiter, 1992).

The "tower-of-knowledge" and quadrivium curricular theories may have affected the thinking of some Harvard College administrators about the role of mathematics in College curricula but it is likely that most administrators thought of mathematics as something which corresponded to the 600-year-old cyphering tradition (Ellerton & Clements, 2012).

From Kilpatrick (1912) and Pelletreau (1907) we know that cyphering took place in some seventeenth-century colonial schools and it is likely that most of the students entering Harvard College during the seventeenth century would, at some stage of their schooling, have prepared at least one cyphering book. It is interesting to reflect on how immersion in practices associated with that tradition might have affected the thinking of not only the College students but also of those teaching and learning mathematics in schools which were preparing students for higher studies.

So far as the preparation of cyphering books was concerned, we know that students

- were expected to display high-level penmanship and calligraphy.
- were not supposed to enter material in their books until they had gained the approval of their masters. This meant that, in theory at least, all

solutions to exercises that students entered in their books should have been correct.

- had dealt with arithmetic that related to real-to-life situations as defined by the *abbaco* curriculum content (which focused on developing understanding of the Hindu-Arabic numeration system and especially its relation to commercial applications).

The first two of these bullet points explains, perhaps, why so many families have been prepared to keep cyphering books for centuries. The E-C cyphering book collection includes 549 North American cyphering books prepared between 1607 and 1865. When we examined those cyphering books we found it hard to avoid the conclusion that most of those who prepared them were very proud of them. They represented the students' best thinking and their most concentrated efforts. This quest for excellence was something which came to be associated with the study of mathematics in the United States during the period 1607–1865. Furthermore, the tradition included the idea that a cyphering book was prepared for beyond-school reference. It was valued because of the expectation that it could be consulted when the need to solve real-world problems arose in the future.

Much has been written (see, e.g., Smith & Ginsburg, 1934) about the failure of the United States to produce a world-class mathematician during the period to 1870. From our perspective, that judgment is unfair to Nathaniel Bowditch and Benjamin Peirce. It also fails to give due attention to the peculiar conditions in the 13 colonies, and in particular to the colonialist policies of the British government with respect to higher education. The British government was pleased to maintain some form of control over the developing forms of higher education in its colonies. An outcome of that policy was that mathematics textbooks written by English authors (like Edward Cocker, John Ward, John Hill, Thomas Dilworth, William Hawney, Daniel Fenning, John Bonnycastle and Charles Hutton) were preferred to books by home-grown authors. From the British perspective books were a convenient export to balance monies expended on the import by the "home" country of tobacco and other "needed" objects or products.

The rush toward French education methods and the adoption of textbooks by French authors was based on a questionable assumption—fueled by the admiration of French mathematicians and French mathematics educators by leaders at the United States Military Academy at West Point, and by John Farrar, Hollis Professor at Harvard College between 1807 and 1836—that French approaches to the teaching and learning of mathematics were preferable to British approaches. Powerful persons—like, for example, Thomas Jefferson—came to believe that mathematics curricula at U.S. colleges should be supported by textbooks written by Continental, especially French, scholars, and not British scholars. All of that contributed to a feeling among U.S. college authorities that mathematics in North American educational institutions was second-class, and so too were the levels of scholarship of teachers of mathematics in the schools and colleges.

What has been written in the last paragraphs could help explain the curious phenomenon of self-educated geniuses like Nathaniel Bowditch and David Rittenhouse, who managed to achieve mathematical excellence despite their not having access to serious mathematical training in colleges. One could argue, nevertheless, that the cyphering tradition had generated an appreciation of the peculiar beauty and power of mathematics in the minds of some young people who were moved to work hard to achieve excellence in the field.

Indeed, one should ask—what other explanation could there be for what transpired with Bowditch and Rittenhouse? And, how could it have been that Benjamin Franklin, who had not been attracted to mathematics at school, somehow stumbled across the remote, mathematically undignified topic of magic squares, and was then captivated by the beauty and order of patterns which were embodied in numerical relationships? Such instances make it clear that untapped and unshaped mathematical talent existed in the United States and was waiting to be recognized in order that great things might happen. Almost certainly, much potential existed in many young minds, but was not recognized or nurtured.

In the early 1760s at the College of William and Mary, in Virginia, there was a young man named Thomas Jefferson who fortuitously came into contact with a generous and knowledgable mathematical mentor, William Small (Ganter, 1947). And Jefferson's inquisitive mind would, lead him to create and implement the world's first fully-decimalized system of currency which would change the monetary systems of the world, and would lead Jefferson to conceptualize a decimalized system of measuring weights and measures even before the French mathematicians formally developed the metric system in the early 1790s.

After 1776, mathematical developments in U.S. colleges were held back because the thinking of those responsible for curriculum development was dominated by the idea that a college curriculum should have classical languages and literature as its base. The colleges were slow to demand of prospective students a knowledge of mathematics beyond low-level *abbaco* arithmetic, and it was not until the nineteenth century that beginning college students would be asked to demonstrate some knowledge of elementary algebra. The fact that hardly any of the persons appointed as professors of mathematics wrote mathematics textbooks or scholarly articles of any kind would also have had an effect. Some of them—like John Winthrop IV at Harvard—did contribute to improvements in applied mathematics, especially in relation to navigation and astronomy, but in general, the professors of mathematics felt no pressure to carry out high-level original research or to contribute to discussions about mathematical developments which were taking place among European scholars. From their perspective, that could be left to the Europeans. That would change after 1800, when Bowditch, Adrain, and Peirce would show the way.

At the government level, there seemed to be a lack of recognition of any need to improve the quality of mathematics in the colleges. Certainly, as previously argued, the creation of USMA at West Point was an important initiative in the right direction, but it was put in the hands of administrators who could not look beyond the need to reproduce "superior" French forms of mathematics. That said, the emphasis at West Point, and at Yale, on using the recitation approach to assess the attained curriculum resulted in some North American students gaining deeper understandings of important aspects of mathematics, especially with respect to the concept of proof.

Answer to Research Question 6: What are the Implications of the Answers to the Five Questions (Above) for Those Investigating the History of "Higher" Mathematics in North America? What Future Research is Needed, and to What Extent will it be Feasible to Conduct that Research?

In providing our answers to these final research questions we wish to emphasize that our fundamental finding is that the history of mathematics between 1607 and 1865 in the mainland part of the present United States of America (excluding Canada and Alaska) should be seen to be about much more than the history of contributions by U.S. mathematicians who carried out "weight-bearing" research into what are regarded as important themes within 21st-century higher mathematics. Today's so-called "Mathematics Subject Classification" (MSC) is produced by the staff of the review databases "Mathematical Reviews" and "Zentralblatt MATH" (Lange et al., 2012), and some mathematics journals ask authors to label their papers with MSC subject codes. The MSC divides mathematics into almost 100 areas, with further subdivisions being provided for sub-areas. Although we do not deny the value of such an exercise for mathematics researchers of the 21st century, we would hasten to add that in writing this present book, which offers a history of North American mathematics between 1607 and 1865, we were *not* guided by MSC.

Our study has been framed by the need to identify, describe, and interpret events which occurred between 1607 and 1865 and which moved the United States closer to an enactment of the principle of "mathematics for all." Clearly, decisions made in the seventeenth, eighteenth and early nineteenth centuries to establish elementary schools in New World, and to include "cyphering" within the intended curriculum of those schools are a highly significant part of our historical story. Linked to that, is the force of a *cyphering tradition* which was initially translated from Europe to North America, and from the outset controlled intended, implemented and attained mathematics curricula in North American educational institutions. And obviously, too, the creation of the colleges in the seventeenth, eighteenth and nineteenth centuries provide time markers for historians, as did the establishment of public high schools from the 1820s onward. The intended, implemented, and attained mathematics curricula of those institutions obviously generated crucially important data for our mathematics-for-all story.

Chapter 2 took up issues associated with the history of early childhood mathematics education in North America. The hornbook was the key artifact in that story. We have argued that it was not until the early 1820s that issues associated with mathematics for young children (i.e., children less than 10 years of age) began to be taken seriously in North America, with Warren Colburn providing the lead. Colburn was much influenced by Johann Heinrich Pestalozzi, the great Swiss educator, and his desire to initiate a decent mathematics education for young children, especially girls, redirected mathematics education in the United States of America. In the 1830s the challenge would be taken up, internationally, by Friedrich Froebel with his kindergarten movement (Downs, 1978). During the period 1820–1865 discourse patterns in mathematics classes would become a matter for discussion with Colburn, Pestalozzi and Froebel all calling for teachers to move toward whole-class instruction in which real-life artifacts were utilized. For the first time, teaching practices in mathematics were put under the spotlight and became an issue for serious discussion.

Our concept of "all" in the expression "mathematics for all" includes European-background and Asian settlers, Native-American indigenous peoples, African-American slaves and their children, European-background indentured servants and their children, and indeed every person living in North America (excluding Canada and Mexico) during the colonial era and then the United States of America. We would like to have given a wider coverage of the history of mathematical content taught to, and applied by, those with non-European-backgrounds. That is a task for future researchers (but see Eels, 1913; James, 2013; Norrell & Myers, 2017; Paraide, Owens, Clarkson, Owens, & Muke, 2022).

Another key aspect of "mathematics for all" is the meaning to be given to "mathematics." That issue was too large for us to consider in detail in this book (but see Courant & Robbins, 1941). In Chapter 6 we showed how some administrators and mathematics teachers in the early academies and colleges were pleased to include *applied* forms of mathematics, such as navigation, surveying, gauging, and mensuration, in their intended and implemented mathematics curricula. Indeed, mathematics was assumed to have a legitimate pure-applied division. Any history of mathematics in North America should take that division as seriously as those administering the academies and colleges did.

The year 1865 simultaneously marked both the end and the beginning of eras in mathematics in North America. The Civil War was over, the Emancipation Act was passed, and what was called for were forms of education which would address the needs of *all* persons—including pre-schoolers, elementary-school children, high-school children, college students, mathematicians, and adults who wished to study mathematics outside the umbrellas of formal education institutions. The power of mathematics would become a reality when the nation would be prepared to adopt a mathematics-for-all attitude—something which is easy to say, but has been extraordinarily difficult to achieve (Clements et al., 2013). Reasons why it was so difficult to achieve are hidden deep in layers of history of society and of mathematics in the North America over the period 1607–1865. Readers are invited to reflect on the following questions:

1. Did leaders of education in North America ever seriously reflect on issues to be associated with a mathematics-for-all mentality?
2. What *needs* to be done, now, to further the cause? What *can* be done?

Limitations and Possibilities for Further Relevant Research

Much more can and should be researched with respect to indigenous forms of mathematics known and used by indigenous Americans, especially, during the seventeenth century. Relevant data are hard to find on that theme, but it is important that issues associated with the gradual extinctions of the indigenous forms—like, for example, the shaping and building of canoes from tree trunks—and with the roles that formal education institutions played in those extinctions, not only be explored but also problematized (see, for example, Paraide et al., 2022).

We have considered elsewhere the importance of the lack of congruence between intended and implemented curricula (Ellerton & Clements, 2012), but much more needs to be done on why that occurred, and its effects. Given the absence of pencil-and-paper achievement test data until well into the nineteenth century, it will be difficult to obtain convincing data for generating answers to "attained" curricular questions. Nevertheless, any attempt to write a history of mathematics education in the United States from the perspective of "student understanding of what was being studied" would be extremely valuable. A good starting point, perhaps, would be working in Yale University's archives and USMA archives to investigate aspects of the conic-sections disputes in the 1820s with respect to assessing students' understanding achieved through the system of recitation, and the effects on learning of the introduction of blackboards. Analysis of early pencil-and-paper mathematics achievement data, from the mid-1840s onward, held in the archives of state departments of education, would also be useful so far as attained curricula are concerned.

From an equity perspective, the 1862 Emancipation Act which was passed during the Civil War led to many African-American children of slaves being given the opportunity to study mathematics formally for the first time. The effects of that momentous change on mathematics and mathematics education in the United States over time is a challenging, but hugely important, matter for further study. Scholars need to identify qualitative, quantitative, and, indeed, any readily accessible data which will make that research a feasible proposition.

One of the intriguing findings by researchers into the history of mathematics in North America has been that some of the best early mathematics research was carried out by those who had not studied much mathematics in formal institutions—like Benjamin Franklin, Nathaniel Bowditch and David Rittenhouse. Issues associated with the history of "out-of-school" mathematics need to be researched—but fruitful data sources may not be easy to locate, and the development of appropriate research methodologies will be challenging.

Effects of colonialist thinking on modern mathematics education practices are worthy of exploration from historical vantage points. Post-modern, critical research methodologies would probably be most appropriate for such investigations. One question which would be of interest is "What remnants of colonialist thinking are still to be found in twenty-first century practices in mathematics education?"

Both authors of this present book attended one-room schools in rural settings in Australia when they were young. In those schools, students aged from 5 to 16 years were in the same room when being taught all subjects, including mathematics, and usually all students in the room were taught by the same teacher. Altogether, more than 200,000 one-room schoolhouses were built in the rural areas of the United States of America, including more than 90,000 in the Midwest states (Mydland, 2011). During the period 1800–1950 the "public" one-room "common schools" were, typically, supported by local taxes and were administered by district committees which appointed and paid teachers, and organized working bees and social events which, among other things, incorporated spelling and mental arithmetic contests. Both authors remember how, when they were students in one-room schools in Australia, they were able to follow and take interest in the mathematics that older students were being asked to study. A historical investigation into the effects of one-room schooling on mathematics learning could generate intriguing results.

From about 1845 onward, pencil-and-paper test mathematics data were gathered by state departments of education from persons wishing to qualify to enter district high-schools, and such data, if they still exist, and can be found, could be useful for those interested in equity issues, and also in attained curricula with respect to different topics (e.g., fractions, and algebra). Similar data sets may be available in college archives with respect to students' performance on pencil-and-paper college-entrance examinations.

The study of calculus in district high schools and colleges came later in the United States of America than in many other "advanced" nations (Zuccheri & Zudini, 2014). One wonders why that was the case. Data on that question, and on other related curricular questions will be buried deep in college archives and in district and state education offices, as well as in the College Entrance Examination Board archives. It should be a challenge for researchers to find persons who know where those data are located.

The present book considered data from an unusual time period 1607–1865. Within that period there are a number of potentially very important "marker" years—such as 1607, 1619, 1620, 1625, 1700, 1776, 1788, 1800, 1820, 1845, 1862, 1865—and the idea of "slicing" the overall period into potentially important sub-periods (e.g., 1776–1820) to investigate changes in some aspect of mathematics or mathematics education during that sub-period, could be fruitful.

Concluding Comments

This book began with the story of a band of just over 100 European-background males who settled in and near what would become known as Jamestown, Virginia. Most of them came with hope in their hearts and a desire to make money within a challenging environment. It could be argued that most of them lacked the skills and knowledge needed to establish a viable pioneering frontier settlement and were unprepared for the fact that on landing they would immediately be confronted by the rightful owners of the territories in which they chose to settle. Evertheless, from that small ill-prepared band, and other like groups who would settle in other regions, would emerge in less than 200 years an independent mix of people who, between 1776 and 1783 would, somehow, combine to defeat the army and navy of one of the most powerful nations on earth.

Among the others who came soon after 1607, was a group seeking religious freedom who, in 1620, landed hundreds of miles north of Jamestown at a place they called Plymouth. In 1625 Dutch-background pioneer settlers established New Netherland, with New Amsterdam as its capital. Forty years later New Amsterdam was captured by the British and renamed New York, but it was soon handed back to Dutch. Finally, in 1674 it was traded to the English who once again called it New York, and a part of it, New Jersey. Other groups would come from various parts of Europe, and increasingly, indentured servants from Great Britain and slaves from Africa would be transported to North America to provide a cheap labor force (Blackburn, 1997). In time, cities would be built and colleges, mostly modelled on the famous British universities at Cambridge and Oxford, created. As in European universities, different forms of mathematics, both pure and applied, would be taught in the colonial colleges.

In this book we have investigated the European-background forms of mathematics and mathematics education which were translated from Europe to the education institutions created in the North American New World. By and large, attempts were made to make the mathematics studied in the New World identical with the mathematics at "home." Whether such attempts were wise could be debated, but they *were* made. Naturally enough, those charged with administering the new education institutions, including those who taught mathematics in them, faced enormous challenges as a consequence of large cultural and educational differences among their students, as well as a serious lack of education-relevant, personnel and financial, resources. In Europe in the early 1600s, John Napier and Henry Briggs, in Great Britain, Simon Stevin in Holland, *Joost Bürgi* in Switzerland, and René Descartes in France were weaving their magic, and having findings of their groundbreaking mathematical research published. In each of the leading European nations there were scholars who identified as professional mathematicians, but none of the best known among them would ever relocate to the North-American New World. Given the circumstances, was it reasonable to expect those in Jamestown, in Plymouth, in the Dutch-background settlement of New Amsterdam, and in other settlements which were established, to study and further develop the same kind of mathematics being created in "home" education institutions?

But, as we said, the attempt *was* made, and we maintain that what was done was amazingly successful. Certainly, for over two centuries "mathematical standards" in the New World lagged well behind those at "home"—in the schools. in the colleges, and in society in general. In order to create satisfactory facilities to educate their children, and to train pastors for their desired religious establishments, the settlers decided to build both from the "bottom up" and from the "top-down." Hornbooks were brought from the homelands, and these helped educate the young children; local schools were funded by local taxes—and Harvard College was created in the late 1630s, with the University of Cambridge as its model. Incredibly, almost immediately, Cambridge agreed to accept qualifications from Harvard as equivalent to those at Cambridge. Even so, Harvard students had to learn that "a colonial college student was ranked not by popularity, athletic prowess, or even intellectual ability, but by the dignity and position of his family" (Morison, 1932, p. 2).

So far as mathematics education in the New World was concerned, one of the most difficult problems the inhabitants of the New World faced during the period 1607–1865 was that of gaining access to a body of teachers who knew their mathematics well enough to be able to teach it in a satisfactory manner. A few very capable mathematics scholars, like Pieter Venema (from The Netherlands) and Alexander Malcolm (from Scotland), made their ways to New York to teach, but even they found the going tough. Often school mathematics was taught by college students wishing to earn enough so that they could survive until they gained a position in law, or medicine, or the church. For most of the period there was a tendency to cling to the curricula and pedagogies of the long-established cyphering tradition—with students, both male and female, preparing cyphering books which incorporated implemented curricula based on the *abbaco* sequence, and teachers talking with students during one-on-one recitation sessions. From the early 1800s, blackboards began to be used in recitations, and "understanding" rather than mere memorization began to be more emphasized. Even the concept of proof began to be taken seriously, especially at West Point (USMA) and Yale College. That trend was accelerated from 1840 onward, when normal schools were created and charged with the specific task of improving teaching methodologies in schools (Harper, 1939).

Between 1820 and 1834, Warren Colburn, fired with the zeal of Johann Heinrich Pestalozzi, encouraged teachers to help young children learn to generalize. Colburn challenged teachers to teach mathematics in a new way by which whole-class and small-group discourse patterns would be more important than ever before. By 1865, however, textbooks written by North American authors like Joseph Ray, Benjamin Greenleaf, and Charles Davies had taken control of intended and implemented curricula and, following the initiative of Horace Mann, written tests were increasingly used for the purpose of grading and ranking students. A "new normal" gripped school mathematics and it would remain in place for the next 150 years (and more).

Increasingly, algebra, geometry, and trigonometry would come to dominate intended curricula of secondary schools, and it would be assumed that beginning college students should and would have basic competence with such subjects. That said, at the close of the period covered by this book, 1607–1865, analytic geometry (with Cartesian graphs) and calculus, were conspicuous be their absence from implemented curricula in secondary schools. Initially, Benjamin Peirce's efforts in the 1830s to elevate the importance of the study of functions was not well received at Harvard. However, Ferdinand Hassler's (1826) attempt to re-interpret the study of trigonometry through a functions perspective which emphasized ratios of sides of a right triangle—rather than the directed-line-segment approach—was better received. Ten editions of Hassler's textbook on the subject would be published between 1826 and 1843 (Karpinski, 1980).

Throughout the seventeenth century hardly any school or college students were expected to tackle applied forms of mathematics like gauging, surveying, or navigation. In fact, in the schools many students did not study any mathematics and even those who did rarely proceeded beyond elementary cyphering for which the content was the lowest level of *abbaco* arithmetic.

The challenge of becoming aware of, catching up with, keeping abreast with, and ultimately surpassing the breathtaking mathematics achievements by European scholars such as Isaac Newton, Gottfried Leibniz, and Pierre-Simon Laplace, or with the high mathematical standards achieved by James Hodgson with 12- to 16-year-olds at the Royal Mathematical School within Christ's Hospital, in London (Ellerton & Clements, 2017; Zitarelli, 2019), was not something taken seriously by administrators of most colonial education establishments. During most of the period 1607–1865 any idea that mathematics learning and research in the colonies (or in the United States) could reach the standards achieved in "home" education institutions was regarded, within Europe, and even within North America, as nothing more than wishful thinking. *Of course* they could not, and no reasonable person should have expected otherwise—there was simply too much catching up to do. But, toward the end of this period the gap was narrowing, and later there would come a time when the gap would be completely closed. Historians of U.S. mathematics need to recognize, and celebrate, the speed and scope of the achievement.

There were, of course, some colonial leaders who thought that mathematics should have an important place in formal education curricula. One such person was Thomas Jefferson (Clements & Ellerton, 2015), the Third U.S. President. For him, mathematicians were expected to honor, philosophically, what they reckoned to be true, even if that would also involve them in recognizing that some of their own beliefs, actions, and life situations were not consistent with what they saw as truth (Miller, 1977). The soaring rhetoric of the axiomatic, Euclid-based Declaration of Independence about all men being "created equal" was not consistent with the fact that Jefferson himself was a slave owner and indeed, was the father of children conceived as a result of his union, apparently over many years, with Sally Hemmings, a black woman in his Monticello entourage, someone much younger than he (Berlin, 1998; Cogliano, 2008; Wiencek, 2012). Yet, in 1776 Jefferson, in a draft version of the Declaration, penned paragraphs, which blamed King George for the adoption of slavery, and for the transport of the associated transatlantic "slave trade," to North America (Cohen, 1969).

In his initial draft, Jefferson described slavery "as a crime against humanity." He prepared the following strongly-worded paragraph—which, ultimately, was rejected by a majority of the other members of the committee charged with the task of preparing the Declaration:

> He [i.e., King George] has waged cruel war against human nature itself, violating its most sacred rights of life & liberty in the persons of a distant people who never offended him, captivating & carrying them into slavery in another hemisphere or to incur miserable death in their transportation thither.

Jefferson referred to slavery as an outcome of "piratical warfare," and as "execrable commerce." He then criticized King George for "exciting those very people to rise in arms among us, and to purchase that liberty of which he has deprived them, by murdering people who had also been deprived of liberties" (quoted in Wiencek, 2012).

Jefferson was driven by the demands of logical thinking in creating and expressing his draft-Declaration judgments on slavery (Quarles, 1961; Tewell, 2012). He knew, surely, that since he himself was a slaveholder, he would be opening himself up to almost unanswerable accusations of hypocrisy. Jefferson had inherited, from his father, his situation with respect to holding slaves (Wiencek, 2012; Wilson, 1992) He must have known that his role as a leader in state and national affairs would likely be seriously questioned if his words on slavery were included in the Declaration. So, although he is credited with infusing ideals of equality and freedom into the nation's founding document, that document would ultimately remain silent on the issue of slavery. One can only speculate what might have flowed from a Declaration which included the remonstration about slavery, and how such words might have changed the paths of educational opportunities for generations of students. How much was the trajectory of mathematics in the United States set back by the omission of that statement? Is the United States still in "catch-up" mode, where equality of opportunity to study mathematics is still a challenge (Berlin, 1998).

One of our aims in writing this book was to reveal how the development of a "mathematics-for-all" way of thinking in America was profoundly influenced by the cyphering tradition, and by an assumption, so common among the ruling classes in colonial North America, that only boys from well-to-do families should have the opportunity to study any form of mathematics beyond *abbaco* arithmetic. Recent analysis of substantial collections of cyphering books prepared in North America has drawn attention to the pedagogical practices and content intimately associated with implemented mathematics curricula in schools during the period 1607–1865. Throughout most of that period, implemented curricula over-emphasized the importance of memorizing, *without* understanding, commercially-oriented aspects of arithmetic, and elementary geometry. Historical perspective suggests that during the first half of the nineteenth century the situation qualitatively changed as a result of the development of the recitation process, assisted by the increasing availability and use of blackboards and, after 1820, greater emphasis on group work. The forward-looking initiatives of educators like Warren Colburn, Catharine Beecher, Charles Davies, Joseph Ray, and Horace Mann—and instructors in colleges and normal schools like Nicholas Tillinghast and Richard Edwards—prepared the way for students ultimately to become capable of contributing actively to genuine research in mathematics.

Investigations of why it became possible for more advanced research agendas in mathematics to be investigated within the nation from about 1875 onward need to take account of how changing societal norms had an impact on implemented and attained mathematics curricula in educational institutions. Following the lead of key individuals, like Warren Colburn, Nathaniel Bowditch, Benjamin Peirce, James Sylvester, and Eliakim Hastings Moore, new emphases in school and college mathematics, and a greater emphasis on the need to create an academic culture which facilitated mathematics research in the United States of America, paved the way for a mathematics research community finally to emerge (Dauben & Parshall, 2014; Parshall, 2003; Zitarelli, 2019).

Just over 40 years ago the Mathematical Association of America (MAA) organized a National Science Foundation-funded conference "to review the progress of the Association and to formulate a plan of action" (MAA, 1978, p. 7). The first conference recommendation was that "new efforts should be made to define, or redefine, the essential mathematical skills that are needed by every citizen" (p. 9). Clearly, in the late 1970s North American mathematicians had come to recognize and accept the importance of a mathematics-for-all concept. But, as James Baldwin (1998) stated—please reread the Baldwin quotation inserted before the start of Chapter 1 in this book—"the great force of history comes from the fact that we carry it within us, are subconsciously controlled by it in many ways, and it is literally present in all that we do" (p. 722).

This book has revealed the main components of the historical effort to achieve "mathematics for all" in North America. Today, almost every child in the United States is studying, or will study, mathematics. So, in a minimalistic sense, "mathematics for all" is being achieved. But, one can argue, education in mathematics should help *all* U.S. children to learn enough mathematics not only to be able to survive with dignity, but also to thrive, within their present and likely future life situations, and there are data which suggest that that aim has not yet been achieved (Cothran, 2018). One can argue that mathematics education should *empower* all people so that whenever it is appropriate they can use quantitative and other mathematics-related techniques to pose and solve problems, thereby enabling them to do more than merely "survive" with dignity (Guasco, 2014). Part of that argument is based on the proposition that "mathematics for all" should be more than a mere slogan.

In the 2020s the mathematics-for-all pathway is, therefore, still under construction. The desired destination has changed from what it was in 1865, but there can be no doubt that there are now many more people traveling further along the pathway than ever before. Even so, the aim now is for *all* people to derive *as many benefits as possible* from their mathematical journeys. As we write this final chapter the world is gripped by a COVID-19 pandemic. Every day, newscasters, politicians and medical professionals are talking of the need to "flatten the curve," of the "slopes of the curves," and of how the "ratios" for one state, or nation, are changing at a slower (or faster) rate than corresponding ratios for another state or nation. There is never-ending talk of models, variables, relationships and graphs, with mathematical terminology such as "percentage increase," "exponential change" and "logarithmic change" being often used. We are sure that in 1865 only a tiny proportion of the U.S. population would have been able to comprehend such language—and, indeed, we wonder what proportion of today's population is able to interpret it appropriately. Nevertheless, we recognize that significant progress has been made, and continues to be made, toward achieving the mathematics-for-all objective.

Although the importance of achieving "mathematics for all" is now accepted by most people, it is not clear what "mathematics" should mean, in that context, or how the *value* of an education in mathematics can be heightened for *all* people (Paraide et al., 2022).

References

Ackerberg-Hastings, A. (2014). Mathematics teaching practices. In A. Karp & G. Schubring (Eds.), *Handbook on the history of mathematics education* (pp. 525–540). Springer. https://doi.org/10.1007/978-1-4614-9155-2_26

Ames, S. M. (1957). *Reading, writing and arithmetic in Virginia, 1607–1699*. Williamsburg, VA: Virginia 350th Anniversary Celebration Corporation.

Australia and Van Diemen's Land (1868). *Chambers' information for the people* (Vol. 1, 1417–1448). New York, NY: United States Publishing Company.

Baldwin, J. (1998). *Collected essays*. New York, NY: Library of America.

Berlin, I. (1998). *Many thousands gone: The first two centuries of slavery in North America*. Cambridge, MA: Harvard University Press. https://doi.org/10.2307/20049174

Blackburn, R. (1997). *The making of new world slavery: From the Baroque to the modern 1492–1800*. London, England: Verso.

Biber, E. (1831). *Henry Pestalozzi, and his plan of education*. London, England: John Souter School Library.

Burton, J. D. (1996). *Puritan town and gown: Harvard College and Cambridge, Massachusetts, 1636–1800*. PhD dissertation, College of William and Mary.

Butchart, R. E. (2010). *Schooling the freed people: Teaching, learning, and the struggle for Black freedom, 1861–1876*. Chapel Hill. NC: University of North Carolina.

Cajori, F. (1890). *The teaching and history of mathematics in the United States* (Circular of Information No. 3, 1890). Washington, DC: Bureau of Education.

Chessman, R. (1965). *Bound for freedom*. New York, NY: Abelard-Schuman.

Clements, M. A., & Ellerton, N. F. (2015). *Thomas Jefferson and his decimals 1775–1810: Neglected years in the history of U.S. school mathematics*. New York, NY: Springer. https://doi.org/10.1007/978-3-319-02505-6

Clements, M. A., Keitel, C., Bishop, A. J., Kilpatrick, J., & Leung, F. (2013). From the few to the many: Historical perspectives on who should learn mathematics. In M. A. Clements, A. J. Bishop, C. Keitel, J. Kilpatrick, & F. Leung (Eds.), *Third international handbook of mathematics education* (pp. 7–40). New York, NY: Springer. https://doi.org/10.1007/978-1-4614-4684-2_1

Cogliano, F. D. (2008). *Thomas Jefferson: Reputation and legacy*. Charlottesville, VA: University of Virginia Press.

Cohen, P. C. (2003). Numeracy in nineteenth-century America. In G. M. A. Stanic & J. Kilpatrick (Eds.), *A history of school mathematics* (pp. 43–76). Reston, VA: National Council of Teachers of Mathematics.

Cohen, W. (1969). Thomas Jefferson and the problem of slavery, *Journal of American History 56*(3), 503–526.

Colburn, W. (1821). *An arithmetic on the plan of Pestalozzi, with some improvements*. Boston, MA: Cummings and Hilliard.

Cothran, J. (2018, April 10). Baltimore City is failing its children. *The Sun* (Baltimore).

Courant, R., & Robbins H. (1941). *What is mathematics? An elementary approach to ideas and methods*. London, England: Oxford University Press.

Cremin, L. A. (1970). *American education: The colonial experience 1607–1783*. New York, NY: Harper & Row.

Cremin, L. A. (1977). *Traditions of American education*. New York, NY: Basic Books.

Dauben, J. W., & K. H. Parshall (2014). Mathematics education in North America to 1800. In A. Karp & G. Schubring (Eds.), *Handbook on the history of mathematics education* (pp. 175–185). New York, NY: Springer.

Doar, A. K. (2006). *Cipher books in the Southern Historical Collection*. Master of Science thesis, Wilson Library, University of North Carolina at Chapel Hill.

Dobyns, H. F. (1983). *Their number become thinned: Native American population dynamics in Eastern North America*. Knoxville TN, University of Tennessee Press.

Downs, R. B. (1978). *Friedrich Froebel*. Boston, MA: Twayne.

Dürr, K. (1951). *The propositional logic of Boethius*. Amsterdam, The Netherlands: North-Holland Publishing Company.

Eels, W. C. (1913). Number systems of the North American Indians. *The American Mathematical Monthly*, *20*(10), 293–299. https://doi.org/10.1080/00029890.1913.11997985

Ellerton, N. F., Aguirre-Holguin, V., & Clements, M. A. (2014). He would be good: Abraham Lincoln's early mathematics, 1819–1826. In N. F. Ellerton & M. A.

Clements (Eds.), *Abraham Lincoln's cyphering book, and ten other extraordinary cyphering books* (pp. 123–186). New York, NY: Springer. https://doi.org/10.1007/978-3-319-02502-5_6

Ellerton, N. F., & Clements, M. A. (1994). *Fractions: A weeping sore in mathematics education.* Item No. 10 in issue No. 2 *Set* (Research information for teachers published jointly by the Australian Council for Educational Research and the New Zealand Council for Educational Research). https://doi.org/10.18296/set.0951

Ellerton, N. F., & Clements, M. A. (2012). *Rewriting the history of mathematics education in North America 1607–1861*. New York, NY: Springer. https://doi.org/10.1007/978-94-007-2639-0

Ellerton, N. F., & Clements, M. A. (2014). *Abraham Lincoln's cyphering book and ten other extraordinary cyphering books*. New York, NY: Springer. https://doi.org/10.1007/978-3-319-02502-5

Ellerton, N. F., & Clements, M. A. (2017). *Samuel Pepys, Isaac Newton, James Hodgson and the beginnings of secondary school mathematics: A history of the Royal Mathematical School at Christ's Hospital 1673–1868*. New York, NY: Springer. https://doi.org/10.1007/978-3-319-46657-6

Ellerton, N. F., & Clements, M. A. (2021). *Cyphering books prepared in the North American colonies (but not Canada), or in the United States of America*. Perth, Australia: Meridian Press.

Elliott, C. A., & Rossiter, M. W. (Eds.). (1992). *Science at Harvard University: Historical perspectives*. Bethlehem, PA: Lehigh University,

Fowle, W. B. (1849). *The child's arithmetick, or the elements of calculation, in the spirit of Pestalozzi's method, for the use of children between the ages of three and seven years*. Boston, MA: Lemuel N. Ide.

Galenson, D. W. (1984).The rise and fall of indentured servitude in the Americas: An economic analysis. *The Journal of Economic History, 4*(1), 1–26. https://doi.org/10.1017/S002205070003134X

Ganter, H. L. (1947). William Small, Jefferson's beloved teacher. *William and Mary Quarterly, 4*, https://doi.org/10.2307/1919640

Geiger, R. L. (2014). *The history of American higher education*. Princeton, NJ: Princeton University Press.

Gilman, D. C., Peck, H. T., Colby, F. M., & Moore, F. (Eds.). (1905). *New international encyclopedia* (pp. 584–586). New York: NY: Dodd, Mead.

Grendler, P. F. (1989). *Schooling in Renaissance Italy literacy and learning, 1300–1600*. Baltimore, MD: Johns Hopkins University Press.

Guasco, M. (2014). *Slaves and Englishmen: Human bondage in the early modern Atlantic world*. Philadelphia, PA: University of Pennsylvania Press. https://doi.org/10.9783/9780812209884

Harper, C. (1939). *A century of public teacher education: The story of the state teachers colleges as they evolved from the normal schools*. Washington, DC: American Association of Teachers Colleges.

Hassler, F. R. (1826). *Elements of analytic trigonometry: Plane and spherical.* New York, NY: Author.

Hertel, J. (2016) Investigating the implemented mathematics curriculum of New England navigation cyphering books. *For the Learning of Mathematics, 36(3)*, 4–10.

Høyrup, J. (2014). Mathematics education in the European Middle Ages. In A. Karp & G. Schubring (Eds.), *Handbook on the history of mathematics education* (pp. 109–124). New York, NY: Springer. https://doi.org/10.1007/978-1-4614-9155-2_6

James, R. Jr. (2013). *Root and branch: Charles Hamilton Houston, Thurgood Marshall, and the struggle to end segregation.* New York, NY: Bloomsbury Publishing.

Karpinski, L. C. (1980). *Bibliography of mathematical works printed in America through 1850.* New York, NY: Arno Press.

Kilpatrick, W. H. (1912). *The Dutch schools of New Netherland and colonial New York.* Washington, DC: United States Bureau of Education.

Lange, C., Ion, P., Dimou, A., Bratsas, C., Sperber, W., Kohlhase, M., & Antoniou, I. (2012). Bringing mathematics to the web of data: The case of the Mathematics Subject Classification. In E. Simperl, P. Cimiano, A. Polleres, O. Corcho, & V. Presutti (Eds.), *The semantic web: Research and applications. Lecture notes in computer science* (pp. 763–777). Berlin, Germany, Springer. https://doi.org/10.1007/978-3-642-30284-8_58

Littlefield, G. E. (1904). *Early schools and school-books of New England.* Boston, MA: The Club of Odd Volumes.

Lopenzina, D. (2012). *Red ink: Native Americans picking up the pen in the colonial period.* Albany, NY: State University of New York Press.

MAA (Mathematical Association of America). (1978). *Prime 80: Proceedings of a conference on prospects in mathematics education in the 1980s.* Washington, DC: The Mathematical Association of America.

Mayo, A. D. (1898). Horace Mann and the American common school. In *Report of the Commissioner of Education for 1896–97* (pp. 715–767). Washington, DC: Education Bureau.

Miller, J. C. (1977). *The wolf by the ears: Thomas Jefferson and slavery.* New York, NY: The Free Press.

Monaghan, E. J. (2007). *Learning to read and write in colonial America.* Amhurst, MA: University of Massachusetts Press.

Morison, S. E. (1932). *Precedence at Harvard College in the seventeenth century.* Worcester, MA: American Antiquarian Society

Mydland, L. (2011). The legacy of one-room schoolhouses: A comparative study of the American Midwest and Norway. *European Journal of American Studies*, *6*(1), 1–23. https://doi.org/10.4000/ejas.9205

Norrell, R., & Myers, A. H. (Ed.). (2017). *Historians in service of a better South: Essays in honor of Paul Gaston.* Montgomery, AL: NewSouth Books.

Paraide, P., Owens, K. D., Clarkson, P. C., Owens, C., & Muke, C. (2022). *Mathematics education in Papua New Guinea: A case study of colonial and postcolonial influences on mathematics education*. New York, NY: Springer.

Parshall, K. H. (2003). Historical contours of the American mathematical research community. In G. M. A. Stanic & J. Kilpatrick (Eds.), *A history of school mathematics* (Vol. 1, pp. 113–158). Reston, VA: National Council of Teachers of Mathematics.

Pelletreau, W. S. (1907). *Historic homes and institutions and genealogical and family history of New York*. New York, NY: Lewis Publishing.

Plimpton, G. A. (1916). *The hornbook and its use in America*. Worcester, MA: American Antiquarian Society. https://doi.org/10.5479/sil.258315.39088004241238

Quarles, B. (1961). *The Negro in the American Revolution*. Chapel Hill, North Carolina: University of North Carolina Press.

Ray, J. (1834). *Introduction to Ray's eclectic arithmetic. The little arithmetic, elementary lessons in intellectual arithmetic on the analytic and inductive methods of instruction, being an introduction to the author's Eclectic Arithmetic*. Cincinnati, OH: Truman, Smith and Co.

Reisch, G. (2002). *Margarita philosophica* (Reprint with an introduction (in Italian) by Lucia Andreini, Salzburg: Institut für Anglistik und Amerikanistik, Universität Salzburg, Austria.

Roberts, D. L. (2014). History of tools and technologies in mathematics education. In A. Karp & G. Schubring (Eds.), *Handbook on the history of mathematics education* (pp. 565–578). New York, NY: Springer. https://doi.org/10.1007/978-1-4614-9155-2_28

Ruter, M. (1828). *The juvenile arithmetick and scholar's guide; wherein theory and practice are combined and adapted to the capacities of young beginners; containing a due proportion of examples in Federal money; and the whole being illustrated by numerous questions similar to those of Pestalozzi*. Cincinnati, OH: N. and G. Guilford.

Schrader, D. V. (1967). The arithmetic of medieval universities. *Mathematics Teacher, 60*, 264–275. https://doi.org/10.5951/MT.60.3.0264

Silverman, D. J. (2006). *Faith and boundaries: Colonists, Christianity and community among the Wampanoag Indians of Martha's Vineyard, 1600–1871. History, 91*(304). 583–584. https://doi.org/10.1111/j.1468-229X.2006.379_7.x

Simons, L. G. (1936). *Bibliography of early American textbooks on algebra*. New York, NY: Scripta Mathematica, Yeshiva College.

Smith, D. E., & Ginsburg, J. (1934). *A history of mathematics in America before 1900*. Chicago, IL: The Mathematical Association of America. https://doi.org/10.1090/car/005

Smith, R. C. (1827). *Practical and mental arithmetic on a new plan in which mental arithmetic is combined with the use of the slate: Containing a complete system for all practical purposes; being in dollars and cents*. Boston, MA: Richardson & Lord.

Stamper, A. W. (1909). *A history of the teaching of elementary geometry, with reference to present-day problems*. PhD dissertation, Columbia University, New York.

Tewell, J. J. (2012). Assuring freedom to the free: Jefferson's Declaration and the conflict over slavery, *Civil War History*, *5*(1), 75–96.

Van Sickle, J. (2011). *A history of trigonometry education in the United States: 1776–1900*. PhD dissertation, Columbia University.

Voigt, J. (1995). *Thematic patterns of interaction and sociomathematical norms*. London, England: Taylor & Francis.

Ward, J. (1719). *The young mathematician's guide: Being a plain and easie introduction to the mathematicks*. London, England: Thomas Horne.

Webber, S. (1801). *Mathematics combined from the best authors and intended to be the textbook of the course of private lectures on these sciences in the University of Cambridge*. Boston, MA: Thomas and Andrews. https://doi.org/10.5962/bhl.title.17251

Wiencek, H. (2012). *Master of the mountain: Thomas Jefferson and his slaves*. Minneapolis, MN: Lerner Publishing.

Wilson, D. L. (1992). Thomas Jefferson and the character issue. *The Atlantic Monthly*, *270*(3), 37–74.

Zitarelli, D. A. (2019). *A history of mathematics in the United States and Canada. Volume 1: 1492–1900*. Washington, DC: American Mathematical Society. https://doi.org/10.1090/spec/094

Zuccheri, L., & Zudini, V. (2014). *History of teaching calculus*. In A. Karp & G. Schubring (Eds.), *Handbook on the history of mathematics education* (pp. 493–513). New York, NY: Springer.

Combined Reference List

(Note that cyphering books mentioned in the text were *not* considered for the purposes of this list.)

A Member of the Royal Institution. (1812). *A vindication of Mr. Lancaster's system of education*. London, England: Longman & Co.

Ackerberg-Hastings, A. (2000). *Mathematics is a gentleman's art: Analysis and synthesis in American college geometry teaching, 1790-1840*. PhD dissertation, Columbia University.

Ackerberg-Hastings, A. (2010). John Farrar and curricular transitions in mathematics education. *International Journal for the History of Mathematics Education, 5*(2), 17–30.

Ackerberg-Hastings, A. (2014). Mathematics teaching practices. In A. Karp & G. Schubring (Eds.), *Handbook on the history of mathematics education* (pp. 525–540). New York, NY: Springer. https://doi.org/10.1007/978-1-4614-9155-2_26

Adams, C., & Russell, T. (Eds.). (1965). *The West Point Thayer papers 1808–1872*. West Point, NY: Association of Graduates.

Adams, D. (1801). *The scholar's arithmetic—Or Federal accountant*. Leominster, MA: Adams and Wilder.

Adams, D. (1802). *The scholar's arithmetic—Or Federal accountant* (2nd ed.). Leomister, MA: Adams & Wilder.

Adams, D. (1817). *Scholar's arithmetic or Federal accountant: Containing, 1. Common arithmetic, the rules and illustrations, II. Examples and answers with blank spaces, sufficient for their operation by the scholar, III. To each rule a supplement, comprehending, 1. Questions in the nature of the rule, its use and manner of its operations, 2 Exercises. IV. Federal money, with rules for all the various operations in it to reduce Federal to the Old Lawful, and the Old Lawful to Federal money. V. Interest cast in Federal money, with compound*

N. F. Ellerton, M. A. (Ken) Clements, *Toward Mathematics for All*, History of Mathematics Education,
https://doi.org/10.1007/978-3-030-85724-0

multiplication, compound division and practice, wrought in Old Lawful and Federal money; the same questions being put in separate columns on the same page in each kind of money, these two modes of account being contrasted, and the great advantage gained by reckoning in Federal money easily discerned. VI. Demonstrations by engravers of the reason and nature of the various steps in the extraction of the square and cube roots, not to be found in any other treatise on arithmetic. VII. Forms of notes, deeds, bonds and other instruments of writing. The whole in a form and method altogether new, for the ease of the master and the greater progress of the scholar (9th ed.). Keene, NH: John Prentiss.

Adams, D. (1848). *Adams's new arithmetic*: *Arithmetic in which the principles of operating by numbers are analytically explained and synthetically explained, thus combining the advantages to be derived both from the inductive and synthetic mode of instructing*. New York, NY: Collins & Brother.

Ahmed, M. M. (2004). How many squares are there, Mr Franklin? Constructing and enumerating Franklin Squares. *American Mathematical Monthly, 111*, 394–410.

Albree, J. (2002). Nicolas Pike's Arithmetic (1788) as the American *Liber Abbaci*. In D. J. Curtin, D. E. Kullman, & D. E. Otero (Eds.), *Proceedings of the Ninth Midwest History of Mathematics Conference* (pp. 53–71). Miami, FL: Miami University.

Allen, J. (1822). *Euclid's elements of geometry, the first six books are added, elements of plain and spherical trigonometry, a system of conick sections, elements of natural philosophy as far as it relates to astronomy, according to the Newtonian system, and elements of astronomy*. Baltimore, MD: Cushing and Jewett.

Anderson, C. (1962). *Technology in American education: 1650–1900* (Report No. OE-34018). Washington, DC: Office of Education, U.S. Department of Health, Education, and Welfare.

Anderson, D. (1997). *The radical enlightenments of Benjamin Franklin*. Baltimore, MD: Johns Hopkins University Press.

Andrews, C. M. (1912). *The colonial period*. New York, NY: Henry Holt and Company.

Angulo, A. J. (2012). The polytechnic comes to America: How French approaches to science instruction influenced mid-nineteenth century American higher education. *History of Science, 50*(168), 315–338. https://doi.org/10.1177/007327531205000304

Archibald, R. C. (Ed.). (1925). *Benjamin Peirce*. Oberlin, OH: The Mathematical Association of America. https://doi.org/10.1080/00029890.1925.11986401

Ascher, M. (1992). Ethnomathematics: A multicultural view of mathematical ideas. *The College Mathematics Journal, 23*(4), 353–355. https://doi.org/10.2307/2686959

Association of Masters of the Boston Public Schools. (1844). *Remarks on the Seventh Annual Report of the Hon. Horace Mann, Secretary of the Massachusetts Board of Education*. Boston, MA: Charles C. Little & James Brown.

Australia and Van Diemen's Land. (1868). *Chamber's information for the people* (Vol. 1, pp. 1417–1448). New York, NY: United States Publishing Company.

Babcock, T. H. (1829). *The practical arithmetic; in which the principles of operating by numbers are analytically explained and synthetically applied*. New York, NY: G. and C. and H. Carvill.

Bailey, E. (1833). *First lessons in algebra, being an easy introduction to that science; designed for the use of academics and common schools*. Boston, MA: Carter, Hendee and Co.

Bailey, M. L (2013). Hornbooks. *The Journal of the History of Childhood and Youth, 6*(1), 3–14.

Baldwin, J. (1998). *Collected essays*. New York, NY: Library of America.

Baldwin, J. (1908). *Barnes's elementary history of the United States*. New York, NY: American Book Company.

Barbin, E., & Menghini, M. (2014). History of teaching geometry. In A. Karp & G. Schubring (Eds.), *Handbook on the history of mathematics education* (pp. 473–492). New York, NY: Springer. https://doi.org/10.1007/978-1-4614-9155-2_23

Barnard, F. A. P. (1875–1876). Progress of the exact sciences. *Harper's New Monthly Magazine, 52*, 82–100.

Barnard, H. (1851) *Normal schools, and other institutions, agencies and means designed for the professional education of teachers*. Hartford, CT: Case, Tiffany & Co.

Barnard, H. (1856). Graduation of public schools with special reference to cities and large villages. *American Journal of Education, 2*, 455–464.

Barnard, H. (Ed.). (1859). *Life, educational principles, and methods of John Henry Pestalozzi, with biographical sketches of several of his assistants and disciples*. New York, NY: F. C. Brownell.

Barnard, H. C. (2008). *Education and the French Revolution*. Cambridge, England: Cambridge University Press.

Barrème, N. (1744). *L'arithmétique du Sr Barrème ou le livre facile*. Paris, France: Gandouin.

Barrow, I. (1659). *Euclidis elementorum*. London, England: R. Daniel.

Bartlett, J. R. (1933). *Letter of instructions to the captain and the supercargo of the brig "Agenoria," engaged in a trading voyage to Africa*. Philadelphia, PA: Howard Greene and Arnold Talbot.

Beadie, N. (2010). Education, social capital and state formation in comparative historical perspectives: Preliminary investigations. *Paedagogica Historica, 46*(1–2), 15–32. https://doi.org/10.1080/00309230903528439

Beckers, D. (2006). Elementary mathematics education in the Netherlands ca. 1800: New challenges, changing goals. *Bulletin of the Belgian Mathematical Society, 13*. 937–940. https://doi.org/10.36045/bbms/1170347816

Beecher, C. E. (1828). *Arithmetic, explained and illustrated, for the use of the Hartford Female Seminary*. Harford, CT: P. Canfield.

Behforooz, H. (2012). Weighted magic squares. *Journal of Recreational Mathematics, 36*(4), 283–286.

Bell, E. T. (1945). *The development of mathematics* (2nd ed.). New York, NY: McGraw-Hill Book Company.

Berlin, I. (1998). *Many thousands gone: The first two centuries of slavery in North America*. Cambridge, MA: Harvard University Press.

Biber, E. (1831). *Henry Pestalozzi, and his plan of education*. London, England: John Souter School Library.

Bishop, A. J. (1988). *Mathematical enculturation*. Dordrecht, The Netherlands: Reidel. https://doi.org/10.1007/978-94-009-2657-8

Blackburn, R. (1997). *The making of new world slavery: From the Baroque to the modern 1492–1800*. London, England: Verso.

Blank, B. E. (2001). *What is mathematics: An elementary approach to ideas and methods. Book review. Notices of the AMS, 48*(11), 1325–132.

Blinderman, A. (1976). *Three early champions of education: Benjamin Franklin, Benjamin Rush, and Noah Webster*. Bloomington, IA: Phi Delta Kappa Educational Foundation.

Bonnycastle, J. (1806). *An introduction to algebra: With notes and observations*. Philadelphia, PA: Joseph Crukshank.

Bonnycastle, J. (1818). *An introduction to mensuration and practical geometry* (2nd ed.). Philadelphia, PA: Kimber and Sharpless.

Bonnycastle, J. (1822). *An introduction to algebra: With notes and observations*. New York, NY: Evert Duyckinck and George Long.

Bonnycastle, J. (1831). *An introduction to algebra: With notes and observations*. New York, NY: Collins and Hannay.

Bordley, J. B. (1789). *On monies, coins, weights and measures*. Philadelphia, PA: Daniel Humphreys.

Bourdon, L. P. M. (1831). *Elements of algebra by Bourdon: Translated from the French for colleges and schools*. Boston, MA: Hilliard, Gray, Little & Wilkins.

Bowditch, N. I. (1840). *Memoir of Nathaniel Bowditch*. Boston, MA: C. C. Little & Brown.

Bowditch, N. (1797). *Journal of a voyage from Salem to Manila in the ship Astrea, E. Prince, Master, in the years 1796 and 1797*. Handwritten manuscript held in the Bowditch Collection, Boston Public Library.

Bowditch, N. (1802). *The new American practical navigator; being an epitome of navigation; containing all the tables necessary to be used with the Nautical almanac. . . . Also, the demonstration of the most useful rules on trigonometry;*

with many useful problems in mensuration, surveying and gauging; and a directory of sea-terms ... Newbury-Port, MA: Edmund Blunt.

Boyd, J. P. (Ed.). (1950a). *The papers of Thomas Jefferson, Volume 1, 1760–1776*. Princeton, NJ: Princeton University Press. https://doi.org/10.1515/9780691184661-006

Boyd, J. P. (Ed.). (1950b). *The papers of Thomas Jefferson, Volume 2, January 1777 to June 1779*. Princeton, NJ: Princeton University Press.

Boyd, J. P. (Ed.). (1953). *The papers of Thomas Jefferson, Volume 7, March 1784 to February 1785*. Princeton, NJ: Princeton University Press.

Boyd, J. P. (1961). Report on weights and measures: Editorial note. In J. P. Boyd (Ed.), *The papers of Thomas Jefferson 16, November 1789 to July 1790)* (pp. 602–617). Princeton, NJ: Princeton University Press.

Braden, W. W. (1983). *Theo tradition in the South*. Baton Rouge, LA: Louisiana State University Press

Bradley, A. D. (1949). Pieter Venema, teacher, textbook author and free thinker. *Scripta Mathematica, 15*(1), 13–16.

Brasch, F. E. (1939, October). *The Newtonian epoch in the American colonies (1680–1783)*. Paper presented to the American Antiquarian Society.

Breed, F. S., Overman, J. R., & Woody, C. (1937). *Child-life arithmetics. Grade Five*. Chicago, IL: Lyons & Carnahan.

Brekke, B. F. (1977). *The copper coinage of Imperial Russia, 1700–1917*. Malmoe, Sweden: Forlagshuser.

Bressoud, D. (2010). Historical reflections on teaching trigonometry. *Mathematics Teacher, 104*(2), 106–112. https://doi.org/10.5951/MT.104.2.0106

Brewer, J., & Porter, R. (2003). *Consumption and the world of goods*. New York, NY: Routledge.

Bridge, B. (1832). *A treatise on the elements of algebra*. Philadelphia, PA: Key, Mielke and Biddle.

Briggs, H. (1617). *Logarithmorum chilias prima*. London, England: Author.

Britton, J. P., Proust, C, & Shnider, S. (2011). Plimpton 322: A review and a different perspective. *Archive for History of Exact Sciences, 65*(5), 519–566. https://doi.org/10.1007/s00407-011-0083-4

Brooks, E. (1879). *Normal methods of teaching containing a brief statement of the principles and methods of the science and art of teaching*. Philadelphia, PA: Normal Publishers.

Broome, E. C. (1903). *A historical and critical discussion of college admission requirements*. New York, NY: Columbia University.

Brown, C. (1721). *The geography of the ancients so far described as it is contain'd in the Greek and Latin classicks*. London, England: Author.

Burke, C. B. (1982). *American collegiate populations. A test of the traditional view*. New York, NY: NYU Press.

Burton, J. D. (1996). *Puritan town and gown: Harvard College and Cambridge, Massachusetts, 1636–1800*. PhD dissertation, College of William and Mary.

Burton, W. (1833). *The district school as it was, by one who went to it*. Boston, MA: Carter, Hendee and Company.

Butchart, R. E. (2010). *Schooling the freed people: Teaching, learning, and the struggle for Black freedom, 1861–1876*. Chapel Hill. NC: University of North Carolina.

Buzbee, L. (2014). *A personal history of the classroom*. Minneapolis, MN: Gray Wolf Press.

Cairns, W. D. (1934). "The Elements of Euclid, Thomas L. Heath." *The American Mathematical Monthly, 41*(6), 383. https://doi.org/10.2307/2301562

Cajori, F. (1890). *The teaching and history of mathematics in the United States* (Circular of Information No. 3, 1890). Washington, DC: Bureau of Education.

Cajori, F. (1907). *A history of elementary mathematics with hints on methods of teaching*. New York, NY: The Macmillan Company.

Cajori, F. (1928). *A history of mathematical notations*. La Salle, IL: Open Court Publishing Company.

Cajori, F. (1980). *The chequered career of Ferdinand Rudolph Hassler*. New York, NY: Arno Press.

Caldwell, O. W., & Courtis, S. A. (1925). *Then and now in education, 1845–1923*. Yonkers-on-Hudson, New York, NY: World Book Company.

Callum, G. W. (1891). *Biographical register of the officers and graduates of the U.S. Military Academy at West Point, N.Y.: From its establishment, in 1802, to 1890, with the early history of the United States Military Academy*. Boston, MA: Houghton Mifflin

Campbell, F. (1968). Latin and the elite tradition in education. *The British Journal of Sociology, 19*(3), 308–325. https://doi.org/10.2307/588835

Chambers' Information for the People. (1868). *Mechanics' institutions* (Vol. 1, pp. 713–744). New York, NY: United States Publishing Company.

Chateauneuf, A. O. (1930). *Changes in the content of elementary algebra since the beginning of the high school movement as revealed by the textbooks of the period*. PhD dissertation, The University of Pennsylvania.

Chessman, R. (1965). *Bound for freedom*. New York, NY: Abelard-Schuman.

Chevigne, L. I. M. (1809). *Mathematical manual for the use of colleges and academies*. Baltimore, MD: G. Doblin & Murphy.

Clarke, J. F., & Hale, E. E. (1892). School days in New England. In K. Munroe & M. H. Catherwood (Eds.), *School and college days* (Vol. VII, pp. 265–280). Boston, MA: Hall and Locke Company.

Clements, M. A., & Ellerton, N. F. (2006). Historical perspectives on mathematical elegance—To what extent is mathematical beauty in the eye of the beholder? In P. Grootenboer, R. Zevenbergen & M. Chinnappan (Eds.), *Identities, cultures and learning spaces* (pp. 147–154). Adelaide, Australia: Mathematics Education Research Group of Australasia.

Clements, M. A., & Ellerton, N. F. (2015). *Thomas Jefferson and his decimals 1775–1810: Neglected years in the history of U.S. school mathematics*. New York, NY: Springer. https://doi.org/10.1007/978-3-319-02505-6

Clements, M. A., Keitel, C., Bishop, A. J., Kilpatrick, J., & Leung, F. (2013). From the few to the many: Historical perspectives on who should learn mathematics. In M. A. Clements, A. J. Bishop, C. Keitel, J. Kilpatrick, & F. Leung (Eds.), *Third international handbook of mathematics education* (pp. 7–40). New York, NY: Springer. https://doi.org/10.1007/978-1-4614-4684-2_1

Cobb, L. (1835). *Cobb's ciphering book, No. 1, containing all the sums and questions for theoretical and practical exercises in Cobb's Explanatory Arithmetic No 1*. Elmira, NY: Birdsall & Huntley.

Cocker, E. (1677). *Cocker's arithmetick: Being a plain and familiar method suitable to the meanest capacity for the full understanding of that incomparable art, as it is now taught by the ablest school-masters in city and country*. London, England: John Hawkins.

Cocker, E. (1678). *Cocker's arithmetic: Being a plain and familiar method suitable to the meanest capacity* ... London, England: H. Tracey.

Cocker, E. (1685). *Cocker's decimal arithmetick,* ... London, England: J. Richardson.

Cocker, E. (1697). *Cocker's arithmetic: Being a plain and familiar method suitable to the meanest capacity* ... London, England: Eben Tracey.

Cocker, E. (1719). *Cocker's arithmetic: Being a plain and familiar method suitable to the meanest capacity* ... London, England: H. Tracey.

Cocker, E. (1720). *Decimal arithmetic, wherein is shewed the nature and use of decimal fractions in the usual rules of arithmetic, . . .* (5th ed.). London, England: J. Darby for M. Wellington.

Cogley, R. W. (1999). *John Eliot's mission to the Indians before King Philip's War*. Cambridge, MA: Harvard University Press.

Cogliano, F. D. (2008). *Thomas Jefferson: Reputation and legacy*. Charlottesville, VA: University of Virginia Press.

Cohen, A. M. (1998). *The shaping of American higher education*. San Francisco, CA: Jossey-Bass Publishers.

Cohen, P. C. (1982). *A calculating people: The spread of numeracy in early America*. Chicago, IL: The University of Chicago Press.

Cohen, P. C. (1993). Reckoning with commerce. Numeracy in 18th-century America. In J. Brewer, & R. Porter (Eds.), *Consumption and the world of goods* (pp. 320–334). New York, NY: Routledge.

Cohen, P. C. (2003). Numeracy in nineteenth-century America. In G. M. A. Stanic & J. Kilpatrick (Eds.), *A history of school mathematics* (Vol. 1, pp. 43–76). Reston, VA: National Council of Teachers of Mathematics.

Cohen, W. (1969). Thomas Jefferson and the problem of slavery. *Journal of American History, 56*(3), 503–526. https://doi.org/10.2307/1904203

Colburn, W (1821). *An arithmetic on the plan of Pestalozzi, with some improvements*. Boston, MA: Cummings and Hilliard.

Colburn, W. (1822). *Arithmetic upon the Inductive method of instruction being a sequel to intellectual arithmetic*. Boston:, MA: Cummings and Hilliard.

Colburn, W. (1825). *An introduction to algebra upon the inductive method of instruction*. Boston, MA: Cummings, Hilliard, and Co.

Colburn, W. (1826). *Arithmetic upon the Inductive method of instruction being a sequel to intellectual arithmetic*. Boston:, MA: Hilliard, Gray, Little and Wilkins.

Colburn, W. (1827). *Arithmetic upon the inductive method of instruction: Being a sequel to Intellectual* Arithmetic (3rd ed.). Boston, MA: Hilliard, Ray, Little and Wilkins.

Colburn, W. (1830/1970). Teaching of arithmetic. In J. K. Bidwell & R. G. Clason (Eds.), *Readings in the history of mathematics education* (pp. 24–37). Washington, DC: National Council of Teachers of Mathematics.

Colburn, W. (1835). *Colburn's first lessons, intellectual arithmetic upon the inductive method of instruction*. Hallowell, ME: Glazier, Masters and Smith.

Connor, R. D. W. (1951). Genesis of higher education in North Carolina. *North Carolina Historical Review, 28,* 1–14.

Coon, C. L. (1915). *North Carolina schools and academies, 1790–1840*: *A documentary history*. Raleigh, NC: Edwards and Broughton Printing Co.

Cotter, C. (n.d.). John Hamilton Moore and Nathaniel Bowditch. *Forum, 30,* 323–326. https://doi.org/10.1017/S0373463300044003

Cowley, W. H., & Williams, D. (1991). *International and historical roots of American higher education*. New York, NY: Garland Publishing, Inc.

Courant, R., & Robbins H. (1941). *What is mathematics? An elementary approach to ideas and methods*. London, England: Oxford University Press.

Crackel, T. J., Rickey, V. F., & Silverberg, J. S. (2017). Provenance lost? George Washington's books and papers lost, found, and (on occasion) lost again. *The Papers of the Bibliographical Society of America, 111*(2), 203–220. https://doi.org/10.1086/691826

Cremin, L. A. (1970). *American education: The colonial experience 1607–1783*. New York, NY: Harper & Row.

Cremin, L. A. (1977). *Traditions of American education*. New York, NY: Basic Books.

Crilly, T. (2008). Arthur Cayley and mathematics education. *HPM, 68*, 1–4.

Crozet, C. (1821). *A treatise on projective geometry, for the use of the cadets of the United States Military Academy. Part 1*. New York, NY: A. T. Goodrich and Co.

Cubberley, E. P. (1920). *The history of education*. Boston, MA: Houghton Mifflin Company.

da Ponte, J. P., & Guimarães, H. M. (2014). Notes for a history of the teaching of algebra. In A. Karp & G. Schubring (Eds.), *Handbook on the history of mathematics education* (pp. 323–334). New York, NY: Springer.

Daboll, N. (1804). *Daboll's schoolmaster's assistant: Being a plain, practical system of arithmetic; adapted to the United States* (3rd ed.). New London, CT: Samuel Green.

Daboll, N. (1813). *Daboll's schoolmaster's assistant: Improved and enlarged being a plain practical system of arithmetic; adapted to the United States* (7th ed.). New London, CT: Samuel Green

Daboll, N. (1818). *Daboll's schoolmaster's assistant: Improved and enlarged being a plain practical system of arithmetic; adapted to the United States* (10th ed.). New London, CT: Samuel Green

Daboll, N. (1820a). *Daboll's schoolmaster's assistant: Improved and enlarged being a plain practical system of arithmetic: Adapted to the United States*. New London, CT: Samuel Green.

Daboll, N. (1820b). *Daboll's practical navigator: Being a concise, easy, and comprehensive system of navigation; calculated for the daily use of seamen, and also for an assistant to the teacher: Containing plane, traverse, parallel, middle latitude, and Mercator's sailing; with all the necessary tables: Concise rules are given, with a variety of examples in every part of common navigation; also, a new, scientific, and very short method of correcting the dead reckoning; with rules for keeping a complete reckoning at sea, applied to practice, and exemplified in three separate journals, in which may be seen all the varieties which can probably happen in a ship's reckoning*. New London, CT: Samuel Green.

Danna, R. (2019). *The spread of Hindu-Arabic numerals in the European tradition of practical mathematics (13th–16th centuries)*. Cambridge, England: Department of History. University of Cambridge.

Dantzig, T. (1930). *Number: The language of science*. London, England: The Macmillan Company. https://doi.org/10.2307/2224269

Dartmouth College Library (n.d.). *Guide to the papers of Benjamin Greenleaf, 1807–1865*. Manuscript MS-1108). Hanover, NH: Author.

Darwin, G. H. (1913). Opening address. In E. W. Hobson & A. E. H. Love (Eds.), *Proceedings of the Fifth International Congress of Mathematicians* (pp. 33–36). Cambridge, England: Cambridge University Press.

Dauben, J. W., & K. H. Parshall (2014). Mathematics education in North America to 1800. In A. Karp & G. Schubring (Eds), *Handbook on the history of mathematics education* (pp. 175–185). New York, NY: Springer.

Davies, C. (1826). *Elements of descriptive geometry, with their application to spherical trigonometry, spherical projections, and warped surfaces*. Philadelphia, PA: H. C. Carey and I. Lea.

Davies, C. (1833). *The common school arithmetic, prepared for the use of academies and common schools in the United States, and also for the use of the young gentleman who may be preparing to enter the Military Academy at West Point*. New York, NY: N. & J. White.

Davies, C. (1835). *Elements of algebra: Translated from the French of M. Bourdon*. New York, NY: Wiley & Long.

Davies, C. (1836). *Elements of the differential and integral calculus*. New York, NY: Wiley and Long.

Davies, C. (1837). *Elements of algebra: Translated from the French of M. Bourdon*. New York, NY: Wiley & Long.

Davies, C. (1838a). *Elements of geometry and trigonometry from the works of A. M. Legendre* (revised ed.). New York, NY: A. S. Barnes & Co.

Davies, C. (1838b). *Mental and practical arithmetic designed for the use of academies and schools*. Hartford, CT: A. S. Barnes & Co.

Davies, C. (1840). *First lessons in arithmetic*. Hartford, CT: A. S. Barnes & Co.

Davies, C. (1844). *Arithmetic, designed for academies and schools, uniting the reasoning of the French with the practical methods of the English with full illustrations of the method of cancellations*. New York, NY: A. S. Barnes and N. L. Burr.

Davies, C. (1847). *Key to Davies' arithmetic*. New York, NY: A. S. Barnes and Burr.

Davies, C. (1852). *School arithmetic, analytical and practical*. New York, NY: A. S. Barnes & Co.

Davis, S. (1826). *The pupil's arithmetick*. Boston, MA: Lincoln and Edmands.

Day, J. (1814). *An introduction to algebra, being the first part of a course of mathematics, adapted to the method of instruction in the American colleges*. New Haven, CT: Howe & Deforest.

Day, J. (1815). *A treatise of plane trigonometry . . . being the second part of a course of mathematics, adopted to the method of instruction in the American colleges*. New Haven, CT: Yale College.

Day, J. (1817). *The mathematical principles of navigation and surveying, with the mensuration of heights and distances*. New Haven, CT: Yale College.

Day, J. (1836). *The teacher's assistant in the "course of mathematics"*. New Haven, CT: Yale College.

De Bellaigue, C. (2007). *Educating women: Schooling and identity in England and France 1800–1867*. Oxford, England: Oxford University Press. https://doi.org/10.1093/acprof:oso/9780199289981.001.0001

De Bock, D., & Vanpaemel, G. (2019). *Rods, sets and arrows: The rise and fall of modern mathematics in Belgium*. Cham, Switzerland: Springer. https://doi.org/10.1007/978-3-030-20599-7

Defense Mapping Agency, Hydrographic/Topographic Center. (1995). *The American practical navigator: An epitome of navigation*. Bethesda, MD: Author.

Deming, C. (1904). Yale wars of the conic sections. *The Independent, 56*, 667–669.

Denniss, J. (2012). *Figuring it out: Children's arithmetical manuscripts 1680–1880*. Oxford, England: Huxley Scientific Press.

Dewees, W., & Dewees, S. B. (1899). *Centennial history of Westtown Board School 1799–1899*. Philadelphia, PA: Sherman & Co.

Dexter, F. B. (1887). Estimates of population in American colonies. *Proceedings of the American Antiquarian Society* (pp. 22–50). Boston, MA. American Antiquarian Society.

Dickens, C. (1850). *David Copperfield*. London, England: Bradbury & Evans. https://doi.org/10.1093/oseo/instance.00121331

Dilworth, T. (1773a). *The schoolmasters assistant: Being a compendium of arithmetic, both practical and theoretical* (17th ed.). Philadelphia, PA: John Dunlap.

Dilworth, T. (1773b). *The schoolmasters assistant: Being a compendium of arithmetic, both practical and theoretical* (17th ed.). Philadelphia, PA: Joseph Crukshank.

Dilworth, T. (1797). *The schoolmasters assistant: Being a compendium of arithmetic, both practical and theoretical: The latest edition*. New London, CT: S. Green.

Dilworth, T. (1806). *The schoolmasters assistant: Being a compendium of arithme-2109 tic, both practical and theoretical*. New York, NY: McFarllane & Long, for Evert Duyckinck.

Doar, A. K. (2006). *Cipher books in the Southern Historical Collection*. Master of Science thesis, Wilson Library, University of North Carolina at Chapel Hill.

Dobyns, H. F. (1983). *Their number become thinned: Native American population dynamics in Eastern North America*. Knoxville TN: University of Tennessee Press.

Downs, R. B. (1978). *Friedrich Froebel*. Boston, MA: Twayne.

Drake, S. (1963). *The American dream and the Negro: 100 years of freedom*. Chicago, IL: Roosevelt University.

Draper, J. (1772). *The young student's pocket companion*. Newcastle-Upon-Tyne, England: Newcastle Literary and Philosophical Society.

Dunlap, L. A. (1959). Lincoln's sum book. *Lincoln Herald, 61*(1), 6–10.

Durkin, J. T. (1942). Journal of the Revd. Adam Marshall, schoolmaster, *U.S.S. North Carolina*, 1824–1825. *Records of the American Catholic Historical Society of Philadelphia, 53*(4), 152–168.

Dürr, K. (1951). *The propositional logic of Boethius*. Amsterdam, The Netherlands: North-Holland Publishing Company.

Dwight, T. (1903). *Memories of Yale life and men*. New Haven, CT: Dodd, Mean and Co.

Earle, A. M. (1899). *Child-life in colonial days*. New York, NY: Macmillan.

Edmonds, M. J. (1991). *Samplers and samplermakers: An American schoolgirl art, 1700–1850*. New York, NY: Rizzoh.

Edson, T. (1856). *Memoir of Warren Colburn, written for the American Journal of Education*. Boston, MA: Brown, Taggart & Chase.

Education. (1802, March 29). *The Salem Register*, p. 1.

Education in the Southern States. (1867, November 9). *Harper's Weekly, 11*(567), 706–707.

Educational Unit (U.S. Merchant Marine Cadet Corps). (1966). *Americans who have contributed to the history and traditions of the United States merchant marine*. Washington, DC: U.S. Cadet Corps.

Edwards, R. (1857). *Memoir of Nicholas Tillinghast, first Principal of the State Normal School at Bridgewater, Massachusetts*. Boston, MA: James Robinson & Co.

Edwards, R. (1902). My schools and schoolmasters. *Educational Review, 23*, 385–399.

Eels, W. C. (1913). Number systems of the North American Indians. *The American Mathematical Monthly*, *20*(10), 293–299. https://doi.org/10.1080/00029890.1913.11997985

Eggleston, E. (1888). *A history of the United States and its people*. New York, NY: American Book Company.

Egloff, K., & Woodward, D. (1992). *First people: The early Indians of Virginia*. Charlottesville, VA: The University Press of Virginia.

Ellerton, N. F., Aguirre, V., & Clements, M. A. (2014). He would be good: Abraham Lincoln's early mathematics, 1819–1826. In N. F. Ellerton & M. A. Clements, *Abraham Lincoln's cyphering book, and ten other extraordinary cyphering books* (pp. 123–186). New York, NY: Springer. https://doi.org/10.1007/978-3-319-02502-5_6

Ellerton, N. F., & Clements, M. A. (Eds.). (1989). *School mathematics: The challenge to change*. Geelong, Australia: Deakin University.

Ellerton, N. F., & Clements, M. A. (1994). *Fractions: A weeping sore in mathematics education*. Item No. 10 in issue No. 2 *Set* (Research information for teachers published jointly by the Australian Council for Educational Research and the New Zealand Council for Educational Research). https://doi.org/10.18296/set.0951

Ellerton, N. F., & Clements, M. A. (2009). Theoretical bases implicit in the *abbaco* and cyphering-book traditions. In M. Tzekaki, M. Kaldrimidou, & H. Sakonidis (Eds.), *Proceedings of the 33rd Conference of the International Group for the Psychology of Mathematics Education* (Vol. 3, pp. 9–16). Thessaloniki, Greece: International Group for the Psychology of Mathematics Education.

Ellerton, N. F., & Clements, M. A. (2011a, March 13). *Beyond witches: Salem, MA—The cradle of North American mathematics*. Paper presented to the

History and Pedagogy of Mathematics (HPM) Americas conference held at the American University, Washington, DC.

Ellerton, N. F., & Clements, M. A. (2011b). Unique mathematics books from a lost tradition. *The Guild of Book Workers' Journal. 11*, 28–39.

Ellerton, N. F., & Clements, M. A. (2012). *Rewriting the history of mathematics education in North America, 1607–1861*. New York, NY: Springer. https://doi.org/10.1007/978-94-007-2639-0

Ellerton, N. F., & Clements, M. A. (2014). *Abraham Lincoln's cyphering book and ten other extraordinary cyphering books*. New York, NY: Springer. https://doi.org/10.1007/978-3-319-02502-5

Ellerton, N. F., & Clements, M. A. (2017). *Samuel Pepys, Isaac Newton, James Hodgson and the beginnings of secondary school mathematics: A history of the Royal Mathematical School at Christ's Hospital 1673–1868*. New York, NY: Springer. https://doi.org/10.1007/978-3-319-46657-6

Ellerton, N. F., & Clements, M. A. (2019, September 22). *Major influences on U.S. school mathematics in the nineteenth century*. Paper presented to a meeting of the HPM/AMS Sectional Meeting in Madison, Wisconsin.

Ellerton, N. F., & Clements, M. A. (2021). *Cyphering books prepared in the North American colonies (but not Canada), or in the United States of America*. Perth, Australia: Meridian Press.

Elliott, C. A., & Rossiter, M. W. (Eds.). (1992). *Science at Harvard University: Historical perspectives*. Bethlehem, PA: Lehigh University.

Else-Quest, N., Hyde, J. S., & Linn, M. (2010). Cross-national patterns of gender differences in mathematics: A meta-analysis. *Psychological Bulletin, 136*(1), 103. https://doi.org/10.1037/a0018053

Emerson, F. (1822). *Female education ... To which is added, the little reckoner, consisting principally of arithmetical questions for infant minds*. Boston, MA: Samuel T. Armstrong and Crocker and Brewster.

Emerson, F. (1829). *The North American arithmetic*: *Part First, containing elementary lessons*. Boston, MA: Lincoln & Edmands.

Emerson, F. (1832). *The North American arithmetic*: *Part Second, uniting oral and written exercises in corresponding chapters*. Boston, MA: Lincoln & Edmands.

Emerson, F. (1834). *The North American arithmetic*: *Part Third, for advanced scholars*. Boston, MA: Russell, Odiorne, & Metcalf.

Emerson, F. (1835) *The North American arithmetic: Part third, for advanced scholars*. Boston, MA: Russell, Odiorne & Metcalf.

Emerson, F. (1838). *Key to the North American arithmetic: Part Second and Part Third for the use of teachers*. Boston, MA: G. W. Palmer & Co.

Euler, L. (1797). *Elements of algebra*. London, England: J. Johnson.

Euler, L. (1818). *An introduction to the elements of algebra designed for the use of those who are acquainted only with the first principles of arithmetic, selected from the algebra of Euclid*. Boston, MA: Hilliard, Gay, Little & Wilkins.

Executive Committee of the State Normal School, New York. (1846). *Annual report*. Albany, NY: Author.

F. D. B. (1911). Blackboards. In P. Monroe (Ed.), *A cyclopedia of education* (Vol. 1, pp. 390–394). New York, NY: The Macmillan Company.

Fanning, D. F., & Newman, E. P. (2011, July 24). More on the *American Accomptant* and the first printed dollar sign. *The E-Sylum, 14,* Numosmatic Bibliomania Society.

Fauvel, J. (1999, April 15). *Thomas Jefferson and mathematics*. Lecture given at the University of Virginia.

Fenzi, G. (1905). *The rubles of Peter the Great*. Moscow, Russia: Open Library.

Finegan, T. E. (1917). Colonial schools and colleges in New York. *Proceedings of the New York State Historical Association, 16,* 165–182.

Flint, A. (1804). *A system of geometry and trigonometry together with a treatise on surveying*. Hartford, CT: Oliver and Cooke.

Folmsbee, B. (1942). *A little history of the hornbook*. Boston, MA: The Horn Book Inc.

Fowle, W. B. (1826). *The child's arithmetick, or the elements of calculation, in the spirit of Pestalozzi's method, for the use of children between the ages of three and seven years*. Boston, MA: Thomas Wells.

Fowle, W. B. (1849). *The child's arithmetick, or the elements of calculation, in the spirit of Pestalozzi's method, for the use of children between the ages of three and seven years*. Boston, MA: Lemuel N. Ide.

Franklin, B. (1749). *Proposals relating to the education of youth in Pensilvania*. Philadelphia, PA: Author.

Franklin, B. (1793). *The private life of the late Benjamin Franklin, LL.D*. London, England: J. Parsons.

Franklin, B. (1917). *The autobiography of Benjamin Franklin*. Boston, MA: Houghton Mifflin Co.

Franklin, B. (1964). *The autobiography of Benjamin Franklin* (2nd ed.). New Haven, CT: Yale University.

Frost, J. (1846). *Lives of American merchants*. New York, NY: Saxon & Miles.

Fuess, C. M. (1917). *An old New England school: A history of Phillips Academy Andover*. Boston, MA: Houghton Mifflin Company.

Fulton, J. F., & Thomson, E. H. (1947). *Benjamin Silliman 1779–1864: Pathfinder in American science*. New York, NY: Henry Schuman.

Galenson, D. W. (1984).The rise and fall of indentured servitude in the Americas: An economic analysis. *The Journal of Economic History, 4*(1), 1–26. https://doi.org/10.1017/S002205070003134X

Ganter, H. L. (1947). William Small, Jefferson's beloved teacher. *William and Mary Quarterly, 4,* 505–511. https://doi.org/10.2307/1919640

Garcia, R., Meyer, S., Sanders, S., & Seitz, A. (2009). Construction and enumeration of Franklin circles. *Involve, 2*(3), 357–370. https://doi.org/10.2140/involve.2009.2.357

Garrett, J., & Guth, R. (2003). *100 greatest U.S. Coins*. Atlanta, GA: H.E. Harris & Co.

Gaydos, T., & Kampas, B. (2010). *American and Canadian ciphering books, n.d., 1727–1864*. Salem, MA: Phillips Library at the Peabody Essex Museum.

Geiger, R. L. (2014). *The history of American higher education*. Princeton, NJ: Princeton University Press.

Geiger, R. L. (2016). *The history of American higher education*. Princeton, NJ: Princeton University Press.

George Washington to Nicolas Pike, June 20, 1788 (in Electronic Text Center, University of Virginia Library). Retrieved December 7, 2006, from https://etext.virginia.edu/etcbin/toccernew2?id=WasFi30.xml&ima

Gibson, R. A. (1785). *The theory and practice of surveying*. Philadelphia, PA. Joseph Crukshank.

Gibson, R. A. (1814). *A treatise on practical surveying*. New York, NY: Duyckinck.

Gies, J., & Gies, F. (1969). *Leonardo of Pisa and the new mathematics of the Middle Ages*. New York, NY: Thomas Y. Crowell.

Gilman, D. C., Peck, H. T., Colby, F. M., & Moore, F. (Eds.). (1905). *New international encyclopedia* (pp. 584–586). New York, NY: Dodd, Mead.

Glaisher, J. W. L. (1873). On the introduction of the decimal point into arithmetic. *Report of the Meeting of the British Association for the Advancement of Science, 43,* 13–17.

Goldsmith, O. (1837). *The Grecian history, from the earliest state to the death of Alexander the Great*. Hartford, CT: Judd, Loomis & Co.

Goldstein, J. (2000). A matter of great magnitude: The conflict over arithmetization in 16th-, 17th- and 18th-century English editions of *Euclid's Elements* Books I through VI (1561–1795). *Historia Mathematics, 27*, 36–53. https://doi.org/10.1006/hmat.1999.2263

Goodchild, L. F., & Wechsler, H. S. (1989). *The history of higher education* (2nd ed.). Boston, MA: Pearson Custom Publishing.

Goodell, W. (1853). *The American slave code in theory and practice*. New York, NY: American & Foreign Anti-Slavery Society.

Goodfriend, J. D. (2017). *Who should rule at home? Confronting the elite in British New York City*. Ithaca, NY: Cornell University Press. https://doi.org/10.7591/cornell/9780801451270.001.0001

Goodrich, S. G. (1833). *Peter Parley's method of teaching arithmetic to children: With numerous engravings*. Boston, MA: Carter, Hendee, & Co.

Goodwin, J. (2003). *Greenback: The almighty dollar and the invention of America*. New York, NY: Henry Holt and Company.

Gough, J. (1788). *A treatise of arithmetic in theory and practice containing everything important in the study of abstract and applicant numbers, adapted to the commerce of Great Britain and Ireland*. Philadelphia, PA: B. Workman.

Green, A. (2015, September 15). The Yale chalkboard rebellion of 1830.

Green, R. W. (1839). *Gradations in algebra, in which the first principles of algebra are inductively explained*. Philadelphia, PA: Thomas, Cowperthwait and Co.

Greenleaf, B. (1850). *The national arithmetic, on the inductive system, combining the analytic and synthetic methods. In which the principles of arithmetic are explained*. Boston, MA: R. S. Davis.

Greenwood, I. (1729). *Arithmetick, vulgar and decimal, with the application thereof to a variety of cases in trade and commerce*. Boston, MA: Kneeland & Green.

Grew, T (1853). *The description and use of the globes, celestial and terrestrial, with variety of examples for the learner's exercise: intended for the use of such persons who would attain to the knowledge of those instruments; but chiefly designed for the instruction of young gentlemen at the Academy of Philadelphia. To which is added rules for working all the cases in plain and spherical triangles without a scheme*. Germantown, PA: Christopher Sower.

Grove, M. J. (2000). *Legacy of one-room schools*. Morgantown, PA: Masthof Press.

Guasco, M. (2014). *Slaves and Englishmen: Human bondage in the early modern Atlantic world*. Philadelphia, PA: University of Pennsylvania Press. https://doi.org/10.9783/9780812209884

Guralnick, S. (1975). *Science and the antebellum American college*. Philadelphia, PA: American Philosophical Society.

Gwynne-Thomas, E. H. (1981). *A concise history of education to 1900 A.D*. Washington, DC: University Press of America, Inc.

Hadden, R. W. (1994). *On the shoulders of merchants: Exchange and the mathematical conception of nature*. New York, NY: SUNY Press.

Halwas, R. (1990). *American mathematics textbooks 1760–1850*. London, England: Author.

Hardy, G. H. (2004). *Mathematician's apology*. Cambridge, England: Cambridge University Press.

Harney, J. H. (1840). *An algebra upon the inductive method of instruction*. Louisville, KY: Morton and Griswold.

Harper, C. (1935). *Development of the teachers college in the United States with special reference to Illinois State University*. Bloomington, IL: McKnight & McKnight.

Harper, C. (1939). *A century of public teacher education: The story of the state teachers colleges as they evolved from the normal schools*. Washington, DC: American Association of Teachers Colleges.

Harper, E. P. (2010). Dame schools. In T. Hunt, T. Lasley, & C. D. Raisch (Eds.), *Encyclopedia of educational reform and dissent* (pp. 259–260). Thousand Oaks, CA: SAGE Publications.

Harris, M. (1997). *Common threads: Women, mathematics and work*. Stoke on Trent, England: Trentham Books.

Harris, P. (1981). Measurement in tribal Aboriginal communities. *The Australian Journal of Indigenous Education, 9*(2), 53–61. https://doi.org/10.1017/S0310582200011482

Harris, P. (1991). *Mathematics in a cultural context; Aboriginal perspectives on space, time and money*. Geelong, Australia: Deakin University.

Hassler, F. R. (1826). *Elements of analytic trigonometry: Plane and spherical*. New York, NY: Author.

Hawney, W. (1775). *The complete measurer: Or, the whole art of measuring. in two parts. The first part teaching decimal arithmetick, with the extraction of the square and cube roots ... The second part teaching to measure all sorts of superficies and solids, by decimals; by cross multiplication, and by scale and* compasses (14th ed.). London, England: J. & F. Rivington.

Hay, C. (1988). (Ed.). *Mathematics from manuscript to print 1300–1600*. Oxford, England: Clarendon Press.

Heal, A. (1931). *The English writing-masters and their copy-books 1570–1800*. Cambridge, England: Cambridge University Press.

Henrich, C. J. (1991). Magic squares and linear algebra. *American Mathematical Monthly, 98*(6), 481–488. https://doi.org/10.1080/00029890.1991.11995746

Henry, J. (1843). An address upon education and common schools. *The Common School Journal, 6*(2), 26–29.

Hepburn, A. B. (1915). *A history of currency in the United States with a brief description of the currency systems of all commercial nations*. New York, NY: The Macmillan Company.

Herbst, J. (2004). The Yale Report of 1828. *International Journal of the Classical Tradition, 11*(2), 213–231. https://doi.org/10.1007/BF02720033

Hertel, J. (2016) Investigating the implemented mathematics curriculum of New England navigation cyphering books. *For the Learning of Mathematics, 36*(3), 4–10.

Hildreth, G. (1936). *Learning the three R's: A modern interpretation*. Minneapolis, MN: Educational Publishers Inc.

Hill, T. (1880). Benjamin Peirce. *The Harvard Register, 6*(1), 91–92.

Hirsch, D., & Van Haften, D. (2015). *Abraham Lincoln and the structure of reason*. El Dorado Hills, CA: Savas Beatie.

Hirsch, D., & Van Haften, D. (2019a). *The tyranny of public discourse*. El Dorado Hills, CA: Savas Beatie.

Hirsch, D., & Van Haften, D. (2019b). *The ultimate guide to the Declaration of Independence*. El Dorado Hills, CA: Savas Beatie.

Hirsch, D., & Van Haften, D. (2019c). *The ultimate guide to the Gettysburg address*. El Dorado Hills, CA: Savas Beatie.

Hobson, E. W., & Love, A. E. H. (Eds.). (1913). *Proceedings of the Fifth International Congress of Mathematicians.* Cambridge, England: Cambridge University Press.

Hodder, J. (1714). *Arithmetick, or that necessary art made most easie* (520th ed.). London, England: N. & M. Baddington.

Hogan, E. R. (1981). Theodore Strong and ante-bellum American mathematics. *Historia Mathematics, 8,* 439–455. https://doi.org/10.1016/0315-0860(81)90052-5

Honeywell, R. J. (1931). *The educational work of Thomas Jefferson.* Cambridge, MA: Harvard University Press. https://doi.org/10.4159/harvard.9780674337299

Howsam, L., & Raven, J. (Eds.). (2011). *Books between Europe and the Americas: Connections and communities, 1620–1860.* Basingstoke, England: Palgrave Macmillan. https://doi.org/10.1057/9780230305090

Howson, G. (1982). *A history of mathematics education in England.* Cambridge, England: Cambridge University Press. https://doi.org/10.1057/9780230305090

Howson, G. (2010). Mathematics, society, and curricula in nineteenth-century England. *International Journal for the History of Mathematics Education, 5*(1), 21–51.

Howson, G., & Rogers, L. (2014). Mathematics education in the England. In A. Karp & G. Schubring (Eds.), *Handbook on the history of mathematics education* (pp. 257–282). Dordrecht, The Netherlands: Springer. https://doi.org/10.1007/978-1-4614-9155-2_13

Høyrup, J. (2005). Leonardo Fibonacci and *abbaco* culture: A proposal to invert the roles. *Revue d'Histoire des Mathématiques, 11*, 23–56.

Høyrup, J. (2008). The tortuous ways toward a new understanding of algebra in the Italian *Abbacus* School (14th –16th centuries). In O. Figueras, J. L Cortina, A. Alatorre, T. Rojano & S. Sepulveda (Eds.), *Proceedings of the joint meeting of PME 32 and PME-NA XXX* (Vol. 1, pp. 1–20). Morelia, Mexico. International Group for the Psychology of Mathematics Education.

Høyrup, J. (2014). Mathematics education in the European Middle Ages. In A. Karp & G. Schubring (Eds), *Handbook on the history of mathematics education* (pp. 109–124). New York, NY: Springer. https://doi.org/10.1007/978-1-4614-9155-2_6

Hu, J. C. (2016, November 4). Why are there so few women mathematicians? *The Atlantic*.

Hughes, R. G. (1932, February). *Joseph Ray*, the mathematician, and the man. *West Virginia Review*.

Hurd, D. H. (1890). *History of Middlesex County, Massachusetts, with biographical sketches of many of its pioneers and prominent men*. Philadelphia, PA: J. W. Lewis & Co.

Hutton, C. (1764). *The schoolmaster's guide: Or, A complete system of practical arithmetic*. London, England: R. Baldwin.

Hutton, C. (1831). *A course of mathematics for the use of academies*, as well as private tuition. New York, NY: W. E. Dean.

Ifrah, G. (2000). *The universal history of numbers from prehistory to the invention of the computer*. New York, NY: John Wiley & Sons. Inc.

Iliffe, R. (1997). Mathematical characters: Flamsteed and Christ's Hospital Royal Mathematical School. In F. Willmoth (Ed.), *Flamsteed's stars: New perspectives on the life and work of the first Astronomer Royal (1646–1719)* (pp. 115–144). Woodbridge, England: The Boydell Press.

Israel, J. (2002). *Radical enlightenment: Philosophy and the making of modernity, 1650–1750*. London, England: Oxford University Press.

Jackson, A. (2002). Teaching math in America: An exhibit at the Smithsonian. *Notices of the American Mathematical Society, 49*(9), 1082–1083.

Jackson, L. L. (1906). *The educational significance of sixteenth century arithmetic from the point of view of the present time*. New York, NY: Columbia Teachers College.

Jacoby, H. (1939). *Navigation*. New York, NY: The Macmillan Company.

James, R. Jr. (2013). *Root and branch: Charles Hamilton Houston, Thurgood Marshall, and the struggle to end segregation*. New York, NY: Bloomsbury Publishing.

Jefferson, T. (1784a). *Notes*. In W. Peden (Ed.), *Notes on the State of Virginia*. Chapel Hill, NC: University of North Carolina Press for the Institute of Early American History and Culture, Williamsburg, Virginia.

Jefferson, T. (1784b). *Notes on the establishment of a money unit and of a coinage for the United States* (handwritten manuscript). Washington, DC: Library of Congress.

Jefferson, T. (1784c) *Some thoughts on a coinage* (handwritten manuscript). [ca. March 1784]. Founders Online, National Archives (Jefferson/01-07-02-0151-0004, ver. 2014-05-09). In J. P. Boyd (Ed.), *The papers of Thomas Jefferson, Vol. 7, 2 March 1784–25 February 1785*, Princeton, NJ: Princeton University Press, 1953, pp. 173–175. Also in T. Jefferson (1785). *Notes on the establishment of a money unit and of a coinage for the United States*. Paris, France: Author. The notes are reproduced in P. F. Ford (Ed.). (*The works of Thomas Jefferson* (Vol. 4, pp. 297–313). New York, NY: G. P Putnam's Sons. This was published in 1904.

Johnston, S. (1996). The identity of mathematical practitioners in 16th-century England. In I Hantsche (Ed.), *Der "mathematicus": Zur entwicklung und beturdung einer neuen berufsgruppe in der zeit Gerhard Mercators* (pp. 93 – 120). Bochum, Germany: Brockmeyer.

Jones, C. (2015). *The sea and the sky: The history of the Royal Mathematical School of Christ's Hospital*. Horsham, England: Author.

Jones, H. (c. 1752). *The reasons and rules and uses of octave computation or natural arithmetic*. London, England: Author (held in the British Museum).

Jones, P. S., & Coxford, A. F. (1970). From discovery to an awakened concern for pedagogy: 1492–1821. In National Council of Teachers (Ed.), *Thirty-second handbook* (pp. 11–23). Washington, DC: National Council of Teachers of Mathematics.

Kamens, D. H., & Benavot, A. (1991). Elite knowledge for the masses. The origins and spread of mathematics and science education in national curricula. *American Journal of Education, 99,* 137–180.

Kanbir, S., Clements, & Ellerton, N. F. (2017). *Using design research and history to tackle a fundamental problem with school algebra*. New York, NY: Springer. https://doi.org/10.1007/978-3-319-59204-6

Karp, A., & Schubring, G. (Eds.). (2014). *Handbook on the history of mathematics education*. New York, NY: Springer. https://doi.org/10.1007/978-1-4614-9155-2

Karpinski, L. C. (1925). *The history of arithmetic*. Chicago, IL: Rand McNally & Company.

Karpinski, L. C. (1940). *Bibliography of mathematical works printed in America through 1850*. Ann Arbor, MI: University of Michigan Press.

Karpinski, L. C. (1980). *Bibliography of mathematical works printed in America through 1850*. New York, NY: Arno Press.

Katz, M. B. (1968). *The irony of early school reform: Educational innovation in mid-19th century Massachusetts*. Cambridge, MA: Harvard University Press.

Keith, T. (1809). *Hawney's complete measurer; or, the whole art of measuring* (3rd ed.). London, England: Johnson, Rivington & Walker.

Kern, F. (1982). *Captain William Cooke Pease: U.S. Coast Guard pioneer*. Bethesda, MD: Alised Enterprises.

Kersey, J. (1689). An appendix containing choice knowledge in arithmetick, both practical and theoretical. In E. Wingate, Mr. Wingate's arithmetick containing a plain and familiar method for attaining the knowledge and practice of common arithmetick (9th ed., pp. 303–544). London, England: James Wingate

Ketcham, J. H. (1901). *The life of Abraham Lincoln*. New York, NY: Perkins Book Company.

Kidwell, P. A., Ackerberg-Hastings, A., & Roberts, D. L. (2008). *Tools of American mathematics teaching, 1800–2000*. Baltimore, MD: Johns Hopkins University Press.

Kiely, E. R. (1947). *Surveying instruments: Their history and classroom use* (*19th yearbook*). New York, NY: Teachers College Columbia University/National Council of Teachers of Mathematics.

Kilpatrick, J. (1992). A history of research in mathematics education. In D. A. Grouws (Ed.), *Handbook of research on mathematics teaching and learning* (pp. 3–38). New York, NY: Macmillan Publishing Company.

Kilpatrick, J. (2013). Warren Colburn and the inductions of reason. In K. Bjarnadóttir, F. Furinghetti, J. Prytz, & G. Schubring (Eds.), *"Dig where you stand" 3. Proceedings of the Third International Conference on the History of Mathematics Education* (pp. 219–232). Uppsala, Sweden: Uppsala University.

Kilpatrick, J. (2014). Mathematics education in the United States and Canada. In A. Karp & G. Schubring (Eds.), *Handbook on the history of mathematics education* (pp. 323–334). New York, NY: Springer. https://doi.org/10.1007/978-1-4614-9155-2_16

Kilpatrick, J., & Iszák, A. (2008). Historical perspectives on algebra in the curriculum. In C. E. Greenes & R. Rubinstein (Eds.), *Algebra and algebraic thinking in school mathematics: Seventieth yearbook* (pp. 3–18). Reston, VA: National Council of Teachers of Mathematics.

Kilpatrick, W. H. (1912). *The Dutch schools of New Netherland and colonial New York*. Washington, DC: United States Bureau of Education.

Klein, H. S. (2012). *A population history of the United States*. Cambridge, England: Cambridge University Press. https://doi.org/10.1017/CBO9781139059954

Kline, M. (1972). *Mathematical thought from ancient to modern times*. New York, NY: Oxford University Press.

Krause, D. A. (2000). "Among the greatest benefactors of mankind": What the success of the chalkboard tells us about the future of computers in the classroom. *Computers and the Future of the Humanities, 33*(2), 6–16. https://doi.org/10.2307/1315198

Kraus, J. W. (1961). The development of the curriculum in the early American colleges. *History of Education Quarterly, 1*(2), 64–76. https://doi.org/10.2307/367641

Kraushaar, O. F. (1976). *Private schools from the Puritans to the present*. Bloomington, IA: Phi Delta Kappa Educational Foundation.

Kullman, D. E. (1998). Joseph Ray: The McGuffey of mathematics. *Ohio Journal of School Mathematics, 38*, 5–10.

Lacroix, S. F. (1818a). *An elementary treatise of arithmetic, and an introduction to the elements of algebra, comprehending the mathematics required for admission to the University of Cambridge, New England*. Cambridge, MA: Hilliard and Metcalf.

Lacroix, S. F. (1818b). *An elementary treatise on arithmetic, taken principally from the arithmetic of S. F. Lacroix and translated into English with such alterations and additions as were found necessary in order to adapt it to the use of the American student*. Cambridge, New England: Hilliard and Metcalf, at the University Press.

Lacroix, S. F. (1818c). *An elementary treatise of arithmetic, and an introduction to the elements of algebra, comprehending the mathematics required for admission to the University of Cambridge, New England*. Cambridge, MA: Hilliard and Metcalf.

Lacroix, S. F., & Bézout, É (1826). *An elementary treatise on plane and spherical trigonometry, and on the application of algebra to geometry from the mathematics of Lacroix and Bézout, for the use of students of the University at Cambridge, New England* (2nd ed.). Cambridge, MA.: Hilliard and Metcalf.

Lancaster, J. (1805). *Improvement in education, as it respects the industrious classes of the community*. London, England: Darton & Harvey.

Landis, J. H. (1854). Examination of teachers' certificates. *Pennsylvania School Journal, 2*(1), 11–12.

Lange, C., Ion, P., Dimou, A., Bratsas, C., Sperber, W., Kohlhase, M., & Antoniou, I. (2012). Bringing mathematics to the web of data: The case of the Mathematics Subject Classification. In E. Simperl, P. Cimiano, A. Polleres, O. Corcho, & V. Presutti (Eds.), *The semantic web: Research and applications. Lecture notes in computer science* (pp. 763–777). Berlin, Germany: Springer. https://doi.org/10.1007/978-3-642-30284-8_58

Leacock, S (1970). *Feast of Stephen*. Toronto, Canada: McClelland & Stewart, Inc.

Lean G. A. (1992). *Counting systems of Papua New Guinea and Oceania*. PhD dissertation, Papua New Guinea University of Technology (Lae, Papua New Guinea).

Lee, C. (1797). *The American accountant; being a plain, practical and systematic compendium of Federal arithmetic* . . . Lansingburgh, NY: William W. Wands.

Legendre, A-M (1794). *Élements de géométrie*. Paris, France: Fermin Didot.

Lemprière, J. (1834). *Lemprière's classical dictionary for schools and academies*. Boston, MA: Carter, Hender & Co.

Levin, J. E., & Levin, M. R. (2010). *Abraham Lincoln's Gettsysburg address*. New York, NY: Simon & Schuster.

Lewis, E. (1848). *A treatise on plane and spherical trigonometry including the construction of the auxiliary tables; a concise tract on the conic sections, and the principles of spherical projection* (2nd ed.). Philadelphia, PA: E. H. Butler & Co., and H. Orr.

Lial, M. L., Miller, C. D., & Hornsby, E. J. (1992). *Beginning algebra*. Boston, MA: Addison Wesley.

Libois, P. (1951). *Les espaces*. Liège, Belgium: Thone.

Linklater, A. (2003). *Measuring America: How the United States was shaped by the greatest land sale in history*. New York, NY: Plume.

Littlefield, G. E. (1904). *Early schools and school-books of New England*. Boston, MA: The Club of Odd Volumes.

Lloyd, H. A. (2012). Commonwealth and empire: Robert Recorde in Tudor England. In G. Roberts & F. Smith (Eds.), *The life and times of a Tudor mathematician* (pp. 145–164). Cardiff, Wales: University of Wales Press

Long, P. O., McGee, D., & Stahl, A. M. (Eds.). (2009). *The book of Michael of Rhodes: A 15th century maritime manuscript*. Cambridge, MA: MIT Press.

Looby, C. (1984). Phonetics and politics: Franklin's alphabet as a political design. *Eighteenth-Century Studies*, *18*(10), 1–34. https://doi.org/10.2307/2738304

Lopenzina, D. (2012). *Red ink: Native Americans picking up the pen in the colonial period*. Albany, NY: State University of New York Press.

Lucas, C. J. (1994). *American higher education: A history*. New York, NY: St. Martin's Griffin.

Lucas, S. E. (1989). Justifying America: The Declaration of Independence as a rhetorical document. In T. W. Benson (Ed.), *American rhetoric: Context and criticism* (pp. 67–130). Carbondale, IL: Southern Illinois University Press.

MAA (Mathematical Association of America). (1978). *Prime 80: Proceedings of a conference on prospects in mathematics education in the 1980s*. Washington, DC: The Mathematical Association of America.

Macintyre, S., & Clark, A. (2004). *The history wars*. Melbourne, Australia: Melbourne University Press.

Macfarlane, A. (1916). *Lectures on ten British mathematicians of the nineteenth century*. New York, NY: John Wiles & Sons.

Malcolm, A. (1730). *A new system of arithmetick, theoretical and practical wherein the science of numbers is demonstrated from its first principles through all the parts and branches thereof, either known to the ancients, or owing to the improvements of the moderns: The practice and application to the affairs of life and commerce being also fully explained: so as to make the whole a complete system of theory, for the purposes of men of science; and of practice, for men of business*. London, England: J. Osborn and T. Longman.

Mann, H. (1845/1925). Boston grammar and writing schools. *The Common School Journal*. In O. W. Caldwell & S. A. Courtis, *Then and now in education 1845–1923* (pp. 237–272). New York, NY: World Book Company.

Marshall, P. J. (Ed.). (2001). *Oxford history of the eighteenth century*. Oxford, England: Oxford University Press.

Martin, G. H. (1897). *The evolution of the Massachusetts public school system: A historical sketch*. New York, NY: D. Appleton and Company.

Martines, L. (1979). *Power and imagination: City states in renaissance Italy*. New York, NY: Alfred A. Knopf.

Mattoon, C. H. (1850). *Common arithmetic upon the analytic method of instruction*. Medary, OH: Steam Press of S. Medary.

Matz, F. P. (1895). B. O. Peirce: Biography. *American Mathematical Monthly, 2*, 173–179. https://doi.org/10.1080/00029890.1895.11998647

Mayo, A. D. (1898). Horace Mann and the American common school. In *Report of the Commissioner of Education for 1896–97* (pp. 715–767). Washington, DC: Education Bureau.

McCusker, J. J. (1992). *Money and exchange in Europe and America 1600–1775*. Chapel Hill, NC: University of North Carolina Press.

McDonald, A. (1785). *The youth's assistant* . . . Norwich, CT. John Trumbull.

Meacham, J. E. (2018). *The soul of America: The battle for our better angels*. New York, NY: Penguin Random House.

Menninger, K. W. (1969). *Number words and number symbols: A cultural history of numbers*. Boston, MA: MIT Press. https://doi.org/10.2307/2799719

Merchant, A. M. (1824). *The first lines of arithmetic: Made easy and adapted to the capacity of junior learners*. New York, NY: John C. Totten.

Meriwether, C. (1907). *Our colonial curriculum 1607–1776*. Washington, DC: Capital Publishing.

Meyer R. (1968). Opponents of classical learning in America during the revolutionary period. *Proceedings of the American Philosophical Society, 112*, 221–234.

Michalowicz, K. D., & Howard, A. C. (2003). Pedagogy in text: An analysis of mathematics texts from the nineteenth century. In G. M. A. Stanic & J. Kilpatrick (Eds.). *A history of school mathematics* (Vol. 1, pp. 77–112). Reston, VA: National Council of Teachers of Mathematics.

Middlekauf, R. (1963). *Ancients and axioms: Secondary education in eighteenth-century New England*. New Haven, CT: Yale University Press.

Miller, J. C. (1977). *The wolf by the ears: Thomas Jefferson and slavery*. New York, NY: The Free Press.

Minnick, J. H. (1921). The recitation in mathematics. *The Mathematics Teacher, 14*(3), 119–123. https://doi.org/10.5951/MT.14.3.0119

Minto, W. (1788). *An inaugural oration, on the progress and importance of the mathematical sciences*. Manuscript, held in the Clements Library, The University of Michigan.

Miter S. (1896, September 8). Our London letter. *The American Stationer, 40*(10), 367–368 and 379. https://doi.org/10.15281/jplantres1887.10.117_367

Molloy, P. M. (1975). *Technical education and the young Republic: West Point as America's Ecole Polytechnique, 1802–1833*. PhD Dissertation, Brown University.

Monaghan, E. J. (2007). *Learning to read and write in colonial America*. Amherst, MA: University of Massachusetts Press.

Monge, G. (1811). *Géométrie descriptive* (New edition). Paris, France: J. Klostermann.

Monroe, W. S. (1911). Warren Colburn. In P. Monroe (Ed.), *A cyclopedia of* education (Vol. 2, p. 48). New York, NY: Macmillan. https://doi.org/10.1086/454173

Monroe, W. S. (1912). A chapter in the development of arithmetic teaching in the United States. *The Elementary School Teacher, 13*(1), 17–24.

Monroe, W. S. (1917). *Development of arithmetic as a school subject*. Washington, DC: Government Printing Office.

Monroe, W. (Will) S. (1969). *History of the Pestalozzian movement in the United States*. New York, NY: Arno Press & *The New York Times*.

Moore, J. H. (1796). *The new practical navigator, being an epitome of navigation, explaining the different methods of working the lunar observations, and all the requisite tables used with the nautical almanac, in determining the latitude and longitude, and keeping a complete reckoning at sea; illustrated by proper rules and examples; the whole exemplified in a journal kept from England to the island of Teneriffe: also, the substance of the examination, every candidate for a commission in the Royal Navy, and Officer in the Honourable East India Company's service, must pass through previous to their being appointed; this with the sea terms, are particularly recommended to the attention of all young gentlemen designed for, or belonging to the sea* (12th ed.). Tower Hill, England: E. Law.

Moore, J. H. (1799). *The new practical navigator*. Newbury-Port, MA:. Edmund March Blunt.

Morison, S. E. (1921). *Maritime history of Massachusetts 1783–1860*. Boston, MA: Houghton Mifflin Company.

Morison, S. E. (1932). *Precedence at Harvard College in the seventeenth century*. Worcester, MA: American Antiquarian Society

Morison, S. E. (1935). *The founding of Harvard College*. Cambridge, MA: Harvard University Press.

Morison, S. E. (1956). *The intellectual life of colonial New England*. Ithaca, NY: Great Seal Books.

Morison, S. E. (1971). *The European discovery of America*. New York, NY: Oxford University Press.

Morrice, D. (1801). *The young midshipman's instructor (designed to be a companion to Hamilton Moore's Navigation*): *With useful hints to parents of sea youth, and to captains and schoolmasters of the Royal Navy*. London, England: Knight and Compton.

Morris, R. (1782, January 15). Robert Morris to the President of Congress, January 15, 1782. In J. P. Boyd (Ed.), *The papers of Thomas Jefferson 16, March 1784 to February 1785* (pp. 160–169). Princeton, NJ: Princeton University Press.

Morrow, G. R. (1970). *Proclus's commentary on the first book of Euclid's Elements*. Princeton, NJ: Princeton University Press.

Muttappallymyalil, J, Mendis S, John L. J., Shanthakumari N., Sreedharan J, & Shaikh R. B. (2016). Evolution of technology in teaching: Blackboard and beyond in medical education. *Nepal Journal of Epidemiology, 6*(3), 588–592. https://doi.org/10.3126/nje.v6i3.15870

Mydland, L. (2011). The legacy of one-room schoolhouses: A comparative study of the American Midwest and Norway. *European Journal of American Studies, 6*(1), 1–23. https://doi.org/10.4000/ejas.9205

Myers, C. & Nash, J. D. (2006). Private education. In W. S. Powell (Ed.), *Encyclopedia of North Carolina*. University of North Carolina Press.

Norrell, R., & Myers, A. H. (Eds.). (2017). *Historians in service of a better South: Essays in honor of Paul Gaston*. Montgomery, AL: NewSouth Books.

Napier, J. (1614). *Mirifici logarithmorum canonis descriptio*. Edinburgh, Scotland: Andrew Hart.

Napier, J. (1619). *The wonderful canon of logarithms*. Edinburgh, Scotland: William Home Lizars, 1857 [English translation by Herschell Filipowski].

Navigation Division, Defense Mapping Agency (1995). *American practical navigator, Bowditch*. Hydrographic/Topographic Center, ASIN: B001B4884O.

Nietz, J. A. (1966). *The evolution of American secondary school textbooks*. Rutland, VT: Charles E. Tuttle.

Nishikawa, S. (1987). The economy of Chōshū on the eve of industrialization. *The Economic Studies Quarterly, 38*(4), 209–222.

Núñez, R., & Cooperrider, K. (2013). The tangle of space and time in human cognition. *Trends in Cognitive Sciences, 17*(5), 220–229. https://doi.org/10.1016/j.tics.2013.03.008

Ogg, F. A. (1927). *Builders of the Republic*. New Haven, NJ: Yale University Press.

Otis, J. (2017). "Set them to the cyphering schoole": Reading, writing, and arithmeticale, circa 1540–1700. *Journal of British Studies, 56*(3), 453–482. https://doi.org/10.1017/jbr.2017.59

Owens, K., Lean, G., Paraide, P., & Muke, C. (2018). *History of number: Evidence from Papua New Guinea and Oceania*. New York, NY: Springer. https://doi.org/10.1007/978-3-319-45483-2

Page, D. P. (1877). *Theory and practice of teaching: The motives and methods of good school-keeping* (90th ed.). New York, NY: A. S. Barnes & Company.

Paraide, P., Owens, K. D., Clarkson, P. C., Owens, C., & Muke, C. (2022). *Mathematics education in Papua New Guinea: A case study of colonial and postcolonial influences on mathematics education*. New York, NY: Springer.

Parshall, K. H. (2003). Historical contours of the American mathematical research community. In G. M. A. Stanic & J. Kilpatrick (Eds.), *A history of school mathematics* (Vol. 1, pp. 113–158). Reston, VA: National Council of Teachers of Mathematics.

Pasles, P. C. (2007). *Benjamin Franklin's numbers: An unsung mathematical odyssey*. Princeton, NJ: Princeton University Press.

Patterson, E. C. (2012). *Mary Somerville and the cultivation of science, 1815–1840*. Boston, MA: Kluwer.

Patterson, R. (1819). *A treatise of practical arithmetic intended for the use of schools* (2 parts). Pittsburgh, PA: R. Patterson and Lambdin.

Peabody Essex Museum. (n.d.). *Salem: Maritime Salem in the age of sail*. Salem, MA: Author.

Peden, W. (Ed.). (1955). *Notes on the State of Virginia by Thomas Jefferson*. Chapel Hill, NC: University of North Carolina Press.

Peirce, B. (1837). *An elementary treatise on algebra; to which are added elementary equations and logarithms*. Boston, MA: James Munroe and Company.

Peirce, B. (1841). *An elementary treatise on curves, functions, and forces (Volume First): Analytic geometry and the differential calculus*. Boston, MA: James Munroe and Company.

Peirce, B. (1846). *An elementary treatise on curves, functions, and forces (Volume Second): Calculus and imaginary quantities, residual calculus, and integral calculus*. Boston, MA: James Munroe and Company.

Pelletreau, W. S. (1907). *Historic homes and institutions and genealogical and family history of New York*. New York, NY: Lewis Publishing.

Perkins, G. R. (1845). *The elements of algebra*. Utica, NY: B. S. Merrill.

Peterson, S. R. (1955). Benjamin Peirce: Mathematician and philosopher. *Journal of the History of Ideas, 16,* 89–112. https://doi.org/10.2307/2707529

Phalen, H. R. (1946). The first professorship of mathematics in the colonies. *The American Mathematical Monthly, 53*(10), 579–582. https://doi.org/10.1080/00029890.1946.11991755

Phillips, C. J. (2015). An officer and a scholar: Nineteenth-century West Point and the invention of the blackboard. *History of Education Quarterly, 55*(1), 82–108. https://doi.org/10.1111/hoeq.12093

Phillips, J. D. (1947). *Salem and the Indies: The story of the great commercial era of the city*. Boston, MA: Houghton Mifflin Company.

Pike, N. (1788). *The new and complete system of arithmetic, composed for the use of the citizens of the United States*. Newbury-Port, MA: John Mycall.

Pike, N. (1793). *Abridgement of the new and complete system of arithmetick composed for the use, and adapted to the commerce of the citizens of the United States*. Newbury-Port, MA: John Mycall, Isaiah Thomas.

Pike, S. (1811). *The teacher's assistant; or a system of practical arithmetic; wherein the several rules of that useful science, are illustrated by a variety of examples, a large proportion of which are in Federal money. The whole is designed to abridge the labour of teachers, and to facilitate the instruction of youth.* Philadelphia, PA: Johnstonn (sic.) and Warner.

Pike, S. (1822). *The teacher's assistant or a system of practical arithmetic; wherein the several rules of that useful science, are illustrated by a variety of examples, a large proportion of which are in Federal money. The whole is designed to abridge the labour of teachers, and to facilitate the instruction of youth.* Philadelphia, PA: Benjamin Warner.

Pillans, J. (1856). *First steps in the physical and classical geography of the ancient world.* London, England: Longman, Brown, Green & Longmans.

Pioariu, R. (2011). Cross-cultural issues in teaching English to Romanian students. In T. Popescu, R. Pioariu, & C. Herteg (Eds.), *Cross-disciplinary approaches to the English language: Theory and practice* (pp. 150–160). Newcastle-on-Tyne, England: Cambridge Scholars Publishing.

Playfair, J. (1814). *Elements of geometry: Containing the first six books of Euclid, with a supplement on the quadrature of the circle and the geometry of solids; to which are added, elements of plane and spherical trigonometry*. New York, NY: Collins & Hannay.

Playfair, J. (1822). *Elements of geometry: Containing the first six books of Euclid, with a supplement on the quadrature of the circle and the geometry of solids; to which are added, elements of plane and spherical* trigonometry (6th ed.). Edinburgh, Scotland: Bell & Bradfute.

Plimpton, G. A. (1912, August 1). Hornbooks. *The Independent, 72*(3322), 264–268. https://doi.org/10.2307/3474145

Plimpton, G. A. (1916). *The hornbook and its use in America*. Worcester, MA: American Antiquarian Society. https://doi.org/10.5479/sil.258315.39088004241238

Plimpton, G. A. (1928). The history of elementary mathematics in the Plimpton Library. *Science, 68(1765),* 390–395. https://doi.org/10.1126/science.68.1765.390

Powell, S., & Dingman, P. (n.d.). Arithmetic is the art of computation: The collation. https://collation.folger.edu/2015/09/arithmetic-is-the-art-of-computation. Folger Shakespeare. This website was viewed for the first time, April 7, 2020.

Preveraud, T. (2015). American mathematics journals and the transmission of French textbooks to the United States. In K. Bjarnadottir, F. Furinghetti, J. Prytz, & G. Schubring (Eds.), *"Dig where you stand 3"* (pp. 309–325). Uppsala, Sweden: Uppsala Universitet.

Prévost, G. (1677). *Briefve méthode et instruction pour apprendre l'arithmétique*. Tournay, France: Jacques Coulon.

Price, D. A. (2003). *John Smith, Pocahontas, and the start of a new nation*. New York, NY: Alfred A. Knoff.

Pycior, H. (1997). *Symbols, impossible numbers and geometric entanglements: British algebra through the commentaries on Newton's Universal Arithmetic*. New York, NY: Cambridge University Press. https://doi.org/10.1017/CBO9780511895470

Quarles, B. (1961). *The Negro in the American Revolution*. Chapel Hill, NC: University of North Carolina Press.

Quincy, J. (1860). *The history of Harvard University*. Boston, MA: Crosby, Nicholls, Lee & Co.

Rappleye, C. (2010). *Robert Morris: Financier of the American Revolution*. New York, NY: Simon & Schuster.

Rawley, J. A. (1981). *The trans-Atlantic slave trade*. New York, NY: W. W. Norton.

Ray, J. (1834a). *Introduction to Ray's eclectic arithmetic. The little arithmetic, elementary lessons in intellectual arithmetic on the analytic and inductive*

methods of instruction, being an introduction to the author's Eclectic Arithmetic. Cincinnati, OH: Truman, Smith and Co.
Ray, J. (1834b). *The little arithmetic*. Cincinnati, OH: Truman, Smith and Co.
Ray, J. (1838). *Ray's eclectic arithmetic on the inductive and analytic methods of instruction* (4th ed.). Cincinnati, OH: Truman and Smith.
Ray, J. (1848). *Ray's algebra: Part first*. Cincinnati, OH: Van Antwerp Bragg & Co.
Ray, J. (1985). *Ray's new practical arithmetic*. Milford, MI: Mott Media, Inc.
Record, R. (1658). *Grounde of the arts:* Teaching the worke and practise, of arithmeticke. London, England: R. Wolff.
Reisch, G. (2002). *Margarita philosophica* (Reprint with an introduction (in Italian) by Lucia Andreini, Salzburg: Institut für Anglistik und Amerikanistik, Universität Salzburg, Austria.
Reisner, E. H. (1930). *The evolution of the common school*. New York, NY: Macmillan.
Richard, C. J. (1994). *The founders and the classics: Greece, Rome, and the American enlightenment*. Cambridge, MA: Harvard University Press.
Rickey, V. F., & Shell-Gellasch, A. (2010, July). Mathematics education at West Point: The first hundred years—Teaching at the Academy. *Convergence* (Publication of the Mathematical Association of America).
Rickey, F., & Shell-Gellasch, A. (2019). Mathematics education at West Point: The first 100 years—Claudius Crozet. https://www.maa.org/book/export/html/116859 (viewed December 5, 2019).
Ring, B. (1993). *Girlhood embroidery: American samplers and pictorial needlework, 1650–1850*. New York, NY: Knopf Publishers.
Roach, J. (1971). *Public examinations in England, 1850–1900*. Cambridge, England: Cambridge University Press. https://doi.org/10.1017/CBO9780511896309
Roberts, D. L. (2014). History of tools and technologies in mathematics education. In A. Karp & G. Schubring (Eds.), *Handbook on the history of mathematics education* (pp. 565–578). New York, NY: Springer. https://doi.org/10.1007/978-1-4614-9155-2_28
Roberts, D. L. (2019). *Republic of numbers: Unexpected stories of mathematical Americans through history*. Baltimore, MD: Johns Hopkins University Press.
Roberts, G., & Smith, F. (2012). *Robert Recorde: The life and times of a Tudor mathematician*. Cardiff, Wales: University of Wales Press.
Robin Halwas Limited (1997). *American mathematical textbooks 1760–1850*. London, England: Author.
Robinson, H. (1870). *The progressive higher arithmetic for schools, academies, and mercantile colleges*. New York, NY: Ivison, Blakeman, Taylor & Co.
Root, E. (1795). *An introduction to arithmetic for the use of common schools*. Norwich, CT: Thomas Hubbard.

Rotherham, W. (1852). *The algebraical equation and problem papers, proposed in the examinations of St. John's College, Cambridge from the year 1794 to the present time*. Cambridge, England: Henry Wallis.

Rudolph, F. (1990). *The American college and university: A history*. Athens, GA: The University of Georgia Press.

Rush, B. (1806), *Essays, literary, moral and philosophical* (2nd ed.). Philadelphia, PA: Thomas and William Bradford.

Ruter, M. (1827). *The juvenile arithmetick, and scholar's guide*. Cincinnati, OH: N and G. Guilford.

Ruter, M. (1828). *The juvenile arithmetick and scholar's guide; wherein theory and practice are combined and adapted to the capacities of young beginners; containing a due proportion of examples in Federal money; and the whole being illustrated by numerous questions similar to those of Pestalozzi*. Cincinnati, OH: N. and G. Guilford.

Sanford, V. L. (1957). Robert Recorde's "Whetstone of witte," 1557. *The Mathematics Teacher, 50*(4), 258–266. https://doi.org/10.5951/MT.50.4.0258

Sarjeant, T. (1788). *Elementary principles of arithmetic with their application to the trade and commerce of the United States of America*. Philadelphia, PA: Dobson and Lang.

Schlesinger, A. M. (Ed.). (1983). *The almanac of American history*. New York, NY: G. P. Putnam & Sons.

Schrader, D. V. (1967). The arithmetic of medieval universities. *Mathematics Teacher, 60*, 264–275.

Seaman, W. H. (1902, March). How Uncle Sam got a decimal coinage. *School Science*, 232–236. https://doi.org/10.1111/j.1949-8594.1902.tb00443.x

Seltman, M., & Goulding, R. (Eds.). (2007). *Thomas Harriot's Artis Analyticae Praxis*. London, England: Springer.

Sesiano, J. (2019). *Magic squares: Their history and construction from ancient times to AD 1600. Studies in the history of mathematics and physical sciences*. Gewrestrasse, Switzerland: Springer Nature.

Seybolt, R. F. (1917). *Apprenticeship and apprenticeship education in colonial New England and New York*. New York, NY: Teachers College Columbia University.

Seybolt, R. F. (1921). *The evening schools of colonial New York City*. Albany, NY: The University of the State of New York.

Seybolt, R. F. (1935). *The private schools of colonial Boston*. Cambridge, MA: Harvard University Press. https://doi.org/10.5951/MT.60.3.0264

Seybolt, R. F. (1969). *The public schools of colonial Boston 1635–1775*. New York, NY: Arno Press.

Sherwin, T. (1842). *An elementary treatise on algebra*. Boston, MA: Benjamin B. Mussey.

Sherwin, T. (1845). *The common school algebra*. Boston, MA: Phillips and Sampson.

SHM. (1856). Written examinations. *Rhode Island Schoolmaster, 2*, 156.

Silverman, D. J. (2006). Faith and boundaries: Colonists, Christianity and community among the Wampanoag Indians of Martha's Vineyard, 1600–1871. *History, 91*(304). 583–584. https://doi.org/10.1111/j.1468-229X.2006.379_7.x

Simons, L. G. (1923). A Dutch textbook of 1730. *The Mathematics Teacher, 16*(6), 340–347. https://doi.org/10.5951/MT.16.6.0340

Simons, L. G. (1924). *Introduction of algebra into American schools in the 18th century*. Washington, DC: Department of the Interior Bureau of Education.

Simons, L. G. (1931). The influence of French mathematicians at the end of the eighteenth century upon the teaching of mathematics in America. *Isis, 15*, 104–123.

Simons, L. G. (1936a). *Bibliography of early American textbooks on algebra*. New York, NY: Scripta Mathematica, Yeshiva College.

Simons, L. G. (1936b). Short stories in colonial geometry. *Osiris, 1*, 584–605. https://doi.org/10.1086/368442

Simson, R. (1756). *The elements of Euclid*. Glasgow, Scotland: R. & A. Foulis.

Simson, R. (1806). *The elements of Euclid, viz, the first six books, together with the eleventh and twelfth, the errors by which Theon and others have long ago vitiated these books, and corrected, and some of Euclides' demonstrations are restored. Also the books of Euclid's data, like manner corrected*. Philadelphia, PA: Mathew Carey.

Sinclair, N. (2008). *The history of the geometry curriculum in the United States*. Charlotte, NC: Information Age Publishing.

Smith, A. E. (1947). *Colonies in bondage: White servitude and convict labor in America 1607–1776*. Chapel Hill, NJ: University of North Carolina.

Smith, D. E. (1911). *The teaching of geometry*. Boston, MA: Ginn and Company.

Smith, D. E., & Ginsburg, J. (1934). *A history of mathematics in America before 1900*. Chicago, IL: The Mathematical Association of America. https://doi.org/10.1090/car/005

Smith, D. E., & Karpinski, L. C. (1911). *The Hindu-Arabic numerals*. Boston, MA: Ginn & Co.

Smith, R. C. (1826). *Practical and mental arithmetic on a new plan*. Boston, MA: S. G. Goodrich and Richardson and Lord.

Smith, R. C. (1827). *Practical and mental arithmetic on a new plan in which mental arithmetic is combined with the use of the slate: Containing a complete system for all practical purposes; being in dollars and cents*. Boston, MA: Richardson & Lord.

Smith, R. C. (1850). *Practical and mental arithmetic on a new plan*. New York, NY: Cady & Burgess.

Solberg, T (1905). *Copyright in Congress 1789–1904*. Washington, DC: Government Printer.

Stamper, A. W. (1909). *A history of the teaching of elementary geometry, with reference to present-day problems*. PhD dissertation, Columbia University, New York.

State of Massachusetts. (1848). *Report No. 12 to the Massachusetts School Board by Horace Mann*. Boston, MA: Author.

State of Massachusetts. (1855). *Eighteenth annual report of the Secretary of the Massachusetts Board of Education*. Boston, MA: Author.

State Street Trust Company. (1917). *Some events of Boston and its neighbors*. Boston, MA: Author.

Stedall, J. (2012). *The history of mathematics: A very short introduction*. Oxford, England: Oxford University Press. https://doi.org/10.1093/actrade/9780199599684.001.0001

Sterry, C., & Sterry, J. (1790). *The American youth: Being a new and complete course of introductory mathematics, designed for the use of private students*. Providence, RI: Authors.

Sterry, C., & Sterry, J. (1795). *A complete exercise book in arithmetic, designed for the use of schools in the United States*. Norwich, CT: John Sterry & Co.

Stevin, S. (1585). *De Thiende*. Leyden, The Netherlands: The University of Leyden.

Stewart, L. (1999). Other centres of calculation, or where the Royal Society didn't count: Commerce, coffee-houses and natural philosophy in early modern London. *British Journal of the History of Science, 32*, 132–153. https://doi.org/10.1017/S0007087499003556

Stocker, H. E. (1922). *A history of the Moravian Church in New York City*. New York, NY: Author.

Stoeckel, A. (1976). Presidents, professors, and politics: The colonial colleges and the American revolution. *Conspectus of History, 1*(3), 45–56.

Struik, D. J. (2012). *A concise history of mathematics* (4th ed.). New York, NY: Dover Publications.

Swan, S. B. (1977). *American women and their needlework 1700–1850*. New York, NY: Holt, Rinehart and Winston.

Sylvester, J. J. (1870, January 6). A plea for the mathematician. *Nature, 1*, 261–263. https://doi.org/10.1038/001261a0

Taylor, E. G. R. (1966). *The mathematical practitioners of Hanoverian England 1714–1840*. Cambridge, England: Cambridge University Press.

Tebbel, J. (1972). *A history of book publishing in the United States*. New York, NY: R. R. Bowke.

Tewell, J. J. (2012). Assuring freedom to the free: Jefferson's Declaration and the conflict over slavery, *Civil War History*, *5*(1), 75–96. https://doi.org/10.1353/cwh.2012.0033

Tharp, P. (1798). *A new and complete system of Federal arithmetic*. Newburgh, NY: D. Denniston.

Thomson, J. B. (1843). *Elements of algebra, being an abridgment of Day's algebra, adapted to the capacities of the young, and the method of instruction, in schools and academies*. New Haven, CT: Durrie & Peck.

Thornton, T. P. (1996). *Handwriting in America: A cultural history*. New Haven, CT: Yale University Press. https://doi.org/10.5149/northcarolina/9781469626932.001.0001

Thornton, T. P. (2016). *Nathaniel Bowditch and the power of numbers: How a nineteenth-century man of business, science and the sea changed American life*. Chapel Hill, NC: University of North Carolina Press.

Tillinghast, N. (1841). *Elements of plane geometry for the use of schools*. Boston, MA: Lewis & Sampson.

Todd, J., Jess, Z., Waring, W., & Paul, J. (1800). *The American tutor's assistant, or a compendium system of practical arithmetic*. ... Philadelphia, PA: Zachariah Poulson.

Todhunter, I. (Ed.). (1955). *Euclid's elements*. London, England: J. M, Dent and Sons.

Truth, S. (1850). *The narrative of Sojourner Truth* (edited by Olive Gilbert). Boston, MA: Author.

Tuer, A. M. (1896). *History of the horn-book*. London, England: Leadenhall Press.

Turner, W. L. (1953). The charity school, the Academy and the College Fourth and Arch Streets. *Transactions of the American Philosophical Society, 43*(1), 179–186. https://doi.org/10.2307/1005670

Tyler, L. G. (1897). Education in colonial Virginia. Part II, private schools and tutors. *William and Mary Quarterly, 6*, 1–6. https://doi.org/10.2307/1914792

United States Census Bureau. (2004). *Colonial and pre-Federal statistics*. Suitland, MD. Author.

United States Department of Education. (1985). *Early American textbooks*. Washington, DC: Author.

Uzes, F. D. (1980). *Illustrated price guide to antique surveying instruments and books*. Rancho Cordova, CA. Landmark Enterprises.

Van Egmond, W. (1976). *The commercial revolution and the beginnings of Western mathematics in Renaissance Florence, 1300–1500*. PhD dissertation, Indiana University.

Van Egmond, W. (1980). *Practical mathematics in the Italian Renaissance: A catalog of Italian abbacus manuscripts and printed books to 1600*. Firenze, Italy: Istituto E Museo di Storia Della Scienza.

Van Sickle, J. (2011). *A history of trigonometry education in the United States: 1776–1900*. PhD dissertation, Columbia University.

Varcoe, K. E. (2002). A historical review of curriculum in American higher education: 1636–1900. PhD Course paper, Nova Southeastern University.

Venema, P. (1725). Unpublished "precursor." Held in the Ellerton-Clements textbook collection, Bloomington, IL.

Venema, P. (1730). *Arithmetica of Cyffer-Konst, volgens de Munten Maten en Gewigten te Nieu-York, gebruykelyk als mede een kort Ontwerp van de Algebra*. New York, NY: Jacob Goelet.

Vickers, D. (2008). *A companion to colonial America*. New York, NY: Wiley-Blackwell.

Viète, F. (1579). *Canon mathematicus seu ad triangula cum appendicibus*. Paris, France: Jean Mettayer.

Wadsworth, G. (2005). *Benjamin Banneker: Pioneering scientist*. Minneapolis, MN: Lerner Publishing.

Walkingame, F. (1785). *The tutor's assistant being a compendium of arithmetic and a complete question book* (21st ed.). London, England: J. Scratcherd & I. Whitaker.

Wallis, P., Wallis, R., Ransom, P., & Fauvel, J. (1991). *Mathematical tradition in the north of England*. Newcastle-Upon-Tyne, England: The University of Newcastle.

Wallis, R. (1997). Edward Cocker (1632?–1676) and his arithmetick: De Morgan demolished, *Annals of Science, 54*, 507–522. https://doi.org/10.1080/00033799700200471

Wallis, R. (2004). Kersey, John, the Elder. *Oxford dictionary of national biography*. Oxford University (online).

Walsh, M. (1801). *A new system of mercantile arithmetic: Adapted to the commerce of the United States, in its domestic and foreign relations; with forms of accounts, and other writings usually occurring trade*. Newbury-Port, MA: Edmund M. Blunt.

Walsh, M. (1804). *A new system of mercantile arithmetic: Adapted to the commerce of the United States, in its domestic and foreign relations; with forms of accounts, and other writings usually occurring trade* (4th ed.). Newbury-Port,-MA: Edmund M. Blunt.

Ward, H. E. (2007). To all who know their ABCs, greeting: A history of the ABCs (Lilly Library, Indiana University). *Indiana Libraries, 26*(3), 58–61.

Ward, J. (1719). *The young mathematician's guide: Being a plain and easie introduction to the mathematicks*. London, England: Thomas Horne.

Ward, J. (1758). *The young mathematician's guide: Being a plain and easy introduction to the mathematicks*. London, England: C. Hitch and L. Hawes.

Wardhaugh, B. (2012). *Poor Robin's prophecies: A curious almanac, and the everyday mathematics of Georgian Britain*. Oxford, England: Oxford University Press.

Wardley, P., & White, P. H. (2003). The Arithmeticke Project: A collaborative research study of the diffusion of Hindu-Arabic numerals. *Family and Community History, 6*(1), 5–17. https://doi.org/10.1179/fch.2003.6.1.002

Wareing, J. (1985). *Emigrants to America: Indentured servants recruited in London 1718–1733*. Baltimore, MD: Genealogical Publishing Co.

Watson, F., & Kandel, I. L. (1911). Examinations. In P. Monroe (Ed.), *A cyclopaedia of education* (Vol. 2, pp. 532–538). New York, NY: The Macmillan Company.

Wayland, F. (1842). *Thoughts on the present collegiate system*. Boston, MA: Gould, Kendall & Lincoln.

Webber, S. (1801). *Mathematics compiled from the best authors and intended to be the textbook of the course of private lectures on these sciences in the University of Cambridge*. Boston, MA: Thomas and Andrews. https://doi.org/10.5962/bhl.title.17251

Webber, S. (1808). *Mathematics compiled from the best authors and intended to be the textbook of the course of private lectures on these sciences in the University at Cambridge* (2nd ed.). Cambridge, MA: William Hilliard.

Webster, N. (1787). *The American speller*. Boston, MA: Isaiah Thomas.

Wecter, D. (1937). *The saga of American society: A record of social aspiration 1697–1937*. New York, NY: Charles Scribner's Sons.

Weems, M. L. (1820). *The life of Benjamin Franklin*. Baltimore, MD: Author.

Wells, R. V. (1975). *Population of the British colonies in America before 1776: A survey of census data*. Princeton, NJ: Princeton University Press.

Welsh, C. (1902). Hornbooks and battledores. *The Literacy Collector, 3*(2), 33–37. https://doi.org/10.1038/scientificamerican02011902-37ebuild

Westbury, I. (1980). Change and stability in the curriculum: An overview of the questions. In H. G. Steiner (Ed.), *Comparative studies of mathematics curricula: Change and stability 1960–1980* (pp. 12–36). Bielefeld, Germany: Institut für Didaktik der Mathematik-Universität Bielefeld.

Whalin, W. T. (1997). *Sojourner Truth: American abolitionist*. Uhrichsville, OH: Barbour Publishing.

Whaples, R. (1995). Where is there consensus among American economic historians? *The Journal of Economic History, 55*(1), 139–154. https://doi.org/10.1017/S0022050700040602

White, E. E. (1886). *The elements of pedagogy*. New York, NY: American Book Company.

Wickersham, J. P. (1886). *A history of education in Pennsylvania*. Lancaster, PA: Inquirer Publishing Company.

Wiencek, H. (2012). *Master of the mountain: Thomas Jefferson and his slaves*. Minneapolis, MN: Lerner Publishing.

Wikipedia contributors. (2021a, April 8). Jeremiah Day. In *Wikipedia, The Free Encyclopedia*. Retrieved July 14, 2021, from https://en.wikipedia.org/w/index.php?title=Jeremiah_Day &oldid=1016597452

Wikipedia contributors. (2021b, May 9). Florian Cajori. In *Wikipedia, The Free Encyclopedia*. Retrieved July 11, 2021 from https://en.wikipedia.org/w/index.php?title=Florian_Cajori &oldid=1022259870

Wikipedia contributors. (2021c, July 4). Sojourner Truth. In *Wikipedia, The Free Encyclopedia*. Retrieved from https://en.wikipedia.org/w/index.php?title=Sojourner_Truth&oldid=1031850304

Wilkins, J. (1668). *Essay towards a real character and a philosophical language*. Held in the Balliol Library, University of Oxford, England.

Williams, J. J. (2011). *Robert Recorde: Tudor polymath, expositor, and practitioner of computation*. London, England: Springer-Verlag. https://doi.org/10.1007/978-0-85729-862-1

Wilson, D. L. (1992). Thomas Jefferson and the character issue. *The Atlantic Monthly, 270*(3), 37–74.

Wilson, J., Fiske, J., & Klos, S. L. (1889a). Benjamin Greenleaf. In J. Wilson, J. Fiske & S. L. Klos (Eds.), *Appleton's cyclopedia of American biography* (p. 399). New York, NY: D. Appleton & Co.

Wilson, J. G., Fiske, J., & Klos, S. L. (Eds.). (1889b). *Appleton's cyclopedia of American biography*. (6 volumes). New York, NY: D. Appleton and Company.

Wilson, N. G. (Ed.). (2006). *Encyclopedia of ancient Greece*. New York, NY: Routledge

Windschuttle, K. (1996). *The killing of history: How literary critics and social theorists are murdering our past*. San Francisco, CA: Encounter Books.

Wingate, E. (1624). *L'usage de la règle de proportion en arithmétique*. Paris, France: Author.

Wingate, E. (1630). *Arithmétique made easie*. London, England: Stephens and Meredith.

Wingate, E. (1689). *Mr. Wingate's arithmetick containing a plain and familiar method for attaining the knowledge and practice of common arithmetick* (9th ed.). London, England: Author.

Woloch, N. (1992). *Early American women*. Belmont, CA: Wadsworth Publishing Company.

Workman, B. (1789). *The American accountant or schoolmaster's new assistant ...* Philadelphia, PA: John M'Culloch.

Workman, B. (1793). *The American accountant or schoolmaster's new assistant ... Revised and corrected by Robert Patterson*. Philadelphia, PA: W. Young.

Wunsch, J. (2007). Magic squares and circles. *Nature, 450*(1162). https://doi.org/10.1080/0046760X.2011.584573

Wylie, C. D. (2012). Teaching manuals and the blackboard: Accessing historical classroom practices. *History of Education, 41*(2), 257–272.

Yale College. (1828). *Reports on the course of instruction in Yale College by a committee of the Corporation, and the academical faculty*. New Haven, CT: Author.

Yeldham, F. A. (1926). *The teaching of arithmetic through four hundred years (1535–1935)*. London, England: George A. Harrap.

Yeldham, F. A. (1936). *The story of reckoning in the Middle Ages*. London, England: George A. Harrap.

Zuccheri, L., & Zudini, V. (2014). *History of teaching calculus*. In A. Karp & G. Schubring (Eds.), *Handbook on the history of mathematics education* (pp. 493–513). New York, NY: Springer.

Zigler, R. (2011). *The Parrys of Philadelphia and New Hope: A Quaker family's lasting impact on two historic towns*. Bloomington, IA: iUniverse.

Zitarelli, D. E. (2019). *A history of mathematics in the United States and* Canada *(Vol. 1, 1492–1900)*. Providence, RI: MAA Press and American Mathematical Society. https://doi.org/10.1090/spec/094

Author Index

N. F. Ellerton, M. A. (Ken) Clements, *Toward Mathematics for All*, History of Mathematics Education,
https://doi.org/10.1007/978-3-030-85724-0

Subject Index

N. F. Ellerton, M. A. (Ken) Clements, *Toward Mathematics for All*, History of Mathematics Education,
https://doi.org/10.1007/978-3-030-85724-0